DISASTER PLANNING AND CONTROL

William M. Kramer

PennWell®

Fire Engineering®

Disclaimer

The recommendations, advice, descriptions, and the methods in this book are presented solely for educational purposes. The author and publisher assume no liability whatsoever for any loss or damage that results from the use of any of the material in this book. Use of the material in this book is solely at the risk of the user. This book is written using a variety of technical resources that are current at the time of writing. The authors have included pertinent Web sites and electronic sources to ensure that readers have the most current information in the field of emergency operations planning. The authors have presented this material for the readers learning experience only. Legal review of plans and procedures should be conducted by the jurisdiction's legal counsel prior to implementation.

PennWell Corporation
1421 South Sheridan Road
Tulsa, Oklahoma 74112-6600 USA

800.752.9764
+1.918.831.9421
sales@pennwell.com
www.pennwellbooks.com
www.pennwell.com

Marketing: Jane Green
National Account Executive: Barbara McGee

Director: Mary McGee
Managing Editor: Stephen Hill
Production Manager: Sheila Brock
Production Editor: Tony Quinn
Cover Designer: Kelly Cook

Library of Congress Cataloging-in-Publication Data

Kramer, William M. (William Michael), 1944-
Disaster planning and control / William M. Kramer.
p. cm.
Includes bibliographical references and index.
ISBN 978-1-59370-189-5
1. Emergency management—United States—Handbooks, manuals, etc. 2. Fire departments—United States—Handbooks, manuals, etc. 3. Fire extinction—United States—Handbooks, manuals, etc. I. Title.
HV551.3.K73 2009
363.34'60973—dc22

2008052899

Printed in the United States of America

1 2 3 4 5 13 12 11 10 09

I dedicate this text book to Cari and Jennie, my two daughters who as children often spent time away from me due to my dual role as a fire service practitioner and educator, and who nonetheless stayed close and grew up to be wonderful young women. I also dedicate this book to the memory of the late Charles "Chuck" Bahme, who coauthored with me the last edition of *Fire Officer's Guide to Disaster Control,* and to my protégé Randall W. Hanifen, who, served as an "associate author" and is responsible for updating much of the information in this book.

CONTENTS

I DISASTER ANTICIPATION AND PREPARATION

II MANAGING DISASTERS

IV GROWING THREATS, GLOBAL CONCERNS

V NATURAL DISASTERS AND RECOVERY

FOREWORD

From the time I was eight years old I knew I wanted to be a member of the fire service. What I didn't know was the transformation of the service I would witness as I climbed the ranks, moved around the country, and accepted continuously high levels of responsibility. In assuming command of the Washington D.C. fire department, I found myself in charge of an agency that still handles routine house fires well, delivers pre-hospital emergency care in exemplary fashion, can handle hazardous materials and perform technical rescues in a community with a daytime population of more than 1,000,000 people.

That's not all. This same department can be called upon to mitigate a disaster with a magnitude unimagined several decades ago. Along the peaceful Potomac River sits the Pentagon, transformed on September 11, 2001 to become a symbol of a new dangerous world. History changed radically in a few minutes, not just in not just in New York, Pennsylvania, and our own nation's capital, but across the United States and across the free world. This world looks to the fire service for its first line of defense and emergency operations plans designed for all types of disasters have become essential.

A predecessor book *Fire Officer's Guide to Disaster Control* by Bill Kramer and Chuck Bahme was an essential part of a fire chief's library. This new text, put together by Bill Kramer with input from Randy Hanifen and some of the brightest minds in the country, was adopted for use by the National Fire Academy while still in the galley stages. It is sure to become a prominent reference work, not only us who are fires service leaders, but police commanders, and emergency management leaders, all of whom must now potentially work together in a unified command to handle natural and manmade disasters, the true extent of which remains to be seen.

Few of us could have envisioned the extent to which the fire service role would be expanded and our resource tables lengthened and broadened to accommodate the expanding role of the fire service. All of us in the emergency service arena, regardless of rank or assignment could at a moment's notice be involved in mitigating a world-shaking emergency.

As my career took me to increasing levels of responsibility in Fairfax, Virginia; Mesa, Arizona; Dothan, Alabama; Norfolk, Virginia; and Atlanta, Georgia, I saw the need for coordinated disaster planning in communities everywhere regardless of size or location. The difference between being a brand-new firefighter and a fire chief is just the luminous responsibility of taking care of the members in the community. As the bar is raised and this level of care grows in magnitude, this book will help us all.

Dennis L. Rubin, Chief of Fire and Emergency Medical Services, Washington, D.C.

PREFACE

New Name and Focus

The world is continually evolving. As with any area of study, disaster planning and control has encountered a new focus based on historical events. The past decade has seen more changes in disaster planning and control than many other areas of study. The events of 9/11 will never be forgotten and have left a mark on the study of disasters. This mark is the increased breadth of disaster planning. In the predecessor version of this text, *Fire Officer's Guide to Disaster Control, Second Edition*, the authors viewed disaster planning from the fire department's perspective, ensuring that the partnerships needed for successful planning and control were included.

Based on knowledge gained in our profession over the past decade, we have recognized that all disciplines of public service, as well as private industry, need to have a basis of educational knowledge that is consistent. It is important that every individual involved in a disaster, from the frontline worker to the highest elected officials, understand how the disaster is planned for and controlled. The slightest misunderstanding by any individual involved in the disaster can lead to a response that is criticized by the public, possibly resulting in their distrust of the capability of the responders to handle subsequent disasters.

The author has consulted personnel from many disciplines and backgrounds to bring together and review a text that can be used by firefighters, police officers, public safety directors, elected officials, and emergency planners at all levels.

As no one person can deliver all of the information needed in a particular subject area, the author and associate author have searched to find relevant links to Web pages that can further your knowledge of a subject area in the text.

It is hoped that you will find this book to be an informative text that you can employ as a reference book for disaster planning and control.

Best wishes as you begin the journey of planning for and controlling the world's future disasters.

TRIBUTE TO CHARLES BAHME

Bill Kramer and Randy Hanifen

Charles Bahme

As this book goes to press, a special tribute is owed to a pioneering great who is in large measure responsible for the book's existence. In 1978 Charles "Chuck" Bahme, deputy fire chief for the city of Los Angeles, published a small but powerful volume, *Fire Officer's Guide to Disaster Control.* This influential text served as a bible for many command officers as they faced some of the most severe natural and man-made calamities. Chief Bahme's manual was commonplace on the front seat of staff cars everywhere and was the forerunner of the more comprehensive disaster planning manuals that exist today.

In 1992, author Bill Kramer teamed up with Charles Bahme for a successor edition entitled *Fire Officer's Guide to Disaster Control*, second edition. This more comprehensive volume included all of the key principles espoused by Chuck in the first edition, together with much new information on incident management, emergency operations planning, resource listing, and similar topics. The second volume was enhanced with more contemporary examples of actual disasters and how emergency commanders coped with them. *Disaster Planning and Control* has captured the essence of the two editions of the *Fire Officer's Guide to Disaster Control* and has updated and broadened them. The result is a manual that is useful across the board not only for fire officers, but also for others in law enforcement, emergency management, and similar fields, all of whom must work together in tackling increasing global threats and the typical array of natural and man-made disasters that have plagued mankind for centuries.

Thank you, Chuck Bahme, for your inspiration that led to the book in front of us now and for the timeless words of wisdom that found their way into the first edition of the *Fire Officer's Guide to Disaster Control* and that continue as nuggets of wisdom still embedded in this current text.

PASSING THE TORCH

Randall Hanifen

On these two adjacent pages are two of the brightest minds to ever have contributed to fire service literature. Charles Bahme was the original author of *Fire Officer's Guide to Disaster Control*. This and many other books were used to train professionals in the emergency services field during the 1970s and 1980s. Chuck passed the torch to William Kramer, who became a coauthor of the revised edition of the *Fire Officer's Guide to Disaster Control*, and who now provides an enhanced version to serve all members of the emergency services community.

Contributing heavily to this new text is the next generation to whom William Kramer will pass the torch. Randall Hanifen was identified by author William Kramer as a rising star in the fire service. As a company officer, he is characterized by the chiefs under whom he serves as "a doer and an achiever." William Kramer has incorporated the very latest technical information available from Randall regarding CBRN (chemical, biological, radiological, and nuclear) threats and many technical areas undergoing updates. Thanks to Randy, these are included throughout the many chapters of this text.

As a member of an active urban search and rescue task force who has served in several deployments, and as an adjunct instructor at the University of Cincinnati who incorporates the latest technology in the classes he teaches, Randall Hanifen has ensured that this text contains the most relevant information available to readers. Randall has helped to ensure that this textbook will remain current long after it is published by including pertinent Web sites. These sites are updated regularly and should ensure that the readers of this book will have access to current information regardless of how long after the publication date they choose to read the text and use its contents.

Author William Kramer expresses a sincere thank-you to the next generation for accepting the torch, lighting the way, and enhancing this textbook with their illumination.

ABOUT THE AUTHOR

Author Bill Kramer has been in emergency services his entire life, both as a practitioner and as an educator. At the age of 18 he began a fire service career as a volunteer firefighter in suburban Cincinnati, later joining the Cincinnati Fire Department where he worked through the ranks. Retiring as Assistant Fire Chief, he proceeded to serve as Fire Chief for the Indianapolis International Airport Facility and later as Fire Chief for Deerfield Township in Warren County, Ohio.

All the while he earned two bachelors degrees, two masters degrees, and a Ph.D. He stayed busy educating firefighters most notably as Director of Fire Science at the University of Cincinnati. He teamed up with Los Angeles Deputy Chief Chuck Bahme in a predecessor book entitled "Fire Officers Guide to Disaster Control." In that text, Bill and Chuck researched and wrote about the latest techniques used by emergency responders to handle disasters and calamities of all types around the world. Many of these were chronicled by Bill in *American Heat* and *Working Fire* videos which have been viewed by firefighters and emergency response personnel over the past several decades.

At the University of Cincinnati, Bill not only leads the ever-popular Distance Learning Bachelor's Degree in fire science, but also a newly developed Master's Degree in Emergency Management. Bill is called upon to serve as a consultant in many communities working to improve fire protection and planning for domestic preparedness and homeland security affairs. Bill has two grown daughters, Cari and Jennie, and his fiancé Deborah in Cincinnati, but travels widely due to his expertise in the disaster and emergency preparedness arena.

Bill Kramer and "Tiller"

1

Disaster Anticipation and Preparation

DISASTER PLANNING 1

THE RAPIDLY CHANGING WORLD

Whether you began your career 30 years ago or 30 days ago, it is guaranteed that you have seen changes within the emergency services profession. Each day we turn on the news, a new crisis is developing in the world, many with the potential to include the United States. Each month we open a fire service trade journal to see that another aspect of our job is changing. The speed with which technology and world events are changing makes anyone wonder what our fire service jobs will encompass in the next year, much less the next 20 years.

One aspect will remain the same: we will be expected to respond. When the authors began their careers, no one would have thought we would fall under the Department of Homeland Security. However, we now receive our federal funding and many of our training mandates from this department. We have many funding opportunities thanks to this new mission, but we also see many new unfunded mandates. If, as a fire officer, you balked at responding to hazmat incidents because you were just a firefighter, I can only imagine how you must feel today. This is likely not to change in the future. As funding becomes more competitive, we will either take on a new mission, or take on a new mission with no funding. Throughout history, the fire service has been the organization at the front lines of all disasters. We must remember that the response to all the disasters began with one call to the local 911 center. This is true of the bombing of the Alfred P. Murrah Federal Building in Oklahoma City and the bombing of the World Trade Center in New York City on September 11, 2001 ("9/11"), and it will be the case in future disasters.

Despite the rapidly changing world, many aspects of disasters and their responses have remained the same. In floods, water still rises. In tornadic events, the wind still spins at high rates of speed. Therefore, we can still learn the basic principles of disaster response and have a high probability of success. However, any good fire officer knows that the minute you stop learning and feel that you know it all is the exact time you should retire. By getting this far in the text, you have committed to make the next disaster response better than the last. It is because of the dedicated men and women involved that the fire service personnel will remain the ones that the public will call when they are in need, whether that need involves an illness or a terrorist event.

DISASTERS OCCUR DAILY

Hardly a day passes in which another new disaster does not occur. Whether it is the day after Christmas in Indonesia, involving a large tsunami, or a clear September day in the United States, people have been shocked repeatedly. Train wrecks, plane crashes, hazardous materials spills and releases, tank truck collisions, overturned buses, and other transportation accidents occur on a regular basis. Many of these events occur in densely populated areas of the country and involve a large number of victims. Many disasters are reoccurring conflagrations. Each year, thousands of acres of forest burn, and hurricanes continue to strike the East and Gulf coasts. Tornadoes occur in the Midwest each spring. Earthquakes occur on a regular basis, with some reaching catastrophic levels in urban areas, thus destroying entire cities. According to the Federal Emergency Management Agency (FEMA) in 2006, a total of 406 disasters had been declared in the United States since the beginning of the year 2000.[1] This was an average of 58 disasters a year, including a wide range of events from fires in Alaska to large snowstorms in South Dakota.

In 2005, the United States saw a rash of very powerful hurricanes and subsequent flooding. Many of these hurricanes repeatedly affected the same areas, giving little time for recovery. The largest and most notable of these storms, Hurricane Katrina, showed that despite vast improvements in federal planning, gaps in response and recovery missions can still occur. Second, the U.S. military has been engaged in war and warlike activities since 2001, thus lessening the ability of an armed forces response to homeland events. As this book is being written, North Korea boasts that it has nuclear capabilities, and there are tensions both in the war areas of Iraq and within the political parties of the United States. This shows that the trend of both foreign and domestic disasters will continue to occur at the same pace or faster than they have historically, due to man-made disasters speeding the pace.

Even the insurance profession realizes the role of the fire department. More than four decades ago, the American Insurance Association released *Special Interest Bulletin No. 310*, stating, "Disasters can and do occur anywhere. It is the duty of large and small fire departments, well equipped and poorly equipped to be thoroughly prepared to assist the operations at the scene of a disaster."[2] This is more of a truth today than ever before, as all of the state and federal response plans indicate the local jurisdiction will retain authority over the incident, thus indicating the fire department's lead role in disaster response.

One of the most important tasks that fire chiefs complete is that of disaster planning. As stated previously, most disasters begin with the response of the local fire department. This means that the success and subsequent calls for assistance will not work in a coordinated effort if the written plans are not present and the responding personnel are not trained to the current plans. Finally, no plan is effective unless it is tested through tabletop and full-scale exercises.

The *American Heritage Dictionary* defines *disaster* as "an occurrence causing widespread destruction and distress; a catastrophe."[3]

Based on this definition, disasters obviously can vary greatly in their particulars. One instance may be a tornado that levels a town. Another may involve a large loss of wildlife from a ruptured oil tanker. No matter the particulars of the disaster, one can be assured that the fire service will become involved in the event's response and recovery.

Often the terms *calamity* and *catastrophe* are used synonymously (as seen in the *American Heritage* definition) with the term *disaster.* Some of the world's greatest disasters are described in the text, indicating the great variety of disasters that a response agency could respond to in its existence. The important part of considering these disasters is to determine which ones are likely to occur in the jurisdiction under concern. Although it is possible for many to occur, wise planners will devote time to those events with a high probability of occurrence.

The disasters that will be examined all involve the possibility for loss of life. These disasters can have varying degrees of impact based on the responder's relationship to the event. If an explosion of a railcar destroys half of a small town or a bus crash kills 30 schoolchildren from the local school, it is likely to severely impact a responder who lives in the area. It may not have an impact in the life of a responder who lives on the other side of the country. However, if another terrorist event involves a large city and is played out on television, as were the events of 9/11, the entire country will feel the effects of this disaster. The point of the varying degrees is that responders will have to respond to the emergency, and based on the size and complexity, those responders can have a very intimate relationship with the victims.

Disasters can also be quite dependent on the availability of forces to deal with the event. Consider a fire in an apartment building located within a small village with a volunteer fire department. For example, a fire could involve an apartment building housing many people who work second shift, and the fire may start during the day. The volunteer forces may be occupied at their places of employment, and the fire could make good headway, possibly killing many residents prior to its extinguishment. This could certainly be termed a *disaster*.

On the other hand, if this apartment building is located in the heart of any urban area, the on-duty forces will be dispatched and will respond immediately. With incoming information from the dispatcher, a second and third alarm would be transmitted prior to the arrival of the first company. Thus a response would be created with numerous firefighting forces to evacuate and extinguish the fire prior to its spread to any adjacent apartments. This event might not even make the nightly news for the given day.

Disasters can also depend on built-in protection. In the 1980s, 1990s, and into the 21st century, there has been a greater emphasis on safety. This built-in safety means that due to the lessons learned from former disasters, many disasters can be averted today. Take for example a tank truck carrying anhydrous ammonia falling off the interstate and spreading its contents over a populated area (as was the case in Houston in 1976). This disaster may not occur today because of the Department of Transportation laws governing the transportation of hazardous cargo. These regulations are the reason hazardous cargo signs are not seen around urban areas. Many hazardous materials events are no longer disasters due to built-in protection.

If a chlorine leak occurred and the resulting leak blanketed a nearby office building, the result would likely involve numerous casualties. However, if the same leak occurred in a plant that had detectors and alarms to alert the workers at the factory, the information could be relayed to responding units. These units could call for an evacuation or shelter in place procedure, and the chemical would dissipate into the air, causing little harm to anything but the environment.

The final determining factor for whether a disaster is remembered and recorded on various lists is the events present in the world at the time. According to a Geocities Web page produced by Don McBride, who chronicles the world's disasters, the FBI siege of Randy Weaver at Ruby Ridge only involved the death of three people, but almost everyone has heard of Ruby Ridge due to the high publicity. According to the same source, 80,000 people were killed the same year in a civil war in El Salvador. Because of the commonality of wars in other countries, the event that killed nearly 80,000 times as many people is not even recalled.

This book is designed to allow the fire officer to study these events and the relevant methods to respond to and deal with the various types of events. By learning from the past and understanding the concepts presented, fire officers can make a difference in the overall outcome of future events.

It has often been stated in the fire service that what occurs in the first five minutes will dictate what is done in the next five hours. With this phrase in mind, it is obviously important that all officers understand the concepts of responding to disasters.

SIMILARITIES AND DIFFERENCES BETWEEN COMMUNITY PUBLIC SAFETY PRACTICES AND DISASTER CONTROL

Today, more than ever, the fire service is involved in delivering disaster control. Throughout the book, it will be shown through examples and case studies how the fire service has been involved with responding to various disasters.

Although many citizens think only of the fire department responding to their house if it is on fire, the public's perception of the duties of the fire service is slowly changing with high-profile disaster events being televised. The one major event in which the fire service's role change was evident was that of 9/11. The outpouring of support to all firefighters post-9/11 showed how much respect the public has for the fire profession and that they understand the fire service's involvement in responding to all disasters. This solidified the fact that the fire service will become involved in all disasters, no matter their cause.

Similarities between disasters and public safety situations

Immediate response. As many departments have recognized in recent years, they will receive the first calls for response to a disaster. In many communities, the fire service is poised to respond at a moment's notice to any type of emergency. In communities that do not have on-duty forces, it is the fire service that will likely still be the first arriving rescuers to the situation.

Mission of the fire department. In many departments, the mission statement is consistently being updated. At one time, fire departments were only organizations that responded to fires. However, progressive fire service leaders soon realized that the public needed more than just a fire suppression force. The public instead needed an organization that could prevent and respond to all types of emergencies in order to save lives and property. Many departments' missions now include emergency medical services, hazardous materials response, technical rescue response, and initial terrorist event response.

Personnel requirements. The fire service is one of the few organizations to have many personnel in a readiness state, downtime to train and maintain a variety of equipment, and by nature of their daily duties, personnel who adapt well to whatever problem is present. Many of the risks that are inherent in firefighting are the same as those in disaster response, thus personnel are already informed about safety and response expectations for the various types of disasters.

Incident management system (command and control). The core responsibility of fire departments is to take an emergency, make sense of what is occurring, and apply personnel and equipment to stop the chain of current events. During disaster response, this is essentially the same goal.

The fire service has long been known for mass organization at a moment's notice. The fire service's utilization of the incident management system (IMS), even on routine calls, allows personnel to quickly divide and conquer an event. Because of the adaptability and expandability of the IMS, it is suited for a one-car accident or the events of 9/11. Each of these incidents will have the same operations, planning, logistics, and finances, just at different levels. The National Incident Management System (NIMS), the National Response Framework (NRF), and specifically the incident command system (ICS) is discussed in detail in other chapters of the text.

Preincident planning. The fire service is often planning and preventing events from occurring, such as creating building plans for use in fire situations, or planning for the neighborhood festival and providing fire and EMS services for this event. Consequently, it only makes sense that the fire service can expand this to planning and exercising for disasters, such as hazardous materials leaks and man-made and natural disasters. The planning and exercise fundamentals are again the same, whether planning for a leaking anhydrous ammonia pipe in an industrial complex or a bombing in a sports arena.

Mutual aid. Fire officers have long recognized that regardless of the number of paid or volunteer firefighters in an organization, some emergencies will exceed their capabilities either through size or complexity, thus requiring mutual aid. Because of this recurring fact, the fire service has always informally borrowed each other's forces.

According to an article released on March 22, 2006, the International Association of Fire Chiefs "is currently developing two projects that will help the fire service's ability to support the National Response Plan and implement concepts presented in the National Incident Management System. The first project is the National Fire Service Mutual Aid System (IMAS), which is addressing intrastate mutual aid (within states). The second is focusing on interstate mutual aid (between states), and is known by the name of the IAFC task force that is formulating recommendations on the subject: the Mutual Aid System Task Force (MASTF)."[4] This shows that the fire service is still the leader in collaborative responses to disasters. The fire service realizes the need for an even greater number of forces working together, even those from outside the state, and understands that planning for their response will ensure a smooth operation when needed.

Differences between disasters and fire situations

Despite the large differences once seen by officers and firefighters from disaster response to firefighting response, a quick examination of the differences is warranted. Recent events and training initiatives show that there are far more similarities than differences.

Scope of emergencies. Usually disasters are not contained to one property and will become spread out over a larger area. Except for the largest fires, often termed *disasters*, operations are limited to one or two buildings, thus making tracking of and communication with personnel much easier.

Types of emergencies. Disasters can be a multitude of events, some of which spawn secondary events. Even progressive fire departments deal with a small range of daily calls for service. Disasters, on the other hand, present themselves for a variety of reasons, such as natural and man-made disasters.

Planning effort. Typical fire department planning involves only local problems, buildings, and assets. In order to plan for disasters, a global and collaborative approach must be taken to include all of the resources necessary to mitigate the disaster.

Agencies and levels of government involved. In a typical day of emergency response, the fire department is likely to have contact with the local police department, the road department, and possibly the government's administration. However, when dealing with disasters, numerous agencies are needed to either provide service or lend technical expertise to the situation. Within the past 10 years, county, state, and federal response teams for hazardous materials, weapons of mass destruction, and urban search and rescue have emerged and become great assets in a time of disaster.

In addition, all levels of government agencies are involved, up to and including the federal government. This can add levels of cooperation and interoperability that are not present in everyday responses, and there must be a plan or flexibility on the behalf of the responders to overcome these additional challenges.

Fire department response to disasters. As mentioned previously, it is inevitable that the fire department will be one of the first agencies summoned to a disaster. Therefore the similarities pointed out will aid the fire service in responding, and the differences described are what the fire officer must anticipate and plan to overcome in response to a disaster. This planning and understanding, as well as the basic concepts of disaster response, will allow a safe and successful response to a large-scale disaster.

What seems like a routine call can be just the beginning of a disaster response. The well-trained fire officer will identify this quickly and begin to employ the concepts of disaster response. If not, he or she could be struggling alone while victims continue to perish. This normal response turning into a disaster response occurred with the Sycamore Township Fire Department (a Cincinnati, Ohio suburb) on the morning of April 9, 1999. The department responded with one fire company for wires down in the intersection of Kemper and Montgomery roads. Upon arrival, they found damage indicative of a tornado. In fact, the tornado covered two states and killed five people. The worst area of damage was in Blue Ash, Montgomery, and Sycamore Township. The tornado in this area was an F4 on the Fujita scale.

Had the first arriving officer not realized that he had arrived at the scene of a possible disaster, he could have called for the utility company and awaited their arrival to tend to the wires down in this intersection. Rather, an incident command system was established and soon fire companies, US&R teams, the Red Cross, and medical disaster teams were summoned from the three adjacent counties. The response and recovery to one of the worst recent tornados in Cincinnati was quickly underway.

Many problems were encountered at this event, such as no common radio frequencies for the adjacent county's responders and an overloaded countywide radio system and dispatch center. In addition, there were no street signs or landmarks due to the destruction, and there was also a shift change by most responding departments. Despite all of these problems, the actions by the first arriving fire company set into motion a well-coordinated disaster response thanks to planning and exercises previously performed by many of the responders.

CAUSES AND EFFECTS OF DISASTERS

In the broadest sense, disasters can be termed *accidental* or *intentional.* Accidental disasters include many natural disasters, such as tsunamis, volcanic eruptions, hurricanes, floods, and mudslides. Accidental disasters also include many transportation accidents, such as plane crashes, bus accidents, and train derailments.

On the other side of the spectrum are intentional or man-made disasters. These include bombings, terrorist activities, and even some of the large forest fires that have been deemed arson.

Whether accidental or intentional, all disasters have in common their large loss of life or property, or both, and possible long-term effects to the environment. Some of the effects include contaminated drinking water, famine, power outages, and civil unrest. The latter has been seen in recent disasters such as Hurricane Katrina.

CIVILIAN AND MILITARY ROLES IN DISASTER CONTROL

Americans live in a free society, with freedom not generally controlled by the military. Furthermore, America is known for its huge commitment to volunteering in a time of need. These two facets must be considered when planning a response to disasters. Two recent enhancements deserve recognition in this section.

The first is the Community Emergency Response Teams. This program is organized by the Citizens Corps. According to Citizens Corps,

> The Community Emergency Response Team (CERT) program helps train people to be better prepared to respond to emergency situations in their communities. When emergencies happen, CERT members can give critical support to first responders, provide immediate assistance to victims, and organize spontaneous volunteers at a disaster site. CERT members can also help with non-emergency projects that help improve the safety of the community...
>
> Over the next two years, the CERT program aims to double the number of participants, with over 400,000 individuals completing the 20 plus hours of training. Train-the-Trainer sessions will be held in all 56 states and territories over the next year to expand the program throughout the United States.[5]

The second is the use of National Guard and the United States Coast Guard for response to disasters. In the aftermath of Hurricane Katrina, these two agencies were instrumental in rescuing trapped people and quelling days of civil unrest. The Coast Guard used helicopters and boats to make rescues, while the National Guard drove large amphibious vehicles through the floods to deliver needed supplies.

According to the Army National Guard's official Web site, "The Army National Guard's state mission is perhaps the most visible and well known. Nearly everyone has seen or heard of Guard units responding to battle fires or helping communities deal with floods, tornadoes, hurricanes, snowstorms or other emergency situations. In times of civil unrest, the citizens of a state can rest assured that the Guard will be ready to respond, if needed."[6] Figure 1–1 shows the military operating at a disaster.

Fig. 1–1. Military operating at site of a disaster

Despite these missions, the response system of the United States is designed for these military forces to be assets rather than to provide command and control of an incident. This may have been confusing in the response to Hurricane Katrina because President George W. Bush placed Vice Admiral Thad W. Allen in charge of the federal response as he removed former FEMA Director Michael Brown. The federal response plan and the military's function within that plan is further discussed in chapter 4.

FEDERAL EMERGENCY MANAGEMENT AGENCY

Throughout this text, the authors will refer to the authority of emergency management, which is the Federal Emergency Management Agency. Found on the World Wide Web at www.fema.gov, this agency now resides under the Department of Homeland Security. The mission statement of FEMA states, "On March 1, 2003, the Federal Emergency Management Agency (FEMA) became part of the U.S. Department of Homeland Security (DHS). The primary mission of FEMA is to reduce the loss of life and property and protect the nation from all hazards, including natural disasters, acts of terrorism, and other man-made disasters, by leading and supporting the nation in a risk-based, comprehensive emergency management system of preparedness, protection, response, recovery, and mitigation."[7] Figure 1–2 shows Department of Homeland Security (DHS) Secretary Janet Napolitano speaking to the press about FEMA activities.

Fig. 1–2. DHS Secretary Janet Napolitano answers questions from the press at FEMA headquarters. *DHS/NyxoLyno Cangemi*

The statutory authority for FEMA is derived through the Robert T. Stafford Act. The organizational structure of FEMA is broken into three divisions: response, recovery, and mitigation. These divisions operate out of 10 regional offices located around the country.

THE STAFFORD ACT

According to FEMA, "Robert T. Stafford Disaster Relief and Emergency Assistance Act, PL 100-707, signed into law November 23, 1988; amended the Disaster Relief Act of 1974, PL 93-288."[8] The Stafford Act constitutes the statutory authority for most federal disaster response activities, especially as they pertain to FEMA and FEMA programs. It is designed to bring an orderly and systematic means of federal natural disaster assistance for state and local governments in carrying out their responsibilities to aid citizens.

The Stafford Act is a 1988 and subsequent 2007 amended version of the Disaster Relief Act of 1974 (Public Law 93-288). It created the system in place today by which a presidential disaster declaration of an emergency triggers financial and physical assistance through FEMA. The Act gives FEMA the responsibility for coordinating government wide relief efforts.[9] The Federal Response Plan it implements includes the contributions of 28 federal agencies and non-governmental organizations, such as the American Red Cross.

The Stafford Act is explained and summarized best by Keith Bea in the publication *Federal Stafford Act Disaster Assistance: Presidential Declarations, Eligible Activities, and Funding.*[10] This document explains that there are five types of declarations available to the President of the United States. These are major disaster, emergency, fire suppression, defense emergency, and predeclaration activities.

Types of declarations

According to *Federal Stafford Act Disaster Assistance*,

> The president issues a major disaster declaration after receiving a request from the governor of the affected state. Major disaster declarations may be issued after a natural catastrophe or "regardless of cause, fire, flood, or explosion." A declaration authorizes DHS [Department of Homeland Security] to administer various federal disaster assistance programs for victims of declared disasters. Each major disaster declaration specifies the type of incident covered, the time period covered, the types of disaster assistance available, the counties affected by the declaration, and the name of the federal coordinating officer.[11]

A fire officer will not be able to directly tell the governor to make the request for a presidential declaration. Rather, he or she will have to work through local and state channels in order to make the request a reality. This was partly to blame for the delayed response during Hurricane Katrina. The mayor of New Orleans and the governor of Louisiana both claimed they did their part, but this lack of request is partially to blame for the delay in federal assistance. Figure 1–3 illustrates a typical progression of disaster resources.

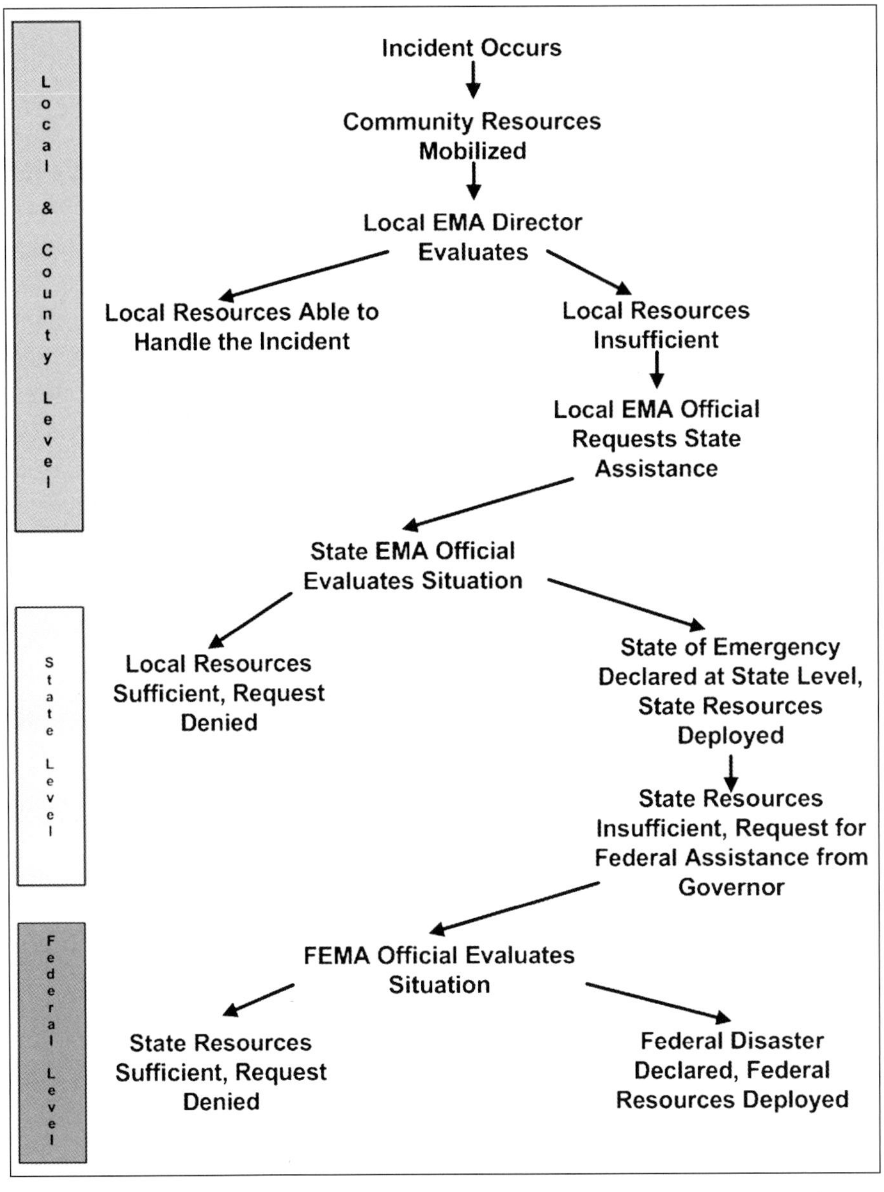

Fig. 1–3. Disaster resource progression. This flowchart illustrates typical progression of resources at a disaster and the processes by which these resources are planned for and obtained.

According to Keith Bea, "The declaration process for emergencies is similar to that used for major disasters; the President may, however, issue an emergency declaration without a gubernatorial request if the primary responsibility rests with the federal government . . . Under an emergency declaration, the federal government funds and undertakes emergency response activities."[12]

This is usually only used for events that occur in federal property, due to the federal government not wanting to pay for disasters that state funding could support. This may come into play if another terrorist attack occurs due to the political drive to respond to and combat terrorist activities.

Under a fire suppression declaration, the Department of Homeland Security is authorized to supplement fire suppression activities when a fire will rise to the magnitude of a disaster. This is common in many wildland fires, as these events frequently escalate to a disaster status.

During a defense emergency, "Upon request from the governor of an affected state, the President may authorize the Department of Defense to carry out emergency work for a period not to exceed 10 days."[13] This work is limited to actions that save lives and preserve property. This could become enacted during any terrorist event due to the large number of victims affected. Second, this declaration is becoming more of a reality, as many divisions within the various branches of the military develop units that mimic those response agencies such as the FEMA US&R system and the DMATs.

During predeclaration activities, "The Secretary of DHS may place agency employees on alert. DHS...deploys teams and resources to maximize the speed and effectiveness of the anticipated federal response and, when necessary, performs preparedness and preliminary damage assessments."[14] Predeclaration activities are now the normal operation in preparation for larger hurricanes. The DHS places many FEMA assets on alert and goes even further by predeploying forces in order to deliver the needed response in a quick time frame.

Types of assistance and eligibility

The second function of the Stafford Act is to establish types of assistance and eligibility available. FEMA has established three categories of assistance. They are individual, public, and hazard mitigation assistance.

Individual assistance covers such items as "immediate temporary shelter, cash grants...home repair grants...[and] unemployment assistance due to the disaster."[15] Public assistance covers "repair, reconstruction, or replacement of infrastructure and recreational facilities."[16] Hazard mitigation covers "assistance to reduce future disaster losses."[17]

A combination of these types of assistance helps restore an area to its predisaster state. Unfortunately, this is not a short process and can take many years to complete. Furthermore, the hazard mitigation assistance is subject to political review. One example is the aftermath of Hurricane Katrina; the decision had to be made to rebuild a levee for an area that was guaranteed to flood if the levee broke again in the future.

DISASTER PLAN AND EMERGENCY OPERATIONS PLAN

A disaster plan is usually called an "emergency operations plan," and is defined as an attempt to anticipate the effects of a particular disaster and develop responses to alleviate the effects on the community. As proven during all past disasters, planning is but one key aspect to successful mitigation and recovery from a disaster. As stated by President Dwight D. Eisenhower, "Plans are worthless, but planning is everything."

Written plans are the first portion of a good plan. Although no two communities are exactly the same, it is found that a generic format and basis for an emergency plan is a good approach because many aspects of different communities are the same. An example is that many cities have police, fire, service, and administration departments within their government, and the duties of these departments are similar from city to city. Fire departments must remember that a written plan is only the first step toward disaster preparedness. Testing of, training to, and revisions of the plan are all portions of plan development that should not be overlooked.

The authors suggest that there are 10 good reasons to develop a written plan for disaster control. These 10 reasons are as follows:

1. Fulfill the responsibility of command.
2. Support the plans of a higher authority.
3. Point out the gaps that need additional study. An idea that seems complete is often found to lack important details when the first attempt is made to write it down.
4. Define the objective of the plan.
5. Define lines or scope of authority and fix responsibility.

6. Avoid the weakness inherent in a "one-man show." The man who has the plan in his head is extremely valuable as long as he is around. If he should be out of town, or have a sudden life-ending event prior to the emergency, his knowledge is nontransferable without a written plan.
7. Reveal the need for coordination of collaborative units.
8. Accurately determine available forces and equipment.
9. Bring together in one document all relevant data necessary for a comprehensive picture of a disaster control operation, from the smallest to the largest unit.
10. National Incident Management System compliance.

The Federal Emergency Management Agency has an extensive collection of emergency operation planning information (www.fema.gov). The *State and Local Guide (SLG) 101: Guide for All-Hazards Emergency Response Planning* Web page offers explanations and links to a number of emergency plan components. According to *SLG 101*:

> This State and Local Guide (SLG) provides emergency managers and other emergency services personnel with information on FEMA's concept for developing risk-based, all-hazard emergency operations plans.
>
> This Guide clarifies the preparedness, response, and short-term recovery planning elements that warrant inclusion in State and local EOPs. It offers FEMA's best judgment and recommendations on how to deal with the entire planning process—from forming a planning team to writing the plan. It also encourages emergency managers to address all of the hazards that threaten their jurisdiction in a single EOP instead of relying on stand-alone plans.[18]

The need for an EOP

As discussed many times, the local government response organizations will be the first to respond to disasters. Therefore, a plan is necessary for dealing with the disaster. A local EOP should focus on measures that are essential to protecting the public. These include warning, evacuation, and sheltering. Next, a process for progressing the incident through the various levels of government is needed.

The local elected officials are legally responsible for ensuring that necessary action to protect the public takes place when disasters and emergencies occur. The only way to ensure that the elected officials have their bases covered is through the use of an emergency operations plan.

Although the EOP is the centerpiece of planning for a disaster, there are limitations to any plan. If these limitations are recognized and responding personnel and command officials understand these limitations prior to use, the plan will work. Often, a lack of this understanding is not recognized prior to usage of the plan. Thus, when the plan is utilized and these limitations become apparent, the users of the plan feel that the plan is not right. They either spend numerous hours rewriting the plan or refuse to utilize the plan because of their prior experience with it.

Some of the limitations include the following:

1. It will not give step-by-step directions on in-field operations.
2. Most plans are written vaguely so that they can be adapted to a variety of variables.
3. The plans are focused on overall strategic functions, especially the basic plan.

4. Plans are divided into basic plans, functional annexes, and hazard specific appendices.

5. Plans are not standard operating procedures (SOPs). SOPs are more of a direct "what to do" in a given situation.

Planning process

A prepackaged plan will not work for any community. Instead, emergency managers must use a systematic process for developing, testing, and revising the plan in order to have a functional plan rather than space filler on a bookshelf. Based on the FEMA *SLG 101* document, the planning process involves the following components:

1. Principles
2. Research
3. Development
4. Validation
5. Maintenance

Any emergency manager or fire officer who has seen the jurisdiction's EOP has probably been amazed at the sheer size of the binder. Any fire officer charged with the development of an EOP probably felt overwhelmed due to this size. The first principle of plan development is "Don't reinvent the wheel." As discussed, a plan for one community will likely have much of the needed content in order to work for another community, due to the commonalities that exist between communities. Also, if the community already has a plan, revision is much easier than an entire rewrite. Figure 1–4 is an outline of the planning process. Readers can refer to this outline throughout the process.

Emergency Operations Planning Process

1. Principles

- Don't Reinvent the wheel
- Elicit the help and opinions of others

2. Research

- Review existing laws, regulations, mutual aid agreements, and high level government organizations

3. Development

- Comprehensive Hazard Analysis
  - Identify Hazards
  - Determine Possible Consequences
  - Prioritize Risks
  - Create and Apply Scenarios
  - Investigate Sources of Information
- Survey the Resource Base
- Research Special Facets of the Planning Environment
- Write the Plan
  - Rough Draft
  - Distribute for Review, Ensure the CEO is kept abreast
  - Incorporate Review

4. Validation

- Ensure that the plan meets all laws and regulations
- Test the plan via exercises to include tabletop, functional, and full-scale
- Incorporate changes found during exercises

5. Maintenance

- Review Periodically

Fig. 1–4. Emergency operations plan development process. Having an outline of the plan development process ensures the planning committee will remain focused and can understand all of the components prior to commencing with the process.

Getting help and other opinions will aid the planner in the daunting task of EOP development. Because of the variety of problems any EOP addresses, a variety of viewpoints during the development of the plan will prevent a one-sided view that will not address all of the issues the plan needs to address upon its completion. The authors suggest developing a committee that is comprised of the different response agencies affected by the plan. It is also a good idea to have all components of the plan reviewed by a subcommittee of differing ranks of those response organizations that will use and/or abide by the plan. This review ensures that the plan will work from a theoretical perspective prior to other real-world testing.

Research phase

The next step is the first step of the actual plan development. This first step is the research phase. This should begin with a review of existing laws, regulations, mutual aid agreements, and higher level government organizations as the first portion of this phase. It would do the committee little good to develop a plan that violated state or local laws. Therefore, this background research will ensure that the committee develops a plan in conjunction with laws and higher authority plans rather than contradict these items.

Comprehensive hazard analysis

The next portion of this phase is an extensive process of conducting a hazard analysis. This is a systematic approach that will ensure the plan highlights the hazard needs of the community. It would be considered a waste of time if the planning committee focused on earthquake response when the community was not located near any fault lines. Although an earthquake is possible, it is better to have a more comprehensive plan to deal with disasters that have a greater likelihood of occurrence for the area.

Comprehensive hazard analysis is a process that can be used to find the hazards of the area being served. The following are components of the comprehensive hazard analysis:

- **Identify hazards.** A place to begin this process is to interview the emergency managers of the jurisdiction in question. Many will remember disasters of significance they have responded to in their careers. This should be coupled with an examination of all hazards and their possibility of occurrence. Once the hazards are written down, they should be examined to determine their possibility for occurrence in the jurisdiction.

- **Determine possible consequences.** Once the applicable disaster list is created, examination of the consequences of that disaster is the next step. A table of the consequences can be created with the hazards on the y-axis of the chart and the following categories placed on the x-axis: frequency, magnitude, location, spatial extent, duration, seasonal pattern, speed of onset, and availability of warning.

- **Compare and prioritize risks.** Upon filling in the columns of the chart, a matrix is created that will help to determine which disasters the plan should focus upon. Developing a numerical value system for each of the x-axis categories will aid in numerically prioritizing the hazards.

- **Create and apply scenarios.** For the top-rated disasters, scenarios should be developed. These scenarios should emphasize events that will occur as a result of the disasters, such as collapsed buildings, loss of communications, etc.

- **Sources.** Sources of information on historical data relevant to hazards should be investigated and collected. Sources could include records from FEMA, USGS maps, and the local emergency planning committee (LEPC). Verifying that the prioritized list is relevant is an important checks-and-balances system.

Resource base

The next phase of plan development involves a survey of the resource base. Querying the emergency managers will tell the planning committee what resources each response agency can bring to the disaster. One of the FY 2007 NIMS compliance objectives was to categorize and type the resources of each agency based on a specified resources typing list. This ensures that planners are using an apples-to-apples comparison when planning for the resources. Even within the fire service, there is a problem determining what type of service a resource can provide. For an example, some departments call a rescue a *paramedic ambulance*, while others refer to a rescue as a *heavy rescue vehicle* capable of extrication from cars, buildings, etc.

Special facets

The final phase of the research step is to consider any special facets of the planning environment. Special facets could include terrain or dependence upon one major industrial facility to maintain the presence of the town. Other facets may involve communication barriers or special problems associated with the response agencies, such as the need to follow more than one emergency operations plan due to joint districts, etc.

Plan development

The next step in the planning process is development. This involves the actual writing of the plan. The authors should begin with a rough draft. This draft should be distributed for review, ensuring that the chief executive official is kept abreast of the progress. Once drafts are distributed for review, a meeting should be scheduled in order to incorporate changes. Although computer software is available for editing a collaborative text document, this process should occur in person, as conversation spawned from the various additions and deletions will be of value to the developers.

Validation of the plan

The validation step is next in the plan development. During the validation process, two phases occur. The first phase is to ensure that the plan meets all regulations and laws. The second phase is the testing process. The plan should be used in a tabletop exercise first, followed by a full-scale exercise. The plan should be updated between these two events in order to incorporate lessons learned from the tabletop exercise.

Plan maintenance

The last step of the plan development process is the plan maintenance step. As with any text, the EOP will likely have areas that could be outdated by the time it is published. Thus the EOP should be considered a living document. If it is created with the intention of having it remain on a shelf in an emergency manager's office, it will not work very well. Plans should be reviewed periodically. The time frame will vary by jurisdiction, but the plan developers should consider updating the plan after completion of drills or incidents involving the use of the plan, as well as when hazards change within the jurisdiction.

Plan components

According to the standard template provided by FEMA, emergency operation plans are divided into the basic plan, functional annexes, and hazard-specific appendices. This division allows the user to use only the portions of the plan needed for the given disaster. Fire service personnel would not desire to read numerous pages about response issues specific to tornadoes if they are standing before a flood. In addition, a member of the structural collapse response team would not want to read pages of the plan that deal with feeding the public.

Basic plan. According to *SLG 101*, "The Basic Plan...provides an overview of the jurisdiction's approach to emergency operations. It details emergency response policies, describes response organizations, and assigns tasks."[19] This portion of the plan is primarily written for the jurisdiction's agency heads. This lays the groundwork for the functional annexes and the hazard-specific appendices. Some of the components of the basic plan are given in the following:

1. **Promulgation document.** This document gives the EOP its official status and provides the authority and responsibility for each agency to perform its tasks.

2. **Dated title page and record of changes.** This allows the user to ensure that he or she has the current version of the plan. Records of all changes, including the dates, should be included in this section.

3. **Record of distribution.** This records that each of the affected agencies has received a copy of the plan. This serves as a record that each of the participating agencies has read the plan and understands its responsibilities within the plan.

4. **Table of contents.** This allows the user of the plan to easily find the desired section. The division of the contents should be user-friendly and methodical. If information is not easy to find, it will likely be overlooked in the time of need.

Although one might feel that an agency executive should know how his or her response agency works collaboratively with the other agencies in the jurisdiction, there are two main reasons to ensure that the basic plan is written and updated periodically. The first is continuity of operations. As stated previously, the agency's chief executive may be unavailable. Second, aside from some of the largest cities in America, disasters are not an everyday occurrence. By having a plan, each agency can refresh training with the plan, and all agencies understand the roles of the other agencies. Essentially, everyone is playing from the same sheet of music.

Functional annex. *SLG 101* states, "Annexes are the parts of the EOP that begin to provide specific information and direction. Annexes should focus on operations: what the function is and who is responsible for carrying it out...annexes should emphasize responsibilities, tasks, and operational actions that pertain to the function being covered."[20] Some of the annexes that should be included in the plan are as follows:

- Command and control
- Communications
- Warning
- Emergency public information
- Evacuation
- Mass care
- Health and medical services
- Resource management

Although each agency may have some role in each of the functions listed in the annexes, it should appoint a lead agency for each function. By tasking an agency with the overall responsibility, it will allow coordination and a final authority for decisions as they relate to the functions. Anyone in the business or government arena has heard the term "committed to death." Collaborative efforts among the various agencies are important, but a final word must be assigned to one agency executive. It is wise to assign the overall responsibility of a given function to an agency that specializes in the particular function. The fire service is an obvious leader of command and control, as well as evacuation.

Hazard specific appendices. *SLG 101* states that "the content of a hazard-specific appendix focuses on the special planning needs generated by the hazard and should not duplicate the information in the functional annex. The appendix contains unique and regulatory response planning details that apply to a single hazard."[21] Typical content of hazard-specific appendices to core functional annexes, provided by FEMA, shows some of the hazard-specific appendices that are associated with the various functional annexes. These appendices are geared toward the task-oriented responders such as urban search and rescue teams.

THE NATIONAL RESPONSE FRAMEWORK AND THE NATIONAL RESPONSE PLAN (SUPERSEDED)

According to the DHS document *Quick Reference Guide for the National Response Plan*,

> As required by Homeland Security Presidential Directive (HSPD)-5, the National Response Plan (NRP) established a single, comprehensive approach to domestic incident management to prevent, prepare for, respond to, and recover from terrorist attacks, major disasters, and other emergencies . . . The NRP can be partially or fully implemented in the context of a threat, anticipation of a significant event, or in response to an incident requiring a coordinated Federal response. This includes events with potential national or long-term implications such as a public health emergency or a cyber incident.[22]

In January 2008, the National Response Framework superseded the NRP. According to the document *National Response Framework*, the NRF

> guides how the Nation conducts all-hazards response. The ***Framework*** documents the key response principles, roles, and structures that organize national response. It describes how communities, tribes, states, the federal government, and private-sector and nongovernmental partners apply these principles for a coordinated, effective national response. And it describes special circumstances where the federal government exercises a larger role, including incidents where federal interests are involved and catastrophic incidents where a State would require significant support. It allows first responders, decision makers, and supporting entities to provide a unified national response.[23]

The National Response Framework consolidates many former national plans and streamlines the federal government's operation as it relates to disasters of all kinds. The NRF is comprised of a base framework, functional annexes, and support annexes. The base framework is comprised of five chapters, titled as follows:

- Roles and Responsibilities
- Response Actions
- Response Organization
- Planning
- Additional Resources

Does this mean that every community must follow this framework? According to DHS,

> The National Response Plan [now updated to the National Response Framework] applies to all incidents requiring a coordi-

nated Federal response as a part of an appropriate combination of Federal, State, local, tribal, private-sector, and nongovernmental entities. The NRP [now NRF] is applicable to all Federal departments and agencies that have primary jurisdiction for or participate in operations requiring a coordinated Federal response. The NRP [NRF] also applies to the American Red Cross, which functions as an Emergency Support Function (ESF) primary organization in coordinating the use of mass care resources.[24]

Therefore, although the local communities are not required to live by this plan on a daily basis, it only makes logical sense to prepare local, regional, and state emergency response plans that will dovetail with this plan. In addition, this plan is always in effect and spells out responsibilities for local, tribal, and state governments.

Plan structure

The structure of the NRF is similar to the structure of the emergency operations plans discussed previously in this chapter. The structure of the NRF, according to the introductory pages of the framework document, contains the following components:

- **Core document.** Contains five chapters, titled as follows:
  - Roles and Responsibilities
  - Response Actions
  - Response Organization
  - Planning: A Critical Element of Effective Response
  - Additional Resources
- **Emergency support functions annexes.** Groups federal resources and capabilities into functional areas that are frequently needed in a national response (e.g., transportation, firefighting, mass care).
- **Support annexes.** Provide guidance for the functional processes and administrative requirements.
- **Incident annexes.** Address the unique aspects of the response to seven broad incident categories (e.g., biological, nuclear/radiological, cyber, and mass evacuation).
- **Partner guides.** Provide ready references describing key roles and actions for local, tribal, state, federal, and private sector response partners.

This is the same format with which the EOPs are created.

Roles and responsibilities of local and tribal authorities

In order for any plan to work effectively, there must be roles and responsibilities assigned within the plan. Second, the persons or organizations responsible for the roles and responsibilities must know, understand, and train for their responsibilities. According to the NRF document, "It is important that each level of government adapt and apply the general roles outlined in the *Framework*."[25] This means that the local level must provide a plan and a commensurate level of resources based on threats present in the jurisdiction. In addition, the plan must identify how to obtain assistance from other local jurisdictions, as well as regional and state resources.

Role of private sector and nongovernmental organizations

Although the private sector is not often a response agency, such as the local or state level, private sector organizations aid the response in almost every major disaster situation. Recognizing this, the federal government

> works with private-sector groups as partners in emergency management. Private sector organizations play a key role before, during, and after an incident. First, they must provide for the welfare and protection of their employees in the workplace. In addition, emergency managers must work seamlessly with businesses that provide water, power, communication networks, transportation, medical care, security, and numerous other services upon which both response and recovery are particularly dependent.[26]

This is the case in many preplanning activities, such as the design of large sports arenas. Although the owner of the arena will not respond to an incident, the creation of the structure around large-scale incident response will aid in the time of need. One example would be to build in a decontamination deluge system at the entrances to the stadium, thus decontaminating the victims as they leave the stadium during a weapons of mass destruction (WMD) attack.

Role of states, territories, and tribal governments

According to the NRF document, the primary "role of State government is to supplement and facilitate local efforts before, during, and after incidents."[27] In addition, "the State provides direct and routine assistance to its local jurisdictions through emergency management program development and by routinely coordinating in these efforts with Federal officials. States must be prepared to maintain or accelerate the provision of commodities and services to local governments when capabilities fall short of demands."[28] Many states (for example, Ohio) are coordinating the response of resources in order to relieve local governments of the pressure of trying to coordinate massive numbers of resources during an emergency. This planning effort has streamlined the delivery of resources. Many states are also formalizing and funding response teams, such as urban search and rescue teams. Lessons learned from prior incidents show that the federal government assistance will take upwards of 24 to 72 hours to actually arrive at the incident, and the states have planned for assets to fill this void.

Role of the federal government

The federal government divides responsibilities based on HSPD-5. Based on this directive, the following departments within the federal government system have the following roles and responsibilities:

Secretary of Homeland Security. This person is responsible for "coordination of Federal resources utilized in the prevention of, preparation for, response to, or recovery from terrorist attacks, major disasters, or other emergencies."[29]

FEMA administrator. "The principle advisor to the President, the Secretary [of Homeland Security], and the Homeland Security Council on all matters regarding emergency management."[30] This position "helps the Secretary [of Homeland Security] in meeting these HSPD-5 responsibilities."[31]

Attorney general. Responsible to lead "criminal investigations of terrorist acts or terrorist threats by individuals or groups inside the United States or directed at U.S. citizens or institutions abroad."[32] Generally working through the FBI, the attorney general is responsible to coordinate the "activities of the other members of the law enforcement community to detect, prevent, and disrupt terrorist attacks against the United States."[33]

Department of Defense. The Department of Defense has significant resources that can aid in the response to a disaster. The Secretary of Defense authorizes support of domestic incidents, so long as they are consistent with military readiness operations and are appropriate under the circumstances and the law. It is noted that the Secretary of Defense retains command of military forces providing civil support.

Department of State. The Department of State, headed by the Secretary of State, is responsible for coordinating "international preparedness, response, and recovery activities relating to domestic incidents and the protection of U.S. citizens and U.S. interests overseas."[34]

Response actions

According to the NRF document, "The *Framework* is intended to strengthen, organize, and coordinate response actions across all levels. The doctrine of tiered response emphasizes that response to incidents should be handled at the lowest jurisdictional level capable of handling the work."[35]

In order to properly respond, preparedness is needed. The preparedness cycle, which is a template to assure proper capability, includes the following: planning; organizing, training, and equipping; exercising; and evaluating and improving. This cycle must continue to occur on a perpetual basis.

Once the incident occurs, organizations must begin the response process. In order to have a successful response, four key actions must occur. These are as follows:

- Gain and maintain situational awareness
- Activate and deploy resources and capabilities
- Coordinate response actions
- Demobilize

Three of the four actions (demobilization excluded) occur in a cyclical process, much like the preparedness cycle.

In order to gain and maintain situational awareness, information channels must be established prior to the incident. Once established, the transfer must occur in a timely manner. The fire service has in the past possessed a limited amount of situational awareness as it relates to terrorism. This fact is changing currently through an IAFC initiative to allow the fire service access to situation reports that relate to terrorist activities. The second method to allow proper situational awareness is the utilization of national reporting. Often planners and responders want information on past incidents to create a quantitative basis for their actions but find that the emergency services fail to report useful information that has consistent measurements. The last suggestion of the NRF is to link operational centers. Utilizing the synergistic effect that is created by having numerous organizations providing information, linking operations centers allows response executives to obtain information from a variety of sources.

The next step in the response process is to activate and deploy resources and capabilities. This begins with the responders assessing the situation, identifying and prioritizing requirements, and activating available resources and capabilities. This process leads to the completion of an incident action plan. In order to ensure proper resources are present, response leaders must activate people, resources, and capabilities and be able to summon additional resources and capabilities. They must also identify needs and preposition resources, based on the intelligence received on the situation.

The third action involves coordinating response actions, which includes managing emergency functions, coordinating initial actions, coordinating requests for initial support, identifying and integrating resources and capabilities, and coordinating information. Failure to coordinate this action will result in resources being wasted, decreased safety due to lack of appropriate resources, and misinformation being relayed to the public.

The last action is demobilization. The safe return of all resources does not occur without a great deal of planning.

Response organization

The NRF provides structures for implementing national-level policy and operational coordination for domestic response. How do NIMS and the NRF interact? How does the NRP coincide with NIMS? NIMS provides a core set of common concepts, principles, terminology, and technologies in the following areas:

- Incident command system
- Multiagency coordination system
- Unified command
- Training
- Identification and management of resources
- Mutual aid and assistance
- Situational awareness
- Qualifications and certification
- Collection, tracking, and reporting of incident information
- Crisis action planning
- Exercises[36]

The various levels of response agencies must work to employ their command structures. However, as noted in prior disasters, these command structures must be integrated. The local responders will usually begin with a simple command consisting of an incident commander, as well as command staff and general staff positions. These positions and their functions are covered in detail in chapter 4.

The next level of command structure that could be utilized is the concept of *area command*, which allows one command element to oversee many commands. Full explanation of this concept is explained in chapter 4.

If the incident escalates to a point at which multiple agency coordination is needed, the implementation of a local EOC is suggested. The "EOCs help form a common operating picture of the incident, relieve on-scene command of the burden of external coordination, and secure additional resources. The core functions of an EOC include coordination, communications, resource allocation and tracking, and information collection, analysis, and dissemination."[37]

If the event exceeds the capabilities of the local responders, the state is the next level at which assistance is gained. According to the *National Response Framework* document,

> State EOCs are activated as necessary to support local EOCs. Therefore, the State EOC is the central location from which off-scene activities supported by the State are coordinated. Chief elected and appointed officials are located at the State EOC, as well as personnel supporting core functions. The key function of State EOC personnel is to ensure that those who are located at the scene have the resources (e.g., personnel, tools, and equipment) they need for the response.[38]

In order to coordinate information being disseminated to those affected by the disaster, the establishment of a joint information center is recommended by the NRF. The *Framework* document states that "the JIC serves as a focal point for coordinated and timely release of incident-related information to the public and the media. Information about where to receive assistance is communicated directly to victims and their families in an accessible format and in appropriate languages for those with limited English proficiency."[39]

The next level involved in a disaster is the federal level. The *Framework* document states,

> The Governor is responsible for requesting Federal assistance for incidents within his or her State. Overall, Federal incident support to the State is generally coordinated through a Joint Field Office (JFO). The JFO provides the means to integrate diverse Federal resources and engage directly with the State. Using unified command principles, a Unified Coordination Group comprised of senior officials from the State and key Federal departments and agencies is established at the JFO. This group of senior officials provides the breadth of national support to achieve shared objectives.[40]

Even though the federal government is now involved in the incident, the state still maintains a presence in the event. According to the *Framework* document,

The *SCO* [State Coordinating Officer] plays a critical role in managing the State response and recovery operations following Stafford Act declarations. The Governor of the affected State appoints the SCO, and lines of authority flow from the Governor to the SCO, following the State's policies and laws. For certain anticipated events in which a Stafford Act declaration is expected, such as an approaching hurricane, the Secretary of Homeland Security or the FEMA Administrator may pre-designate one or more Federal officials to coordinate with the SCO to determine resources and actions that will likely be required, and begin deployment of assets. The specific roles and responsibilities of the SCO include:

- Serving as the primary representative of the Governor for the affected State or locality with the RRCC or within the JFO once it is established.
- Working with the Federal Coordinating Officer to formulate State requirements, including those that are beyond State capability, and set priorities for employment of Federal resources provided to the State.
- Ensuring coordination of resources provided to the State via mutual aid and assistance compacts.
- Providing a linkage to local government.
- Serving in the Unified Coordination Group in the JFO.

Governor's Authorized Representative. As the complexity of the response dictates, the *Framework* recognizes that the Governor may empower a Governor's Authorized Representative to:

- Execute all necessary documents for disaster assistance on behalf of the State, including certification of applications for public assistance.
- Represent the Governor of the impacted State in the Unified Coordination Group, when required.
- Coordinate and supervise the State disaster assistance program to include serving as its grant administrator.
- Identify, in coordination with the SCO, the State's critical information needs for incorporation into a list of Essential Elements of Information (critical items of specific information required to plan and execute an operation and to support timely, logical decisions).[41]

The next level of organization present at a disaster is the federal level. This is present when the state affected by the disaster is resource deficient after state supplementation.

Based on the policy set forth in the *Framework*,

The President leads the Nation in responding effectively and ensuring the necessary coordinating structure, leadership, and resources are applied quickly and efficiently to large-scale incidents. The Homeland Security Council (HSC) and National Security Council (NSC) advise the President on national strategic and policy during large-scale incidents. The HSC and NSC ensure coordination for all homeland and national security-related activities among executive departments and agencies and promote effective development and implementation of related policy. The HSC and NSC ensure unified leadership across the Federal Government.

The Assistant to the President for Homeland Security and Counterterrorism and the Assistant to the President for National Security Affairs coordinate interagency policy for domestic and international incident management, respectively, and convene interagency meetings to coordinate policy issues. Both Councils use well-established policy development structures to identify issues that require interagency coordination.

To support domestic interagency policy coordination on a routine basis, HSC and NSC deputies and principals convene to resolve significant policy issues. They are supported by the following two Policy Coordination Committees at the assistant secretary level:

- *Domestic Readiness Group (DRG).* The DRG is an interagency body convened on a regular basis to develop and coordinate preparedness, response, and incident management policy. This group evaluates various policy issues of interagency importance regarding domestic preparedness and incident management and makes recommendations to senior levels of the policymaking structure for decision. During an incident, the DRG may be convened by DHS to evaluate relevant interagency policy issues regarding response and develop recommendations as may be required.
- *Counterterrorism Security Group (CSG).* The CSG is an interagency body convened on a regular basis to develop terrorism prevention policy and to coordinate threat response and law enforcement investigations associated with terrorism. This group evaluates various policy issues of interagency importance regarding counterterrorism and makes recommendations to senior levels of the policymaking structure for decision.[42]

From an operational standpoint, "The Secretary of Homeland Security is the principal Federal official responsible for domestic incident management. This includes coordinating Federal operations and resource deployments within the United States to prepare for, respond to, and recover from terrorist attacks, major disasters, or other emergencies."[43]

The *Framework* document also states,

The FEMA Administrator is the principal advisor to the President, the Secretary of Homeland Security, and the Homeland Security Council regarding emergency management. The FEMA Administrator's duties include operation of the National Response Coordination Center, the effective support of all Emergency Support Functions, and, more generally, preparation for, protection against, response to, and recovery from all-hazards incidents. Reporting to the Secretary of Homeland Security, the Administrator also is responsible for management of the core DHS grant programs supporting homeland security.

Other DHS agency heads have a lead response role or an otherwise significant role, depending upon the type and severity of the event. For example, the U.S. Coast Guard Commandant has statutory lead authority for certain mass migration management scenarios and significant oil/hazardous substance spill incidents in the maritime environment.

The DHS Director of Operations Coordination is the Secretary's principal advisor for the overall departmental level of integration of incident management operations and oversees the National Operations Center. Run by the Director, the National Operations Center is intended to provide a one-stop information source for incident information sharing with the White House and other Federal departments and agencies at the headquarters level.[44]

In order to ensure that the federal response is coordinated, certain response support structures are present. One of the support structures is the National Operations Center (NOC). The *Framework* states,

---

The NOC is the primary national hub for situational awareness and operations coordination across the Federal Government for incident management. It provides the Secretary of Homeland Security and other principals with information necessary to make critical national-level incident management decisions.

The NOC is a continuously operating multiagency operations center. The NOC's staff monitors many sources of threat and hazard information from across the United States and abroad. It is supported by a 24/7 watch officer contingent, including: (1) NOC managers; (2) selected Federal interagency, State, and local law enforcement representatives; (3) intelligence community liaison officers provided by the DHS Chief Intelligence Officer; (4) analysts from the Operations Division's interagency planning element; and (5) watch standers representing dozens of organizations and disciplines from the Federal Government and others from the private sector.[45]

The *Framework* document also addresses the NOC operational components that provide mission support, including the National Response Coordination Center (NRCC). According to the *Framework*, "The NRCC is FEMA's primary operations management center, as well as the focal point for national resource coordination. As a 24/7 operations center, the NRCC monitors potential or developing incidents and supports the efforts of regional and field components."[46]

Figure 1–5 shows this coordination center in action.

Fig. 1–5. An NRCC operating during a disaster situation. Note the wide variety of organizations represented.

According to the *Framework*, another support structure is the *National Infrastructure Coordinating Center (NICC)*, which "monitors the Nation's critical infrastructure and key resources on an ongoing basis. During an incident, the NICC provides a coordinating forum to share information across infrastructure and key resources sectors through appropriate information-sharing entities such as the Information Sharing and Analysis Centers and the Sector Coordinating Councils."[47]

In addition, the various departments within the federal government have supporting federal operations centers that must maintain situational awareness of their functional areas. According to the *Framework*, one of these support centers is the *National Military Command Center (NMCC)*, which "is the Nation's focal point for continuous monitoring and coordination of worldwide military operations."[48] Other support centers mentioned by the *Framework* are

- ***National Counterterrorism Center (NCTC).***
  - The NCTC serves as the primary Federal organization for integrating and analyzing all intelligence pertaining to terrorism and counterterrorism and for conducting strategic operational planning by integrating all instruments of national power.
- ***Strategic Information and Operations Center (SIOC).***
  - The FBI SIOC is the focal point and operational control center for all Federal intelligence, law enforcement, and investigative law enforcement activities related to domestic terrorist incidents or credible threats, including leading attribution investigations.[49]

Emergency support functions

At the scene of any incident, there are certain categories of tasks that need to be completed to bring the event under control and recover from the event. Based on this premise, the federal government has broken down its response series into emergency support functions (ESFs): "The ESFs provide the structure for coordinating federal interagency support for a federal response to an incident. They are mechanisms for grouping functions most frequently used to provide federal support to states and federal-to-federal support, both for declared disasters and emergencies under the Stafford Act and for non-Stafford Act incidents."[50]

According to the *Framework*, ESFs and their coordinators are listed as follows:

- ***ESF #1—Transportation***
  - ESF Coordinator: U.S. Department of Transportation
    - o Aviation/airspace management and control
    - o Transportation safety
    - o Restoration and recovery of transportation infrastructure
    - o Movement restrictions
    - o Damage and impact assessment
- ***ESF #2—Communications***

- ESF Coordinator: DHS (National Communications System)
    - o Coordination with telecommunications and information technology industries
    - o Restoration and repair of telecommunications infrastructure
    - o Protection, restoration, and sustainment of national cyber and information technology resources
    - o Oversight of communications within the Federal incident management and response structures

- ***ESF #3—Public Works and Engineering***
  - ESF Coordinator: Department of Defense (U.S. Army Corps of Engineers)
    - o Infrastructure protection and emergency repair
    - o Infrastructure restoration
    - o Engineering services and construction management
    - o Emergency contracting support for life-saving and life-sustaining services

- ***ESF #4—Firefighting***
  - ESF Coordinator: Department of Agriculture (U.S. Forest Service)
    - o Coordination of Federal firefighting activities
    - o Support to wildland, rural, and urban firefighting operations

- ***ESF #5—Emergency Management***
  - ESF Coordinator: DHS (FEMA)
    - o Coordination of incident management and response efforts
    - o Issuance of mission assignments
    - o Resource and human capital
    - o Incident action planning
    - o Financial management

- ***ESF #6—Mass Care, Emergency Assistance, Housing, and Human Services***
  - ESF Coordinator: DHS (FEMA)
    - o Mass care
    - o Emergency assistance
    - o Disaster housing
    - o Human services

- ***ESF #7—Logistics Management and Resource Support***
  - ESF Coordinator: General Services Administration and DHS (FEMA)
    - o Comprehensive, national incident logistics planning, management, and sustainment capability
    - o Resource support (facility space, office equipment and supplies, contracting services, etc.)

- ***ESF #8—Public Health and Medical Services***
  - ESF Coordinator: Department of Health and Human Services
    - o Public health
    - o Medical
    - o Mental health services
    - o Mass fatality management

- ***ESF #9—Search and Rescue***
  - ESF Coordinator: DHS (FEMA)
    - o Life-saving assistance

- Search and rescue operations

- ***ESF #10—Oil and Hazardous Materials Response***
  - ESF Coordinator: Environmental Protection Agency
    - Oil and hazardous materials (chemical, biological, radiological, etc.) response
    - Environmental short- and long-term cleanup

- ***ESF #11—Agriculture and Natural Resources***
  - ESF Coordinator: Department of Agriculture
    - Nutrition assistance
    - Animal and plant disease and pest response
    - Food safety and security
    - Natural and cultural resources and historic properties protection
    - Safety and well-being of household pets

- ***ESF #12—Energy***
  - ESF Coordinator: Department of Energy
    - Energy infrastructure assessment, repair, and restoration
    - Energy industry utilities coordination
    - Energy forecast

- ***ESF #13—Public Safety and Security***
  - ESF Coordinator: Department of Justice
    - Facility and resource security
    - Security planning and technical and resource assistance
    - Public safety and security support
    - Support to access, traffic, and crowd control

- ***ESF #14—Long-Term Community Recovery***
  - ESF Coordinator: DHS (FEMA)
    - Social and economic community impact assessment
    - Long-term community recovery assistance to States, tribes, local governments, and the private sector
    - Analysis and review of mitigation program implementation

- ***ESF #15—External Affairs***
  - ESF Coordinator: DHS
    - Emergency public information and protective action guidance
    - Media and community relations
    - Congressional and international affairs
    - Tribal and insular affairs[51]

Field support structure

Due to some disasters being "known" events, the federal government may dispatch assets within hours of the event. According to the *Framework*, one of these support units is the:

> Incident Management Assistance Teams (IMATs), which are interagency, regionally based response teams that provide a forward Federal presence to improve response to serious incidents.
>
> The IMATs support efforts to meet the emergent needs of State and local jurisdictions, possess the capability to provide initial situational awareness for Federal decision makers, and support the establishment of Federal coordination efforts with the State.[52]

Two other teams and support units are the Hurricane Liaison Team (HLT) and the Urban Search and Rescue (US&R) Task Forces. According to the *Framework*,

- **Hurricane Liaison Team (HLT).** The HLT is a small team designed to enhance hurricane disaster response by facilitating information exchange between the National Hurricane Center in Miami and other National Oceanic and Atmospheric Administration components, as well as Federal, State, tribal, and local government officials.

- **Urban Search and Rescue (US&R) Task Forces.** The National US&R Response System is a framework for structuring local emergency services personnel into integrated response task forces. The 28 National US&R Task Forces, complete with the necessary tools, equipment, skills, and techniques, can be deployed by FEMA to assist State, tribal, and local governments in rescuing victims of structural collapse incidents or to assist in other search and rescue missions.[53]

Another support unit is the *Mobile Emergency Response Support (MERS)*. According to the *Framework*, "The primary function of MERS is to provide mobile telecommunications capabilities and life, logistics, operational, and power-generation support required for the on-site management of response activities. MERS support falls into three broad categories: (1) operational support elements, (2) communications equipment and operators, and (3) logistics support."[54]

Other specialized response teams provided by the federal government are:

- Emergency Response Teams (ERT)—to be replaced by the Federal Incident Response Support
- Teams (FIRST) and Incident Management Assistance Teams (IMAT)
- Damage Assessment Teams
- Nuclear Incident Response Team (NIRT)
- Disaster Medical Assistance Teams (DMATs)
- Department of Health and Human Services' Incident
- Response Coordination Team—formerly the Secretary's Emergency Response Team
- Department of Labor/Occupational Safety and Health Administration's Specialized Response Teams
- National Veterinary Response Teams (NVRT)—formerly the Veterinarian Medical Assistance Teams (VMATs)
- Disaster Mortuary Operational Response Teams (DMORTs)
- Medical Emergency Radiological Response Team (MERRT)

- National Medical Response Teams (NMRTs)
- Scientific and Technical Advisory and Response Teams (STARTs)
- Donations Coordination Teams
- Urban Search and Rescue (US&R) Task Forces
- Incident Management Teams (IMTs)
- Domestic Emergency Support Team (DEST)
- Domestic Animal and Wildlife Emergency Response Teams and Mitigation Assessment Team[55]

Joint field office

In order to ensure that local, state, and federal response efforts are coordinated, a joint field office (JFO) is put in place. The JFO is the primary federal incident management field structure. The JFO is a temporary federal facility that provides a central location for the coordination of federal, state, tribal, and local governments and private sector and nongovernmental organizations with primary responsibility for response and recovery. The JFO structure is organized, staffed, and managed in a manner consistent with NIMS principles and is led by the Unified Coordination Group. Although the JFO uses an ICS structure, the JFO does not manage on-scene operations. Instead, the JFO focuses on providing support to on-scene efforts and conducting broader support operations that may extend beyond the incident site.

According to the *Framework*,

> The JFO is led by the Unified Coordination Group, which is comprised of specified senior leaders representing State and Federal interests, and in certain circumstances tribal governments, local jurisdictions, the private sector, or NGOs. The Unified Coordination Group typically consists of the Principal Federal Official (if designated), Federal Coordinating Officer (FCO), State Coordinating Officer, and senior officials from other entities with primary statutory or jurisdictional responsibility and significant operational responsibility for an aspect of an incident (e.g., the Senior Health Official, Department of Defense representative, or a Senior Federal Law Enforcement Official if assigned). Within the Unified Coordination Group, the FCO is the primary federal official responsible for coordinating, integrating, and synchronizing Federal response activities.[56]

Additional federal field structure

Presidential directives and other authorities outline the primary areas of responsibility that guide federal support at national, regional, and field levels. The field structures are designed to implement these lines of responsibility and provide coordination to ensure an effective response. According to the *Framework*, in addition to the JFO, these include:

- **Joint Operations Center (JOC).** The JOC is an interagency command post established by the FBI to manage terrorist threats or incidents and investigative and intelligence activities. The JOC coordinates the necessary local, State, and Federal assets required to support the investigation, and to prepare for, respond to, and resolve the threat or incident.
- **Joint Information Center (JIC).** The JIC is responsible for the coordination and dissemination of information for the public and media concerning an incident. JICs may be established locally, regionally, or nationally depending on the size and magnitude of the incident.[57]

Field support roles

The federal team that assembles to provide unified coordination is composed of multiple senior leaders performing supporting roles tailored to the specific event. Not all of these will be deployed at every incident that involves a federal response.

- **Principal federal official (PFO).** According to the *Framework*, "By law and by Presidential directive, the Secretary of Homeland Security is the principal Federal official responsible for coordination of all domestic incidents requiring multiagency Federal response."[58]

- **Federal coordinating officer (FCO).** According to the *Framework*, "For Stafford Act incidents (i.e., emergencies or major disasters), upon the recommendation of the FEMA Administrator and the Secretary of Homeland Security, the President appoints an FCO. The FCO is a senior FEMA official trained, certified, and well experienced in emergency management, and specifically appointed to coordinate Federal support in the response to and recovery from emergencies and major disasters."[59]

- **Federal resource coordinator (FRC).** According to the *Framework*, "In non–Stafford Act situations, when a Federal department or agency acting under its own authority has requested the assistance of the Secretary of Homeland Security to obtain support from other Federal departments and agencies, DHS may designate an FRC."[60]

- **Defense coordinating officer (DCO).** The *Framework* states that the "DOD has appointed 10 DCOs and assigned one to each FEMA region. If requested and approved, the DCO serves as DOD's single point of contact at the JFO for requesting assistance from DOD."[61]

- **Senior federal law enforcement official (SFLEO).** The *Framework* states that "the SFLEO is an official appointed by the Attorney General during an incident requiring a coordinated Federal response to coordinate all law enforcement, public safety, and security operations with intelligence or investigative law enforcement operations directly related to the incident."[62]

- **Joint task force (JTF) commander.** According to the *Framework*, "Based on the complexity and type of incident, and the anticipated level of DOD resource involvement, DOD may elect to designate a JTF to command Federal (Title 10) military activities in support of the incident objectives."[63]

Proactive federal response

In some disaster circumstances, a proactive deployment of forces is the best option to deliver the needed services in a timely manner. As learned by the highly publicized response to Hurricane Katrina, much of the ESF resources take a considerable amount of time to mount, thus predeployment will speed the needed response.

Putting all of the elements together

Upon notification of the incident, the responsible jurisdiction will initiate the NIMS and designate an incident commander. As many of the incidents are multifaceted, the utilization of the unified command portion of the NIMS is often the best option to manage these incidents. As the incident size or complexity, or both, are realized, the local command will notify the appropriate emergency operations center. Often this is a regional EOC. The EOC will then coordinate multiple agencies and deal with communication issues and resource dispatch and tracking. Once the local and state EOCs determine the need for additional assets, they notify the national operations center (NOC). This often requires the declaration of the disaster from the affected state's governor. Simultaneously, the federal EOC goes online.

Upon receipt of an incident, the NOC assesses the overall situation and makes the initial determination to initiate the coordination of federal information-sharing and incident management activities. When the incident or threat has the potential to become an incident of national significance, the NOC reports to the DHS secretary, who will determine what components of the NRF are warranted.

Within the NRF are the guidelines by which the DHS secretary determines if the event will be managed by the DHS secretary, in terms of federal response, or whether the event should be handled by responsible jurisdictions and agencies. The latter would involve other established authorities and existing plans executed in coordination with the NRF's comprehensive framework of incident annexes.

LEVELS OF PLANNING

As disasters begin, the local responders race to the scene to employ control and mitigation practices in order to deal with the aftermath of the disaster. In many cases, the local responders will realize that an incident of this magnitude exceeds their capabilities and will begin to request assistance in the form of mutual aid. In order to utilize the many resources needed to combat the disaster, the local jurisdiction will request assistance from the state in which the jurisdiction resides. If upon evaluation by the state officials, they realize it is beyond the resources and financial backing that is available from the state, the state will request federal assistance. This continual increase in assistance would not be possible if planning did not occur at various levels. The levels of planning that are common from the emergency management community are local, regional, state, and federal.

National Response Framework

The National Response Framework updated the National Response Plan, learning from past national disasters, to streamline the framework in which the nation responds to disasters, both natural and man-made. The National Response Plan incorporated the many plans that were created by such agencies as the Environmental Protection Agency and FEMA. The National Response Plan addressed the many issues of how all of the federal assets would respond in conjunction and established lines of authority for a wide variety of disasters.

State plans

A state plan provides two areas of support response. The first is a coordination of all of the state's assets. Much like the local plans discussed previously, these plans establish lines of responsibility and provide agencies a plan of how and what operations will be carried out and how to respond to various types of emergencies. Second, state plans address how to provide assistance to the local jurisdiction requesting assistance, as well as how to integrate with the local jurisdiction's EOP. State plans also discuss the opposite side of the assistance chain, dealing with how to request and incorporate assistance from the federal government.

Regional plans

One of the most famous regional plans is the Firefighting Resources of California Organized for Potential Emergencies (FIRESCOPE) program, which grew out of the disastrous California brush fires in 1970. The plan was created to be a coordinated effort of all firefighting agencies in California, joined in one organization to fight wildland brush fires. Administrative complications can arise as an incident commander attempts to unite many large organizations into one cohesive organization. FIRESCOPE agencies include the U.S. Forest Service, California Department of Forestry, California Office of Emergency Service, and numerous fire departments. The project utilizes computers to determine the location of fire crews and equipment and how long it will take them to travel to a given location, and it considers individual capabilities and other operational conditions.

The plan was designed to reduce the cost of fighting fires, as well as losses in structures and watershed. It enables the firefighters to know quickly where the fire will spread, how fast it will travel, where structures are located that need protection, and what resources are available at any time or place. The FIRESCOPE system is the predecessor to the incident command system.

As fire and police departments cope with additional responsibilities such as clandestine lab response and technical rescue response, they have formed regional plans and response units to deal with these highly technical, expensive, and infrequent events. It only makes sense to utilize experts from many departments rather than trying to spend money to train all of a department's personnel.

Local plans

Local plans, which are relatively the same from community to community, establish a plan of response for the local jurisdiction's response agencies. As discussed previously in the disaster plans section, these plans consist of a basic plan, functional annexes, and hazard-specific appendices. This plan coordinates about a dozen emergency divisions such as fire, police, health, utilities, public works, housing and shelter, transportation and communication, property and procurement, personnel, and recruitment. Other areas addressed are succession of command or continuity of operations, emergency procurement of equipment and supplies, and logistical issues. Although these plans are often created by committee, the fire department is often tasked with taking the lead on the plan's development and revisions due to the extensive background these personnel have in responding to disasters and emergencies.

Comprehensive emergency management

Even with a good review of history and the best research to predict future disaster, no planner can account for every possible scenario and know what resources, strategies, and tactics will be required in order to combat the event. Thus planners should focus on comprehensive emergency management. At one time in history, only local volunteer responders and the isolated paid fire department focused on emergency management. In today's environment, everyone from the corporate leader of a Fortune 500 company to the director of a volunteer program to feed the homeless is concerned with emergency management and development of emergency plans. This is due in part to the highly publicized, large magnitude disasters, as well as a recognition that a business that suffers a great disaster, such as a fire destroying its entire operation, is likely not to recover from such an event.

The FEMA publication *Emergency Management Guide for Business and Industry: A Step-by-Step Approach to Emergency Planning, Response, and Recovery for Companies of All Sizes* describes how to research, develop, and implement a comprehensive emergency management plan and program. Some of the components of this program are:

- The four steps in the planning process
  - Establish a planning team
  - Analyze capabilities and hazards
  - Develop the plan
  - Implement the plan
- Emergency management considerations
  - Direction and control
  - Communications
  - Life safety
  - Property protection
  - Community outreach
  - Recovery and restoration
  - Administration and logistics
- Hazard-specific information
- Information sources[64]

As probably noted, much of the content and process is identical to an emergency operation plan development for the local jurisdiction. Due to the success of this process in the public emergency management sector, its adoption to the business and industry (private) sector only made sense. Not only does this ensure that a consistent process is in place for both public and private sector businesses, but it allows a common understanding and cohesiveness in operations between the businesses and public agency heads. The Emergency Management Institute (EMI) hosts numerous classes that explain the plan development process. (Readers are encouraged to visit the EMI Web site for additional training: http://training.fema.gov/index.asp.) Figure 1–6 shows the building in which EMI is located.

Fig. 1–6. EMI headquarters building. This site hosts campus-based courses in emergency management.

CONCLUSION

This chapter introduces the concepts of disaster management. The chapter shows the similarities and the few differences that disasters have in common with daily emergencies to which the emergency services respond. The chapter covered extensively the development of disaster and emergency operation plans. The chapter also covered many of the regulations that are related to disasters, such as the Stafford Act. Understanding plans and how they are developed creates the foundation for the successive chapters on the command and control functions, as well as the strategies and tactics for responding to various types of disasters.

REFERENCES

1. U.S. Department of Homeland Security, Federal Emergency Management Agency. 2006. FEMA home page. Retrieved November 3, 2006 from: http://www.fema.gov.

2. Bahme, C. W. 1978. *Fire Officer's Guide to Disaster Control.* 1st ed. Boston, MA: NFPA.

3. Dictionary.com. 2006. Disaster—definitions from Dictionary.com. Retrieved November 5, 2006 from: http://dictionary.reference.com/browse/Disaster.

4. International Association of Fire Chiefs. 2006. Clarifying the IAFC's role in mutual aid initiatives. March 22. Retrieved November 8, 2006 from: http://www.iafc.org/displayindustryarticle.cfm?articlenbr=29847.

5. Citizen Corps. 2006. Programs and partners. *Community Emergency Response Team (CERT).* Retrieved November 8, 2006 from http://www.citizencorps.gov/programs/cert.shtm.

6. Army National Guard. 2006. Our state mission. Retrieved November 8, 2006 from: http://www.arng.army.mil/aidingamerica.aspx.

7. U.S. Department of Homeland Defense, Federal Emergency Management Agency. 2006. About FEMA. Retrieved November 12, 2006 from: http://www.fema.gov/about/index.shtm#0.

8. Ibid.

9. U.S. Department of Homeland Security, Federal Emergency Management Agency. 2007. Stafford Act. June. Retrieved July 15, 2008 from: http://www.fema.gov/pdf/about/stafford_act.pdf.

10. Bea, K. 2005. *Federal Stafford Act Disaster Assistance: Presidential Declarations, Eligible Activities,*

and Funding. August 29. pp. CRS-1–CRS-12. Retrieved November 8, 2006 from: http://fpc.state.gov/documents/organization/52793.pdf.

11. Ibid. pp. CRS-1–CRS-2.
12. Ibid. p. CRS-2.
13. Ibid.
14. Ibid. pp. CRS-2–CRS-3.
15. Ibid. p. CRS-3.
16. Ibid.
17. Ibid. p. CRS-4.
18. U.S. Department of Homeland Security, Federal Emergency Management Agency. 2006. *State and Local Guide (SLG) 101: Guide for All-Hazard Emergency Operations Planning*. Chapter 4. Basic plan content. Retrieved November 12, 2006 from: http://www.fema.gov/plan/gaheop.shtm.
19. U.S. Department of Homeland Security, Federal Emergency Management Agency. 2006. *State and Local Guide (SLG) 101: Guide for All-Hazard Emergency Operations Planning*. Chapter 4. Basic plan content. p. 4-1. Retrieved November 13, 2006 from: http://www.fema.gov/pdf/plan/4-ch.pdf.
20. U.S. Department of Homeland Security, Federal Emergency Management Agency. 2006. *State and Local Guide (SLG) 101: Guide for All-Hazard Emergency Operations Planning*. Chapter 5. Functional annex content. p. 5-1. Retrieved November 13, 2006 from: http://www.fema.gov/pdf/plan/5-ch-0.pdf.
21. U.S. Department of Homeland Security, Federal Emergency Management Agency. 2006. *State and Local Guide (SLG) 101: Guide for All-Hazard Emergency Operations Planning*. Chapter 6. Hazard-unique planning considerations. p. 6-2. Retrieved November 13, 2006 from: http://www.fema.gov/pdf/plan/6-ch-0.pdf.
22. U.S. Department of Homeland Security. 2006. *Quick Reference Guide for the National Response Plan*. Version 4.0. May 22. p. 1. Retrieved October 14, 2008 from: http://www.apco911.org/new/commcenter911/downloads/Homeland_Security_Quick_Reference_Guide.pdf.
23. U.S. Department of Homeland Security. 2008. *National Response Framework*. Appendix 6. Glossary and acronyms. January. Retrieved October 14, 2008 from: http://www.fema.gov/pdf/emergency/nrf/nrf-glossary.pdf.
24. U.S. Department of Homeland Security. 2006. *Quick Reference Guide for the National Response Plan*. Version 4.0. May 22. p. 1. Retrieved October 14, 2008 from: http://www.apco911.org/new/commcenter911/downloads/Homeland_Security_Quick_Reference_Guide.pdf.
25. U.S. Department of Homeland Security. 2008. *National Response Framework*. January. p. 5. Retrieved March 17, 2008 from: http://www.fema.gov/pdf/emergency/nrf/nrf-core.pdf.
26. Ibid. p. 18.
27. Ibid. p. 6.
28. Ibid. p. 21.
29. Ibid. p. 25.
30. Ibid.
31. Ibid.
32. Ibid.
33. Ibid.
34. Ibid. p. 26.
35. Ibid. p. 27.
36. Ibid. pp. 48–49.
37. Ibid. p. 50.
38. Ibid. p. 51.
39. Ibid. p. 52.
40. Ibid.

41. Ibid. pp. 51–52.
42. Ibid. pp. 53–54.
43. Ibid. p. 54.
44. Ibid. p. 55.
45. Ibid.
46. Ibid. p. 56.
47. Ibid.
48. Ibid.
49. Ibid. p. 57.
50. http://www.fema.gov/pdf/emergencies/nrf/nrf-esf-intro.pdf.
51. Ibid. pp. 58–59.
52. Ibid. p. 62.
53. Ibid.
54. Ibid.
55. http://www.whitehouse.gov/infocus/homeland/nshs/2007/sectionVII.html.
56. U.S. Department of Homeland Security. 2008. p. 64.
57. Ibid. p. 65.
58. Ibid. p. 66.
59. Ibid. p. 67.
60. Ibid. p. 68.
61. Ibid.
62. Ibid.
63. Ibid.
64. U.S. Department of Homeland Security, Federal Emergency Management Agency. 1993. *Emergency Management Guide for Business and Industry: A Step-by-Step Approach to Emergency Planning, Response, and Recovery for Companies of All Sizes*. October. Washington, DC: FEMA. Retrieved August 29, 2008 from: http://www.fema.gov/pdf/library/bizindst.pdf.

HISTORICAL LESSONS 2

INTRODUCTION

History is noted to repeat. This can be good and bad simultaneously. The good part of history repeating is that a historical study can be conducted to determine what might happen in the future. From a disaster perspective, it can be noted that hurricanes have active and inactive cycles based on the data collected by the National Hurricane Center over the past decades. With this information, it can be predicted that if this cycle remains consistent and actually does repeat, it should be possible to predict times in which resources and planning need to be top notch in a hurricane-prone area in order to mitigate a hurricane's effects. As another example, history has shown that California is subject to a higher frequency and magnitude of earthquakes than other portions of the country. Thus the focus of disaster management there should involve a good deal of earthquake management. This focus on the possibility of earthquakes can be seen in the building codes enacted in California, requiring buildings there to be constructed to withstand certain levels of seismic activity.

On the negative side of history repeating, there has also been poor disaster response. Since history does not repeat on a predictable basis, in some instances, poor prevention codes have been adopted. There can also be a general lack of enthusiasm on behalf of those who need to be involved in the planning process. This can be seen in many federal grant programs. According to FEMA, "In 1991, the Federal Emergency Management Agency (FEMA) incorporated this concept [the National US&R program] into the Federal Response Plan (now the National Response Plan). Events such as the 1995 bombing of the Alfred P. Murrah building in Oklahoma City [and] the Northridge earthquake...underscore the need for highly skilled teams to rescue trapped victims."[1] However, the period of 1996 through 2001 saw limited funding to the US&R system. Funding had an insurgence after the 9/11 attacks and revamped the composition of the teams, but in the funding years post-9/11, the funding decreased due to that imminent threat not being present.

History has shown continual problems with planning for disasters and responding to their aftermath, thus leading one to believe that the lessons learned are not being applied properly. Furthermore, there is a continual attrition cycle of both emergency responders and emergency planners and executives. These challenges make it imperative that today's emergency service leaders from police, fire, public works, etc. study the past to ensure they are prepared not to make the same mistakes as their predecessors.

COMMUNITY RISK ASSESSMENT

Although many communities around the country share certain similarities, each has its own unique character. This character and image are usually a culmination of many facets of the community. This character is also what needs to be preserved in order for a community to survive a disaster. Emergency planners must understand what is important to the community and what particular aspects of different losses would mean to the community. For example, Cincinnati, Ohio is home to the Proctor & Gamble Co. Figure 2–1 shows the Proctor & Gamble headquarters. Although not every person in the community works for this employer, many of the residents either work directly or indirectly for this company. Loss of Proctor & Gamble would put a drastic strain upon the community. Therefore, emergency planners must place a high emphasis on this company when developing emergency plans.

Fig. 2–1. Proctor & Gamble headquarters

Photo courtesy of associate author Randall W. Hanifen.

Other than just trying to guess at what is important to the community, how can emergency managers figure out what is important in planning? There are numerous risk assessment tools for a community to utilize. The National Oceanic and Atmospheric Administration has a CD-ROM entitled "Community Vulnerability Assessment Tool," or CVAT. This tutorial program instructs emergency managers how to conduct a community vulnerability assessment for a variety of disasters. This assessment process is broken into seven steps. Figure 2–2 depicts this process in a flowchart. The reader is encouraged to copy this figure prior to starting the process. The steps, as listed by the program, are as follows:

1. Hazard identification
2. Hazard analysis
3. Critical facilities analysis
4. Societal analysis
5. Economic analysis
6. Environmental analysis
7. Mitigation opportunities analysis

Hazard identification

Step 1, hazard identification, involves deciding what hazards are a concern and determining how they should be prioritized.[2] Some of the hazards present in this section of the process are storm surge, wind, flood, earthquake, etc. This list can be comprehensive or can involve planning for only one potential disaster at a time. The priorities are established in the same fashion as the hazard identification processes used in the creation of emergency planning, which involves the creation of a matrix. The CVAT process suggests the use of the following factors: frequency, area impacted, and magnitude. By placing a numerical value in each of these categories and multiplying the factor, the relative importance of the hazard can be determined.

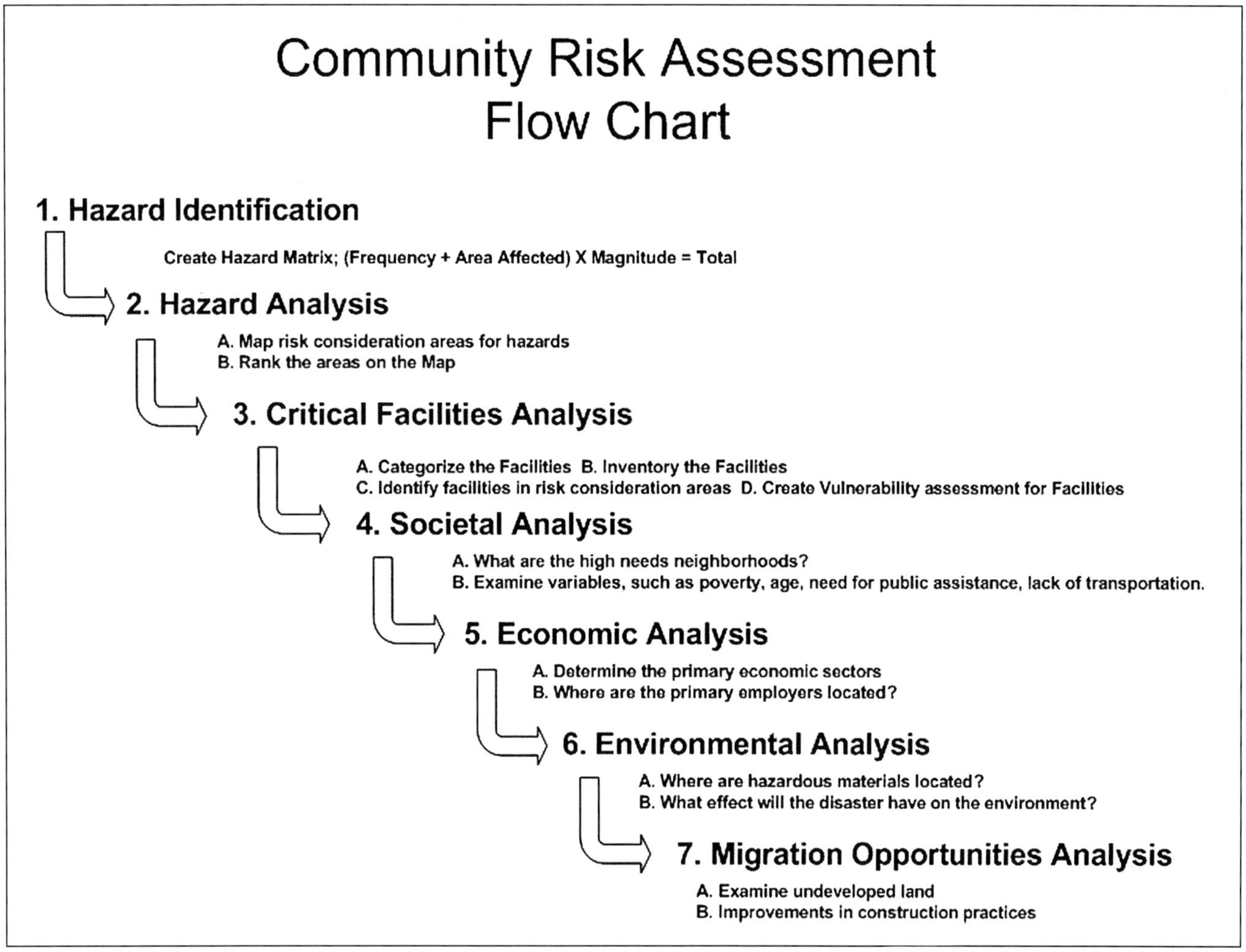

Fig. 2–2. Community risk assessment flowchart

An example of this process can be examined with the impact of Hurricane Katrina on New Orleans. The other hazard example in the matrix is an earthquake in New Orleans. For the hurricane, a frequency number of four can be used. This is arbitrary based on its location on the Gulf Coast, but comprising only a small area in comparison to the whole Gulf Coast. The area impacted is large in relation to population and given the overall width of hurricanes; thus an eight will be assigned to this category. The final category, magnitude, will be assigned a seven, as total destruction may not occur, but it will have a large impact on the city for a great length of time.

For the earthquake example, frequency could be assigned a two, as earthquakes are not prevalent in New Orleans. However, earthquakes can still occur and thus cannot be disregarded entirely. Furthermore, if the New Madrid fault line produces an earthquake above a 6.0 on the Richter scale, it could affect New Orleans. The area affected would depend upon its magnitude. Using a 7.0 on the Richter scale, widespread damage would be possible, as evident based on the maps provided on the *New Madrid Seismic Zone Maps* Web site at http://showme.net/~fkeller/quake/maps.htm. The category of area affected could thus be assigned a six. As to the magnitude, it could be assigned a value of seven as well, due to the lack of seismic planning in the area. Looking at table 2–1, it is evident that due to frequency, a hurricane is the priority hazard.

Table 2.1 Hazard Matrix for New Orleans

Hazard Matrix

Hazard	Frequency +	Affected Area X	Magnitude =	Total
Hurricane	4	8	7	84
Earthquake	2	6	7	56

(Frequency + Area Affected) X Magnitude =Total

Hazard analysis

Step 2, hazard analysis, involves answering the two questions: "Where are your risk consideration areas? How would you prioritize these risk areas?"[3] In order to answer the first question, the step 2a process calls for mapping risk consideration areas for hazards. This will allow the planners to see what areas could be affected by a particular hazard. This works best for flooding or hazardous materials releases but has less accuracy in disasters such as an earthquake or hurricane, in which the entire area the planners are responsible for could be affected simultaneously and uniformly.

The step 2b process involves further ranking the areas outlined in step 2a. An example would involve designating relative levels for the different zones, such as the imminent flood areas, the 100-year flood plain, the 200-year flood plain, etc. This helps further limit the areas that will provide the biggest planning needs, as well as identify levels of need based on the severity of the disaster. In the flood example, a given number of resources are needed if the floodwaters rise 10 feet. However, if they rise 20 feet, this need could triple due to the increase in area affected.

Critical facility analysis

Step 3 calls for examination of the critical facilities. This step of the CVAT process from NOAA asks for evaluation of two questions: "What are your critical facilities and where are they located? How vulnerable are they to physical and operational impacts from hazards?"[4] This step contains four sections. The first section involves categorizing the critical facilities. Some categories could be nursing homes, criminal institutions, fire stations, police stations, or communication centers.

Step 3b involves inventorying the critical facilities in the various categories. One possible way to speed this operation is to query the fire department to see if its occupancy files show the occupancies by type. Many of the computer databases used by fire departments have this capability, which could save an immense amount of time compared to physically inventorying these facilities. Some of the information that the planners should collect are full address, hours of operation, and emergency contacts.

In step 3c, the planners need to identify which facilities are within the zones created in step 2. The facilities in the zones that have the highest probability of being affected should receive priority in full planning, as the probability of a catastrophic event happening at these facilities is highest. An example of this information aiding disaster planners is to decide which facilities would need evacuation first during a flood. It would be of little benefit to evacuate a facility that will not flood for two days, while leaving people in a facility that is beginning to fill with water.

The final portion of step 3 involves creating a vulnerability assessment for all of the facilities identified.

Societal analysis

The next step in the community risk assessment is an analysis of the society involved in the area being planned. In this section, the NOAA asks the planners to consider the following questions: "What are your high-need neighborhoods? How vulnerable are they to hazard impacts?"[5] Some of the variables that need to be examined are poverty, age, lack of transportation, public assistance, etc. The first process in this step is to identify areas with high-needs neighborhoods. This is more easily accomplished in some cities than others, so planners should elicit the help of volunteer agencies and government services organizations to aid in identifying these areas. The second process in this step is to identify where the high-needs neighborhoods intersect with the high-risk areas for the given hazard. In the example of Hurricane Katrina, many of the people left behind in the Superdome were from high-needs neighborhoods. Many of the people were elderly or had no transportation to evacuate the area. The final process in this step is to take an inventory of all the locations defined as special needs neighborhoods in high-risk areas. This will allow for prioritization of these sites, as well as facilitate planning how to respond to disasters that affect the areas, such as determining how large a population is contained in each of these areas that might need to be evacuated.

Economic analysis

The fifth step is to conduct an economic analysis. Many towns depend on the employment of a major business, such as a paper factory or a steel mill. In fact, many towns exist because of these anchor businesses. An example of one such area is Norwood, Ohio. A General Motors plant operated in the town, employed many of the residents, and provided a tax base that funded three firehouses. When the plant closed, many of the town's residents were left without good paying jobs, including members of the fire department, which was downsized to one firehouse.

According to the NOAA CVAT, the questions to be answered in this step are: "What are your primary economic sectors and how vulnerable are they to hazards? Where are your largest employers and how vulnerable are they to hazard impacts?"[6] In this step, there are five substeps to complete. The first is to identify and locate all of the economic sectors. This can easily be accomplished with the help of the area's chamber of commerce. They will likely have economic facts on the area under consideration. The second process is to identify where these key economic areas intersect the hazard area. Substep 3 is to create an inventory and prioritize these areas based on the hazard locations. The fourth and fifth substeps are to locate and cross-reference the large employers and the hazard areas in order to create a vulnerability assessment of the structures used by the large employers. By having these areas planned through a vulnerability assessment, the emergency manager can assign resources to provide preventive measures to these facilities. In the instance of a flood, it may only take placing sandbags along the perimeter of the large employer's facility so that damage can be averted and the citizens can remain employed in the postdisaster environment. This will provide one less step to full recovery of the town.

Environmental analysis

During the environmental analysis step, planners need to answer the following questions, as suggested by the CVAT in its sixth step: "Where are your hazardous materials located and how vulnerable are they to natural hazards? How vulnerable are critical natural resources to secondary hazard impacts?"[7] In this step, planners need to perform four tasks. Those involve locating the sites that contain hazardous materials and determining which sites intersect with hazard areas. In addition, planners must identify critical natural resources in the same area and conduct a vulnerability analysis on each of these sites.

During this phase, planners should enlist the assistance of the local emergency planning commission in order to speed identification of sites that contain hazardous materials. Another useful partner in this phase is the local fire department. If fire service personnel conduct preincident plans, it is likely they have identified sites that contain hazardous materials in their district. In addition, a state's division of natural resources can aid in identifying critical natural resources in the area and can provide insight on the impacts of their disturbance from a disaster.

Mitigation opportunities analysis

The final step in the community risk assessment is to conduct a mitigation opportunities analysis. According to the CVAT, during this step planners need to ask and be prepared to answer the following question: "Where are your best opportunities for changes in mitigation policy that will have significant impacts on future hazard vulnerability?"[8] During this step, planners should examine any undeveloped land in the areas that are likely to be impacted by disasters, and recommend improvements during construction to make the facilities less vulnerable to disasters. They can also examine any code improvements that could benefit existing facilities or buildings. Although many citizens will not enact their recommendations for improvements due to financial constraints, having the information in a postdisaster state will allow for improved rebuilding, thus reducing the impacts of subsequent disasters.

GEOGRAPHICAL DIFFERENCES

Response to disasters and the preparedness to handle a disaster will vary based on the geographic characteristics of the area in question. For instance, a Mississippi River flood is likely to be planned for by the communities in the areas surrounding the river. However, if the same type of flood occurred in the northern portion of Ohio, the preparedness levels would be decreased due to the infrequent occurrence of such an event. During the planning phase, planners will encounter areas that must be addressed by the plan, but they lack the resources in the immediate vicinity to handle the problems. For example, consider a flash flood. In the California spillways, flash floods are yearly events, and when citizens become trapped, the local rescue service is prepared to make helicopter rescues in order to save lives. However, if this same scenario occurs in an area that may see floods, but rarely flash floods, then responders have to rely upon less than a well-scripted helicopter rescue. Although it would be great for each community to have an asset such as the helicopter rescue team, the initial training, ongoing training, and the equipment far exceed the anticipated use of the service.

This is the same as rural versus urban fire protection. Urban fire protection is based on large amounts of manpower, built-in fire protection, and specialized high-rise techniques. These operational methods would not fare well if the fire instead occurred in a rural setting where reliance is on natural sources of water, tanker shuttles, and summoning of enough volunteer forces. However, contingency plans must be in place. As land becomes more expensive, large manufacturing and distribution facilities are locating in these once-rural areas. In a good portion of the instances where this occurs, the volunteer fire department is left to contend with incidents at these facilities, with little to no funding increase. This again displays the vital importance of working mutual aid agreements.

Many of the types of disasters that will be studied may not have a high frequency of occurrence in some areas of planning. However, having a plan and identifying the resources prior to the disaster will ensure the incident will have the best possible outcome. Study of the subsequent chapters pertaining to the prevalent types of disasters and tailoring the response plan to each community is a recipe for success.

PLANS DEVELOPMENT AND DISASTER PREPAREDNESS

In the past, the fire service spearheaded much of the disaster planning as a natural extension of its job, as discussed in chapter 1. The need for disaster preparedness has grown over the past years and has become a priority for all groups within the municipality's government. This includes the police department, services departments, and in many cases, a newly created emergency management department. This cooperation and desire to create and test plans was not always the case. An IAFC study presented in *Fire Officer's Guide to Disaster Control* stated,

> A survey of fire chiefs conducted by the International Association of Fire Chiefs (IAFC) indicated problems that confront chiefs in plans development and disaster preparedness. Among the most frequent responses were these:
>
> - Lack of interest by other agencies and officials
> - Limited resources—personnel, money, and time—for planning
> - Inadequate cooperation with other organizations
> - Lack of common mutual aid radio frequency with fire and rescue agencies.[9]

This study was conducted by the IAFC in 1980 and still had relevance in 1992. Examination of these problems reveals that some of them are still echoed today. It is true that planning has become a municipality-wide process, and many fire chiefs have direction from their bosses to allocate personnel, money, and time for planning. However, many are still trying to establish cooperation between organizations, and until recently, the interoperability problem was present and no plausible solutions were in the works.

In 2006, there was a major push from the federal government, with associated financial incentives, to create a formalized and consistent disaster management program. Born out of *Homeland Security Presidential Directive/HSPD-5*, the National Incident Management System

> will provide a consistent nationwide approach for Federal, State, and local governments to work effectively and efficiently together to prepare for, respond to, and recover from domestic incidents, regardless of cause, size, or complexity. To provide for interoperability and compatibility among Federal, State, and local capabilities, the NIMS will include a core set of concepts, principles, terminology, and technologies covering the incident command system; multi-agency coordination systems; unified command; training; identification and management of resources (including systems for classifying types of resources); qualifications and certification; and the collection, tracking, and reporting of incident information and incident resources.[10]

This push stemmed from the response to the 9/11 attacks, which, as seen at any previous disaster, demonstrated the need for consistency and a formalized plan in order to deal with an event of significance.

The benefit of disaster planning

One of the obvious benefits of NIMS adoption is eligibility for federal funding. Since NIMS adoption is a contingency of receiving federal assistance of any kind for the community, fire service personnel need to examine the realistic benefits of spending the inordinate amount of time required to create a disaster plan and practice implementing it.

Each day, when a person goes to work, he likely gets out of bed, showers, dresses, gets into his car, and drives to work. It is likely that he has developed a routine in which all of these functions happen in a monotonous, methodical fashion. However, consider what would happen if that same person awoke in the middle of the woods and realized that there was no shower or clothes available, and he needed to be at work in 30 minutes. What would he do? Many people are inventive, but few plan to wake up with no clothes in the middle of the woods on a work day. Most would have to experiment with different plans to discover how to overcome this problem. Some of these experiments may work; others may not.

This scenario is not far removed from what could occur as a community responds to a disaster. Police officers normally deal with traffic patrol and keeping the peace within the community. How will they react upon a civil disturbance situation that has resulted from widespread destruction and the citizens' attempts to procure supplies for their survival? The fire department responds to a variety of small disasters on a regular basis, such as high-rise fires or hazardous materials leaks. But what procedures do they enact when none of the fire apparatus or command officers can talk to each other via the radio system because a tornado has destroyed all of the radio towers?

Questions such as these should not be left to a "Let's decide when the time comes" attitude. Many executives say, "Why plan for the far-reaching events? We need to focus on what we do every day." Although organizations should emphasize what service they deliver day in and day out, they should also have planners at all levels in the organizations' determining processes and procedures that should be enacted for the day that is not the normal service delivery day.

Some might ask, "Can't standard operating procedures be used as a plan? We already follow these procedures every day." It is important to remember that standard operating procedures are just that: standard. Many procedures for caring for the injured and ill are modified greatly during a disaster. Emergency medical crews would not be expected to respond to injuries from a fall, and within five seconds, decide the victim cannot be salvaged, leave the victim behind, and move on to the next call. However, in a time of disaster, EMS crews can make fast triage of a large number of patients, utilizing a system known as *START triage*. In a START triage, this quick assessment and decision is made, thus freeing available resources to tend to patients who have the highest likelihood of survival.

Another issue that planners address is the old adage, "No plan can cover everything." Although this is true, a plan does assign various responsibilities because of the disaster, which allows those services to operate and render care or assistance as the situation dictates. Who would envision the far-reaching effects of just turning off the electricity in a town, as was the case in New York, Pennsylvania, and Ohio on Thursday, August 14, 2003? One of the major issues became the heat. As this loss of power occurred during a heat wave, many elderly were adversely affected by the staggering temperatures. Emergency planners must consider what steps their community would take to aid the elderly in a similar time of need. Planners also have to consider how they would fuel their vehicles. These are some very basic problems that fire service personnel and other emergency responders could encounter with nothing more than a loss of electricity.

Others might ask, "Why plan? We are just a small community, and nothing happens here." Although many of the highly publicized disasters do occur in major metropolitan areas, disasters can occur anywhere and at any time. Take for instance the flood that hit Hogate, Ohio on May 5, 2000. Floodwaters of the town's river overflowed their banks and flooded the entire town. Although only 25 homes were damaged, evacuation had to take place, and shelter had to be provided for those who were evacuated. This disaster likely did not make any national headlines, but many of the same problems that were present in the flooding following Hurricane Katrina were likely encountered in this event. Whereas Katrina had many local, state, and federal responders, the town of Hogate relied upon citizens of the town and the neighboring community to aid with the response to the disaster.

Scope of the plan

Although inventorying assets and response personnel, as well as volunteers, is a good start to the planning process, it is merely the tip of the iceberg. Planners who inventory the local and regional resources have created a picture of what could be called for at that particular time, but perhaps not what will be required two years later. They have to ask questions such as, "What will the assets do when they arrive?" "How will they be contacted?" "What protective equipment do they need?" All of these are issues that need to be addressed in order for the resources to arrive and perform as the planners envision. Planners also must contend with business coming and going, changes in the police department's and fire department's response capabilities, and volunteers moving away. Continual updating is critical if the plan is to function as designed.

How broad or narrow should the plan be? A plan will address the four basic questions of *what*, *when*, *where*, and *how*. During the aftermath of Hurricane Katrina, a plan was devised to utilize America's firefighters to aid in the FEMA assistance distribution. The e-mail from FEMA called for two-person teams from fire departments, preferably career, to fly into Atlanta for training and deployment for a period of 30 days. They were to bring no more with them than what could fit into a backpack. The associate author, Randall W. Hanifen, volunteers for one of FEMA's US&R teams, in which a 10-day deployment requires nearly 80 lbs of personal gear and attire in order for a team member to be self-sustaining for that 10-day period. This is a case of answering the *what*, *when*, *where*, but failing to answer the *how*. This disorganization and half-planning made the fire service headlines for days as these deploying teams removed themselves from service as they learned of their fate for the next 30 days.

Although a plan is essentially a guide, not an operations manual to follow to the letter, it must answer the basic questions and should be tested prior to actual implementation. Had this firefighter assistance program been planned and exercised even through the tabletop stages of the exercising elements, planners would have soon realized a lack of essential elements, such as changes of clothes, medicines, vaccinations, etc.

HISTORICAL BACKGROUND

Disaster planning began with the beginning of mankind. In prehistoric days, disasters were controlled with the use of a club. As mankind progressed, so did his form of disaster planning. Castles were built in high areas, with moats for protection and portholes from which to fire weapons, thus fending off enemies. This type of planning ensured success when battle time arose. If the occupants of the castle had chosen instead to wait until they were under attack to design their fortress, it would not be built in time or have the benefits of the precise planning and building that could take place during a time of peace.

Rather than embark on a long history lesson of disaster planning, this text will only be concerned with the past half century. One trend that is evident through this period is that as disaster planners began to work collaboratively, planning began to take on a more formalized and organized approach.

Disaster planning in the 1960s

According to the government report *The Role of Local Civil Defense in Disaster Planning*, the basic characteristics of disaster planning were identified and studied from the 1960s through the 1970s.[11] Some of its conclusions identify significant events, forces, and developments from earlier historical periods that have affected subsequent disaster planning.

In the government report, the decade from 1960 to 1970 is described as a period of transition in disaster planning. Certain common attributes among cities were identified and some basic trends were listed. The following paragraphs identify characteristics of that period.

In the 1960s, disaster planning was located in three different segments of the community, with minimal contact among the segments. The three segments were the local civil defense office, private health and welfare agencies, such as the American Red Cross and the Salvation Army, and municipal emergency organizations, such as police and fire departments. At the time, the civil defense agencies were concentrating on the threat of nuclear attack. The fire department and other agencies with stronger ties to serving the local community became much more concerned with conflagrations, floods, hurricanes, and tornadoes, which they felt presented more imminent types of threats.

Consequently, most community organizations tended to engage in planning in isolation from other community organizations. Because of the various contradictory pulls that many community organizations felt, each organization tended to develop its own planning, often with little relationship to other organizations within the community.

Planning throughout the 1960s tended to be cyclical and spasmodic. At the beginning of the 1960s, the Berlin crisis, and more importantly, the Cuban Missile Crisis, resulted in increased federal funding for certain types of programs related to civil defense, particularly fallout shelter construction, location, and supply.

Direction of planning in the 1970s

Although the previously mentioned trends seemed to have dominated the planning efforts in the 1960s, many crosscurrents were creating change within American communities. These were the influences for the changes that came about in the 1970s.

The first change was the broadening of the scope of the planning. Natural disasters always received attention in many disaster plans, but now many new hazards needed to be added to the plans. One example occurred with the addition of the national highway system, which allowed the corridors to be filled with hazardous materials shipments. The standard plan that involved fixed-site planning had to be modified due to the variety of chemicals that could be encountered on the highways.

The second change involved a shift in the focus of disaster planning from having a great emphasis on national security to retaining the viability of the local community. The assumption from the 1960s was that the local communities would work towards supporting the federal government's goal of national security. As communities found the national security problems did not come to fruition, they began to focus on the local problems, such as regular flooding or preparing for hazardous materials leaks from key facilities. Although the communities did not discard national security completely, they began a more diverse approach.

The third change was the number of organizations involved in community planning. In the 1960s many organizations did their own ad hoc planning, but as experience in disasters grew, it was recognized that the synergistic effect of planning in unison was far more beneficial than many separate plans. Much of this collaborative effort was led by fire service personnel, who experienced the lack of planning in real-world incidents.

Trends in the 1980s and 1990s

The first new trend was that of the federal focus in planning. Many new regulations, programs, and networking opportunities aided in this shift. The first program was the Integrated Emergency Management System (IEMS), which was a FEMA program developed to facilitate the development of comprehensive emergency management capability at the federal, state, and local levels of government.

The second new trend was the training of the nation's fire chiefs at the National Fire Academy. The National Fire Academy offers a number of courses in disaster planning. This developed consistency in the nation's leading planning executives. The academy, housed with the Emergency Management Institute and offering courses developed in conjunction with the EMI, ensured that the nation's fire chiefs and emergency planners took the same information back to their communities.

The third trend occurred due to the enactment of the Superfund Amendment and Reauthorization Act (SARA) Title III. This national act mandated many planning and response requirements in order to meet the law. SARA Title III is described in detail in chapter 9.

Incorporation and developments in technology

Although technology is often slow to be adopted in the public service arena, mostly due to financial constraints, the 1990s saw a few great technology improvements in the field of disaster response. The first improvement was that of computer-aided dispatching. This program allowed computer tracking of the status of units. Second, based on information obtained from the phone line, the computer could recognize and prompt the correct units based on the type of call. This system also featured automated vehicle locators, which gave the system the ability to dispatch the closest units, based on their real-time location.

The second technology advancement of this decade was the computer database known as Computer-Aided Management of Emergency Operations (CAMEO). This program allowed responders to quickly obtain crucial information about hazardous chemicals at the scene of an incident. This program also is described in detail in chapter 9.

Integrated emergency medical response

As many fire departments took on the role of emergency medical services (EMS), this mass care capability was integrated and controlled by the local government. This control, as well as the cross training of firefighting personnel, allowed a more powerful and coordinated response to mass casualty disasters.

Interagency cooperation

In the 1960s, agencies planned on their own. In the late 1970s, some integration occurred, but in the 1980s, a genuine attempt for harmonious integrated planning among all agencies became evident. New mutual aid pacts were developed to ensure that disasters that crossed jurisdictional borders were handled with coordination. If an event of magnitude occurred in one community, resources would respond from adjacent communities.

CHANGES IN THE NEW CENTURY

A new federal look

During just the first six to seven years of the new century, disaster planning and response saw major changes. In the weeks following the 9/11 attacks, "The President [George W. Bush]—using the maximum legal authority available to him—created the Office of Homeland Security and the Homeland Security Council...as an immediate step to secure the homeland."[12] With this transformation to a Department of Homeland Security, the following divisions were placed under this newly created department:

- Border and Transportation Security
- Emergency Preparedness and Response
- Chemical, Biological, Radiological, and Nuclear Countermeasures
- Information Analysis and Infrastructure Protection

Within this transition, the Federal Emergency Management Agency was placed under the Department of Homeland Security, as well as the National Fire Administration and the Assistance to Firefighters Grant Program. All of the nation's emergency response services have been placed into one department, thus making the attainment of information and training much easier and more condensed.

National Incident Management System

"Homeland Security Presidential Directive 5: *Management of Domestic Incidents* (HSPD-5)... requires DHS to lead a national, coordinated effort with other Federal departments and agencies and State, local, and tribal governments to establish a National Response Plan and the National Incident Management System (NIMS)."[13]

The National Incident Management System requires each federal, state, local, and tribal jurisdiction to adopt NIMS procedures. Adoption and implementation of NIMS is a requirement for continued federal funding. As mentioned previously, *HSPD-5* states that NIMS gives the framework for consistent response to domestic disasters on a nationwide basis. The NIMS components are covered in depth in chapters 3 through 6.

Technology continued into the 21st century

Technology is moving at a record pace in the 21st century. Although the emergency services are slow to adopt emerging technology due to cost, some very useful resources are available for responders, executive staff, and planners. One of the most useful technologies to come into the hands of responders is computerized mapping and its availability. With a notebook computer and an Internet connection (available anywhere a cellular signal is available), the responders can access an aerial view, with street layouts and latitude and longitude. These are available at http://earth.google.com/. This free service allows responders from other communities to respond to and map out areas affected by the particular disaster. GPS systems can integrate with this mapping and identify areas of concern at emergencies that involve large geographical areas.

Technological advances have affected almost all areas of planning, including the use of databases to trend incidents, track resources available, and view lessons learned from other jurisdictions. A Web site that provides a plethora of information related to planning and response is the *Lessons Learned Information Sharing* Web site hosted by the Department of Homeland Security. The site is located at www.llis.gov and is password and authentication protected.

Planning is continuous and cumulative

By the 1990s, planning had become a continuous and cumulative activity in large cities. With the latest federal government initiatives, such as NIMS, it is hoped that this continuous and cumulative planning will trickle down to all jurisdictions regardless of size. The key to ensuring continuous planning is to test the plans in tabletop and full-scale exercises during nonevent times and to ensure their use during real events.

In addition, it is important to examine what went well and what revisions are needed in order to keep a working, realistic plan that all responders can utilize.

Utilization of positions

In many of the former plans, communities would list information naming an individual (such as "Assistant Chief John Doe") who was responsible for a particular task. However, many plans now utilize terminology such as *the operations section chief* to designate responsibility for a particular task. With the implementation of NIMS, the incident command system (a component of NIMS) places certain responsibilities with certain positions. These positions are labeled and used the same in every jurisdiction, and thus the confusion of stating that a certain chief should accomplish a given task is eliminated. This is stressed in chapter 3, which covers the command and control component of NIMS.

USING PLANS/LESSONS LEARNED

In order to develop working plans, one of the planning committee members should review historical lessons learned from other disasters and the associated plans. This careful review should ensure that problems encountered previously are not inadvertently repeated when developing the jurisdiction's own plan. In this section, the common components of disaster response and the associated lessons learned in each area will be reviewed. Although this is not a list of every possible lesson learned, due to lessons being learned and reiterated on almost a daily basis, it will give the plan developer some crucial areas to examine. Some possible solutions to the lessons learned will also be examined.

General actions

As discussed in chapter 1, a disaster always begins with a call to the local emergency services, with the exception of anticipated disasters such as hurricanes or slow-rising floods. Therefore, it will be up to the first public safety officer, usually a fire officer, to recognize the disaster and begin the appropriate response. When the first arriving officer arrives on the scene, he must immediately establish the incident management system. While this first arriving officer may not be an expert in disaster management, he or she can begin to coordinate efforts and gain information about the event.

By establishing the incident management system and working as the incident commander, this first arriving officer has eliminated the possibility of multiple plans of action and duplication of resources. Such duplication could occur if two different officers from the same organization or two different organizations arrive and begin to set into motion competing plans. Not only would this duplication delay useful response due to a lack of coordination, it could be entirely unsafe. Consider a situation in which one fire officer arrives at a large fire and decides to take an offensive interior attack, while another engine company officer arrives and directs his firefighters to squirt water from the exterior, thus pushing the fire onto the interior crew. This interior crew could be severely burnt due to lack of coordination and a failure to implement a single plan of action.

One area with which this "one command officer" concept becomes difficult is when a disaster encompasses many jurisdictions. As mentioned in chapter 1, one of the concepts in the National Incident Management System is *area command.* This ensures that only one commander is operating the entire area, but it allows for field commanders at the separate locations/jurisdictions. Having an area command will improve operations for a disaster such as a widespread tornado. Often a tornado will produce sporadic damage over a large area. If separate command systems are competing for finite resources, such as search and rescue teams, none of the disaster victims is being well served. This overall area command can triage the numerous disaster locations to ensure that the area most in need of a resource gets the priority. In addition, a unified command structure enhances the command and control operations at large, widespread disasters.

Enacting the emergency operations plan

At what point is the emergency termed a *disaster*, thus requiring the invocation of the emergency response plan for the jurisdiction? Some may try to relegate this policy decision by stating that if the flooding involves more than a certain number of homes, or after the evacuation of a certain number of people in an area, the emergency operations plan should be invoked. However, this proves difficult and has some inherent problems.

The first problem involves the commanders who want to keep the incident simple. Many fire departments have an alarm card system that will dispatch a predetermined amount of resources based on the number of the alarm, such as two engines, one ladder, and one chief officer for the first alarm. Any subsequent alarms will alert high-ranking chiefs of the incident, and thus the original commander will only ask for an additional engine or ladder each time he requests in order to not invoke the additional command officers and supporting staff's response.

Instead, it is possible to more generally term the declaration necessity as "any incident involving a larger-than-normal area" or "an event that is expected to impact more than a few residents." This will allow commanders to feel free to enact the disaster response plan for any incident they feel will escalate or will involve more than the ordinary call. Of course, there may be certain commanders who are reluctant to call for invoking the plan, but those officers must be trained and monitored to ensure that they are abiding by the guideline. It must also be supported by the top administration. If invoking the disaster plan turns out not to be warranted, one must remember that it was at least a good drill in starting to operate under the plan.

Communications

In nearly every after action report, communications is listed as one of the top three problems, as observed by the authors. One of the initiatives in the National Incident Management System is interoperability, which is due to the fact that communications cause problems on nearly every incident.

What can be done until the day that every responder can talk to the other responders regardless of their affiliation? The first item to address is a communications policy for each jurisdiction, such as operating frequencies for incidents that are separate from the primary dispatch channel. By establishing policies for the utilization of the given frequencies, all forces will be able to talk to each other with the greatest efficiency within the equipment's limitations.

The next action is to note what radio frequencies exist within the surrounding jurisdictions and to consider the degree of compatibility with the different jurisdictions' radios. Sometimes this is a simple matter, such as reprogramming the radios to include the needed frequencies. Of course, this requires working relationships and cooperation from the different jurisdictions. The associate author works within a county that will not allow its radio frequencies to be utilized outside of the county. This creates many problems as the jurisdiction abuts two other counties.

If the radio frequencies of the surrounding jurisdictions are not compatible, the purchase of a radio relay device that allows different radios to plug into the device should be considered. This relay device will patch the radio frequencies together, and the device is portable, thus making it useful for mobile command units. There are multiple manufacturers of these devices, and anyone interested should check with his or her communications center as a good starting point.

Communications centers are often the forgotten hard workers of any disaster. During the first stages of the disaster, they are answering subsequent 911 calls and attempting to make notifications for the incident commander. Many dispatch centers are also an answering service for the public safety agencies, and thus they may still have to perform this operation. Finally, just because a disaster is occurring does not preclude other emergencies from continuing.

In order to relieve and assist the communications center, the emergency services of the community should enact the emergency operations center. By establishing the emergency operations center early, dispatchers can be relieved of trying to dispatch and track a large amount of resources. Although they will still dispatch and track the resources, they will have direction and assistance in the task. This can allow better deployment of resources by keeping units in their district and not having them pass each other responding to subsequent calls.

Radio talk groups/networks

Obviously, each level within the management has issues that pertain to their tasks. However, many of the different levels of management and staff are working on different tasks, and to one group, hearing what the other group is talking about is irrelevant to the tasks at hand. An example of this could be a commander of a large-scale building collapse who is trying to obtain resources to deal with rescuing people, extinguishing fires, caring for victims, and keeping bystanders at bay. Simultaneously, the rescuer who is trying to shore the building is contacting his rescue squad officer to obtain the correct diameter nails. Having all of this communication on one frequency can lead to gridlock on the radio and wasted time listening to transmissions that are irrelevant to the position's task. This can be eliminated by establishing radio networks. Some of the common radio networks are command, logistics, operations, hazmat, rescue, and emergency medical services. These are, of course, only suggestions, but the communications unit, under the logistics section chief, should develop radio networks that increase the efficiency and limit the number of differing levels on the same radio network.

The final item to note for radio networks is that each level or radio network must be able to communicate vertically within the organizational chart. Consider an example in which a command network is developed that places the incident commander, operations section chief, planning section chief, and administration chief on the channel. Subsequently an operations radio network is created that contains the operations section chief, the hazmat branch officer, the rescue branch officer, and the EMS branch officer on the network. It now is necessary to ensure that the operations section chief can monitor and transmit on both the command network and the operations network. This creates overlapping radio frequencies. The communications units should also designate an emergency frequency, if possible, that can be monitored. If the lack of interoperable radios prevents this, a phone number to call the command post or EOC in the event of an emergency could serve as a good contingency plan due to the number of cellular phones carried by responders.

Deployment of forces

In the beginning of a disaster, deployment of forces will be difficult at best. One of the first tasks during a large-scale disaster is to develop reconnaissance teams. This will allow the incident commander to get a grasp of the size and complexity of the incident. It will also serve as a triage of the incident. It would be a poor deployment decision to send all of the operating forces to the area originally summoning the safety service's help when another section of town has many more lives in need of emergency service. This again would be an instance in which an area command would benefit responders.

When deploying resources, the unit leader must attempt to predict the future. When parking emergency vehicles, the crew must know what the disaster will do before it reaches homeostasis. This could involve the floodwaters rising, the fire moving up the canyon, or the building having a secondary collapse. Crew leaders must always have a plan of escape. Some past mistakes include the following:

- Failing to recognize which roads will flood next due to not paying attention to the topography of the land.
- Failing to recognize the burn pattern of a wildland fire, thus trapping crews and equipment.
- Parking apparatus in front of buildings that have sustained severe structural damage or that have a chance of collapse. When this occurred on 9/11, it wiped out nearly half of the FDNY's responding fleet of fire apparatus.

Relief of personnel

When the span of an incident lasts longer than one operational period (usually 12 hours), personnel need relief. During some incidents, this may involve sending all of the operating personnel back to their own jurisdictions and allowing them to return to their residences. Other types of resources are designed to supply their own downtime sleeping area away from the scene. It is likely that if local and state resources are utilized, they will only be self-sufficient and be able to stay at the scene for one operational period. If federal resources such as federal firefighters or search and rescue teams are utilized, they will likely be self-sufficient for a period of 3 to 10 days. This is an issue the planning team will need to account for when ordering the resource.

When relieving personnel, it may be helpful to make the change during the daylight hours. This could have several benefits, one of the largest being the safety of the operating personnel. Due to the daylight, personnel from the off-going shift will be able to visually point out safety concerns to the oncoming shift. Other reasons involve transporting personnel and logistical issues such as leaving the scene with personal deployment gear.

When an incident exceeds one operational period and personnel will take relief periods near the incident, a base camp must be established. Establishing these in areas that are safe and have an ability to receive supplies is important. During the response to Hurricane Katrina, response teams had concerns for their safety due to the rioting. This could extend to the base camp due to the presence of basic supplies such as food and medical supplies.

During past relief operations, identity of crews became a concern. By utilizing the ICS system within NIMS, this problem is alleviated. It is important to maintain divisions, groups, and branches during relief periods. If possible, the incident action plan, including the organizational chart and communications plan, should be reviewed with the oncoming shift prior to field deployment.

Relief periods have often been a benchmark to iron out some chaotic operations. Many times due to the speed at which personnel are assigned and given tasks at the incident, it becomes chaotic, and less-than-optimum organization occurs. This is a perfect time to shuffle divisions, groups, and the incident action plan to iron out any identified problems.

Feeding

During many disasters, the personnel utilized at the disaster will be in the field and at base camps for many days. Besides accommodations, the responsible jurisdiction must also make provisions for feeding. Work will come to a standstill if no nutrition is provided to the workers. Although food can wait for about six to eight hours, immediate preparations must be in place for hydration. Many disaster workers have heat-related illnesses from long hours of work in hot conditions. Even if it is not hot outside, the body still loses water through hard work.

One area that the jurisdiction's EOP should address is food acquisition. For the immediate incident commander, summoning of the local Red Cross can at least deliver snacks, bottled water, and coffee to the scene. In addition, many support organizations such as fire auxiliary groups perform this function. These are important contacts to place in the jurisdiction's resource guide. Often the local jurisdiction will have an emergency procurement fund that allows agency directors to place a predetermined amount on a credit card during times of emergencies. This can at least bring some fast food to the scene.

Commanders must remember to calculate lead time into the feeding. It is possible to feed one's immediate family by just walking into McDonald's. In contrast, much more preparation time is needed to fix food for 100 responders, many of whom will eat more than a standard meal due to their nutritional demand from the work. The commanders also must know how far away the food is and whether it will be delivered or will require scene personnel to pick it up.

One plausible solution to the problem would be to have the local emergency management agency keep a supply of meals, ready-to-eat (MREs) at a predetermined site. This would allow their distribution during the first operational period. The MREs are designed to replenish nutritional demands of hard-working responders and have a long shelf life. One drawback is the cost (averaging $7 a piece), and they do have expiration dates. In order to overcome expiration issues, they can be built into the full-scale drill required by NIMS, thus rotating new ones to stock and charging the expense to the drill.

Food safety is a concern that will present itself in any long-term disaster. During many of the past disasters, civilians have provided an outpouring of support to the responders in the form of supplies and food. Although many are well intentioned, they provide food that could spoil or have some other food-borne contaminant by the time it is given to responders. If an entire operating group develops a food-borne illness, it can wipe out the entire group and require medical resources to care for these personnel. One solution is to have a doctor look over the food. Many times local and state-sponsored response teams, such as the US&R teams or disaster medical assistance teams, have a doctor on their response matrix. Utilization of these personnel to check over the food prior to feeding the responders may be helpful.

Medical care

Under the logistics section of NIMS ICS is the medical unit. Although many small incidents do not require the establishment of the medical unit, extended incidents with numerous responders will require the implementation of this group. One of the most confusing task assignments for various divisions and groups is the assignment of the medical unit. Many assume that the ambulances summoned to care for victims can be diverted to care for responders and that the EMS branch functions to care for responders. This is not true. During a large incident, the incident commander must begin to think about the health and care of the responders. Much of disaster response work can lead to injuries among the responders.

The responsibilities of the medical unit include care of responders and transporting them to local medical facilities. The medical unit leader should determine the number and areas of operating personnel and whether they have immediate medical care with them. Next, the medical group leader needs to survey the area's medical facilities to determine their capabilities. Waiting to find out that a responder 10 miles away has a severe trauma may delay appropriate care to the point that the responder may die from his injuries.

Record-keeping

America is a very litigious society. Each incident will have a host of problems that will reveal themselves. As will be reiterated in the legal section of this chapter, these records will become valuable to the incident commander as he or she testifies in the many hearings and court proceedings that seem to follow the more recent disasters.

For example, SARA Title III allows for recuperation of funds spent responding to hazardous materials incidents. Many times jurisdictions will send a bill to the company requesting reimbursement, but they have no records of why they used these items. If the company balks at paying the bill, it is likely this lack of record-keeping will soon translate to a lack of funds.

Another area where record-keeping is important is related to responder injuries. These injuries will turn into workers' compensation claims a good portion of the time. Even if responders did not report an injury on scene, chronic problems can be discovered years from the incident. This is the case for many of the responders of the 9/11 attacks. They now have what is termed the *WTC cough*. This is likely the result of breathing the contaminated air within the scene.

It is important for large incidents to establish a finance/administration section and section chief in order to compile all of the records of the incident. Some of the information to compile includes contracts signed and utilized, fuel purchases, and personnel usage. These reports and records will become invaluable postincident information.

Water information

Due to the frequency of disasters involving fire, it is important that planning involve locations of water sources. Although this may seem like an easy task of accounting for the hydrants in the jurisdiction, planners must remember that these hydrants may not be usable during a disaster. In the case of a flood, they will be submersed. In the case of an earthquake, the water mains could break. Therefore, planners must look to identify water main control valves, along with secondary water sources, such as lakes, retention ponds, and storage tanks. Many of the aerial mapping programs such as Google Maps and Google Earth can identify these secondary water sources. However, planners must remember that these photos can be up to three years old.

Mutual aid agreements

In some cases, neighboring communities will respond to a disaster only to have a need in their own community. It is customary that if a mutual aid resource and personnel need to return to their home community, they will contact the command post and request to be released from the incident. Although this has occurred in the past, proper planning and mutual aid agreements should alleviate this problem. Ohio has a statewide mutual aid agreement that will only allow the commitment of resources that the providing department can supply and still have its department be operational.

In times where a disaster is sporadic and provides damage over a large area, this type of problem can occur. Jurisdictions must employ "big picture" thinking during a disaster and utilize area command. The first notified jurisdiction may be in a rush to call all of their neighboring communities for assistance, only for the neighboring community to learn while en route that their own community is affected. Thus the request for assistance is cancelled, further delaying and complicating the disaster. Commanders must work to obtain an accurate reconnaissance of the disaster in order to plan for the needed resources from the correct communities.

Legal problems

In the case of the 9/11 attacks, it is likely that emergency workers will have to sue in order to get workers' compensation to cover their respiratory problems. In the case of Hurricane Katrina, many of the homeowners took legal action to have their insurance pay for their property damage.

Another form of legal problems that takes effect is in the form of authority and mutual aid agreements. Some of the legal problems that arise concern who will pay for the workers if they are injured. Who compensates the personnel? Fortunately, as planners have focused on these problems for the past 10 years, many states now have statewide mutual aid agreements. In addition, the International Association of Fire Chiefs has drafted a strategic plan for mutual aid. Information on this planning is located at www.iafc.org/mutualaid. (Mutual aid, and the programs surrounding this issue, are discussed in length in chapter 4.) Another concern is what authority the jurisdiction has in signing contracts with vendors. If these concerns are not addressed and answered prior to the disaster, the person ordering the services could potentially be sued for the cost of the services.

As legal concerns relate to responses provided by the community, an emerging trend in the legal world is to bring litigation against the local jurisdiction, including the fire department. This comes as the community holds the government to a standard. Although at one time the community was happy with any response the jurisdiction provided, the current trend is that this response must meet certain standards. Within the fire service, the National Fire Protection Association (http://www.nfpa.org) provides standards that organizations are held to during their operations.

Public information

Disseminating information to the public is very crucial in a time of disaster. However, it must be planned. With the presence of the news media and their telescopic lenses, helicopters, and very enthusiastic reporters, either the local government officials will provide timely information to the media, or the media will seek it out and report what they want. For instance, during Hurricane Katrina, the media reported the horrific conditions at the Superdome. Although this was a microcosm of the entire response to the hurricane, it was the portion that the news media reported, thus making the response to the hurricane appear to be more of a disaster than the actual disaster itself.

The National Response Framework addresses the public information officer (PIO) and stipulates that all agencies deliver their information through a coordinated single PIO. Worse than providing no information to the public is providing conflicting or inaccurate information.

Employees of all response groups must be trained to not provide information to reporters unless authorized by their command officers. Some problems that have occurred from this are the identification of deceased responders prior to their families being notified. This only leads to the possibility of a loved one seeing his or her significant other's death on the five o'clock news. Another area is the Health Insurance Portability and Accountability Act (HIPAA) privacy regulations. If emergency workers release information on victims, they could be susceptible to fines due to this law.

It is good to portray a positive picture of the responding forces. However, as proven by former FEMA Director Brown, it is not wise to appear to have a positive outlook on the situation when there are obvious problems. Second, the leadership must provide solutions to problems and appear to have a grasp of the situation. If they appear not to be in touch with the situation or do not present an attitude of authority, the public will grow concerned. The Federal Emergency Management Agency through the Emergency Management Institute provides both basic and advanced public information officer training. Any member of the command organization who will fulfill this position should consider attending these valuable courses.

Evacuations

As seen in many recent disasters such as hurricanes Rita and Katrina, problems can occur with evacuations. Until recently, few planners had given thought to testing how long it would take to evacuate a major city. During Hurricane Rita, an evacuation turned Interstate 45 into a parking lot. "'We've traveled approximately 20 miles in nine hours with two sheepdogs,' said Nick Nichols, who headed out of Houston at 6 a.m. 'The orders were to evacuate from Houston, and...we're out here on a parking lot.'"[14] The problems associated with this evacuation involved too many vehicles trying to leave the city on only a limited number of highways. In response to the traffic congestion, Interstate 45 was altered by the transportation department to turn both directions of travel into all-northbound lanes. This action initially alleviated the problem, but as motorists began to run out of fuel due to long wait times, the stranded vehicles, along with the extreme heat, created a secondary disaster as a result of the evacuation. A second quagmire of evacuations surrounded the Hurricane Katrina evacuations. Late notice and an inability to evacuate the poor and elderly created the scenes aired on the news outlets for days depicting the suffering and dying of the elderly due to lack of food and medical care.

Past incidents have demonstrated the need to ensure that an agency has the authority to evacuate. Past incidents have encountered a communication problem between the police and fire departments not knowing which agency had the authority to evacuate. It should also be spelled out in the EOP which agency will perform evacuations. Past incidents have proven that during most disaster situations, the fire service is busily engaged in disaster response efforts and has little time to evacuate citizens. Therefore, the logical agency to perform evacuations is the police department.

During evacuations, people should be advised of the importance of their evacuation, but it is likely responders or officers will not be able to force them from their homes. It is a good practice to write down the locations where the victims will not evacuate so that areas of rescue can be identified after the disaster has occurred.

Information technology

The amount of technology available to responders makes reconnaissance and plan development activities much more accurate and efficient. One of the biggest problems for commanders is knowing how much area is affected and how extensive the damage is for the given situation. In prior years, printed maps of the area would need to be obtained. Relaying the information to field units was cumbersome, and marking the maps with information proved difficult.

With the advent of advanced mapping, helicopter video feeds, and GPS units, command personnel are able to bring up maps of the area, see live video of the damage streamed into their command post, and overlay GPS data obtained from field crews. This enables them to deliver forces to the most needed areas first and get a grasp on the extent of the incident. All of this technology is available, but planners must understand how to obtain this information for commanders prior to the incident. Having an information technology group supporting the command staff is a true need on large incidents.

One of the emerging problems with mapping is the different datums and coordinate systems of different maps. Planners must ensure they establish which coordinate system will be used at the incident. Some of the popular systems are the U.S. National Grid (USNG), latitude/longitude, Universal Transverse Mercator (UTM), and the Military Grid Reference System (MGRS). These systems might or might not work with the mapping software being used. Planning and policies will solve these issues prior to the incident.

Simultaneous threats

"One disaster will lead to another." This is a motto command and planning personnel must remember. Earthquakes lead to massive fires. Hurricanes lead to flooding. Earthquakes trigger tsunamis. The list could go on, and some of the common compound disasters are listed in figure 2–3. If planners and commanders are only planning for the disaster at hand and do not have the forethought to see the compounding disasters, they will be behind for the entire duration of the response phase. Examples can be seen in the recent rash of hurricane events, such as the flooding of New Orleans, or the building collapse problems seen on 9/11.

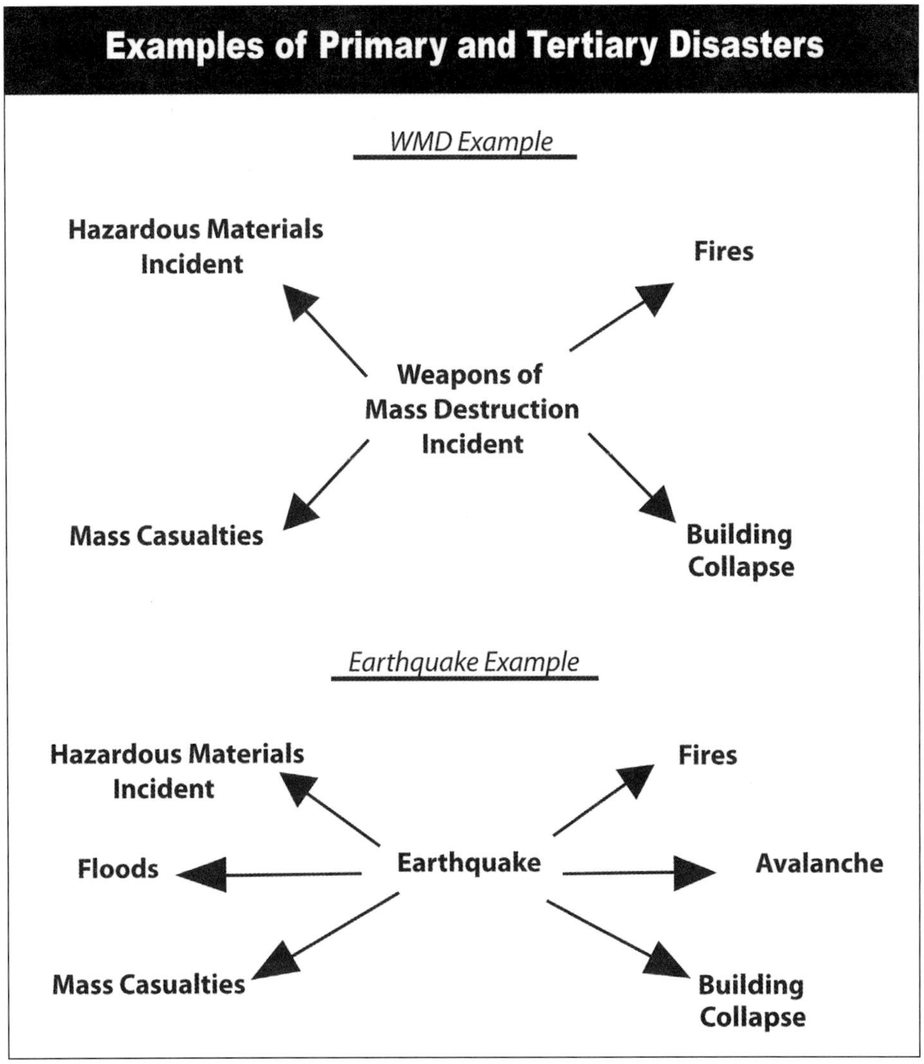

Fig. 2–3. Cause and effect tree for disasters

CONCLUSION

Emergency response planning has taken advantage of historical lessons and developed into a holistic approach. However, as shown by recent examples of disaster lessons learned, it is apparent that responders continue to experience many of the same recurring problems. Therefore, planners and exercise coordinators must focus on these problems and offer solutions. Just being aware of the problem is only half of a solution. Government officials must use political power to address the problem with interoperability and mutual aid plans. Some of this framework is already in place and must be followed through with future government leaders.

REFERENCES

1. U.S. Department of Homeland Security, Federal Emergency Management Agency. 2006. About US&R. Retrieved November 16, 2006 from: http://www.fema.gov/emergency/usr/about.shtm.

2. U.S. Department of Commerce, National Oceanic and Atmospheric Administration. 2006. *Community Vulnerability Assessment Tool.* Step 1—hazard identification. Retrieved November 18, 2006 from: http://www.csc.noaa.gov/products/nchaz/htm/step1.htm.

3. U.S. Department of Commerce, National Oceanic and Atmospheric Administration. 2006. *Community Vulnerability Assessment Tool.* Step 2—hazard analysis. Retrieved November 18, 2006 from: http://www.csc.noaa.gov/products/nchaz/htm/step2.htm.

4. U.S. Department of Commerce, National Oceanic and Atmospheric Administration. 2006. *Community Vulnerability Assessment Tool.* Step 3—critical facilities analysis. Retrieved November 18, 2006 from: http://www.csc.noaa.gov/products/nchaz/htm/step3.htm.

5. U.S. Department of Commerce, National Oceanic and Atmospheric Administration. 2006. *Community Vulnerability Assessment Tool.* Step 4—societal analysis. Retrieved November 18, 2006 from: http://www.csc.noaa.gov/products/nchaz/htm/step4.htm.

6. U.S. Department of Commerce, National Oceanic and Atmospheric Administration. 2006. *Community Vulnerability Assessment Tool.* Step 5—economic analysis. Retrieved November 18, 2006 from: http://www.csc.noaa.gov/products/nchaz/htm/step5.htm.

7. U.S. Department of Commerce, National Oceanic and Atmospheric Administration. 2006. *Community Vulnerability Assessment Tool.* Step 6—environmental analysis. Retrieved November 18, 2006 from: http://www.csc.noaa.gov/products/nchaz/htm/step6.htm.

8. U.S. Department of Commerce, National Oceanic and Atmospheric Administration. 2006. *Community Vulnerability Assessment Tool.* Step 7—mitigation opportunities analysis. Retrieved November 18, 2006 from: http://www.csc.noaa.gov/products/nchaz/htm/step7.htm.

9. International Association of Fire Chiefs. 1980. *Disaster Planning Guidelines for Fire Chiefs.* July. Washington, DC: FEMA. As quoted in Kramer, W. M., and C. W. Bahme. 1992. *Fire Officer's Guide to Disaster Control.* 2nd ed. Saddle Brook, NJ: Fire Engineering Books and Videos, PennWell. p. 40.

10. The White House. 2003. *Homeland Security Presidential Directive/HSPD-5.* February 28. Retrieved November 29, 2006 from: http://www.whitehouse.gov/news/releases/2003/02/20030228-9.html.

11. Dynes, R. R., and E. L. Quarantelli. 1975. *The Role of Local Civil Defense in Disaster Planning.* Washington, DC: GPO. As quoted in Kramer, W. M., and C. W. Bahme. 1992. p. 44.

12. U.S. Department of Homeland Security. 2002. *The Department of Homeland Security: President George W. Bush.* June. p. 4. Retrieved December 3, 2006 from: http://www.dhs.gov/xlibrary/assets/book.pdf.

13. U.S. Department of Homeland Security. 2005. *Interim National Preparedness Goal.* March 31. p. 2. Retrieved December 3, 2006 from: http://www.ojp.usdoj.gov/odp/docs/InterimNational-PreparednessGoal_03-31-05_1.pdf.

14. CNN.com. 2005. Rita turns to threaten Texas, Louisiana. September 22. Retrieved December 9, 2006 from: http://www.cnn.com/2005/WEATHER/09/22/rita/index.html.

TRAINING AND PREPARATION 3

INTRODUCTION

No matter how good the plan for any community may seem, the plan is only as good as the people who carry it out in the time of need. In order to ensure those personnel are the highly competent and experienced disaster responders that the community expects, it is necessary to examine how to properly prepare the personnel who will deliver the disaster mitigation and recovery efforts.

This preparation can take many forms, such as education, training, drills, conferences, tabletop exercises, functional exercises, and full-scale exercises. These different learning ventures are all relevant to the development of great command and planning personnel, not to mention creating great disaster workers.

This preparation was put to the test on September 11, 2001 and created validity that disaster response always starts with the local fire and police departments, when two planes struck the World Trade Center towers. What began as a firefighting and rescue problem contained to one building quickly became one of the largest disasters in U.S. history. If it were not for the concepts of ICS, mutual aid, and standard operating procedures, this disaster could have been far worse, and far more than 343 firefighters could have been sacrificed that day.

The value of learning from past disasters and enhancing operations and training for the future were demonstrated by the disaster on 9/11. The adoption of NIMS, the creation and updating of the National Response Plan/National Response Framework, and the enhancement of FEMA funding to properly equip and train response teams are all positives that came out of such a negative day. As disaster responders, we must honor their loss and the enhancements that have come about as a result of these disasters by dedicating ourselves to becoming better educated to respond to disasters.

This dedication to becoming better prepared was evident at Hurricane Rita. Just over a month prior to the projected landfall of Hurricane Rita, Hurricane Katrina had devastated the Gulf Coast. Lack of evacuation, no food and ice, along with a bureaucratic snafu, created an embarrassing response to the aftermath of Hurricane Katrina. However, because of the lessons learned, an evacuation of the city of Houston went fairly well and resulted in all of the citizens being removed from harm. The associate author can remember the eerie feeling that came about as he drove into Houston at 4 p.m. on a Friday to a deserted downtown. However, it was blatantly obvious that the country had learned from the tragic aftermath of the previous hurricane.

NEED FOR DISASTER CONTROL TRAINING

During the career of a proactive disaster planner, commander, or responder, he or she will spend a significant amount of time attending and completing various types of preoperational activities in order to hone needed skills. There are four basic types of preoperational activities that one can expect to complete. These are the following:

- Educational sessions
- Training sessions
- Conferences/seminar sessions
- Exercise and drill sessions

Educational sessions

During educational sessions, the participant is "provided a broader base that not only allows a fire officer to perform the job more skillfully, but allows broader interaction with the world and the environment. Education is theory based, applies reasoning, evaluates judgment, and examines the how and why."[1] The disaster responder will likely receive this portion of the necessary training through college-level courses or through many of the management-related courses on disaster preparedness and response. This portion of the preparation builds the critical thinking portion of the responder's abilities. By taking concepts and theories, and interpreting and applying their use in these experiences, disaster responders gain the ability to overcome issues that they have not experienced or directly trained for in the past. From its inception, the National Fire Academy's Degrees at a Distance baccalaureate program has included a core course in disaster planning.

Training sessions

Training sessions will likely represent the bulk of the disaster responder's preparation time. Training will directly teach what actions to take provided a given situation. As an example, consider the training in the ICS portion of NIMS. The ICS 100 teaches that an incident commander is required for any incident involving multiple resources. Therefore, the disaster responder knows through training that if two fire engines are responding to an incident, an incident commander is needed. Thus the disaster responder should appoint or retain this position based on his or her role in the response.

One of the new and valuable training sessions created to enhance the response to disasters is the basic emergency rescue technician (BERT) certification. This teaches firefighters and fire officers the basics of responding to technical rescue and disaster events. The curriculum involves building collapse, wide-area search, and NIMS as it relates to large-scale disasters. By training the line responders how to react to these various types of disasters, the recognition and response to disasters is enhanced at all levels. Many other training sessions are offered through the Emergency Management Institute, such as Exercise Design and Evaluation, which

> is designed for emergency management exercise staff to acquire fundamental knowledge and skills necessary to develop, conduct, evaluate, and follow up basic activities and exercises, including exercise orientations, drills, and tabletop exercises.
>
> The content addresses the fundamental requirements of an overall exercise program and its components. A special emphasis

is placed on the evaluation process and the development of sound evaluation tools to assist evaluation staff. Participants are challenged to develop, conduct, and evaluate these activities during the conduct of the course through the use of proficiency demonstrations.[2]

Additional training from the Emergency Management Institute can be found on the Internet at http://www.training.fema.gov/EMIWeb.

Conferences and seminars

Within disaster management and planning are many subdisciplines and areas of interest. In addition, many experts from around the country and the world are willing to share their expertise and experiences through conferences. There are many conferences across a wide range of professions and disciplines that are conducted each year. Some of the more popular conferences are described in the following paragraphs.

Within the firefighting profession, the Fire Department Instructors Conference (FDIC) has many hands-on and conference-style sessions that deal with disaster planning and response. Some of the programs available at the 2007 conference were: Response to Terrorist Bombings, I-400 ICS, Worst Case Scenarios in Rescue, Strengthening Interstate and Intrastate Mutual Aid, WMD Vulnerability Assessments, Fire Service and Counter Terrorism, Plans Section Chief 101, Planning: An All-hazards Approach, and ICS for Dummies. These sessions offered some of the latest thinking and examples available in the industry. Further information on FDIC can be found at http://www.fdic.com/index.html.

Within the emergency management and disaster preparedness profession, many conferences are held annually. Some of the conferences available include the International Disaster Management Conference, the International Emergency Managers Conference, and the annual World Conference on Disaster Management. Some of the past sessions available at the World Conference on Disaster Management were: Creating Community Partnerships to Ensure Business Continuity, Community Recovery, and Strategic Crisis Leadership. More information about the classes offered by the World Conference on Disaster Management can be found at http://www.wcdm.org.

Within the police profession, one of the largest conferences available is the IACP conference, sponsored by the International Association of Chiefs of Police. This six-day conference offers many educational sessions relating to current law enforcement issues. Also covered are issues that arise during disaster situations. More information can be found at http://www.theiacpconference.org.

NIMS REQUIREMENTS BY POSITION

The Federal Emergency Management Agency, through the Emergency Management Institute, has also made receiving NIMS training, as well as other planning and emergency management training, easier than ever. This has been made possible with the advent of distance learning through online courses called the FEMA Independent Study Program. In addition to being able to conduct the training from any computer with Internet access, participants will also receive college credits for completing the course. These courses also receive continuing education credits through the International Association of Continuing Education and Training (IACET). Figure 3–1 shows the control center of EE Net, an online preparedness training, which is a distance learning program.

Fig. 3–1. EE Net distance learning control center. A distance-learning program provided by FEMA's Emergency Management Institute.

FEMA News Photo.

Table 3.1 NIMS Training Requirements by Position

NIMS Incident Command Systems (ICS) Training Requirements By Position

Position	Required Courses
Entry Level Responder or Disaster Worker	• IS-700 • IS-100
Front-Line Supervisor	• IS-700 • IS-100 • IS-200
Middle Management	• IS-700 • IS-800 • IS-100 • IS-200 • IS-300
Command & General Staff	• IS-700 • IS-800 • IS-100 • IS-200 • IS-300 • IS-400

Some of the online courses are now required in order to meet the NIMS implementation guidelines. Based on the FY 07 NIMS implementation requirements for local and tribal entities, the IS-100, 200, 300, 400, and 700 may be required for compliance, based on the positions the personnel within the organization are expected to fill. Table 3–1 is a synopsis of those requirements.

Although this table provides the basics of training requirements for the FY 07 NIMS compliance, a full description can be viewed at http://www.fema.gov/pdf/emergency/nims/imp_mtrx_tribal.pdf. Within the NIMS integration center's Web site at http://www.fema.gov/emergency/nims/index.shtm is a list of all compliance requirements based on the fiscal year and the type of organization involved.

EXERCISES

After a responder has attended many hours of education, training classes, and seminars in order to hone his or her disaster planning and response skills, it is now time to put those skills to the test. According to many industry experts, people perform like they train. Therefore, it is important to test how that training will come together in the form of exercises. According to the FEMA publication *IS-139 Exercise Design and Evaluation*,

There are five main types of activities in a comprehensive exercise program:

- Orientation seminar
- Drill
- Tabletop exercise
- Functional exercise
- Full-scale exercise

As…discussed, these activities build from simple to complex, from narrow to broad, from least expensive to most costly to implement, from theoretical to realistic. When carefully planned to achieve specified objectives and goals, this progression of exercise activities provides an important element of an integrated emergency preparedness system.[3]

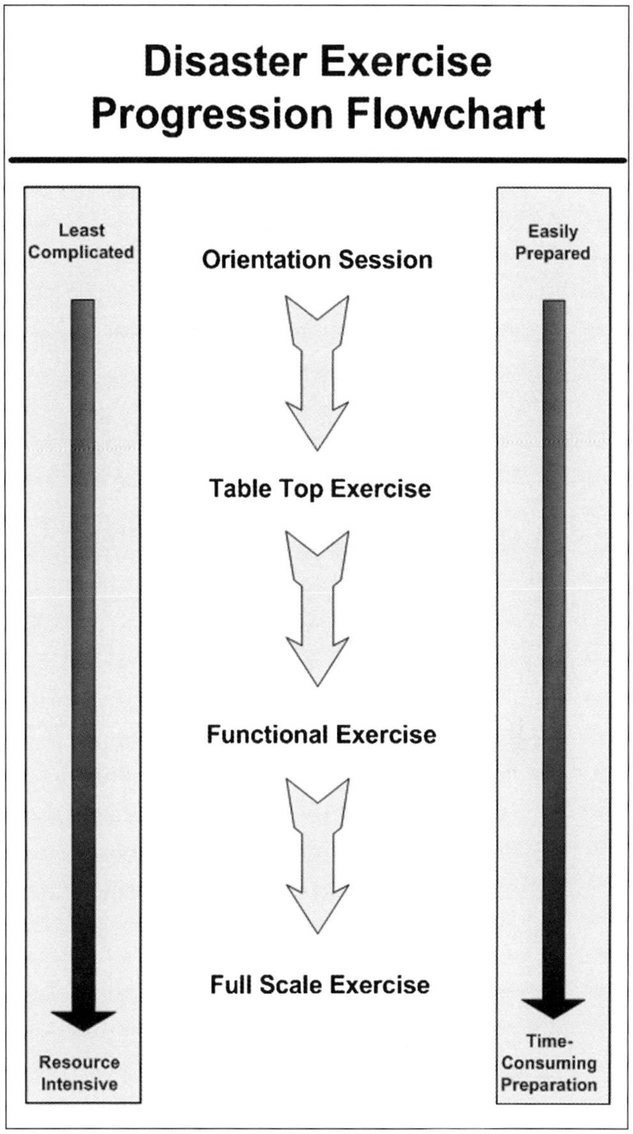

Fig. 3–2. Disaster exercise progression flowchart

Figure 3–2 illustrates how the progression of the different types of exercises occurs. Combining the various types of activities will yield the jurisdiction conducting the activities a better understanding of areas that are sound and those that will need improvement. Utilizing the plans created from chapter 2 in the activities will enable the planners to visualize the portions of their plans that will work well in the real environment, and the portions of the plan that need improvement in order to allow them full utilization in a real-case scenario. It may also shed light on areas that need additional training, based on the participants in the activities.

Orientation seminar

The orientation seminar is an "overview or introduction. Its purpose is to familiarize participants with roles, plans, procedures, or equipment. It can also be used to resolve questions of coordination and assignment of responsibilities."[4] The orientation seminar

> is a very low-stress event, usually presented as an informal discussion in a group setting. There is little or no simulation. (For this reason, orientations do not qualify as FEMA-recognized exercises)...
>
> The orientation seminar can be used for a wide variety of purposes, including:

- Discussing a topic or problem in a group setting.
- Introducing something new (e.g., policies and plans).
- Explaining existing plans to new people (e.g., newly elected officials or executives need an explanation of the EOP and their role at the EOC; new employees need an orientation to operational plans as they relate to emergencies).
- Introducing a cycle of exercises or preparing participants for success in more complex exercises.
- Motivating people for participation in subsequent exercises.[5]

The orientation seminar is designed by the facilitator, who will be the primary speaker, and is attended by jurisdictions that have relevancy in the process. Some of the attendees could be participants in subsequent activities such as tabletop exercises or drills. It is important to include as many entities as would be present during a real disaster. By not including particular entities, it can be assumed that they will not function properly when a real situation occurs.

DRILLS

According to *IS-139 Exercise Design and Evaluation*, a drill is a

> coordinated, supervised exercise activity, normally used to test a single specific operation or function. With a drill, there is no attempt to coordinate organizations or fully activate the EOC. Its role in an exercise program is to practice and perfect one small part of the response plan and help prepare for more extensive exercises, in which several functions will be coordinated and tested. The effectiveness of a drill is its focus on a single, relatively limited portion of the overall emergency management system. It makes possible a tight focus on a potential problem area...
>
> Drills are used to test a specific operation. They are also used to provide training with new equipment, to develop new policies or procedures, or to practice and maintain current skills. Drills are a routine part of the daily job and organizational training in the field, in a facility, or at the EOC. Some examples of drills run by different organizations are listed below:

- EOC: Call down procedures
- Public works: Locating and placing road barriers under time constraints
- Public health and safety: Site assessment and sampling
- Red Cross: Locating specific types of blood within a time constraint

- Military: Activation and mobilization drill
- Airport: Fire Department response to the furthest part of a runway within a given time
- Chemical plant: Evacuation and isolation of spill area and valve system shutoff[6]

Within the public safety professions, such as fire, police, and EMS, drilling is a common practice that occurs on almost a daily basis in many full-time departments. This type of activity is important, especially in the disaster management field. Many fire and police departments respond to numerous calls involving emergency medical services, fire alarms, and fires, but only a few of the members of a fire department have responded as the first responders to a large disaster. Due to the infrequency of large disasters, more drilling is required to maintain proficiency.

A drill needs a leader. This person must be well versed in the function being tested. In many types of drills, some props or simulators are needed in order to create the drill. These must be obtained in advance, and a backup plan for failure of these props is needed. A clear set of objectives should be developed for the drill, so that all participants know what is expected and what the parameters of the drill entail. Many times, creating a drill that involves the command of an incident is not feasible, such as a first response to a train derailment. However, with the advent of technology, there are numerous computer simulation programs available that allow the company officers to arrive at the scene of a disaster, make decisions based on their current training, and see the effects of their decisions. These computer simulators operate in real time, and the situation will progress based on the decisions of the participants. One of the fire service versions of this type of training aid is Digital Combustion, found at the Web site http://www.digitalcombustion.com/.

TABLETOP EXERCISES

According to *IS-139*, a tabletop exercise is a

> facilitated analysis of an emergency situation in an informal, stress-free environment. It is designed to elicit constructive discussion as participants examine and resolve problems based on existing operational plans and identify where those plans need to be refined. The success of the exercise is largely determined by group participation in the identification of problem areas.
>
> There is minimal attempt at simulation in a tabletop exercise. Equipment is not used, resources are not deployed, and time pressures are not introduced…
>
> The exercise begins with the reading of a short narrative, which sets the stage for the hypothetical emergency. Then, the facilitator may stimulate discussion in two ways:

- Problem statements: Problem statements (describing major or detailed events) may be addressed either to individual participants or to participating departments or agencies. Recipients of problem statements then discuss the actions they might take in response.
- Simulated messages: These messages are more specific than problem statements. Again, the recipients discuss their responses.

In either case, the discussion generated by the problem focuses on roles (how the participants would respond in a real emergency), plans, coordination, the effect of decisions on other organizations, and similar concerns. Often maps, charts, and packets of materials are used to add to the realism of the exercise.[7]

Figure 3–3 illustrates a tabletop exercise being conducted.

Fig. 3–3. Tabletop exercise in progress

FEMA News Photo.

The tabletop exercise is the lowest level of coordinated exercises. It brings together many people representing organizations that could have a role in the simulated problem. By having only the paper problem, the issues brought about during the tabletop exercise can be resolved prior to utilizing many participants from a multitude of organizations. Many problems with the responders can be identified and worked through in theory.

The tabletop exercise should elicit discussions from the participants in order to ensure the plans are working in the interest of all participants and that the participants understand their roles when a larger exercise is conducted. This is a form of a dry run. The discussion should be open ended and only be ceased due to time constraints agreed to prior to the tabletop exercise.

FUNCTIONAL EXERCISES

According to *IS-139*, a functional exercise is a

fully simulated interactive exercise that tests the capability of an organization to respond to a simulated event. The exercise tests multiple functions of the organization's operational plan. It is a coordinated response to a situation in a time-pressured, realistic simulation.

A functional exercise focuses on the coordination, integration, and interaction of an organization's policies, procedures, roles, and responsibilities before, during, or after the simulated event...

This is an interactive exercise—similar to a full-scale exercise without the equipment. It simulates an incident in the most realistic manner possible short of moving resources to an actual site. A functional exercise is:

- Geared for policy, coordination, and operations personnel—the "players" in the exercise—who practice responding in a realistic way to carefully planned and sequenced messages given to them by "simulators." The messages reflect ongoing events and problems that might actually occur in a real emergency.

- A stressful exercise because players respond in real time, with on-the-spot decisions and actions. All of the participants' decisions and actions generate real responses and consequences from other players.
- Complex—Messages must be carefully scripted to cause participants to make decisions and act on them. This complexity makes the functional exercise difficult to design.

Functional exercises make it possible to test several functions and exercise several agencies or departments without incurring the cost of a full-scale exercise. A functional exercise is always a prerequisite to a full-scale exercise.

In some instances, taking part in a functional exercise may serve as a full-scale exercise for a participating organization (e.g., a hospital may conduct its own full-scale exercise as part of a community-based functional exercise).

Functional exercises are complex in their organization of leadership and the assignment of roles. The following general roles are used:

- Controller: Manages and directs the exercise
- Players: Participants who respond as they would in a real emergency (Players should include policy makers; may include coordinators and operational personnel directing field activities.)
- Simulators: Assume external roles and deliver planned messages to the players
- Evaluators: Observers who assess performance[8]

FULL-SCALE EXERCISES

According to *IS-139*, a full-scale exercise

simulates a real event as closely as possible. It is an exercise designed to evaluate the operational capability of emergency management systems in a highly stressful environment that simulates actual response conditions. To accomplish this realism, it requires the mobilization and actual movement of emergency personnel, equipment, and resources. Ideally, the full-scale exercise should test and evaluate most functions of the emergency management plan or operational plan.

A full-scale exercise differs from a drill in that it coordinates the actions of several entities, tests several emergency functions, and activates the EOC or other operating center. Realism is achieved through:

- On-scene actions and decisions.
- Simulated "victims."
- Search and rescue requirements.
- Communication devices.
- Equipment deployment.
- Actual resource and personnel allocation...

The exercise begins with a description of the event, communicated to responders in the same manner as would occur in a real event. Personnel conducting the field component must proceed to their assigned locations, where they see a "visual narrative" in the form of a mock emergency (e.g., a plane crash with victims, a "burning" building, a simulated chemical spill on a highway, or a terrorist attack). From then on, actions taken at the scene serve as input to the simulation taking place at the EOC or operating center.[9]

Design of exercises

Many officials want to test their capabilities and want to push the excitement of a full-scale exercise. However, this may not be the best starting point for any given organization. According to *IS-139*,

The best way to determine whether you need an exercise—and what kind of exercise is needed—is to study your situation systematically.

Conducting a needs assessment will help you define the problems, establish the reasons to do an exercise, and identify the functions to be exercised. A needs assessment should begin with a review of the emergency plan and should address:

- Hazards—the risks that you are most likely to face and the priority levels of those hazards.
- Area(s) most vulnerable.
- Functions most in need of rehearsal.
- Potential participants (agencies, organizations, departments, personnel).
- Exercise requirements and capabilities.[10]

It is often best to work in a progressive fashion towards the full-scale exercise and create the exercise scenario around a realistic event. Trying to run a full-scale exercise of a nuclear attack on a high-rise building is an excellent idea if the jurisdiction of concern is a large metropolitan city. However, if it is a suburb that only contains one high-rise office building, it is not very practical to attempt this same scenario. A suburb may be better suited to deal with a chlorine leak at the local wastewater treatment plant.

In the example of the chlorine leak, it may be best to begin with an orientation session that brings together the police, fire, public works, emergency management agency, and the hazardous materials team. By having a top executive at each of the orientation sessions, the exercise design team leader can brief the individuals on the process by which the tabletop exercise will be conducted, and possibly the progression of this scenario through the functional and full-scale exercise process. Getting the leaders of these organizations together will give them a chance to network and begin to understand and respect each other's responsibilities prior to any exercising.

During the orientation session, the design team leader could suggest drills the individual organizations can conduct within their own organizations in preparation for the upcoming exercises. An example would be for police to work on evacuating citizens within a given radius. The fire department could drill on initial scene actions and research of the initial actions based on chemical information, and the hazardous materials team could drill on the containment of a 1-ton chlorine cylinder. By having the individual agencies focus on these drills, they will have a better understanding of their organization's capabilities and be able to pass this on to the exercise design team.

History certainly is an indicator of the future. According to *IS-139, Exercise Design*,

In doing a needs assessment for a single exercise, an obvious starting point is the evaluations of past exercises:

- Who participated in the exercise and who did not?
- To what extent were the exercise objectives achieved?
- What lessons were learned?
- What problems were revealed, and what is needed to resolve them?
- What improvements were made following past exercises, and have they been tested?

In summary, your needs assessment should reveal the following types of issues if they exist:

- Primary and secondary hazards that the organization faces
- Problems that need to be resolved
- Problems that recur
- Skills that need to be practiced
- Functions that are weak
- Improvements implemented after previous exercises, which now need to be tested
- New facilities, personnel, or equipment that have not been included in an exercise
- Weaknesses (such as gaps, conflicting policies, or vague procedures) in the emergency plan or the SOPs
- The need for role clarification
- The need for a certain type of exercise[11]

Once the needs assessment answers are obtained, the design team can have a good understanding of a starting point to creating the exercise. By understanding the past record of participation and outcomes of these exercises, the design team can begin to define the scope of the exercise. As an example, the hazardous materials leak in the suburban setting can be continued. If the jurisdiction is well versed in the delivery of hazardous materials mitigation, the exercise designers could up the ante and progress the chlorine plume towards a nearby retirement home. However, if the jurisdiction has limited manpower and limited experience with hazardous materials response, keeping the scope of the exercise contained to the chlorine leak alone would be in the best interest of all parties. That is not to say that in another year or two, the planners cannot expand the scope, but having a wide scope and limited experience will create a negative experience for all involved, and thus agencies will be reluctant to participate in future exercises.

Once the scope is determined, the purpose or benefit of the drill needs to be cited. If a person does not know where he is supposed to end up, he will not know if he is going in the right direction. In this example, the purpose may be to test the integration of the various response agencies and test the level of hazardous materials mitigation from the local hazmat team. If this team is in its infancy, as many suburban hazmat teams are, this could show strengths and weaknesses of the current operation. It could also show how these response agencies integrate in a NIMS/

ICS situation, possibly testing a unified command structure. The purpose needs to be in writing and should be clearly expressed during the design of the drill. Allowing the local fire department to control the leak prior to a leak needing evacuation would defeat integrating both the police and hazmat team.

The next issue to examine is the location of the exercise. This should be as realistic as possible. It would not create a realistic effect for any of the exercises if the chlorine cylinders were in a vacant parking lot. In this example, the design team should elicit the participation of the wastewater treatment facility personnel. Creating a good map of the area or using staging props, if possible, will add realism to the incident, thus making the participants feel a sense of purpose for the exercise. Most fire service personnel and other responders have been to training and education sessions that they felt were a waste of time due to their irrelevance. It should be the goal of the design team to prevent this from occurring.

It is important to consider all of the possible participants. This should have occurred during the orientation, but a review of that session could reveal that an organization was overlooked. Putting together a list of participants and their contact information is the next step. Ensuring the participation of all of the agencies in the drill will ensure that a large amount of planning will not go unused. In the chlorine leak example, at the very least the fire, police, EMS, and hazmat personnel should be present.

Once the scope, purpose, and participants have been identified, the exercise design can begin. As with any education or training outline, it is important to establish objectives, create a process to meet the objectives, determine evaluation criteria, and prepare a follow-up process for lessons learned. In the chlorine leak example, the objectives may include:

- Establishment of ICS under NIMS guidelines
- Recognizance and initial actions by the fire department
- Ability for police to coordinate evacuation of a particular area
- Ability of the hazmat team to control and contain the chemical leak
- Proper coordination among the agencies present
- Adherence to applicable federal state and local laws and standards related to the response

The process should be a part of the exercise scenario. As discussed previously, factors should lead the response agencies through the objectives. By allowing response agencies to bypass the desired scenario, the exercise would not test what the design team intended. An example would be the first arriving fire company approaching the cylinder and rolling the cylinder to slow the leak. The design team should be prepared to interject with the engine company becoming incapacitated. Although this adds to the scenario, it will prevent the cessation of the scenario before involving other agencies. The objectives should be simple, measurable, achievable, realistic, and task oriented, known in the *IS-139* course as *SMART objectives*.

The evaluation criteria should be in the form of a checklist or short answer sheet in order for the evaluators to capture the needed field data. Evaluators should have a good technical knowledge of the functions being examined in order to provide a quality evaluation of the exercise. Not receiving quality feedback would be an additional reason for the participants to feel a lack of usefulness in the process. According to *IS-139*, the narrative is a brief description of the events that have occurred up to the minute the exercise begins. The narrative has two important functions.

First, it sets the mood for the exercise. Participants need to be motivated to participate. The narrative captures their attention and makes them want to go on.

Second, the narrative sets the stage for later action by providing information that the participants will need during the exercise.

A good narrative:

- Is usually one to five paragraphs long.
- Is very specific.
- Is phrased in present tense.
- Is written in short sentences to lend immediacy and tension.
- May develop the situation chronologically (event with warning time).
- May emphasize the emergency environment.[12]

The author of the narrative must keep in mind that some of the exercise participants may not have the author's background or technical knowledge of the situation.

Once the narrative is composed, events need to be created in order to support the scenario in the narrative. These events should be linked to the functions that are being tested. When the event is presented, the composer should have in mind the response the group participating will provide. Having secondary events in the case of unanticipated field decisions is a good idea to keep the exercise on track.

Aside from the events taking place in the scenario, communication is a vital process of any incident; therefore, it should be included in the exercises. As a way to take advantage of communications, and test what response certain communications elicit, the exercise designers should compose messages to have participants or the controllers read at specific points in the exercise. In the example of the chlorine leak, a responder could succumb to the product and need immediate rescue. By having a controller or participant read this message in a lifelike act, it will test the mental abilities of the commanding personnel.

The final step to creating the exercise is to create a timeline of when the narrative, events, and messages will occur during the exercise. This will keep it on track and provide a more realistic environment for the participants.

Disaster training facilities

Because there are various types of disaster training and exercises, there are a variety of facilities that can be utilized for preparing for disasters. Some of the facilities discussed in this text are drill sites, simulators, EOCs, and exercise facilities. Although different facilities will be examined, the planners of the drills and exercises ultimately know what will best suit their situation. They should strive to make the facility realistic to what the disaster response participants would encounter in a real situation. It would not make sense to create a high-rise collapse structure when exercising for a small suburb with no tall buildings. Likewise, it would make no sense to have a hurricane simulation in a small city in Kansas.

Drill sites

At the basis of every response is the technician. This person could be a police officer, a firefighter, a public works employee, or a highly trained urban search and rescue member. Training for the expertise that these personnel are expected to deliver requires being able to simulate scenarios that they will encounter. The training of firefighters might include training on firefighting that involves natural gas, such as would be the case if an earthquake occurred. In 1989, an earthquake shook the cities of San Francisco and Oakland in California. In the aftermath, many buildings caught fire as a result of ruptured gas lines. From 5:00 p.m. on October 17 through midnight on October 19, there were a reported 33 fires in San Francisco.[13] With this number of fires, most occurring simultaneously, the fire officers had to rely on training in controlling natural gas fires, but also commanding and controlling fires with limited manpower. Chief officers will be spread thin under these conditions.

Police officers must be versed in riot control and evacuation procedures. This can occur in a local park or in an urban setting on a weekend, so as to not alarm the normal business customers. Practicing for riot control takes the coordination of all of the on-duty police and will require specialized equipment. One of the best drills the associate author participated in for riot control involved utilization of the firefighters from the same community to act as rioters. This allowed the firefighters to have a sense of awareness of the duties of their counterparts. Second, it provided a networking opportunity among the fire and police of the community, allowing each other to experience working and interacting together.

Drill sites for specialized operations can be as simple as a field with debris from a disaster, such as sections of homes, to an elaborate parking garage structure that has been strategically collapsed. The Texas Engineering Extension Service (TEEX) has created a training facility known as Disaster City,

> a place where tragedy and training meet... Located in College Station, Texas, this 52-acre training facility, located adjacent to the world renowned Brayton Fire Training Field, delivers the full array of skills and techniques needed by today's emergency response professionals.
>
> Created by the Texas Engineering Extension Service (TEEX), a member of the Texas A&M System, the mock community features full-scale collapsible structures designed to simulate various levels of disaster and wreckage, which can be customized for the specific training needs of any group.
>
> Emergency responders from across the globe venture to Disaster City for unparalleled search and rescue training and exercises.[14]

Exercise facilities

Training for command and support personnel at a disaster involves the exercises listed earlier in the chapter. For many of the exercises, such as the tabletop and the functional exercise, a conference room will serve as the basic training facility for the event. It is optimal, though, to have the functional exercise take place in an EOC, preferably the same one that will be utilized in the event of a disaster. This will allow the participants to obtain a feel for the EOC and identify any deficiencies of the facility while conducting the exercises. If many of the past exercises have been conducted in the designated EOC, it is a good idea to utilize a backup or secondary EOC to ensure that participants do not become too comfortable with a certain facility.

The advent of computer technology is benefiting disaster response training. Through simulations, which command personnel can view as if they were actually present, these personnel can be tested on their ability to think through problems as if they were occurring in real time. A first-class example of this occurs at the previously mentioned Disaster City, sponsored by TEEX.

According to TEEX,

> The 14,000-square-foot Emergency Operations Training Center (EOTC) uses state-of-the-art simulation and computer-based technologies to train first responders and city officials to manage a crisis through a unified command approach. Located adjacent to Disaster City in College Station, the multi-million-dollar training center optically links active and virtual simulations, including direct links to active response in Disaster City. Numerous cameras located throughout Disaster City allow training to be observed from within the EOTC, which can also be configured as an incident command post for full-scale, functional exercises held on the campus. Classrooms can be configured for a variety of simulation exercises and courses.[15]

This setup allows the training of multiple groups of responders in a vast variety of tasks, just as a full-scale exercise would, without the utilization of all of a municipality's resources.

Although it would not be practical to send all of the disaster planning and command personnel, as well as all of the local responders from fire, police, EMS, and public works, it would be advantageous to send a group of representatives to experience this setup and see how to improve the locality's functional and full-scale exercises.

Many other computer-simulated exercises are available from a whole host of vendors. The key to utilizing this technology is to make it practical to the community in which the exercise will occur. Recent versions of the computer-simulated software allow the importing of pictures and video segments of buildings taken by the end-user. One of the popular software programs for use by firefighters is CommandSim, located on the Web at www.commandsim.com.

WHAT SHOULD DISASTER TRAINING ACCOMPLISH?

Ultimately, the end result of any training should be the reproduction of a positive outcome at a real incident. The first item to identify when designing and choosing training and education for disaster responders is to identify what responsibilities they will be expected to handle in the event of a real incident. This should be done by position. Second, once their primary position has been identified, it should be determined what other functions they could be expected to perform. Any worthwhile continuity of operations plan will examine what roles individuals will perform on a normal basis and what roles they will perform in the event of certain individuals being absent from the incident. In the case of a fire department, the battalion chief is normally the command officer at incidents. However, as was the case with the earthquake in San Francisco in 1989, a large number of simultaneous incidents may preclude this from being true. Therefore, it is important to ensure that company officers can at a minimum establish and perform in the command role on a limited basis. If there is a failure to train personnel in their primary and secondary positions, Murphy's Law is guaranteed to occur at the disaster incident.

After determining what functions, primarily and secondarily, the given position is expected to perform, also known as a *job analysis*, it is necessary to match training and education courses that are needed to be able to perform these functions. This is accomplished in the NIMS integration document, reproduced previously in this chapter in table 3–1. Beyond the required NIMS training, jurisdictions should look at different training and education courses offered by local training academies, community colleges and universities, national schools such as the National Fire Academy, and various online learning venues. Next, it should be determined how to integrate these personnel into the various exercises previously mentioned. Once the courses, exercises, and drills needed for this position are determined, a list and check sheet of these courses should be created. This allows each person in this position to examine the required courses and preparation activities that are required of the position.

Based on the required courses, competencies should now be created. Many of the courses and accreditation agencies require a competency-based testing system in order to complete the classes. However, this knowledge requirement is often broad based and is not jurisdiction-specific. Utilizing the class competencies and conforming them to the jurisdiction ensures that the correct knowledge will be present in the time of need. One example of this is in studying emergency operations plans. A class could require the participant to identify the functional annex of the police department and identify core functions of the model EOP, but this model may be different from the EOP used by the jurisdiction.

After the position training and education requirement and the core competencies are met, the individual should be credentialed for that position. This will allow command personnel to know which positions a particular person is qualified to function within. This is currently being worked at on a national basis through NIMS. The credentialing process is underway, but many logistical and planning obstacles are being encountered as the plan is being enacted. Proactive creation of this process will ensure that one's community is ready with trained personnel in the interim.

Lastly, the credentialing should only be valid for a specified period. The creation of refresher training is needed in order to ensure that key concepts and skills are not forgotten in subsequent years of service. Sending a person back to all of the schools required for the position is not necessary, but a competency test is a good indicator of the topics that need to be reviewed. In addition, as the emergency operations plan and/or other response protocols change, the response personnel should receive training on the new plan or protocol.

The only true method to know if personnel can handle their assigned duties is for a disaster to occur. However, with proper design of functional and full-scale exercises, much of the reality that is present and makes individuals perform can be mimicked through these exercises. It is important to note how certain individuals performed and create corrective action plans based on their performance, thus improving the likelihood of success in the future.

LEARNING FROM ACTUAL DISASTERS

Fortunately, many response personnel will never attend a disaster of great proportions. However, this creates a problem in that the first-time response personnel may see a disaster such as a hurricane or tornado; they will need to know how to perform in a top-notch fashion. Because disaster response personnel have the dedication to consistently better themselves and to allow others to learn from their mistakes, many "Lessons Learned" Web sites, publications, and articles are produced each year.

The Department of Homeland Security hosts a Web site of these lessons learned. The Web site requires verification to the responder's employer to ensure that legitimate use of the Web site is occurring. The Web site, located at www.llis.gov, lists numerous resources for emergency response personnel. A best practices section describes how experts have best found to respond to various types of incidents. The Web site allows collaboration with other members of the Web site to ask questions and post thoughts on various subjects.

Because responders who experience disasters should share their experiences with others, the creation of after action reports (AAR) is essential for all personnel, not only from the jurisdiction, but also from jurisdictions who read the AARs. An

AAR is a report on the event that reviews what went correctly, what went wrong, and what actions are required to fix the problems or increase the efficiency of the response. The Office of Domestic Preparedness, through the program Homeland Security Exercise and Evaluation (HSEEP) Program, has produced an AAR template. According to volume III of the HSEEP, "The after action report should be created within one month of the event or exercise."[16] This AAR should be in a format that is easily understandable to government officials, as this after action report will be the basis for the improvement plan.

The improvement plan addresses the issues that arise in the after action plan that have been identified as needing to be improved. An example would be increased ICS training for command personnel. An improvement plan would identify the type and kind of training needed, who needs it, and when the training is expected to be conducted.

Having the plan designers or the command staff complete the AAR can sometimes be a good approach, but including members of all levels will ensure that issues are brought out in full. Often human nature is to only see the good in any situation, but in order to properly learn from incidents and exercises, an outside vision is needed, especially one that is not biased.

CONCLUSION

Second only to plan creation, training is the most important portion of disaster preparedness. This chapter has covered a multitude of various training and educational classes and exercises that are available. Depending upon the function of the various response positions, it is important to customize the training, education, and exercises to that position. It is important to maintain proficiency and meet federal mandates, as described in the NIMS implementation schedule. In addition, credentialing personnel ensures that personnel have met the education and training standards established both by the Department of Homeland Security and the local jurisdiction. Finally, evaluating response capabilities through a series of exercises ensures that the training and education are translated to proper actions. The old adage, "Train like you play," carries a lot of weight in the disaster response community.

REFERENCES

1. Kramer, W. M. 1995. *The Fire Chief's Handbook: Training and Education*. Saddle Brook, NJ: Fire Engineering Books and Videos. pp. 328–329.
2. U.S. Department of Homeland Security, Federal Emergency Management Agency, Emergency Management Institute. 2007. *E/B 132 Exercise Design and Evaluation*. Retrieved January 19, 2007 from: http://training.fema.gov/EMIWeb/CEC/E132.asp.
3. U.S. Department of Homeland Security, Federal Emergency Management Agency. 2003. *IS-139. Exercise Design*. Unit 2: Comprehensive exercise program. March. p. 2.5. Retrieved January 21, 2007 from: http://training.fema.gov/EMIWeb/IS/is139lst.asp.
4. Ibid.
5. Ibid. p. 2.6.
6. Ibid. pp. 2.6–2.7.
7. Ibid. pp. 2.10–2.11.
8. Ibid. pp. 2.12–2.13.
9. Ibid. pp. 2.15–2.16.
10. U.S. Department of Homeland Security, Federal Emergency Management Agency. 2003. *IS-139. Exercise Design*. Unit 4: Exercise design steps. March. p. 4.2. Retrieved January 21, 2007 from: http://training.fema.gov/EMIWeb/IS/is139lst.asp.
11. Ibid. p. 4.3.

12. Ibid. p. 4.27.

13. Virtual Museum of the City of San Francisco. 2007. Fires by location and origin. Retrieved January 30, 2007 from: http://www.sfmuseum.net/quake/fires.htm.

14. Texas Engineering Extension Service. 2007. Disaster City. Retrieved January 31, 2007 from: http://teexweb.tamu.edu/teex.cfm?pageid=USARprog&area=USAR&templateid=1117.

15. Texas Engineering Extension Service. 2007. Emergency Operations Training Center. Retrieved January 31, 2007 from: http://www.teex.com/teex.cfm?pageid=agency&area=teex&templateid=1382.

16. U.S. Department of Homeland Security, Office of Domestic Preparedness. 2004. *Homeland Security Exercise and Evaluation Program*. Vol. III. pp. 33–34. Retrieved January 31, 2007 from: https://hseep.dhs.gov/support/HSEEP%20Volume%20III.pdf.

2

Managing Disasters

ORGANIZATIONAL STRUCTURE AND INCIDENT COMMAND

4

INTRODUCTION

Second to communication problems in after action reports is the difficulty with organization at an incident. In addition, most experts state that a good portion of communication problems stem from an organizational deficiency. Most government organizations are structured in much the same fashion. Whether the organization is the police force, the fire department, or public works, there is a chain of command within the organization. Many of the organizational structures mimic popular business models. Each organization has a chief executive officer, a layer of executive management, a middle management layer, and the workers. During nonemergency operations, if one of the players is unavailable, it is often acceptable to wait for that person's return, or to refer to someone of higher authority when a decision is needed. However, what happens when the worker level person is the first to arrive at a disaster, and higher-ranking officials are unavailable at the time? Should the person wait for consultation? Does the worker continue to only perform his normal duties? Neither will suffice in this instance.

Enter the incident command system. It is recognized by nearly every public service organization that the executives may not arrive first, and that waiting and/or consulting those normally expected to manage a public service organization is not acceptable. Much like the military, public service has to have an organizational structure that will allow success regardless of who is at the helm.

In this chapter, the authors will examine the incident command components of the National Incident Management System (NIMS) and will discuss organizational structure, specifically the components of a successful structure in addition to the standard stipulated by NIMS.

USING AND EXPANDING EMERGENCY SERVICES STRUCTURES

Fire departments are the group of public responders who are most familiar with the incident command system (ICS) due to their frequent use of the system. The ICS system came to the forefront in the 1980s with FIRESCOPE and was later regulated by SARA Title III, which regulates hazardous materials response. Due to the fire department's frequent use of ICS, its personnel often serve as consultants and trainers on the ICS portion of the NIMS compliance. However, the basic incident command system has many components. One of the benefits of the system is its scalability. This means that not every component of the ICS must be staffed and utilized on each incident. Inversely, it is expandable by placing deputies with the section chiefs, thus allowing the organization to grow beyond that depicted in a full ICS organizational chart.

Because this system is scalable, the emergency services may only use certain portions on a daily basis and will have some refresher training when establishing very large organizational charts. The fire department commonly uses an incident commander and a few groups/divisions. The incident commander position is required on every call, and this position retains the responsibilities of the other section chiefs until their establishment. The groups/divisions are often the individual fire companies and are responsible for fire suppression, ventilation, rescue, and overhaul. Often the incident commander talks with each group/division leader.

However, what ICS components are needed during the aftermath of a tornado that has struck a good portion of the town, including a 500-bed nursing home, which is partially collapsed? No longer can the incident commander be responsible for all of the strategic and tactical objectives at this incident, nor can the few groups/divisions deal with the numerous problems in the field. In this case, there is a need for widespread reconnaissance to determine the amount of damage and a full-scale mass casualty from the number of potential victims. There is a potential fire and hazardous materials incident based on the items that could be in a nursing home, as well as a large-scale building collapse rescue. In addition, there will be the planning, logistics, and finances associated with responding to this disaster. Furthermore, what happens to the remainder of the town while resources are responding to the nursing home? Is it acceptable to let them wait, possibly perishing in the interim? In a large-scale disaster, use of the incident command system is vital.

NATIONAL INCIDENT MANAGEMENT SYSTEM

On February 28, 2003, the president of the United States issued *Homeland Security Presidential Directive/HSPD-5*, which directed the Secretary of Homeland Security to develop and administer a National Incident Management System (NIMS). According to HSPD-5:

"This system will provide a consistent nationwide approach for Federal, State, and local governments to work effectively and efficiently together to prepare for, respond to, and recover from domestic incidents, regardless of cause, size, or complexity. To provide for interoperability and compatibility among Federal, State, and local capabilities, the NIMS will include a core set of concepts, principles, terminology, and technologies covering the incident command system; multi-agency coordination systems; unified command; training; identification and management of resources (including classifying types of resources; qualifications and certification; and the collection, tracking, and reporting of incident information and incident resources."

Concepts and principles

To provide this framework for interoperability and compatibility, the NIMS is based on an appropriate balance of flexibility and standardization.

- **Flexibility.** The NIMS provides a consistent, flexible, and adjustable national framework within which government and private entities at all levels can work together to manage domestic incidents, regardless of their cause, size, location, or complexity.

- **Standardization.** The NIMS provides a set of standardized organizational structures, such as the incident command system (ICS), multiagency coordination systems, and public information systems, as well as requirements for processes, procedures, and systems designed to improve interoperability among jurisdictions and disciplines in various areas. These include training, resource management, personnel qualification and certification, equipment certification, communications and information management, technology support, and continuous system improvement.

NIMS components

NIMS is comprised of various components that when used together provide the framework for the coordinated effort to prepare for, prevent, respond to, and recover from domestic incidents, regardless of their cause, size, or complexity. These various components are dispersed throughout the text as they relate to the various chapters' contents. As with the concepts in the text, all of the concepts used in conjunction will allow a coordinated and methodical approach to planning for and responding to disasters of all types.

- **Command and management.** Three key organizational systems provide the command and management portion of NIMS. They are:
  - *The ICS.* Discussed in this chapter, the ICS defines the operating characteristics, interactive management components, and structure of incident management and emergency response organizations engaged throughout the life cycle of an incident.
  - *Multiagency coordination systems.* These define the operating characteristics, interactive management components, and organizational structure of supporting incident management entities engaged at the federal, state, local, tribal, and regional levels through mutual aid agreements and other assistance agreements.
  - *Public information systems.* These refer to processes, procedures, and systems for communicating timely and accurate information to the public during crisis and emergency situations.
- **Preparedness.** The old adage, "The best defense is a good offense," certainly plays a part in effective emergency management. NIMS focuses on certain activities that allow jurisdictions, state, and federal assets to maintain a high degree of readiness. Those activities are:
  - *Planning.* As discussed in chapter 1, a well-coordinated response begins with a solid plan.
  - *Training.* As covered in chapter 3, training is one of the most important predisaster activities. It ensures that personnel know what action to take in the event of an emergency. It ensures that personnel have a cognitive approach to disaster planning and response.
  - *Exercises.* A plan is only good if it works. By conducting exercises, plans can be tested in a controlled environment. Exercises allow personnel responsible for enacting the plan to practice this enactment. It also ensures the plan will function as written.
  - *Personnel qualification and certification.* This activity allows national-level standards measurements to ensure that incident management and emergency responder personnel are appropriately qualified and officially certified to perform NIMS-related functions.
  - *Equipment acquisition and certification.* Incident management organizations and emergency responders at all levels rely on various types of equipment to perform

mission essential tasks. A critical component of operational preparedness is the acquisition of equipment that will perform to certain standards, including the capability to be interoperable with similar equipment used by other jurisdictions.

  - *Mutual aid.* Mutual aid agreements are discussed in chapter 5. As noted in chapter 5, no one agency can handle all events that occur within its borders. Therefore, it is imperative that mutual aid agreements are in place to supplement the resources of the given jurisdiction.

  - *Publications management.* Having standardized forms and publications ensures that all planners and responders are utilizing the same documents. Publications management ensures that documents are updated as needed and that distribution occurs in a timely fashion.

- **Resource management.** NIMS standardizes the mechanisms and establishes requirements for processes to describe, inventory, mobilize, dispatch, track, and recover resources over the life cycle of the incident. This system is described in detail in chapter 5.

- **Communications and information management.** NIMS identifies the requirement for communications, information management, and information sharing at all levels of incident management. Those elements are:

  - Incident management communications

  - Information management

These elements are further discussed in chapter 6 under the heading "NIMS Interoperability Requirements."

- **Supporting technologies.** Technological support is needed for supporting the incident objectives. These technologies include voice and data communications systems, information management systems, and data display systems.

- **Ongoing management and maintenance.** As with emergency operations plans (EOPs), the NIMS document must be constantly evaluated for effectiveness and updated as needed. The NIMS document provides for routine review and refinement to the document.[1]

Components of the ICS

In the discussion of the ICS, the authors have utilized the ICS portion of the NIMS document, as it is the federally mandated system of the United States. This document from the Department of Homeland Security is entitled *National Incident Management System*, and it is the source document for the following discussion.[2]

General requirements of the ICS

In order for the ICS to function properly, there are certain organizational principles and components needed so that all responders can operate under the same system. Those principles and components are:

- Broad in scope
- Standardized units
- Practical to the situation
- Easily understood/common terminology
- Flexible/expandable
- Coordinated
- Properly trained staff
- Principles of organization/unity of command

By having a command system that is broad in scope, it allows all responders to utilize the system and allows these agencies to utilize the system on every call and event, regardless of its nature. The ICS is even being used to organize public safety at large events such as the Olympics and can be utilized by every jurisdiction to provide safety to even the smallest event.

Standardized units are important so that all operating and planning personnel can have the same understanding of what they can expect when they order a particular resource. If this were not the case, one department may term a *task force* as three engines, while another department may send a technical rescue truck. This would be very inefficient.

The ICS system must be a practical system in order not to frustrate all of the users of the system. Fire service personnel and other responders have all worked with systems that looked good on paper but were quite impractical from actual use. These systems fail because human nature is to look for a better system when the one in use is impractical.

By having the system in an easily understood format and providing common terminology, responders who do not use the system on a daily basis can easily integrate into the system. Many local elected officials are vital during a time of disaster, but these officials do not use the system in their daily business. If the system were very technical and complicated, they would be delayed trying to utilize the system in a real time of need.

Flexibility and expandability are important, especially in disaster situations. If the system is not flexible enough to deal with incidents ranging from a traffic stop by a police officer to the 9/11 attacks, responders would have difficulty knowing which system to employ at any given time. In addition, at what point would a responder switch systems? With the ICS system, the same system is flexible enough to use in any incident or disaster.

Coordination ensures that the response is efficient and in the best interest of all victims. In contrast, a lack of coordination in an emergency response can be a safety concern. An example of an uncoordinated response would involve each fire company arriving and setting into motion their own action plan. E-1 wants to rescue only pets; E-2 wants to only extinguish fires; and E-3 wants to operate in command mode and produce orders for E-1 and E-2. In this response to a widespread disaster, all companies are operating on their own and only conducting activities that are in their interest, and no one is rescuing victims of the disaster. With ICS, only one incident commander is responsible for the entire event, and only one incident action plan is operating at a time.

The best system in the world will not work in the field if the end users are not trained how to utilize the system. Training is now mandated under NIMS requirements. Currently, there are various requirements for the various levels of supervision expected to be performed at the incident scene. In table 3–1, presented on page 62, the supervisory levels as well as the required training are listed. As shown from the table, all responders and elected officials must have the IS-700. From this basis, the requirements may include: IS-100 Introduction to ICS, IS-200 ICS Basic, IS-300 Intermediate ICS, and IS-400 Advanced ICS. This standardized training will improve the integration of many disciplines that are present at disasters. By ensuring that a jurisdiction is compliant with the training standards, improved organizational management is assured at the next big disaster.

Having only one supervisor and having certain subordinates ensures that each person knows from whom their orders originated as well as to whom problems and progress should be reported during the operation. This clear line of command ensures that vital information is sent and received. It also eliminates conflicting orders that can delay an operation or diminish safety, or both. This "one boss" concept is known as *unity of command*. It is what makes ICS so workable at the scene of an incident. Many times in nonemergency situations, the popular management theories call for committees, focus groups, etc. in order to develop and institute a plan of action. Unfortunately, at the incident scene, there is not time to place a committee in action and await their best solution. Instead, the time constraint does not allow for interruptions that include differing opinions. That is not to say that unity of command is a complete dictatorship. There are provisions for safety, including the establishment of a safety officer and the ability to question due to a safety concern. However, it is nearly a dictatorship, thus increasing efficiency and safety due to only one plan operating at any given time.

Finally, ICS works because the span of control is addressed in the system. As with any project management, one person can only receive and process information from a limited number of sources. In the nonemergency setting, this number can range in the low teens, but when trying to process critical, time-sensitive information, a person is found to only be able to have three to seven subordinates, with five being optimal. This span of control is addressed throughout the ICS component of NIMS and allows managers of varying degrees of experience to succeed in these positions. Figure 4–1 depicts an expanded ICS organizational chart.

The incident commander

At any incident, with the exception of one managed by unified command, there is only one incident commander. (Unified command is explained in more detail later in this chapter.) This person has the overall responsibility for the incident and must retain any duties he or she does not delegate. At a dumpster fire, the incident commander can be the engine company officer. At a civil disturbance, it may be the chief of police. No matter the daily title of the individual assigned the incident commander responsibility, his or her title at the incident is *command.* By using the positional title within the ICS system, it allows consistency and provides for less confusion than calling the person Chief John Doe, etc. If this same person is reassigned planning section chief, he or she will then have the designation *planning.*

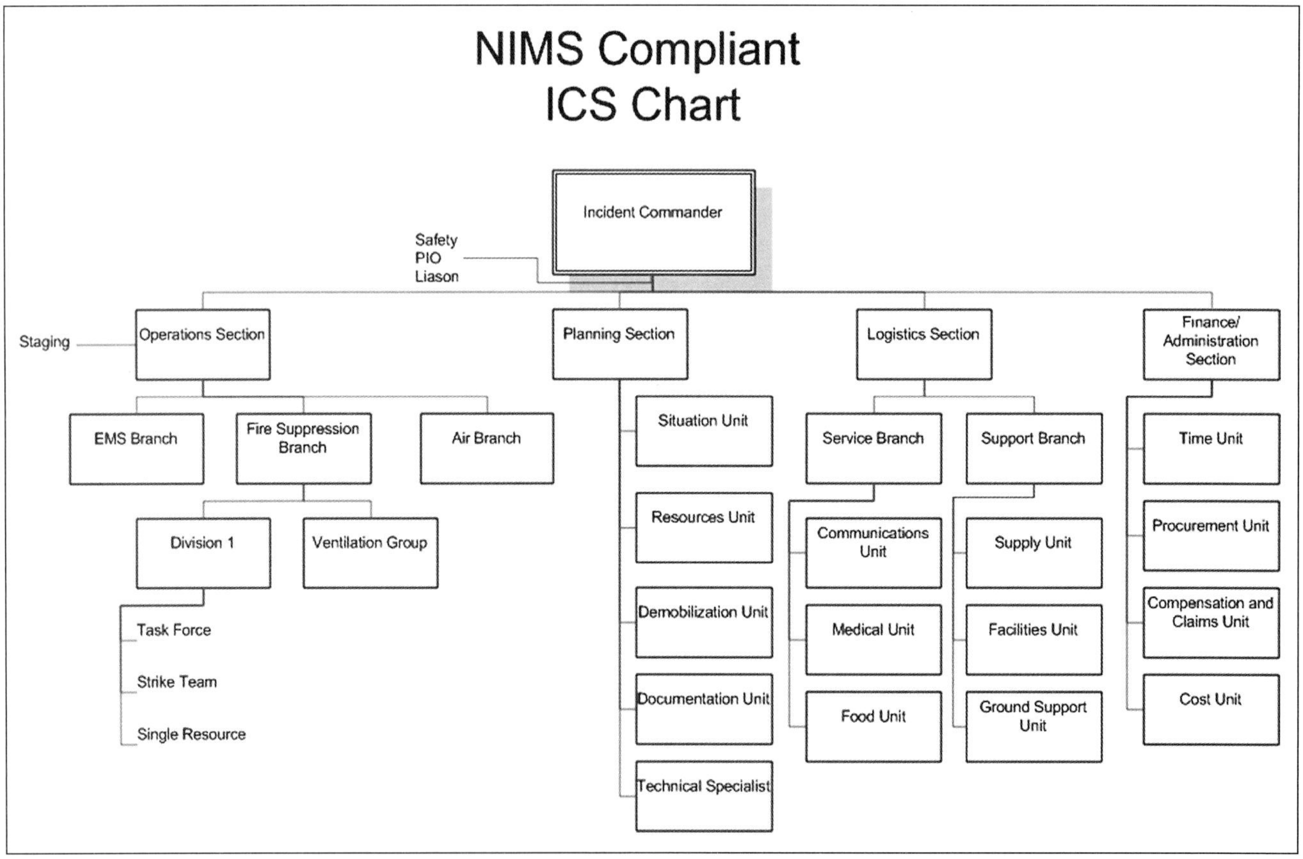

Fig. 4–1. NIMS compliant ICS chart. Ensuring that all documentation related to ICS is NIMS compliant will aid in a seamless, coordinated response.

The incident commander must accomplish the following in order to ensure the incident at hand is dealt with in a timely fashion:

- Development of a strategic plan
- Creation of objectives to fulfill the strategy
- Creation of tasks to meet the objectives

Although each of these functions is a responsibility of command, the incident commander can appoint section chiefs to develop the objectives and tactics of each of these sections. The sections are *operations*, *planning*, *logistics*, and *finance/administration*. Within each of these sections, objectives, tactics, and support decisions will be necessary to support the strategic plan created by the incident commander.

As an example, a hazardous materials spill can have a strategic plan of mitigating the chemical spill and disposing of it in a proper fashion. In order to meet this strategy, some objectives must be determined. The first objective is to identify and contain the product. Second, a decision must be made as to the proper mitigation techniques for the given situation. Finally, it is necessary to consult with outside agencies to determine the proper disposal.

In order to fulfill these objectives, tasks must be created and completed to meet the objectives. The first task may be to send a hazmat team into the scene to identify the chemical and the factors associated with the spill. This will also require a proper decontamination plan and process. Then the hazardous materials team could be tasked with conducting leak control measures to prevent further product escape. The next task could involve discussing the proper disposal of the product with the Environmental Protection Agency representative.

During all of these tasks, planning, support, and financial decisions must occur in order to support these tasks. Some of the planning may involve creation of evacuation plans in case the spill size increases. Logistically, all of the vehicles and equipment to perform these and subsequent tasks must be provided. Financially, a method of funding the equipment, vehicles, and support facilities such as food and restroom facilities must be coordinated and tracked. All of these functions must occur simultaneously, thus it is apparent that the single incident commander would not be able to handle this workload. Instead, he or she will delegate the duties while building an ICS organizational chart. The key for the incident command person to remember is that whatever tasks are not delegated, he or she must perform.

Operations section

The operations section is responsible for managing tactical operations of the incident. The person overseeing this section is the *operations section chief.* This again conforms to common terminology, and is the term used regardless of the person's normal title.

The operations section is responsible for employing the incident commander's strategy. This command can be one person, or in a case of an incident involving multiple jurisdictions, the command may be unified. In the latter case, the operations section chief is blending all of the various needs of the multiple jurisdictions into one operational plan. This plan is not created in isolation but rather takes into account the commander's demands, as well as receiving information from his or her subordinates on the success of the current operation and the conditions that accompany the incident.

Within the operations section can be up to five branches, 25 groups or divisions, and numerous resources. The labor divisions can serve many purposes. The main reason to add divisions, groups, and branches is to retain the span of control. As discussed earlier, the optimal span of control is five. In a high-demand incident, such as a rapidly advancing fire, information will be at a peak, thus having more than five subordinates makes tracking and processing the information difficult.

Branches. Division of labor is the primary reason that extra workgroup supervisors are placed into the ICS organization. Because the ICS organization by design is primarily built from the ground up and from the top down, branches may be one of the last organizational components to be installed into the organizational chart on a routine basis. However, in the following instances, placing branches early will enhance the efficiency of the overall organization of the incident:

- Dividing the operation into functional divisions
- Dividing the operation by jurisdiction

Many large-scale incidents have many different functions that need to take place nearly simultaneously but have different organizations handling these objectives. During a large-scale building collapse, there could be firefighting forces, police investigation forces, emergency medical services, hazardous materials teams, and search and rescue forces all operating at the scene. In order to properly administer each of the interrelated, but separate, operating forces, the incident commander and/or the operations section chief could elect to divide the forces into branches. This will allow a commander who is well versed in the operations of that particular function to oversee that function. In addition, it allows the incident commander and/or the operations section chief to oversee the blending of all of these functions to ensure that the strategic goals of the operation are being met.

Groups and divisions. The level of work division directly below the branch level is the group or division. Within everyday incidents, this workforce division is the most often employed within the ICS organization. The groups and divisions are at the same management level, but have different meanings. *Groups* are designed to divide the incident by functions, such as security, rescue, fire attack, etc. *Divisions*, on the other hand, divide the incident by geographical boundaries. This could be designated as *east*, *west*, *first floor*, *second floor*, etc. Commonly, the divisions within and around the structure follow the pattern of alphabetical descriptors for exterior boundaries. In general, designation begins with **A** at the front of the building and continues through the alphabet in a clockwise direction. This would mean that a rectangular building would be lettered **A** though **D**, with the front as **A** and the rear as **C**. Within the building, the floors above grade are named numerically with the grade level being 1. Any floors below grade level are designated with the prefix *basement.* These are labeled from the grade to the lowest level in ascending numerical order. Thus the first level below grade is B-1, and the levels increase with descent to lower levels. Although these are common namings for divisions, the important factor in the naming is consistency and ensuring that all operating forces are aware of the labels. It has been noted in past firefighter death reports that labeling problems and confusion have been the root cause of at least one death.

Resources. The basic operating forces at any incident are resources. These can be a police car, a road crew, a backhoe operator, or an entire force of search and rescue members. There are three different types of resources that can exist at an incident. These are:

- Single resource
- Strike team
- Task force

The *single resource* is an independent resource. An example would be a fire engine company or a police patrol car. During the initial phases of an incident, these are often how resources arrive and are assigned. An example of a house fire would be to have two engine companies and a ladder company respond to the fire. After their arrival, they could remain as a single resource or be assigned to a division or group.

A *strike team* is a set number of resources of the same kind and type. Typing will be discussed later in this text. This allows numerous resources with the same function to operate together, thus limiting the span of control. The strike team is assigned a strike team leader as its management component.

A *task force* is a group of different resources placed together for a specific mission. The task force has a leader, and the entire task force has common communications. The task force could be as simple as a response task force in a riot situation, which may consist of a fire company and a police unit that utilize face-to-face communication. It could also be as complex as a FEMA US&R team consisting of 80 members, with search, rescue, hazmat, medical, and logistics components utilizing a complex communications system involving their own repeaters and communications system.

The planning section

The planning section is responsible for collecting, evaluating, and disseminating tactical information for the incident. In addition, this section tracks the resources assigned to the incident. This section also prepares the incident action plan and any relevant maps or electronic or hardbound information related to the plan. In addition to working on the current operational period, the planning section focuses on future operations. This section is essential for long-term operations, due to the importance of planning and tracking processes over the ongoing operational periods. Utilization of ICS Form 201, as shown in figure 4–2, will formalize and track the planning process and allow a written, organized plan.

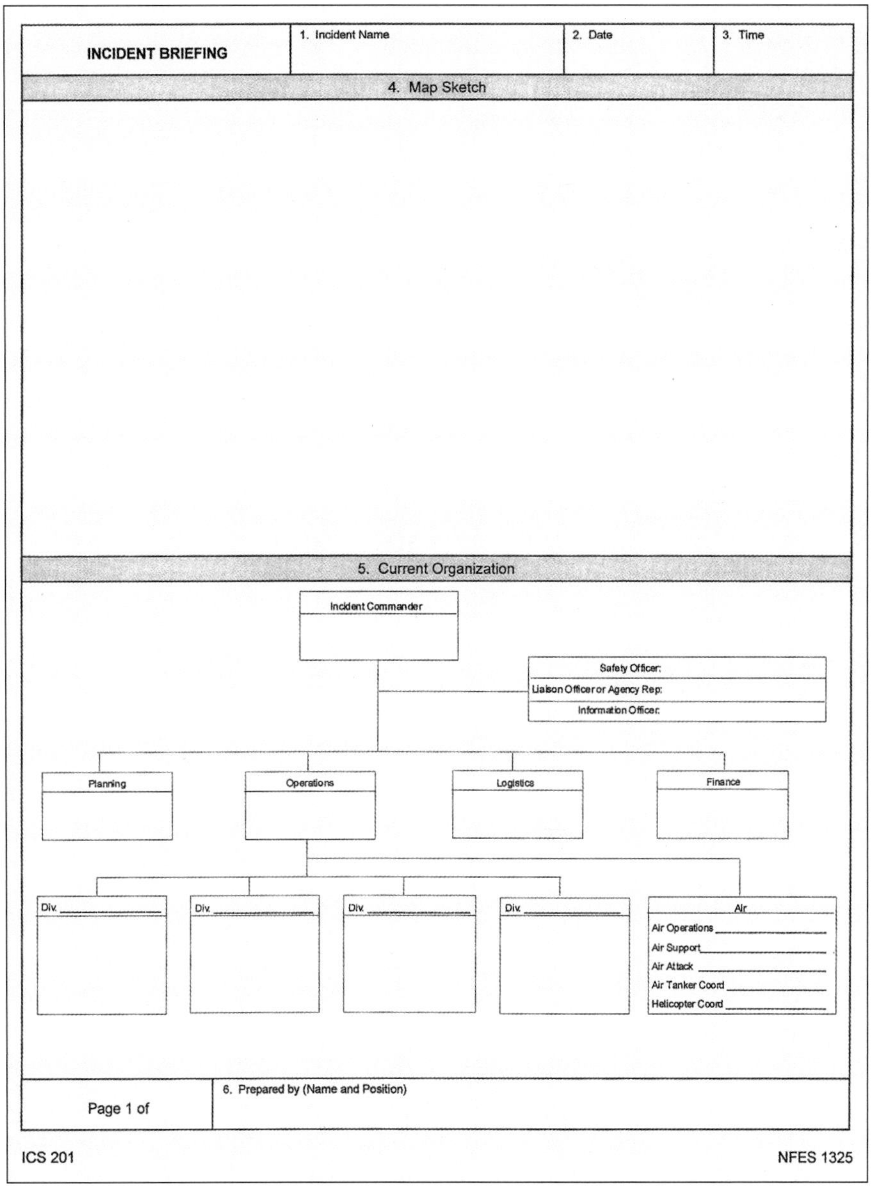
INCIDENT BRIEFING
1. Incident Name
2. Date
3. Time
4. Map Sketch
5. Current Organization
Incident Commander
Safety Officer:
Liaison Officer or Agency Rep:
Information Officer:
Planning
Operations
Logistics
Finance
Div.
Div.
Div.
Div.
Air
Air Operations
Air Support
Air Attack
Air Tanker Coord
Helicopter Coord
Page 1 of
6. Prepared by (Name and Position)
ICS 201
NFES 1325

Fig. 4–2. ICS Form 201. The ICS 201 form's use will ensure that planning follows a formalized process. These documents will also become part of the incident documentation, which could be utilized in a court of law.

The planning section has four primary units operating under the planning section chief, as well as any number of technical specialists who are pertinent to the given operation. The primary units working within the planning section are the following:

- Situation unit
- Resources unit
- Demobilization unit
- Documentation unit
- Technical specialists

The individual responsible for overseeing the planning section is the planning section chief. This person is responsible for overseeing incident-related data gathering and analysis regarding incident operations and assigned resources. He or she develops alternatives for tactical operations, conducts planning meetings, and prepares the incident action plan for each operational period. It is customary for the individual staffing the planning section chief to come from the jurisdiction with the primary incident responsibility. Other jurisdictions can be represented by assigning deputies to the planning section chief.

Resources unit. The resources unit is responsible for ensuring that all resources are checked in at the incident and is also responsible for subsequently tracking all of these resources. Obviously, the unit will need a method of tracking resources. Some of the accountability systems common to disaster services are the T-card system and the passport system. A "resource" can consist of personnel, teams, facilities, supplies, or major items of equipment.

Within the resources unit, there must be a consistent way to identify and track units. A portion of the NIMS compliance is the typing of resources. Typing of resources allows the situation unit to know exactly what the resources are, as well as allowing all operating forces to have the same knowledge. One example of this pertaining to the fire service is the term *rescue unit.* This same unit designation may mean a paramedic ambulance or it could mean a heavy rescue unit for vehicle extrication. This variance in terminology could cause a great deal of delay in an operation if the resource unit thinks they are sending a heavy rescue to the incident, but instead they are sending the operating forces a paramedic ambulance.

The resource unit recognizes that there are three status possibilities for any resource. These are:

- *Assigned.* This status indicates that the resource is checked into the incident and is currently operating to support the incident operation.
- *Available.* This status indicates that the resource is checked into the incident and is available for assignment to incident operations.
- *Out of service.* This status indicates that the unit is assigned to the incident but is unable to be assigned to support incident operations due to mechanical problems, rest, or conduction of the resource.

All changes in status are the responsibility of the resources leader and are to be reported to the resources unit. The unit leader must relay not only the status of the resource but also its location.

Situation unit. The situation unit collects, processes, and organizes ongoing situation information. It prepares situation summaries and develops projections and forecasts of future events related to the incident. The situation unit also prepares maps and gathers and disseminates information and intelligence for use in the IAP. This unit may also require the expertise of technical specialists and operations and information security specialists.

Documentation unit. The documentation unit maintains accurate and complete incident files, including a complete record of the major steps taken to resolve the incident; provides duplication services to incident personnel; and files, maintains, and stores incident files for legal, analytical, and historical purposes. Documentation is part of the planning section primarily because this unit prepares the IAP and maintains many of the files and records that are developed as part of the overall IAP and planning function.

Demobilization unit. The demobilization unit develops an incident demobilization plan that includes specific instructions for all personnel and resources that will require demobilization. This unit should begin its work early in the incident, creating rosters of personnel and resources and obtaining any missing information as check-in proceeds.

Note that many city- and county-provided resources, because they are local, do not require specific demobilization instructions. Once the incident demobilization plan has been approved, the demobilization unit ensures that it is distributed both at the incident and elsewhere as necessary.

Technical specialist. The ICS is designed to function in a wide variety of incident scenarios requiring the use of technical specialists. These personnel have special skills and are activated only when needed. Specialists may serve anywhere within the organization, including the command staff. No minimum qualifications are prescribed, as technical specialists normally perform the same duties during an incident that they perform in their everyday jobs, and they are typically specially certified in their fields or professions.

Technical specialists assigned to the planning section may report directly to its chief or to any function in an existing unit. They also could form a separate unit within the planning section, depending on the requirements of the incident and the needs of the section chief. Technical specialists may also be assigned to other parts of the organization (e.g., to the operations section to assist with tactical matters or to the finance/administration section to assist with fiscal matters). For example, a legal specialist or legal counsel may be assigned directly to the command staff to advise the IC on legal matters, such as emergency proclamations, legality of evacuation orders, and legal rights and restrictions pertaining to media access. Generally, if the expertise is needed for only a short period and normally involves only one individual, that individual should be assigned to the situation unit. If the expertise will be required on a long-term basis and may require several personnel, it is advisable to establish a separate technical unit in the planning section.

Logistics section

The logistics section meets all support needs for the incident, including ordering resources through appropriate procurement authorities from off-incident locations. It also provides facilities, transportation, supplies, equipment maintenance and fueling, food service, communications, and medical services for incident personnel. (See fig. 4–1.) The logistics section is led by a section chief, who may also have a deputy. Having a deputy is encouraged when all designated units are established at an incident site. When the incident is very large or requires a number of facilities with large numbers of equipment, the logistics section can be divided into two branches. The units operating under the logistics section are the following:

- Service branch
  - Communications unit
  - Medical unit
  - Food unit
- Support branch
  - Supply unit
  - Facilities unit
  - Ground support unit

Supply unit. The supply unit orders, receives, stores, and processes all incident-related resources, personnel, and supplies.

Once established, the supply unit also has the basic responsibility for all off-incident ordering, including the following:

- All tactical and support resources (including personnel)
- All expendable and nonexpendable supplies required for incident support

The supply unit provides the support required to receive, process, store, and distribute all supply orders. The unit also handles tool operations, which include storing, disbursing, and servicing all tools and portable, nonexpendable equipment.

Facilities unit. The facilities unit sets up, maintains, and demobilizes all facilities used in support of incident operations. The unit also provides facility maintenance and security services required to support incident operations.

The facilities unit sets up the ICP, incident base, and camps, as well as trailers and/or other forms of shelter for use in and around the incident area. The incident base and camps may often be established in areas having existing structures, which may be used in their entirety or only in part. The facilities unit also provides and sets up necessary personnel support facilities, including areas for the following:

- Food and water service
- Sleeping
- Sanitation and showers
- Staging

This unit also orders, through supply, such additional support items as portable toilets, shower facilities, and lighting units.

Ground support unit. The ground support unit is tasked with the following:

- Maintain and repair primary tactical equipment, vehicles, and mobile ground support equipment
- Record usage time for all ground equipment (including contract equipment) assigned to the incident
- Supply fuel for all mobile equipment
- Provide transportation in support of incident operations (except aircraft)
- Develop and implement the incident traffic plan.

In addition to its primary functions of maintaining and servicing vehicles and mobile equipment, the ground support unit also maintains a transportation pool for major incidents. This pool consists of vehicles (e.g., staff cars, buses, pickups) that are suitable for transporting personnel. The ground support unit also provides up-to-date information on the location and status of transportation vehicles to the resources unit.

Communications unit. The communications unit is responsible for development of the communications plan. This is completed on ICS Form 205. The communications plan should make effective use of the communications equipment and facilities affiliated with the incident. Within the communications plan should be radio networks. These networks make use of the available radio frequencies so that different levels within the ICS organization can communicate effectively.

The communications leader oversees the incident communications center and is essential to all planning meetings so that he or she can establish networks for large-scale, multijurisdictional incidents. By attending these meetings, the communications leader can ensure that the radio networks and communications plan will support the tactical objectives for the next operational period.

In addition to setting up command, air, support, and tactical networks, the communications unit is tasked with establishing landline connections for the incident. The communications unit should now be tasked with establishment of a wireless network at the command post. With the plethora of Internet-based informational sources, an established high-speed Internet connection is necessary for incidents lasting more than one operational period. There is an abundance of information available on the Internet concerning hazardous materials, mapping, and weather conditions and forecasts.

Some of the radio networks that the communications leader is responsible for establishing are the following:

- *Command network.* Links together incident command, command staff, section chiefs, branch directors, divisions and group supervisors
- *Tactical nets.* Several tactical nets may be established in order to connect agencies, departments, geographical areas, or specific functional units. These nets should be established based on need, planning, and logistical functions, but they are the overall responsibility of the communications unit leader.
- *Support net.* This network can be used for reporting resource status changes but can also handle much of the logistical radio traffic.

- *Ground-to-air net.* This network is used to coordinate ground-to-air traffic.
- *Air-to-air net.* This network is used for communications between air resources. These networks are often designated by the air resources and will be assigned for use at the incident.

Food unit. The food unit determines food and water requirements. It plans menus, orders food, provides cooking facilities, cooks, serves, maintains food service areas, and manages food security and safety concerns.

Efficient food service is important, but especially so for any extended incident. The food unit must be able to anticipate incident needs, both in terms of the number of people who will need to be fed and whether the type, location, or complexity of the incident indicates that there may be special food requirements. The unit must supply food needs for the entire incident, including all remote locations (i.e., camps and staging areas), as well as supply food service to operations personnel unable to leave operational assignments. The food unit must interact closely with the following elements:

- Planning section, to determine the number of personnel that must be fed
- Facilities unit, to arrange food-service areas
- Supply unit, to order food
- Ground support unit, to obtain ground transportation
- Air operations branch director, to obtain air transportation

Careful planning and monitoring is required to ensure food safety before and during food service operations, including the assignment, as indicated, of public health professionals with expertise in environmental health and food safety.

Note that feeding victims is a critical operational activity that will be incorporated into the IAP. Feeding activities will normally be conducted by members of an appropriate nongovernmental organization, such as the American Red Cross or similar entities.

Medical unit. The primary responsibilities of the medical unit include the following:

- Develop the incident medical plan (for incident personnel)
- Develop procedures for handling any major medical emergency involving incident personnel
- Provide continuity of medical care, including vaccinations, vector control, occupational health, prophylaxis, and mental health services for incident personnel
- Provide transportation for injured incident personnel
- Ensure that incident personnel patients are tracked as they move from origin, to care facility, to final disposition
- Assist in processing all paperwork related to injuries or deaths of incident-assigned personnel
- Coordinate personnel and mortuary affairs for incident personnel fatalities

The medical unit is responsible for the effective and efficient provision of medical services to incident personnel. The medical unit leader will develop a medical plan that will, in turn, form part of the IAP. The medical plan should provide specific information on medical assistance capabilities at incident locations, potential hazardous areas or conditions, and off-incident medical assistance facilities and procedures for handling complex medical emergencies. The medical unit will also assist the finance/administration section with the administrative requirements related to injury compensation, including obtaining written authorizations, billing forms, witness statements, administrative medical documents, and reimbursement as required.

The medical unit will ensure patient privacy to the fullest extent possible.

Note that patient care and medical services for those who are not incident personnel (victims of a bioterror attack, hurricane victims, etc.) are critical operational activities associated with a host of potential incident scenarios. As such, these activities are incorporated into the IAP as key considerations of the plans and operations sections. These sections should be staffed accordingly with appropriately qualified EMS public health, medical personnel, technical experts, and other professional personnel, as required.

The finance/administration section

When there is a specific need for financial, reimbursement (individual and agency or department), and/or administrative services to support incident management activities, a finance/administration section is established. Under the ICS, not all agencies will require such assistance. In large, complex scenarios involving significant funding originating from multiple sources, the finance/administrative section is an essential part of the ICS. In addition to monitoring multiple sources of funds, the section chief must track and report to the IC the financial "burn rate" as the incident progresses. This allows the IC to forecast the need for additional funds before operations are affected negatively. This is particularly important if significant operational assets are under contract from the private sector. The section chief may also need to monitor cost expenditures to ensure that statutory rules that apply are met. Close coordination with the planning section and logistics section is also essential so that operational records can be reconciled with financial documents.

Note that in some cases, only one specific function may be required (e.g., cost analysis), which a technical specialist in the planning section could provide. The finance/administration section chief will determine, given current and anticipated future requirements, the need for establishing specific subordinate units. In some of the functional areas (e.g., procurement), an actual unit need not be established if it would consist of only one person. In such a case, a procurement technical specialist would be assigned in the planning section instead. Because of the specialized nature of finance functions, the section chief should come from the agency that has the greatest requirement for this support. The section chief may have a deputy.

Time unit. The time unit is primarily responsible for ensuring proper daily recording of personnel time, in accordance with the policies of the relevant agencies. The time unit also ensures that the logistics section records or captures equipment usage time, through the ground support unit for ground equipment and through the air operations support group for aircraft.

If applicable (depending on the agencies involved), personnel time records will be collected and processed for each operational period. The unit leader may require the assistance of personnel familiar with the relevant policies of any affected agencies. These records must be verified, checked for accuracy, and posted according to existing policies. Excess hours worked must also be determined, for which separate logs must be maintained.

Procurement unit. The procurement unit administers all financial matters pertaining to vendor contracts. This unit coordinates with local jurisdictions to identify sources for equipment, prepares and signs equipment rental agreements, and processes all administrative requirements associated with equipment rental and supply contracts.

Note that in some agencies, the supply unit in the logistics section will be responsible for certain procurement activities. The procurement unit will also work closely with local cost authorities.

Compensation and claims unit. Under ICS, a single unit handles injury compensation and claims. The specific activities are, of course, varied and may not always be accomplished by the same person. The individual handling injury compensation ensures that all forms required by workers' compensation programs and local agencies are completed. This individual also maintains files on injuries and illnesses associated with the incident and ensures that all witness statements are obtained in writing. Since the medical unit may also perform certain of these tasks, close coordination between the medical and compensation and claims units is essential. The claims function handles investigations of all civil tort claims involving property associated with or involved in the incident. The compensation and claims unit maintains logs on the claims, obtains witness statements, and documents investigations and agency follow-up requirements.

Cost unit. The cost unit provides cost analysis data for the incident. This unit must ensure that equipment and personnel for which payment is required are properly identified. It must obtain and record all cost data and analyze and prepare estimates of incident costs. The cost unit also provides input on cost estimates for resource use to the planning section. The cost unit must maintain accurate information on the actual costs of all assigned resources.

UNIFIED COMMAND

Unified command is one variation of the incident command system. It is utilized when more that one jurisdiction is responding to an incident, such as an incident that covers more than one political boundary. In addition, the unified command system is employed when multiple agencies respond to and have responsibility for an incident. In this case, a civil disturbance is a good example. During a unified command operation, the command position would no longer involve only one person but would involve a representative from each agency or political subdivision in order to create the strategy and objective for the operations section to employ.

It is important to remember that only the command position is now a committee, not the entire ICS structure. The operations section chief will still only involve one person. During a unified incident, the operations section chief may have a deputy from another agency. In the case of civil disturbance, the police have the primary role, but the fire department must extinguish fires and rescue the injured as a result of the event. In this case, the unified command may involve both the police and fire departments. The operations section chief may be a police official, but the deputy operations section chief would be a fire official. In this unified structure, the use of functional branches helps to coordinate (but keep separate) the various agencies responding to the incident, allowing them to focus on their responsibilities. This is accomplished while maintaining a command structure that accommodates the responsibilities of all responding agencies.

EXPANDED COMMAND STRUCTURES

Within the principles of the ICS system, the command structure should only be as large as necessary to handle the incident. However, many emergency response personnel will respond to incidents that are very large in scale and require an expanded ICS structure. Within the ICS chart, the positions will not change regardless of size, with the exception of an area or unified command. However, employment of certain positions can allow the command system used daily to be expanded to fit the incident.

Within large events, the branch director is a good position to employ in order to break the workforce into areas of responsibilities. The branch director should be a person with expertise in the area of responsibility for the branch. For a hazardous materials incident in which there are numerous victims, an EMS branch and a hazardous materials branch can be employed. This will allow the operations section chief or the incident commander to focus on coordinating the two different but interrelated services needed to mitigate the incident.

The other ICS component that allows for rapid expansion of resources is the use of the strike team and the task force. Each is a group of five resources with a commander and common communication. The strike team is the same kind and type of resource, while the task force is a mixture of kinds and types. The use of the strike team or task force can allow the quintupling of the amount of resources operating at an incident without changing the span of control.

AREA COMMAND

An area command is established when the complexity of the incident and incident management span-of-control considerations so dictate. Generally, the administrator or administrators of the agency having jurisdictional responsibility for the incident make the decision to establish an area command.

There can be a couple of different purposes for an area command. One purpose of an area command could be to oversee the management of multiple incidents that are each being handled by a separate ICS organization. A different purpose might be to oversee the management of a very large or complex incident that has multiple incident management teams engaged.

This type of command is generally used when there are a number of incidents in the same area and of the same type, such as two or more hazmat spills or fires. These are usually the kinds of incidents that may compete for the same resources. When incidents are of different types or do not have similar resource demands, they are usually handled as separate incidents or are coordinated through an EOC. If the incidents under the authority of the area command span multiple jurisdictions, a unified area command should be established. This allows each jurisdiction involved to have appropriate representation in the area command.

Area commands are particularly relevant to public health emergencies, given that these events are typically not site specific or immediately identifiable and could be geographically dispersed and evolve over time ranging from days to weeks. Such events as these, as well as acts of biological, chemical, radiological, and nuclear terrorism, call for a coordinated intergovernmental, private-sector, and nongovernmental organization response. These events require large-scale coordination typically conducted at a higher jurisdictional level. An ICS chart and graphic are provided in figure 4–3 to illustrate this fairly new organizational element.

Responsibilities

The area command does not have operational responsibilities. For the incidents under its authority, the area command is charged with the following responsibilities:

- Set overall agency incident-related priorities
- Allocate critical resources according to the established priorities
- Ensure that incidents are properly managed
- Ensure effective communications
- Ensure that incident management objectives are met and do not conflict with each other or with agency policies
- Identify critical resource needs and report them to the interagency coordination system (generally EOCs)
- Ensure that short-term "emergency" recovery is coordinated to assist in the transition to full recovery operations
- Provide for personnel accountability and a safe operating environment

The area command develops an action plan detailing incident management priorities, needs, and objectives. This plan should clearly state policy, objectives, and priorities; provide a structural organization with clear lines of authority and communications; and identify incident management functions to be performed by the area command (i.e., public communications).

Organization

The area command organization operates under the same basic principles as ICS. Typically, an area command will comprise the following key personnel, all of whom must possess appropriate qualifications and certifications:

1. **Area commander (unified area command).** The area commander is responsible for the overall direction of the incident management teams assigned to the same incident or to incidents in close proximity. This responsibility includes ensuring that conflicts are resolved, that incident objectives are established, and that strategies are selected for the use of critical resources. The area command is also responsible for coordinating with federal, state, local, tribal, and participating private organizations.

2. **Area command logistics chief.** The area command logistics chief provides facilities, services, and materials at the area command level and ensures the effective allocation of critical resources and supplies among the incident management teams.

3. **Area command planning chief.** The area command planning chief collects information from various incident management teams to assess and evaluate potential conflicts in establishing incident objectives, strategies, and priorities for allocating critical resources.

4. **Area command support positions.** The following positions are activated as necessary:

a. *Area command critical resources unit leader.* The critical resources unit leader tracks and maintains the status and availability of critical resources assigned to each incident under the area command.

b. *Area command situation unit leader.* The situation unit leader monitors the status of objectives for each incident or IMT assigned to the area command.

c. *Area command public information officer.* The PIO provides public information coordination between incident locations and serves as the point of contact for media requests to the area command.

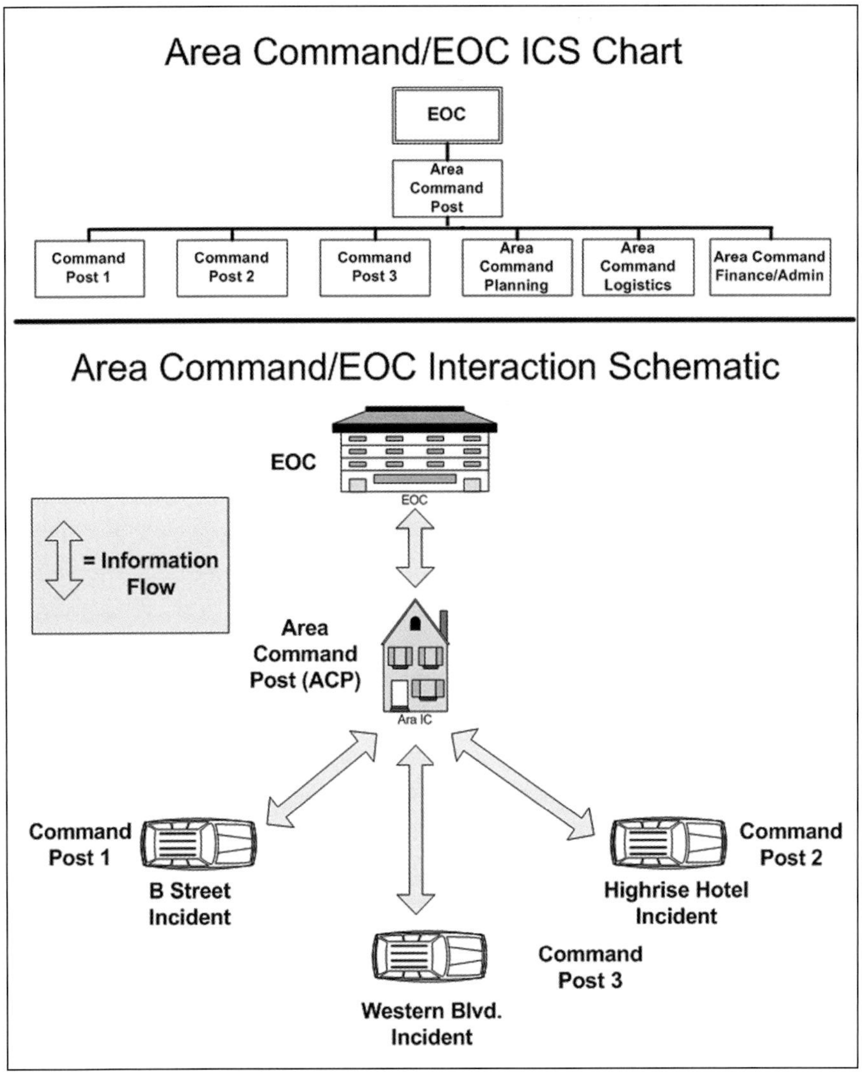

Fig. 4–3. Area command schematic and ICS chart. Large-scale incidents or those spanning a large geographical area will require the use of expanded command structures. This schematic illustrates how to utilize an area command structure.

d. *Area command liaison officer.* The liaison officer helps maintain off-incident interagency contacts and coordination.

e. *Area command aviation coordinator.* An aviation coordinator is assigned when aviation resources are competing for common airspace and critical resources, and works in coordination with incident aviation organizations to evaluate potential conflicts, develop common airspace management procedures, and prioritize critical resources.

Location

The following guidelines should be followed in locating an area command:

- To the extent possible, the area command should be established in close proximity to the incidents under its authority. This makes it easier for the area commander and the ICs to meet and otherwise interact.
- It is, however, best not to collocate an area command with any individual ICP. Doing so might cause confusion with the command and management activities associated with that particular incident.
- Area commands must establish effective, efficient communications and coordination processes and protocols with subordinate ICPs, as well as with other incident management organizations involved in incident operations.
- The facility used to house the organization should be large enough to accommodate a full area command staff. It should also be able to accommodate meetings between the area command staff, the ICs, and agency executives, as well as news media representatives.
- Area commands may be collocated with EOCs.

Reporting relationships

When an area command is involved in coordinating multiple incident management activities, the following reporting relationships will apply:

- The ICs for the incidents under the area command's authority report to the area commander.
- The area commander is accountable to the agency or agencies or to the appropriate jurisdictional executives or administrators.
- If one or more incidents within the area command occur in multiple jurisdictions, a unified area command should be established. In this instance, ICs would report to the unified area commander for their jurisdiction.

EMERGENCY OPERATIONS CENTERS

Many local and regional EOCs are in place or are currently under construction. Realizing the importance of having all of the interested parties in one place, the EOC is becoming a part of the safety infrastructure of many communities. These facilities are the central hub of information gathering and decision making that allows the field units to operate and be informed as well as supported logistically. In figures 4–4, 4–5, and 4–6, the newly constructed Greater Cincinnati Emergency Operations Center is featured. Note the numerous workstations and the large screen displays that can project nearly any information wanted. (The associate author thanks Steve Sierveld for his tour and demonstration of the facility.)

Fig. 4–4. Greater Cincinnati Regional EOC workstations. This is a typical workstation at an EOC. Ensuring that office equipment, such as phones and computers, is available to each EOC participant will speed the time in which decisions can be processed.

Photo courtesy of associate author Randall W. Hanifen.

Fig. 4–5. Greater Cincinnati Regional EOC video monitors. Today's technology permits the streaming of information into the EOC. This real-time information allows EOC participants to make the most informed decisions about the incident.

Photo courtesy of associate author Randall W. Hanifen.

Fig. 4–6. Greater Cincinnati Regional EOC breakout room. Having specialized breakout rooms available at the EOC will allow collaboration of EOC participants who have a common interest. An example could be all of the local fire chiefs deciding how to properly distribute the fire protection forces to the incident.

Photo courtesy of associate author Randall W. Hanifen.

CONCLUSION

The National Incident Management System has provided a consistent baseline to manage a variety of incidents. Some public service departments have used the system with success for years. Now all responding and support personnel must utilize it. Hopefully, this will take effect in much the same fashion as ICS improved the command and control of fire department operations throughout the country.

Area command is a newer concept for many response agencies but has been used in an unnamed fashion during many past disasters. Unified command has enabled many response agencies from the same jurisdiction or multiple jurisdictions to work in conjunction to mitigate an incident.

There are a few key elements to the success of the system. The first is its use for all incidents, planned and unplanned. The second is the assurance that all personnel operating at the incident have been trained in the system. The last key element is that there is a thorough review of incidents as they occur to see what deficiencies occurred in the area of command and control.

REFERENCES

1. U.S. Department of Homeland Security. 2004. *National Incident Management System*. March. Retrieved December 1, 2006 from: http://www.fema.gov/emergency/nims/index.shtm.

2. Ibid.

RESOURCE MANAGEMENT AND AUGMENTATION 5

INTRODUCTION

Disasters within the first decade of the 21st century have proven that no one city or state can provide all of the necessary resources to mitigate these incidents. In 2001, the people of New York City were shocked, and their fire service taxed to its limits, when two planes struck the World Trade Center towers. It was one of the only times in history that personnel of the American fire service can remember an incident being greater than the amount of resources available on hand from FDNY. The status of the FDNY as the largest fire department in the United States, serving the largest U.S. city, caused many to assume that no incident could tax the FDNY beyond its limits. This tragedy, as well as the rash of destructive hurricanes that struck the Gulf Coast in 2005, prove that every community, regardless of size, must create, test, and maintain a plan of how to summon additional resources during a time of disaster. To fail to create this document could be viewed as negligence on behalf of the emergency planners and executives.

A second reason to plan for resources beyond that of one's own community is that disasters are also known to involve large areas. Experience with tornadoes in all parts of the United States and Canada illustrates how damage may spread across more than one community, even into other states and provinces. Thus, it has been accurately stated that disaster control planning must be geared to an area concept.

Merely identifying that more responders, apparatus, food, and supplies are needed does not constitute a mutual aid plan. Coordination and written plans are a necessity for a coordinated effort to occur. Although many proactive communities have had these plans in place for the past few decades, planning is becoming a regional and a state function, as well as a renewed function of the federal government. This text will discuss the National Response Framework (effective March 2008, formerly the Federal Response Plan), and how it is designed to function, as well as examine Ohio's statewide mutual aid plan. The authors will also examine how to enhance the incident command structure with the utilization of mutual aid.

AGENCIES INVOLVED IN DISASTER CONTROL

In the early days of disaster response, it was believed that the response to disasters was a fire department problem. With the latest disaster response thinking, it is realized that it is the entire municipality's responsibility. Each service and support division of the community must work in a coordinated fashion in order to effectively manage and mitigate the disaster. Although it is not the responsibility of the fire department to manage the disaster, it is often the lead or consulting agency in the response due to its amount of experience and the related functions this department performs.

As discussed in chapter 4, some of the variations of the ICS system are applicable to multiagency responses. By utilizing unified and area command, the other agencies involved in the response will be able to coordinate their efforts and ensure that the objectives of each agency responding to the incident are examined and considered.

Often many support agencies can be overlooked due to their sporadic use at incidents. Some of the agencies the authors will examine include the service departments of the municipality, local health agencies, the American Red Cross, the Salvation Army, civil defense agencies, industrial representatives, military organizations, government employees, transportation, utilities, food and shelter, and special services. Keeping a copy of the emergency support functions (ESF) chart, such as the one in figure 5–1, in the EOC or command vehicles can allow decision makers to easily review the common needs at a large-scale disaster.

Departments within the local political entity

As discussed previously, the fire service often takes a leading role in disaster preparedness and response to disasters. This is due to their everyday exposure to and mitigation of smaller scale emergencies. In addition, the services provided by the fire service are often those needed by the citizens most in the postdisaster environment, such as rendering medical care, extinguishing or containing fires, and rescuing citizens trapped because of the disaster.

The police are likely the second in line as the most-utilized force in the disaster response arena. Their ability to restore civil order and prevent further citizen harm makes them a good addition to the medical care and rescue efforts. In the post-9/11 environment, the police will also be utilized to examine the scene while rescue operations are being conducted.

Service or road departments are utilized to support the fire and police operations with such actions as delivering supplies and working to restore roadways, either through repair or by clearing debris. In water emergencies, this department is often tasked with removing the water by operating large pumps. During snow emergencies, this group is the primary response organization, and the police and fire departments can support their operations by assisting them in relief.

American Red Cross, Salvation Army

According to the American Red Cross Web site,

Although the American Red Cross is not a government agency, its authority to provide disaster relief was formalized when, in 1905, the Red Cross was chartered by Congress to "carry on a system of national and interna-

ESF #1 Transportation	
• Federal and civil transportation support • Transportation safety • Restoration / recovery of transportation infrastructure • Movement restrictions • Damage and impact assessment	ESF Coordinator: U.S. Department of Transportation
ESF #2 – Communications	
• Coordination with telecommunications industry • Restoration / repair and temporary provisioning of communications infrastructure • Protection, restoration, and sustainment of national cyber and information technology resources	ESF Coordinator: U.S. Department of Homeland Security / National Communications System
ESF #3 – Public Works and Engineering	
• Infrastructure protection and emergency repair • Infrastructure restoration • Engineering services, construction management • Critical infrastructure liaison	ESF Coordinator: U.S. Department of Defense / U.S. Army Corps of Engineers
ESF #4 – Firefighting	
• Firefighting activities on Federal lands • Resource support to rural and urban firefighting operations	ESF Coordinator: U.S. Department of Agriculture
ESF #5 – Emergency Management	
• Coordination of incident management efforts • Issuance of mission assignments • Resource and human capital • Incident action planning • Financial management	ESF Coordinator: U.S. Department of Homeland Security / Federal Emergency Management Agency
ESF #6 – Mass Care, Housing, and Human Services	
• Mass care • Disaster housing • Human services	ESF Coordinator: U.S. Department of Homeland Security / Federal Emergency Management Agency
ESF #7 – Resource Support	
• Resource support (facility space, office equipment & supplies, contracting services, etc.)	ESF Coordinator: U.S. General Services Administration
ESF #8 – Public Health and Medical Services	
• Public health • Medical • Mental health services • Mortuary services	ESF Coordinator: U.S. Department of Health and Human Services
ESF #9 – Urban Search and Rescue	
• Life-saving assistance • Urban search and rescue	ESF Coordinator: U.S. Department of Homeland Security / Federal Emergency Management Agency

ESF #10 – Oil and Hazardous Materials Response	
• Oil and hazardous materials (chemical, biological, radiological, etc.) response • Environmental safety and short- and long-term cleanup	ESF Coordinator: U.S. Environmental Protection Agency
ESF #11 – Agriculture and Natural Resources	
• Nutrition assistance • Animal and plant disease / pest response • Food safety and security • Natural and cultural resources and historic properties protection, and restoration	ESF Coordinator: U.S. Department of Agriculture
ESF #12 – Energy	
• Energy infrastructure assessment, repair, and restoration • Energy industry utilities coordination • Energy forecast	ESF Coordinator: U.S. Department of Energy
ESF #13 – Public Safety and Security	
• Facility and resource security • Security planning and technical and resource assistance • Public safety / security support • Support to access, traffic, and crowd control	ESF Coordinator: U.S. Department of Justice
ESF #14 – Long-Term Community Recovery	
• Social and economic community impact assessment • Long-term community recovery assistance to States, local governments, and the private sector • Mitigation analysis and program implementation	ESF Coordinator: U.S. Department of Homeland Security / Federal Emergency Management Agency
ESF #15 – External Affairs	
• Emergency public information and protective action guidance • Media and community relations • Congressional and international affairs • Tribal and insular affairs	ESF Coordinator: U.S. Department of Homeland Security

Fig. 5–1. ESF chart. The chart depicts the 15 emergency support functions. This chart can serve as a reminder concerning needs at a disaster.

tional relief in time of peace and apply the same in mitigating the sufferings caused by pestilence, famine, fire, floods, and other great national calamities, and to devise and carry on measures for preventing the same." The Charter is not only a grant of power, but also an imposition of duties and obligations to the nation, to disaster victims, and to the people who generously support its work with their donations...

The Red Cross also feeds emergency workers, handles inquiries from concerned family members outside the disaster area, provides blood and blood products to disaster victims, and helps those affected by disaster to access other available resources.[1]

Fig. 5–2 shows an American Red Cross response vehicle.

Fig. 5–2. American Red Cross response vehicle. These vehicles deploy to the scene of a disaster to support responders and victims.

FEMA News Photo.

According to the Salvation Army Web site,

When The Salvation Army initiates a disaster relief operation, the first aim is to meet the basic needs of those who have been affected, both survivors and first responders (such as firefighters). Even at this level, The Salvation Army's workers are ministering in that they serve as a means of expressing God's love. The Salvation Army's goals are to offer:

- Material comfort
- Physical comfort
- Emotional comfort
- Spiritual comfort[2]

Both of these organizations can provide the support to the frontline forces that is needed to ensure their continued performance. Any planner or commander must remember to take care of those providing the care.

Civil defense agencies

Many planners and responders have difficulty understanding where the military intertwines with disaster response. Many people watched as the military enacted security measures due to the looting in the aftermath of Hurricane Katrina. Widespread media coverage also showed the National Guard rescuing many people. As a constitutional liberty of the United States, the armed forces cannot be used against U.S. citizens. However, as stated in the second edition of the *Fire Officer's Guide to Disaster Control*, "If the military forces are repulsing an enemy attack, that is 'military defense.' If civilians are coping with an emergency, that is civil defense."[3]

Civil preparedness agency

During normal, day-to-day operations, single departments operate under their standard operating procedures and with the necessary resources to conduct their operations. However, when disasters take place, it becomes necessary to coordinate the efforts of many agencies with numerous support resources. The civil preparedness agency or committee will aid in ensuring that the preparedness plan is implemented and will provide the necessary emergency support functions, such as approval of expenditures. With the induction of NIMS, the civil preparedness agency or committee is plugged into positions such as the finance administration section chief and other key logistics positions.

Industrial representatives

Many disasters occur at fixed facilities and are a result of a failure at the facility. When this occurs, the industrial representatives are the primary source of information for responding agencies. In the event of a hazardous materials incident, the on-site maintenance personnel or chemists can fulfill the technical information specialist position in the planning section and can guide the incident commander to successful mitigation of the hazardous materials leak.

According to OSHA publication 3088, *How to Plan for Workplace Emergencies and Evacuations*, "When developing your emergency action plan, it's a good idea to look at a wide variety of potential emergencies that could occur in your workplace. It should be tailored to your worksite and include information about all potential sources of emergencies. Developing an emergency action plan means you should do a hazard assessment to determine what, if any, physical or chemical hazards in your workplaces could cause an emergency. If you have more than one worksite, each site should have an emergency action plan."[4]

Military organizations

Under normal operations, the military is assigned to carry out militant operations, such as the fighting of wars and fulfilling operations around the world that relate to maintaining the freedom of the United States. However, during times of disaster in the United States, the military has provided support and quelled civil disturbances.

According to the U.S. Army Corps of Engineers Web site,

The United States Army Corps of Engineers (USACE) serves the Armed Forces and the Nation by providing vital engineering services and capabilities, as a public service, across the full spectrum of operations—from peace to war—in support of national interests.

Corps missions include five broad areas:

- Water Resources
- Environment
- Infrastructure
- Homeland Security
- Warfighting

These capabilities are complementary and reinforcing. For example, employees working on water resources projects on our Nation's waterways have deployed to Iraq and Afghanistan and applied their technical skills to support the warfighter.[5]

It is also noted that this group develops technical publications for rescues, such as the FEMA *US&R Structures Specialist Field Operations Guide*, which depicts how to build shoring to rescue trapped people from buildings.

Another important source of support is the U.S. Coast Guard. According to the U.S. Coast Guard Web site,

---

The United States Coast Guard is a military, multimission, maritime service within the Department of Homeland Security and one of the nation's five armed services. Its core roles are to protect the public, the environment, and U.S. economic and security interests in any maritime region in which those interests may be at risk, including international waters and America's coasts, ports, and inland waterways.

The Coast Guard provides unique benefits to the nation because of its distinctive blend of military, humanitarian, and civilian law-enforcement capabilities.[6]

The Coast Guard has five basic areas of service to the public:

- Maritime safety
- Maritime security
- Maritime mobility
- National defense
- Protection of natural resources

The Coast Guard was instrumental in saving many lives via helicopter rescue from the floodwaters of Hurricane Katrina.

Government employees

In a time of disaster, all of the employees from the government services should be utilized. Many municipalities use their nonemergency workers from the government services to complete various functions during a disaster. An example would be to place the clerk and the clerk's assistants in charge of the finance administration section of an incident. Who could be better equipped to fill the role of the compensation and claims unit leader than the payroll coordinator for the municipality? Another example of utilization would be to assign the sanitation workers to assist in light debris removal from the streets in a posttornado environment. No matter how they are utilized, the important point is that they are utilized in the plan and trained to operate in the NIMS system.

Transportation

During many disasters, the utilization of mass transit will determine the success of the operation. As seen during Hurricane Katrina, many of the low-income residents of New Orleans did not have transportation to vacate the city in order to avoid the hurricane. The city government subsequently failed to utilize enough mass transit buses in order to vacate the residents, and thus the media termed the planning and response a disaster. As with any form of transportation, the logistics section chief must support the vehicles with fuel and maintenance in order to ensure their reliability.

Property and procurement

Many operations will require heavy equipment or other equipment that is property of the private sector. Ensuring that the procurement unit of the finance/administration section is operating will ensure that rental contracts are signed and the procurement of these resources happens in a timely manner. The creation of a local resource list is a good planning activity that will speed this process. During the creation of the resource list, a member of the planning committee can make contact with the owners of the resources, thus allowing such items as special operators for the equipment to be coordinated prior to the incident.

Utilities control

Many disasters affect the basic utilities of the community. In many neighborhoods, the aboveground wires will be subjected to high winds, rising waters, or fire during disasters, thus causing their failure. In addition, as citizens return to their homes in a postdisaster time, they will likely power their homes with generators, creating a back-feeding problem for downed power lines.

Calling for the electric and gas company may be a futile effort during a widespread disaster due to the sheer volume of work they are facing. However, if it is possible to summon a gas and electric company supervisor to the EOC or command post, the gas and electric company can assist in a coordinated restoration effort. By combining the reconnaissance efforts of the emergency responders with the technical knowledge of the gas and electric company, the joint venture will allow the gas and electric workers to respond to areas where they can feasibly work. The employees will not have to waste time conducting their own reconnaissance work. Second, the incident commander can relay where the restoration of services would best aid the response effort.

Traffic control

Traffic control can often be overlooked during a large-scale disaster. With the police forces being utilized to aid in evacuations or to quell civil disturbance, traffic control is often a low priority. Traffic control became such a problem in the Hurricane Rita evacuation of Houston, Texas, that it caused traffic jams for nearly 100 miles. An effort from the Texas Department of Transportation found that converting the entire highway to one direction alleviated some of the congestion.

It is recommended that some basic planning occur in the establishment of evacuation routes. Although they may need to be tweaked during the actual event, the plan allows all parties responsible for traffic control to have an understanding of the process. Traffic control may be a very good task for the police explorers to undertake. Their uniform represents authority to the drivers, but their lack of certification precludes them from performing actual policing duties.

Health and sanitation

Many of the potential threats for mass destruction and terrorism present aspects that will heavily task the local health department. Such large-scale emergency situations also are likely to overwhelm the emergency care systems in the community, beginning with the EMS and continuing into the hospitals. Working with the local health department during the creation of the emergency operations plan will help the emergency responders understand the problems associated with disaster relocation. Disaster relocation occurs when the EMS personnel transport all of their victims to one hospital or transport contaminated victims to the hospital. In effect, they have just moved the disaster to the hospital. Planners must work with these facilities to understand capacities and understand how (or if) a disaster hospital network operates in their area.

Sanitation facilities are necessary for any event of extended duration. Although emergency workers will remedy the problem in the field as needed, it presents a public perception problem if they continue to do this past the initial hours of a response. Some municipalities may have mobile lavatories in their inventory of resources. If not, many companies supply these for a rental charge. This will be a task of the facilities unit in the logistics section. The national Disaster Medical Assistance Teams (DMATs) can be of great assistance in the form of medical responses. More information on these teams and their missions can be found through the U.S. Department of Health and Human Services at http://www.hhs.gov/news/press/2001pres/20010911c.html.

Food and shelter

Humans are what make up the response capabilities at any disaster. Just as vehicles need fuel, humans need fuel in the form of food. Although the initial operating period can suffice with snacks and fluids for rehydration, workers will soon slow down or cease to work if the proper nutrients are not present within about a six- to eight-hour period. This area will be tasked to the food unit under the logistics section. Often during the first and second operational period, the local Red Cross unit will respond to the scene in order to provide food and beverages for the emergency responders. However, as the incident draws out and the incident base is established, more formalized catering of meals may be appropriate. In order to make this a reality, planning for the proper lead time in order to prepare and deliver the food, as well as obtaining an accurate worker count, is essential.

If the disaster did not affect the government services buildings such as the local firehouses, the sheltering of emergency responders may not take a great deal of planning and logistics. However, as is the case with many widespread disasters, the local government facilities may have been destroyed along with the remainder of the town. In this case, provisions will involve either transporting emergency workers to nearby communities to utilize their facilities or establishing shelter facilities at the incident base. This can be accomplished by summoning the local wildland firefighting unit, if available, or the military forces. Second, some state and all federal US&R teams have the logistical support to provide tents for sheltering emergency responders. This is another asset that the local resource guide should address so that it is available if needed.

Special services

Dependent upon the type of disaster, there are numerous different special services that a community could need for mitigation and recovery from the incident. Some of the special services may include civil defense forces to clear roadways with large heavy equipment or military 6x4 vehicles capable of traversing high water. The latter may be necessary in many flooding situations. Figure 5–3 illustrates heavy equipment removing debris from a disaster area.

Fig. 5–3. Heavy equipment clearing debris from a disaster site

FEMA News Photo.

Weapons of mass destruction incidents will require very close coordination with many of the Department of Defense's special operations teams, such as the civil support teams of the National Guard. According to their mission, "Ohio National Guard Civil Support Team members train to handle a variety of chemical, biological and radiological situations within U.S. borders...According to CST standard procedures, the team must be ready to send a survey team into the contaminated area, or 'hot zone,' within 90 minutes of being called."[7]

Another resource that will need to be integrated during a WMD event is the Federal Bureau of Investigation (FBI). "Today, the FBI is part of a vast national and international campaign dedicated to defeating terrorism. Working hand-in-hand with partners in law enforcement, intelligence, the military, and diplomatic circles, our job is to neutralize terrorist cells and operatives here in the U.S. and to help dismantle terrorist networks worldwide."[8]

Personnel augmentation

Any good emergency operations plan must address the need to get each department up and operating for 24-hour-a-day operations and to sustain those operations for a 72-hour period. Many departments such as the fire and police may consider this a simple task because they already operate on a 24/7 basis. However, these department leaders must still be cognizant of the issues of 24-hour working operations. The human body will not sustain constant work for 24 hours while remaining alert and safe. Adjusting to a schedule involving 12 hours on duty, 12 hours off duty may be a sustainable solution.

Other departments within the municipal government may have to create a plan in which their personnel have emergency sheltering items kept at the government buildings so they may work for extended hours. Another question that arises during these mandatory callback times is how to handle the employees' families. Provisions must be made for child care for children of single parents, etc.

There will be other questions that must be addressed. For instance, how will forces be augmented if the municipal emergency forces are all volunteer personnel? Plans must address the use of neighboring emergency forces so that the response personnel can be alleviated in order to tend to their career positions. A law in Ohio allows volunteers to miss their career employment while on volunteer emergency service calls without penalty. Although they will likely not receive compensation from their career position, they will not be punished for aiding in a time of need.

Another way of augmenting personnel is to maintain a list of viable volunteers within the community. These can be ordinary citizens, but during a disaster, they can provide aid to the community such as passing out rations to the victims or aiding in walking the street to locate victims. This list, like the resource list, must be kept up-to-date or it will not help when needed.

MUTUAL AID IN DISASTER CONTROL

History of mutual aid

Lending assistance is the heart and soul of the emergency services. Therefore, mutual aid has been in use since the beginning of the fire and other emergency services. For many years, the mutual aid system was very informal. It boiled down to "You call, we come." However, as the emergency services became more formalized, governments began to intervene in the management of those services, and the legal aspects of this system became apparent. As the formalizing of the management of incidents occurred through ICS, the emergency services began to formalize their mutual aid systems.

Many of these formalizations were needed. One example is the outpouring of response to the 9/11 attacks. Many fire companies took it upon themselves to respond uninvited, trying to gain access to the site and create their own missions. This does nothing more than to create a safety and logistical nightmare to the persons responsible for management of the incident. Although these kind-hearted companies did not intend to create a safety hazard, they needed to have followed a formal written mutual aid plan.

Mutual aid is the only successful way to combat incidents that occur outside of large urban centers. As an example, consider the 1986 train derailment in Ohio. According to the Ohio State Historical Society,

> On July 8, 1986, fifteen cars of a forty-four-car CSX train derailed near Miamisburg, Ohio. Some of the cars that derailed contained phosphorus. When ignited, phosphorus releases poisonous gas. Shortly after the derailment, city officials ordered

> thousands of Miamisburg residents to evacuate their homes. The next day, officials allowed these people to return, only to have the phosphorus ignite. One Miamisburg resident described the resulting vapor cloud as "real heavy stuff, hugging right on the ground, like fog. People had their lights on, it was so dark." A second evacuation occurred, with approximately twenty thousand people seeking shelter at Dayton, Ohio, at the University of Dayton and the Dayton Convention Center. This was the largest evacuation to occur in Ohio's history...Over three hundred people sought treatment at area hospitals due to respiratory problems.[9]

Miamisburg is a two-station fire department. None of the evacuations, medical transports, and hazardous materials mitigation efforts would have been enough if the city of Miamisburg had not utilized mutual aid for this incident.

The give and take of mutual aid

By definition, *mutual* denotes a sharing between partners. Thus any community cannot expect mutual aid to be a one-way street. Rather, it must be considered a two-way street in which a community must give in order to receive. According to *The Fire Officer's Guide to Disaster Control*, second edition, "Mutual aid, then, is an agreement in which two or more parties agree to furnish resources or facilities to render service to each and every other party of the agreement, based on established criteria."[10] Sharing resources, such as special operations teams depicted in figure 5–4, allows large specialized units at a fraction of the cost to the community.

Fig. 5–4. Special operations team vehicle. These teams are expensive to equip and staff. Utilizing them as a joint resource saves the community money and allows its use when needed.

Photo courtesy of associate author Randall W. Hanifen.

Legal considerations of mutual aid

There are many legal issues the emergency services will face in the scope of their practice. However, many issues related to mutual aid are further complicated due to much of the law existing at a state level. Consider an example in which an ambulance responds to another state due to a hospital fire. During the transport of a patient having chest pains from the disaster, the patient is administered the wrong medication, and the paramedic receives a needle puncture while giving the medication. The issue arises concerning whose protocol is the correct one to operate by in this case. Furthermore, who covers the workers' compensation case?

According to the NGA Center for Best Practices issue brief *Beyond EMAC: Legal Issues in Mutual Aid Agreements for Public Health Practice*,

> Any cross-border mutual aid agreement that envisions the fast and efficient transfer of equipment and/or personnel will require participating states to resolve conflicts or contradictions among applicable laws and regulations. To achieve this compatibility, governors should consider the following strategies:

- Assess whether the laws in their states allow for license portability from other states and consider joining existing multi-state agreements...
- Review their state's policies for the use of volunteer disaster workers, particularly as those policies relate to workers' compensation claims;
- Assess whether sovereign immunity, Good Samaritan laws or other statutes provide liability protections to volunteer health professionals.[11]

A second issue that arises is that "the general rule is that a fire department may not operate outside its jurisdiction unless expressly authorized to do so by statute."[12] Although this giving and receiving of mutual aid often occurs despite these legal issues, the suing of entities, especially government entities, is at an all-time high. If a firefighter contracts an illness while operating at a fire, will the employer cover it? If the department fights the claim and failed to obtain all of the needed documentation as it relates to the mutual aid process, they will likely lose in court, thus incurring the medical bills of the worker and the legal costs associated with the court proceedings.

NIMS RESOURCE MANAGEMENT AND TYPING

Problems can arise when sending resources across state lines due to differences among states pertaining to the qualifications of personnel and due to differences in what resources are termed. As a part of the NIMS integration, resource typing on a national basis is occurring. According to the NIMS Integration Center,

> Resource typing is the categorization and description of response resources that are commonly exchanged in disasters through mutual aid agreements. The National Integration Center (NIC) Incident Management Systems Division has developed and published 120 Resource Typing Definitions...
>
> Standard resource typing definitions help responders request and deploy the resources they need through the use of common terminology. They allow emergency management personnel to identify, locate, request, order and track outside resources quickly and effectively and facilitate the movement of these resources to the jurisdiction that needs them.[13]

Figure 5–5 displays the Department of Homeland Security's resource typing matrix for an EOC management support team.

Furthermore, the credentials of a given responder can cause confusion about the level of training and what tasks can be expected of the person or the crew. In order to address this problem, NIMS has a section dealing with a national credentialing system. "The development of a nationwide credentialing system is a fundamental component of NIMS. A national credentialing system can document minimum professional qualifications, certifications, training and education requirements that define baseline criteria expected of emergency response professionals and volunteers for deployment as mutual aid to disasters."[14]

PROCEDURES—MUTUAL AID REQUESTS

In order for mutual aid to arrive in a timely manner, the procedures to request the assistance of mutual aid must be known to those who will make the request. These procedures should be known to any member of the emergency service organizations with staff members who could be expected to function even temporarily as incident commanders. An emergency cannot await an executive to arrive to carry out the procedure of requesting additional assistance.

Assistance through mutual aid happens at various levels. These levels and systems are discussed in the next sections.

FEMA

U.S. Department of Homeland Security
Federal Emergency Management Agency

RESOURCE:	EOC Management Support Team					
CATEGORY:	Resource Management			KIND:	Team	
MINIMUM CAPABILITIES: COMPONENT	METRIC	TYPE I	TYPE II	TYPE III	TYPE IV	OTHER
Personnel	Information Officer	Yes	Yes	Yes	Yes	
Personnel	Liaison Officer	Yes	Yes	Yes	Yes	
Personnel	Safety Officer	Yes	Yes			
Personnel	Incident Commander See Note 1	Optional	Optional	Optional		
Personnel	Administrative Aide	Yes				

COMMENTS:

Provides support to an Incident Commander. Typically comprised of an Information Officer, Liaison Officer, Safety Officer, and Administrative Aide, although some functions may be optional.

Note 1: An Incident Commander is an optional member of the team, since it is assumed that an Incident Command/lead has already been established under which these support functions will operate. Refer also to "Incident Management Team."

Information Officer: The Information Officer is responsible for developing and releasing information about the incident to the news media, to incident personnel, and to other appropriate agencies and organizations. Only one Information Officer will be assigned for each incident, including incidents operating under Unified Command and multijurisdictional incidents. The Information Officer may have assistants as necessary, and the assistants may also represent assisting agencies or jurisdictions.

Liaison Officer: Incidents that are multijurisdictional, or have several agencies involved, may require the establishment of the Liaison Officer position on the Command Staff. Only one Liaison Officer will be assigned for each incident, including incidents operating under Unified Command and multijurisdictional incidents. The Liaison Officer may have assistants as necessary, and the assistants may also represent assisting agencies or jurisdictions. The Liaison Officer is the contact for the personnel assigned to the incident by assisting or cooperating agencies. These are personnel other than those on direct tactical assignments or those involved in a Unified Command.

Safety Officer: The Safety Officer's function is to develop and recommend measures for assuring personnel safety, and to assess and/or anticipate hazardous and unsafe situations. Only one Safety Officer will be assigned for each incident. The Safety Officer may have assistants as necessary, and the assistants may also represent assisting agencies or jurisdictions. Safety assistants may have specific responsibilities such as air operations, hazardous materials, etc.

Administrative Aide: The Administrative Aide's function is to provide administrative/secretarial support to the EOC Management Support Team. Responsibilities include keeping official minutes of team meetings, receiving phone calls to the EOC, making meeting arrangements, and other duties as needed.

Source: FIRESCOPE, California Department of Emergency Services, 2001; Phoenix Fireground, City of Phoenix Fire Department, 2002

Fig. 5–5. Resource typing matrix for EOC management support team. All of the 120 resources typed by DHS have similar matrices to define the details of the resource, such as its capability.

Levels of mutual aid

Each community will have various levels of mutual aid compacts that could be called upon for response. Some of the forms of mutual aid are:

- Automatic aid from neighboring communities
- Mutual aid from local communities
- Regional, statewide, or national mutual aid

From an everyday basis, a community may operate with an automatic aid agreement for neighboring jurisdictions. This type of aid is designed so that upon receipt of a call, all needed mutual aid companies are contacted simultaneously with the department holding jurisdiction. This type of arrangement can provide reduced response times due to the jurisdictional boundaries not always allowing a fire or police station to be located near every area of responsibility. Second, this type of arrangement can ensure that the proper quantity and types of equipment arrive at an incident. Many communities cannot afford to equip and staff a hazardous materials or technical rescue vehicle or to provide a SWAT team. In disaster situations, these specialized teams and vehicles will prove invaluable. The early arrival of these units will enhance the response, not to mention that the benefit of training and working together on past incidents will improve efficiency and cooperation with the outside personnel.

Mutual aid from local communities takes the form of requesting the particular resource and having a signed agreement between the communities. These written agreements can vary greatly, but the National Emergency Management Agency lists best practices in its publication *Model Intrastate Mutual Aid Legislation*. Some of these practices are

- To the extent permitted by applicable law, include automatic renewal language if agreement instead of legislation.
- Agreements/legislation should mandate joint planning, training and exercise with liability immunity as if it were an actual emergency.
- Develop forms to facilitate requests for aid, record-keeping regarding movement of equipment and personnel and reimbursement…
- A reimbursement provision. This provision may include an off-set for any insurance proceeds applicable to the costs claimed by the responding political subdivision. The insurance claims process should not delay reimbursement. Reimbursement shall not duplicate any state or federal assistance available for the costs.[15]

STATEWIDE MUTUAL AID–INTRASTATE MUTUAL AID COMPACTS (IMAC)

Many states have realized the benefits of creating statewide mutual aid systems. The Ohio Intrastate Mutual Aid Compact became law on December 23, 2002. According to the Ohio Emergency Management Agency, "The purpose of the IMAC is to develop and improve intrastate mutual aid capabilities for providing services and resources across local boundaries in response to and recovery from any disaster resulting in a formal declaration of emergency."[16]

Although Ohio created its IMAC in 2002, many states still do not have any formalized form of mutual aid on a wide-scale basis. The question arises concerning which departments or agencies could safely be called upon in those states. As pointed out previously, the fire service is good about providing assistance when needed. However, sometimes there is a cost associated with utilizing resources, such as a hazardous materials team. These teams can utilize chemical suits that cost thousands of dollars per use. A technical rescue team could utilize pneumatic struts that must be left in place in order to maintain the integrity of the building. If there is not a formalized plan that addresses these issues, then no one is exactly sure who will resupply these units. These units thus may have to be placed in a nonfunctional situation until monies are available for their replacement.

By having a plan that is statewide, jurisdictions can have a sense of understanding of what reimbursement can be expected when they respond to help another community. As well, the requesting community can begin to budget for the services and determine if their capabilities are becoming overwhelmed.

The International Association of Fire Chiefs has begun to address IMACs at the national level. According to the IAFC,

> The goal of a national fire service mutual aid system is to create a venue in each of the 50 states that would ensure a comprehensive, coordinated response to any incident within, or immediately adjacent to, its borders. Creating and implementing such a response system, with an understanding of resources and capabilities, paves the way for a more integrated and efficient response in the event of a national disaster. Specifically, it ensures that the local fire service brings its exceptional skills, unique knowledge and tactical resources to the table to support national strategies such as NIMS and state-based efforts, such as EMAC.
>
> Even with the best of intentions, execution of such agreements can exist at the whim of such issues as personal interpretations or local politics, and can have disastrous effects—particularly at large-scale, multi-jurisdictional incidents.[17]

REGIONWIDE MUTUAL AID—EMAC

Some disasters will overwhelm even the available state resources. Additionally, it may make more logistical sense to utilize a neighboring state's resources rather than await resources from the home state. In these instances, an IMAC agreement may not suffice due to the differing states workers' compensation, certification, and repayment methods.

A case of the need for EMAC occurred in Greendale, Indiana. According to author John M. Buckman III, "275 firefighters from 3 states battled a fire on Thursday, July 7, 2005, and Friday, July 8, 2005, in Greendale, Indiana at the Alternative Plastics Warehouse...The building is a 6-story structure formerly housing a distillery. There are no windows in the structure and there was a sprinkler system."[18]

In this case, the geographic location of Greendale, Indiana made summoning of the Ohio and Kentucky departments that are in the suburbs of Cincinnati more logical than awaiting other Indiana departments. The associate author was rostered for deployment to this fire with a department that was only 33 miles from the fire, but was in Ohio rather than Indiana. Although it was unsure at the time how the mutual aid agreement would work, departments sent equipment from both Ohio and Kentucky.

Many of the large wildfires have caused the western states to adopt interstate mutual aid agreements. FIRESCOPE encompassed this problem. (FIRESCOPE is covered in depth in the second edition of *Fire Officer's Guide to Disaster Control.*)

In order to meet the NIMS objectives, provide an answer to many of the problems previously examined, and to prevent subsequent legal action, the National Association of Emergency Managers created the Emergency Management Assistance Compact (EMAC). The EMAC is "a national Governor's interstate mutual aid compact that facilitates the sharing of resources, personnel and equipment across state lines during times of disaster and emergency. EMAC is formalized into law by member parties."[19]

According to the National Emergency Management Association, the EMAC works in the following steps:

1. Disaster occurs, Governor declares state of emergency
2. State emergency management agency assesses needs...
3. Resources are requested via the EMAC system...
4. Assisting states provide resources[20]

The EMAC is a great agreement to deliver resources during a declared disaster. However, what about sending resources during a nondeclared disaster, such as the Greendale fire in Indiana? The International Association of Fire Chiefs has been diligently working to improve mutual aid throughout the country. According to the IAFC, it is

> currently working on two projects...that will help the fire service's ability to support the National Response Framework (NRF) and implement concepts presented in the National Incident Management System (NIMS). The ultimate goal of the National Fire Service Intrastate Mutual Aid System (IMAS) project is to support the creation of a formalized, comprehensive, exercised intrastate mutual aid plans. In conjunction with the Intrastate Mutual Aid System, the Fire Service Mutual Aid System Task Force (MASTF) is developing a plan for an interstate mutual aid system. Once completed, the Intrastate Mutual Aid System plans will support the Mutual Aid System Task Force plan, creating a mutual aid system that will cover the continental United States.[21]

EXPANDING COMMAND STAFF THROUGH MUTUAL AID

At everyday incidents, departments rarely utilize the numerous components of the ICS system within NIMS. This is due to the scalability of the program. At a car fire or a traffic stop, the crew leader in both cases can easily handle the incident commander, operations sections chief, and group/ division leader positions with just one individual. Second, the on-duty staffing of many departments outside the urban areas will not have enough chief officers to fulfill the multiple section chief, branch manager, and group leader positions that are needed in a time of a disaster.

A plausible solution to this problem is discussed in a *Fire Engineering* article by David C. Comstock Jr., "The Mutual-Aid Chief Officer's Role at Structure Fires." Comstock states,

> Unless you are in charge of a large suburban or metropolitan department, it is unlikely that you will have the 20 or more firefighters needed to complete the tactical assignments, especially during daylight working hours. With staffing shortages, you can't

even begin to worry about "command staff" or general staff positions since the fire must be put out!...

Company officers are usually (or should be) in a position of leadership because of their fire service knowledge, experience, training, and education. The company officer's job is to complete the assigned task while looking out for the safety of the firefighters assigned to his individual company...

The answer to this dilemma is found in the mutual-aid chief officer (MACO). Incident commanders (ICs) must request a sufficient number of chief officers to the scene to cover the command positions present at every fire or emergency incident.[22]

Working with these chief officers at everyday fire situations will improve cohesiveness, which will be apparent when a large disaster hits. It should be considered at each of these fires to mentor the company officers in these needed roles. If the mutual-aid chief officers are not available in the early stages of a disaster, the first arriving company officer can receive adequate experience in section chief positions in order to function properly in times of personnel shortages.

MUTUAL AID DRAWBACKS

Although the pros far outweigh the cons of mutual aid, planners and commanders must plan for and understand the possible problems that can arise from mutual aid.

The first problem that can arise is the dependence upon their presence in a time of disaster. It must be considered that in widespread disasters, the neighboring mutual aid companies will likely be hard at work in their own community and may not be able to send their normal complement. They also could be fighting for scarce resources. This is a time when the use of an area command will serve all affected by the disaster and eliminate the fight over finite resources.

The second problem that can occur is due to operational differences in departments responding to the call. If the jurisdiction requesting the mutual aid has an expectation of that resource upon its arrival based on local operational policies, then the commander must relay this and not just expect it to occur. During mutual aid received instances, the incident commander or the appropriate NIMS supervisor must give explicit directions to the incoming companies. The supervisor must also check with the crew leader to ensure that the crew staffing will support the task assignment. Many departments utilize various levels of staffing, and thus they may only deliver what the requesting agency feels is a partial crew.

The third issue that can arise is the depletion of the resources in the community sending the mutual aid. Many communities can only afford to place a minimal amount of fire, EMS, and law enforcement forces on duty at any time. Provisions must be made by the sending agency to backfill or utilize additional mutual aid to answer subsequent calls in their jurisdiction. Requesting agencies must also make provisions when utilizing smaller jurisdictions for mutual aid, as they may be answering calls within their own community and be unavailable to send resources.

From a financial perspective, the drawback is that the mutual aid system can place a financial strain on the providing jurisdiction, as personnel must be immediately paid for their services. The affected jurisdiction may only be able to afford to reimburse their mutual aid companies upon receipt of reimbursement for the declared disaster.

CONCLUSION

Resource management is the basis for any disaster response. Without personnel and equipment, the disaster would not be mitigated. No jurisdiction can provide enough on-duty and recall personnel or purchase every possible piece of equipment that is needed for every disaster. Therefore, each jurisdiction must prepare plans for attainment of additional forces to deal with these resource-intensive incidents. Emergency services executives and incident commanders must understand how the various levels of emergency operations plans dovetail with each other. From the local plan through the National Response Framework, the processes must be understood in order to ensure that the incident escalates through the various levels of government in order to obtain the needed resources and to properly fund those resources.

REFERENCES

1. The Red Cross. 2006. Disaster services. Retrieved January 4, 2007 from: http://www.redcross.org/services/disaster/0,1082,0_319_,00.html.
2. The Salvation Army. 2006. Our role in disaster services. Retrieved January 7, 2007 from: http://www.salvationarmyusa.org/usn/www_usn.nsf/vw-dynamic-arrays/8214F21A914E09CE80256FE80071ECF1?openDocument.
3. Kramer, W. M., and C. W. Bahme. 1992. *Fire Officer's Guide to Disaster Control*. 2nd ed. Saddle Brook, NJ: Fire Engineering Books and Videos. p. 96.
4. U.S. Department of Labor, Occupational Health and Safety Administration. 2006. *How to Plan for Workplace Emergencies and Evacuations*. Retrieved January 7, 2007 from: http://www.osha.gov/Publications/osha3088.html.
5. U.S. Army Corps of Engineers. 2006. Missions. Retrieved January 4, 2007 from: http://www.usace.army.mil/missions/index.html.
6. U.S. Coast Guard. 2007. Missions. Retrieved January 4, 2007 from: http://www.uscg.mil/top/missions.
7. Ohio National Guard. 2007. At a moment's notice. Retrieved January 6, 2007 from: http://www.ohionationalguard.com/pao/news/2006%20Stories/20060628a.htm.
8. U.S. Department of Justice, Federal Bureau of Investigation. 2007. Counterterrorism. Retrieved January 6, 2007 from: http://www.fbi.gov/terrorinfo/counterrorism/waronterrorhome.htm.
9. Ohio Historical Society. 2007. *Ohio History Central*. Miamisburg, Ohio train derailment. Retrieved January 13, 2007 from: http://www.ohiohistorycentral.org/entry.php?rec=1632.
10. Kramer, W. M., and C. W. Bahme. 1992. p. 104.
11. NGA Center for Best Practices. 2007. *Beyond EMAC: Legal Issues in Mutual Aid Agreements for Public Health Practice*. Retrieved January 13, 2007 from: http://www.nga.org/Files/pdf/05LEGALISSUES.pdf.
12. Kramer, W. M., and C. W. Bahme. 1992. p. 104.
13. U.S. Department of Homeland Security, Federal Emergency Management Agency. 2007. *Frequently Asked Questions. Resource Management and Credentialing*. Resource typing. Retrieved January 14, 2007 from: http://www.fema.gov/emergency/nims/faq/rm.shtm#0.
14. U.S. Department of Homeland Security, Federal Emergency Management Agency. 2007. *Frequently Asked Questions. Resource Management and Credentialing*. Resource credentialing. Retrieved January 14, 2007 from: http://www.fema.gov/emergency/nims/faq/rm.shtm#1.
15. National Emergency Management Association. 2004. *Model Intrastate Mutual Aid Legislation*.

March. p. 9. Retrieved August 25, 2008 from: http://www.emacweb.org/?76.

16. Ohio Department of Public Safety, Ohio Emergency Management Agency. 2007. Ohio Plans Branch. Retrieved January 14, 2007 from: http://www.ema.ohio.gov/plans.asp.

17. Simpson, A. H., and A. Davison. International Association of Fire Chiefs. 2006. Project to help build intrastate mutual aid system. February 1. Retrieved January 14, 2007 from: http://www.iafc.org/displayindustryarticle.cfm?articlenbr=29275.

18. Buckman, J. M. III. 2005. Greendale, Indiana plastics warehouse fire. *Firehouse.com*. July 18. Retrieved January 14, 2007 from: http://cms.firehouse.com/content/article/article.jsp?sectionId=45&id=43390.

19. National Emergency Management Association. 2006. *EMAC Overview*. August 16. Retrieved January 14, 2007 from: http://www.iafc.org/associations/4685/files/EMAC_Briefing_8-16-2006.pdf.

20. Ibid.

21. International Association of Fire Chiefs. 2007. IAFC advances mutual aid. Retrieved January 14, 2007 from: http://www.iafc.org/displayindustryarticle.cfm?articlenbr=29847.

22. Comstock, D. C., Jr. 2005. The mutual-aid chief officer's role at structure fires. *Fire Engineering*. August. 1. Retrieved January 14, 2007 from: http://fe.pennnet.com/articles/article_display.cfm?Section=ARCHI&C=Depar&ARTICLE_ID=235073&KEYWORDS=Mutual%20Aid&p=25.

COMMUNICATIONS AND INFORMATION MANAGEMENT 6

INTRODUCTION

Communication is defined by Dictionary.com as "the imparting or interchange of thoughts, opinions, or information by speech, writing, or signs."[1] In nearly every after action report produced because of a large-scale incident, communications is one of the top problems identified throughout the incident. It is not unusual for there to be a lack of common radio frequencies during normal operation. The additional challenges that arise from the destruction of the communications infrastructure during disasters such as tsunamis, hurricanes, and tornadoes make exchanging information relevant to coordinating the response troubled at best.

This chapter will examine the common communications infrastructure utilized by many emergency response agencies. It will discuss options available for responders when the communications infrastructure is destroyed as a result of the disaster. The personnel and training that allow effective function of the communications equipment will be discussed, as well as the NIMS requirements as they relate to the emergency services. Finally, emerging technology will be addressed, both its usage now and in the future of emergency response.

COMMUNICATIONS PERSONNEL

The most common communications on an everyday basis for emergency services will involve the dispatching or communications center. Communications in the emergency services field is based on the concept of redundancy. Some of the redundancy should involve multiple dispatching consoles, backup of the computer servers that power the communications equipment, power backup for the equipment, and a fully trained complement of dispatchers.

The dispatch centers should work to support the emergency services field operations. The dispatch center is often the link between the citizens and the emergency services. In order to create an effective dispatch center, the management of the emergency services must understand and influence some aspect of the dispatch center. The first is the design and capacity of the dispatch center. Each dispatch center should conduct an analysis of its workload and ensure that the staffing and configuration can support the workload. This may be as simple as assigning another dispatcher and segregating agencies onto two different frequencies. It could be as complicated as upgrading the building, equipment, and number of personnel operating in the center.

At a minimum, the dispatch center should have one dispatcher and a supervisor. In many small rural areas, this is the same person. However, in areas that service a larger population, these duties should be split. The supervisor should understand and be able to diagnose problems with any of the systems used by the dispatch center. This is not to say they must be able to fix all of the equipment, but they should know when the equipment is not functioning properly and be able to contact the correct service personnel. Callback procedures are important procedures to have in place at a dispatch center. The callback procedure should be initiated by the dispatch supervisor very early into the recognition that a disaster is occurring. Dispatchers will become quickly overwhelmed at the beginning of a disaster due to the numerous calls to report the disaster or to inquire about the disaster.

Dispatch center equipment

Once the dispatch center is staffed, equipment access is the second priority. Although some municipalities are reluctant to construct a dispatch center beyond what is used on a daily basis, planning and design should focus on anticipated growth and use during a disaster. Second, a redundancy in the equipment will allow servicing of the equipment while maintaining functionality of the dispatch center. Figure 6–1 depicts the dispatch consoles at a regional EOC. Having these consoles allows the dispatchers to be controlled directly by EOC personnel to provide a seamless operation. Each dispatching console should have access to all of the programs utilized by the center. Some have the philosophy that designing consoles to serve only one jurisdiction or agency is best, such as police consoles and fire-rescue consoles. However, they are designed so that in peak times for either agency, the other agency's dispatcher can support the primary dispatcher if given access to the programs and information.

Fig. 6–1. Remote dispatching consoles at a regional EOC. Having dispatchers in the same area as the EOC allows for a coordinated effort in planning and response.

Photo courtesy of associate author Randall W. Hanifen.

The types of programs and information available to the dispatch personnel will depend upon the design of the information and communications infrastructure of the jurisdiction or jurisdictions that the center services. It could be as antiquated as a tone generator and radio transmitter. Many of the dispatch centers have computer-aided dispatching, interface with the mobile data terminals, geographic information systems, automatic vehicle locators, and video links to traffic and public safety systems.

With the automation of emergency communications centers, it is important to have a backup system in place for the time when the computer systems fail. Current trends are on relying increasingly on computer technology. However, in an earthquake, the power could fail, and the natural gas lines feeding the backup generators also could fail. At this point, if a manual system were not in place and trained for, the dispatching and processing of emergency calls would cease. In addition, it is important that dispatchers should be familiar with this manual system, and thus planned days in which the manual system is utilized should be scheduled.

Personnel training

The next aspect of the dispatch center, and likely one of the most important, is the training of the dispatchers and the supervisors. Initial training of new dispatchers should address the systems and procedures in place at the center. An operations manual should be created for this training. This manual will serve as a backup in a time when technical or operational information is needed.

COMMUNICATIONS NEEDS IN FIRE DEFENSE AND DISASTER OPERATIONS

Communications equipment

During a disaster situation, the communications needs of emergency personnel will be at a peak. All available communications equipment will be needed for both field and command/EOC use. Having a plan of how to deliver extra radios, cellular telephones, and required communications equipment in a timely manner to those needing the equipment is essential to a successful operation. A county in the Cincinnati area has a "hot box" that contains extra portable radios for field units and command personnel at the site of a disaster. Other plans could utilize the local amateur radio personnel to provide needed communications. Plans often call for recalling off-duty personnel; therefore, the plans must also equip them with the needed communications equipment.

NIMS INTEROPERABILITY REQUIREMENTS

Chapter 5 of the *NIMS* document, "Communications and Information Management," states that

> Effective communications, information management, and information and intelligence sharing are critical aspects of domestic incident management. Establishing and maintaining a common operating picture and ensuring accessibility and interoperability are principal goals of communications and information management. A common operating picture and systems interoperability provide the framework necessary to:

- formulate and disseminate indications and warnings;
- formulate, execute, and communicate operational decisions at an incident site, as well as between incident management entities across jurisdictions and functional agencies;
- prepare for potential requirements and requests supporting incident management activities; and
- develop and maintain overall awareness and understanding of an incident within and across jurisdictions.[2]

This common operating picture allows all levels of management to make consistent decisions about the incident. Many times this is reliant upon common communications and data standards. According to the NIMS document,

Communications interoperability in the context of incident management is also critical. Effective communications outside the incident structure—between other levels of government and between government and private entities—for resources and other support is also enhanced by adherence to such standards…Additional progress toward common communications and data standards and systems interoperability will be accomplished over time through a sustained collaborative effort facilitated by the NIMS Integration Center.[3]

InterAgency Board

One of the most common problems encountered by emergency personnel is the inability of various equipment, including the communications equipment, to operate interchangeably. As a result of this known problem in all response disciplines, the InterAgency Board (IAB) was created. According to the IAB Web site,

Sanctioned by the Attorney General of the United States, the InterAgency Board for Equipment Standardization and Interoperability (IAB) was founded by the Department of Defense's Consequence Management Program Integration Office and the Department of Justice's Federal Bureau of Investigation Weapons of Mass Destruction Countermeasures on October 13, 1998...

The Interagency Board (IAB) for Equipment Standardization and Interoperability is designed to establish and coordinate local, state, and federal standardization, interoperability, compatibility, and responder health and safety to prepare for, train and respond to, mitigate, and recover from any incident by identifying requirements for an all-hazards incident response with a special emphasis on Chemical, Biological, Radiological, Nuclear or Explosive (CBRNE) issues.[4]

Because of the work of the InterAgency Board, specifically the standard equipment list in section 6 (SEL 6), there are standards and considerations explained when specifying or purchasing communications equipment. This list can be located on the Internet at http://www.iab.gov/Download/sel_section6_2006.pdf. For example, according to the IAB, a radio tower must meet the following specifications:

- May include omnidirectional antennae.
- Should be constructed with consideration for potential hazards (such as hurricanes, floods, tornadoes, forest fires, etc.).
- All public safety towers must meet current EIA standards for the climates in which they operate. Towers should have engineering inspections to ensure that the tower has not been overloaded with too much equipment for the wind loading expected.[5]

Also according to the IAB, portable radios should meet the following specifications:

- Digital and analog capable.
- Supports 25Khz and 12.5Khz channels.
- Supports conventional systems.
- Project 25 compatible (800 MHz).
- Project 25 required (700 MHz).
- See 47 CFR 90 FCC for applicable standard.

- Portable radios may not be advisable in EOD operations—consider hard-line, or other solutions.
- Can be intrinsically safe.[6]

A standardized equipment list not only provides guidelines for those entities that want to purchase new communications equipment, but it also provides a basis of equipment that is mandated in order to purchase it with a grant. As the cost of much of the communications equipment is beyond affordability for many entities without this financial assistance, the list and standard becomes nearly mandatory. Second, the equipment manufacturers now understand the correct standards for their equipment, and in turn, have an incentive to only produce compliant equipment.

Project 25

As noted in the IAB's requirements for the portable radios, Project 25 is required for radios operating in the 700 MHz spectrum, and Project 25 was recommended for the 800 MHz radios. According to the P25.com Web site,

> Project 25 is a public safety communications standard dedicated to ensuring interoperability in communications. It is designed to ensure fast and secure communications between local, state and federal agencies when protecting the public's welfare.
>
> The Project 25 standard organization is comprised of the Association of Public Safety Communications Officials (APCO), the National Association of State Telecommunications Directors (NASTD) and the U.S. Federal Government. The Telecommunications Industry Association (TIA) offers manufacturers representation in the standard's creation and modification.[7]

In order to receive the designation of "P25 compatible," equipment must meet the following guidelines:

- The Common Air Interface (CAI) specifies how information is coded, transmitted and received over the air.
- It enables users to interoperate and communicate digitally across networks, agencies, and vendors.
- The Improved Multi-Band Excitation (IMBE) vocoder converts speech into a digital bit stream.
- Test panels judged IMBE as the coding scheme most successful at making male and female voices audible against background noises such as moving vehicles, sirens, gunshots, and traffic noise—the conditions of public safety use.[8]

Interoperability scorecards

Now that a standard is established for allowing the various radios to work together, how can personnel get to the same channel in order to talk to one another? Many communities have begun regional efforts to share frequencies and upgrade their radio systems. This interoperability at the tactical level is now being graded by the Department of Homeland Security, through the tactical interoperability communication scorecards. According to the Department of Homeland Security,

> The tactical interoperable communications scorecard assesses the maturity of tactical interoperable communications capabilities in

75 urban/metropolitan areas. These scorecards were developed by subject matter expert panels that reviewed documentation on current communications plans, exercises, and a self-assessment to arrive at consensus findings and recommendations for each region on how to best improve that region's communications capabilities.[9]

According to the Department of Homeland Security, the scorecards cover three areas:

- SOPs—For many of the urban areas, the Tactical Interoperable Communications Plans (TICP) developed through the Interoperable Communications Technical Assistance Program (ICTAP) provided the first formal, regionwide communications interoperability SOPs. Additional steps should be taken to ensure that these procedures (as well as those outlined in the National Incident Management System) are fully instituted at the command and responder levels.
- Usage—The proficiency in the use of communications interoperability equipment and accompanying procedures varies by the types of equipment used and is increasingly complex as additional agencies are included in response efforts. In addition, almost no region had completed a communications-focused exercise before the TICP validation exercise, which meant that the areas had no specific practice using their interoperable communications capabilities.
- Governance—Areas with mature governance structures have advanced further in implementing shared systems/solutions that facilitate regional communications. Regionalized strategic plans are largely not in place and should be developed for communications interoperability with careful consideration for how investments can be shared across the region.[10]

The results of how different urban regions scored are available at http://www.dhs.gov/xlibrary/assets/grants-scorecard-report-0102407.pdf.

DATA TECHNOLOGIES

With the advent of the personal computer and consistent technology improvements throughout a number of professions, data technologies have continued to improve. Given the rapid developments in technology in the areas of data storage, retrieval, and transmission, it is difficult to keep pace with the cutting edge technology. Figure 6–2 shows how all of the currently available data technologies work in conjunction to aid responders with information.

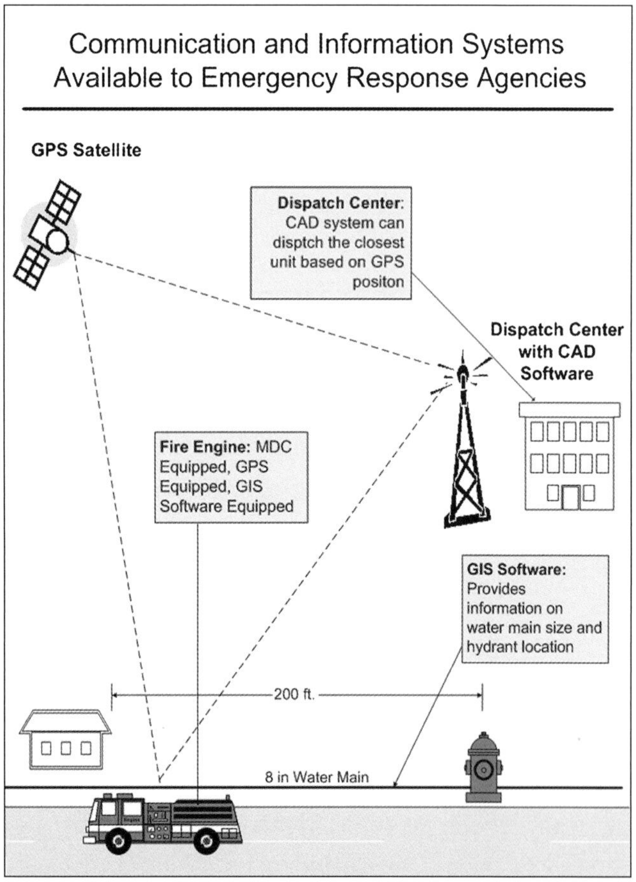

Fig. 6–2. Data technology graphic. This graphic shows how the data technologies are interconnected to deliver information to responders.

The improvements in technology allow numerous forms of data to be transmitted almost anywhere, and into just about any device. Thus, it is necessary to first focus on what data is desired, what data is needed, when the data is needed, and to what device the data will be delivered. This section will cover some of the basic technologies available to the emergency services and what use they can have in a time of disaster.

Mobile data computers

Providing information to responding personnel is of the utmost importance in disaster and nondisaster periods. One of the ways to ensure that personnel in response units, such as fire trucks and police cars, have this information is to equip these units with mobile data computers. Previously, mobile data terminals were installed, but with the reduction in cost, full computers are regularly installed in lieu of a "dummy terminal." Equipping the units with a computer opens the door to a variety of data options.

Mobile data computers (MDCs) are usually interfaced with a computer-aided dispatching program. The dispatching program generates the data, packages it for transmission, and sends it via a medium to one or more mobile data computers that need the information. Some of the information that can be sent over the MDC is as follows:

- Location of the call
- Type of call
- Cross streets
- History of calls to the location
- District and fire defense zone
- Whether a preplan of the location is available
- Additional comments entered by the dispatcher relevant to the call
- Units assigned to the detail
- Complainant information

This is just a sample, but with customizable MDC and computer-aided dispatching programs, the choices are nearly limitless.

Wireless information networks

With computers installed in the response vehicles, it is possible to create a host of wireless information networks. Depending upon the configuration of the medium to transfer the information to the MDC, there are numerous possibilities. These include transferring information such as preincident plans, connecting to a server that contains volumes of information and databases, and the ability to navigate the Internet from the MDC.

Currently, there are two methods by which information can be transferred to the MDC. The first is to dedicate radio frequencies to data and utilize radio modems to transfer the information from radio signals to data readable by the computer. This option relies upon the agency's current radio network and can limit the amount of voice transmissions due to locking out channels solely for data.

The second method for information transfer is to utilize a cell phone provider's broadband wireless card. This card inserts into a laptop and has an antenna that allows the computer to connect to the cell phone's network. All of the large cell phone providers now offer this service. Many claim the same data transmission speeds as a home high-speed network. This allows agencies to retain the full amount of channels for voice transmission and allows their MDCs to function as well.

The major disadvantage to the use of cell phones with wireless cards is the resulting reliance on a third party to provide the connection. During peak hours, the wireless system can become slower due to excessive usage by the business community. Second, if their towers are affected by the disaster, they may not have the ability to set up a field version of their tower in order to get the information medium available for use by emergency responders.

Another disadvantage of using cell phones with wireless cards is the cost involved. As of the writing of this text, the costs can range from $30 to $70 per unit per month. If a department has a fleet of 30 vehicles, this could cost on average $900 to $2,100 per month, or $10,800 to $25,200 per year. This is a large cost to achieve getting the data to the units in the field. Furthermore, this cost would not include the software and hardware that is needed to complete the operation.

Mobile Wi-Fi

With the advancement in technologies, specifically the broadband access cards, it is possible for departments to create their own Wi-Fi hot spots. Much like going to the local coffee shop and using a laptop to connect to the Internet, jurisdictions can now utilize a Linksys Model WRT54G3G-ST router, or a like product known as a *concentrator.* The mobile broadband card is placed into the router, thus creating a mobile hot spot. This could be installed in a supervisor's car in order to make a temporary command post with research capabilities, until the arrival of a full-scale mobile command post or the establishment of an EOC.

GLOBAL POSITIONING SYSTEMS/GEOGRAPHIC INFORMATION SYSTEMS

The wide use of global positioning systems (GPS) has revolutionized how the emergency services operate. Short of the MDC, GPS is likely the best addition and improvement to the emergency services in relation to technology. Many of the computer-aided dispatching software programs utilize GPS coordinating systems. Some of the systems now utilize geographic information systems (GIS), which "integrates hardware, software, and data for capturing, managing, analyzing, and displaying all forms of geographically referenced information."[11]

GPS is the utilization of coordinates to pinpoint a given location. There are many different grids available to report the coordinates, such as latitude/longitude (lat/long), Universal Transverse Mercator (UTM), and the Universal Grid System. Most of the GPS units report in lat/long, as this is the most recognized. However, the U.S. military uses the Military Grid Reference System (MGRS), which is similar to UTM. The important issue to recognize is that the coordinates reported must be utilized with the correct system. Many of the current GPS systems will make the necessary conversions, or the unit can be set to the proper grid system.

GPS is now utilized in many disaster operations to search the areas affected and report incident locations. The current US&R training requires knowledge of GPS units and systems, as this is how the field reconnaissance work is completed in order to determine areas that need to be searched in order to locate victims.

GPS usage on an everyday basis helps to make sure the closest, most appropriate emergency response unit arrives at the scene of an incident in the least amount of time. Many of the computer-assisted dispatch (CAD) systems utilize automatic vehicle locators (AVLs). AVLs triangulate the position of the emergency response units versus the position of the call (at the time of the call) to ensure that the closest unit is dispatched. Many times, based on the nonemergency workload of the emergency services personnel, the units may be closer to the call than the regularly assigned unit.

From a hazardous materials incident standpoint, the units can be programmed with the affected areas, thus allowing workers to evacuate areas and have an alarm if they enter an area that is expected to have the leaked product. Second, the evacuation crews can utilize the devices to mark the homes that were evacuated, thus allowing command personnel to see how many people were evacuated during the leak, and from which locations.

GIS technology allows numerous government infrastructures to be marked and then overlaid onto a map for responding personnel. Because the infrastructures are in layers, responding personnel can turn on and off different items. For example, the fire department may want to see the location of the fire hydrants and water mains, but the police department may want to locate all of the speed limit signs. Each user can turn on only the items they would want to locate.

One of the useful GIS layers for hazardous materials incidents is the locations of the spillways. This allows command personnel to have knowledge of where a leak may migrate to as a result of the various spillways. Thus they can assign crews to create dams at these spillways to contain the incident.

Many of the mapping programs that are utilized by the CAD and MDC programs will create directions for the unit based on their current location. With the increase in automatic and mutual aid response, having a map that encompasses many jurisdictions makes finding the incident in other jurisdictions much easier. This is a significant improvement over the old version of carrying multiple map books and or paper maps. Some of the more sophisticated programs will even give text directions, similar to MapQuest or Yahoo! Maps.

The utilization of GPS and GIS will only improve the information available to field response units in order to deliver disaster and nondisaster emergency services to the community in a more efficient manner. Readers are encouraged to visit Web sites for the current GIS vendors, such as ESRI, which have information on the use of GIS in the public safety/disaster preparation fields. The ESRI Web site can be found at www.esri.com.

CELLULAR AND PAGING TECHNOLOGIES

Cellular or wireless phones are so common today that telephone area codes have been redistricted in the past 10 years in order to accommodate the overwhelming number of cell phone numbers in use today. Cellular phone service can range from just basic cellular phone coverage to a mobile computer that has a phone built into the device. Some of the services available on cellular phones are:

- Voicemail
- Text messaging
- E-mail retrieval
- E-mail answering
- Full Microsoft Office document creation
- Full interactive contact list
- Mobile Internet
- Voice recording
- Picture taking
- Small video recoding
- GPS locators

From a response perspective, cellular phones with many of these newer features are a good addition to command personnel. The command personnel can have a whole host of contacts for response issues with them at all times. Command personnel also can have the ability to talk to other supervisors who may not have a mobile radio with them. In addition, they can have a device that will be able to access the Internet for valuable response-related information, such as the weather, MSDS sheets, reaction calculators, maps of the area, and directions to the scene from any given address. The uses for these devices are limitless. These devices could even be used in a hazardous materials incident to recon the scene and take photos or movies in order to bring to a technical information specialist at the command post.

Although pagers are nearly an item of the past due to the advancement and consolidation by the cellular phone providers, they still are used to alert volunteers and can provide a backup or an alternative to the MDC. Many fire and police departments utilize text pagers to send call information to response crews, thus giving each person a one-way MDC. Many of the computer-aided dispatching systems will create and send the pages as a part of the program. The only drawback to this is the reliance on a third-party carrier for the service. This is why some dispatch centers have entered the business of paging services in order to have their own paging company and not have the reliance on or expense of a third-party carrier.

MOBILE COMMAND CENTERS

For a majority of calls, the emergency services supervisor (fire/police) can handle the command of the incident from the seat or tailgate of the response vehicle. However, for incidents of longer duration that will employ many section chiefs, the utilization of mobile command centers is more appropriate. These vehicles come in many configurations and are available from a variety of manufacturers.

The first consideration for a mobile command center is who will use the vehicle and when it will be used. The current trend is to create a mobile command center that can be used by many jurisdictions within a region, such as a county or metropolitan area. Figure 6–3 is an example of a mobile command center used in the Cincinnati region. Once the vehicle's users are identified, a committee should be formed to develop specifications for the vehicle. If one is already available, the committee can examine the specifications and ensure it will work for the area in which it is utilized, such as ensuring the radios are compatible with those in the area.

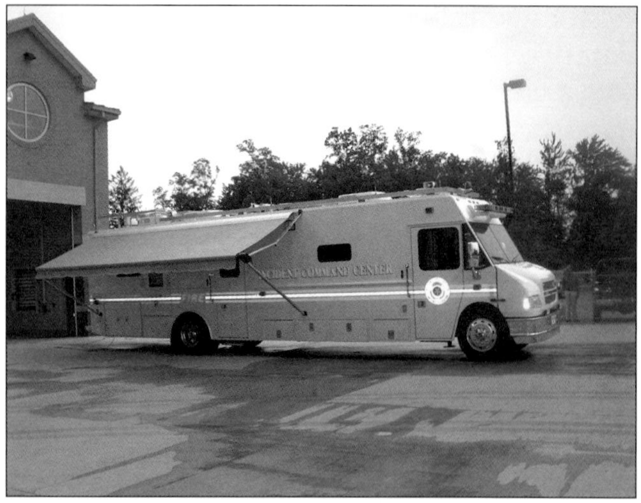

Fig. 6–3. Regional command vehicle. Utilizing mobile command centers that serve more than one jurisdiction can allow for a well-equipped vehicle that is staffed with experts, while not severely burdening one jurisdiction in terms of finances and personnel.

Photo courtesy of associate author Randall W. Hanifen.

The committee should then examine what configuration, equipment, and technology is needed for the vehicle. First, the vehicle's participants should be decided. Will this vehicle be used for planning meetings? Will a unified command use the vehicle? How many command personnel will be expected to be housed in the vehicle? For practicality, at least five operating positions should be available so that the incident commander and the four section chiefs can be located in the vehicle.

The next consideration is the duration of response and the independence the vehicle will be expected to provide. Some of the mobile command centers have a small kitchen and bathroom area available within so that once the personnel are in the vehicle, they do not have to leave for the duration of the operational period. There are other considerations that must be addressed, such as what power source the vehicle will operate from and how long the fuel supply could be expected to last. Many vehicles utilize an onboard generator that is fueled from the vehicle's fuel tank. Many times this generator can run for up to 12 hours on one tank of fuel. If supplies, power, or facilities are needed for the vehicle, the logistical personnel must know this in advance to ensure the requirements are met prior to deploying the vehicle. It would cause a real delay in the command of an incident for the mobile command center to become inoperable in the middle of an incident.

The third consideration is the communications and technology equipment that will be available within the command post. Examining the locations at which the vehicle will be used will dictate the communications equipment the vehicle will need. Equipping the vehicle with the latest 800 MHz and 700 MHz radio systems would not work in an area still operating on a high band frequency. Plans should be made to operate the vehicle within the jurisdictions planned and be able to accommodate mutual aid forces and their communications equipment. Many companies now manufacture patching devices that will allow many different radios, cell phones, and other communications devices to be plugged into the machine. All of the devices are then patched to the same frequency, and thus operating forces utilizing a high-band-frequency radio can hear what the personnel utilizing the 800 MHz radio are saying, and vice versa. The devices are worthwhile during incidents that will summon response forces from a wide area covering many jurisdictions.

Technology is nearly limitless today. Common amenities in new mobile command centers include computer network systems, which allow laptops from all of the operating positions to share information electronically. The network also provides Internet access and can provide a Wi-Fi hot spot. The latter can broadcast outside of the vehicle, allowing branch directors or technical information specialists to utilize the connection from their laptop without having to be inside the command vehicle. One of the latest features is the ability to receive video signals from a helicopter's video camera, such as a police helicopter. In a time of disaster, obtaining a full reconnaissance is difficult, thus flying over the area and broadcasting the signal into the command center can speed the information needed to begin to properly deploy forces to the needed locations. The ability to deliver evacuation instructions and video feed to media outlets will aid in the dissemination of critical information for the victims of the disaster or to citizens who could become affected by the disaster.

The last consideration is who will be responsible for operating the vehicle. This should be a team of logistical support personnel separate from the command staff. Command personnel cannot afford the time to figure out how the various systems on the vehicle operate during an incident. The support team should train regularly with the vehicle to ensure they are proficient in its use. In addition, the support personnel should be tasked with delivery of the vehicle to the needed location.

RADIO PROCEDURES

Each organization should strive to create a policy that dictates the utilization of the radios available to them. As a portion of the *NIMS* document, "Incident communications are facilitated through the development and use of a common communications plan and interoperable communications processes and architectures."[12]

This also translates into the usage of clear text. The usage of codes can add confusion as these codes can vary by jurisdiction. In jurisdiction A, a "10-40" may indicate a working fire, but in jurisdiction B, this code may indicate a frozen fire hydrant.

During normal operations, the utilization of various radio channels will not be of great importance, as long as all units assigned to incident are on the same frequency. However, when incidents grow in size and complexity, the need for a written communications plan comes into play. At this point, it is likely the incident will need more than one frequency. During a disaster, the command personnel do not have the time or the airtime to be able to listen to every crew operating at the incident, nor do they need to, based on the proper use of the ICS system.

When more than one frequency is utilized, the establishment of radio networks is necessary.

Figure 6–4 shows a typical assignment of radio frequencies based on ICS organization. The most common division of the radio frequencies is at the section chief level. The incident commander and the section chiefs will utilize one frequency, often known as the *command network*. The other responders will utilize another frequency known as the *tactical network*. If too many personnel are utilizing the tactical network, splitting the frequencies by branches is the next obvious split, if many branches are established under the operations section chief. The important point of dividing the frequencies on which the personnel are operating is that it is known to all personnel operating at the incident by completing an ICS Form 205, an incident communications form. Information is able to flow vertically and laterally as needed to successfully mitigate the incident.

Noted problems in after action reports indicate that personnel operating at the scene of a disaster need to exercise radio discipline. Under normal responses, response personnel are encouraged to relay all relevant information to their subordinates and supervisors as needed. However, due to the many operations that will occur simultaneously at the scene of a disaster and due to the time it will take to establish many radio networks through a formal communications plan, it is important that only needed information be transmitted over the radio. The leader of a company may find that it is often best in these times to write down notes of items that need to be relayed. When the radio channel is clear, he or she can utilize time wisely by relaying all of the notes as written down. This of course should not be the case when life safety is of concern. At this point, the phrase "emergency

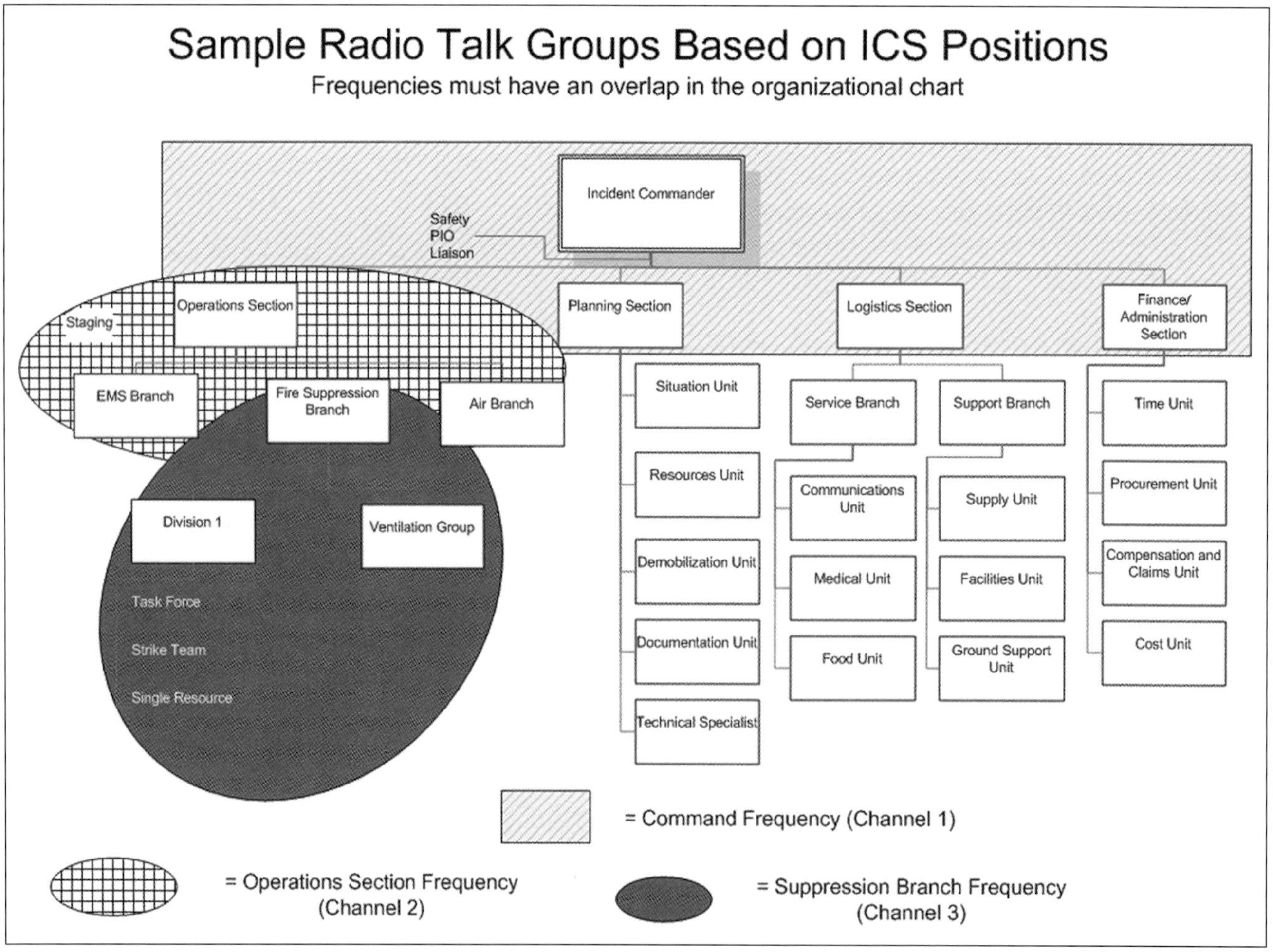

Fig. 6–4. Tactical frequency assignments based on ICS organization. Having personnel with the same level in the organization of the incident on the same frequency helps ensure vital information is heard and transmitted. Note the overlapping of the frequencies.

traffic" can be utilized to indicate to others the urgency of the message.

It is important to remember that communications is a two-way operation. Acknowledging and briefly summarizing the order to the requester allows the person requesting to know the order has been transmitted. Even the best radio systems can experience overload and dropped transmissions, thus orders can be missed or misunderstood. Either can cause a life safety issue in an emergency.

Finally, the communicator should remain calm. Not only is it difficult to understand a person who is yelling into a radio, it also creates apprehension for those listening. If it is a commanding officer who is excited, it can instill a lack of confidence in the subordinates. A commanding officer who remains calm and direct will instill confidence and reduce the number of times a transmission is required in order to be understood.

FIELD COMMUNICATIONS

During many disasters, the standard communications equipment may be knocked off-line. At this time, it is imperative that planning has addressed a way to establish field communications. Often the amateur radio operators will assist the jurisdiction in setting up a field communications unit.

In times of crisis and natural disasters, amateur radio provides emergency communications when wireline, cell phones, and other means of communications fail. Unlike commercial systems, according to *World Book Encyclopedia*, "an amateur radio station basically consists of a transmitter, a receiver, and an antenna."[13] "Amateur radio becomes the principal or only means of communication when disasters destroy normal channels of communication."[14]

Amateur radio operators are also experienced in improvising antennas and power sources, and most equipment can be powered by an automobile battery. Annual field days are held in many countries to practice these emergency improvisational skills. Amateur radio operators can use hundreds of frequencies and can quickly establish networks tying disparate agencies together to enhance interoperability. This was evident in the public service announcement by EmergencyRadio.org, which depicts amateur radio operators helping to establish communication links.[15]

Although these radio operators may have extensive knowledge in setting up emergency communications, they may have limited knowledge or experience in the disaster response disciplines. Therefore, utilizing their services in full-scale exercises and during drills at the various response organizations will ensure they are able to integrate into the operation as if they worked alongside the responders on an everyday basis.

When communications towers are nonfunctional, the utilization of satellite phones is a good alternative. These phones utilize satellites orbiting the Earth rather than land-based antennas. Therefore, these devices will work regardless of damage to the infrastructure. The drawback is the expense of the phone. Some response agencies have arranged to rent the phones in the event of an emergency. Assigning one of the administrative/office personnel from a response agency or government administration will enable the delivery of the phones without delaying response to the disaster.

DISASTER WARNING DISSEMINATION

Second to prevention and planning, the dissemination of information to either affected or unaffected citizens is top priority. Because people do not deal with disasters on a daily basis, and due to the extreme circumstances that surround disasters, citizens often desire and require information in order to respond appropriately. Citizens who are not currently affected will need information on how not to become part of the disaster, such as sheltering or where to travel to get away from expected area of destruction. Although there are numerous ways for people to receive messages, two accepted and intertwined systems are utilized to deliver information to the public in a time of crisis. These are the Emergency Alert System signals and the Emergency Broadcast System.

Emergency Alert System signals

Emergency Alert System signals consist of outdoor warning sirens, television, radio, and electronic media, such as text messaging. Figure 6–5 shows a typical outdoor warning siren. These sirens are to alert those outdoors to seek information from a media source. The outdoor warning sirens are often used for impending weather conditions, such as tornadoes. The sirens are operated by a variety of configurations. Within the Cincinnati area, there are at least five different agencies that have outdoor warning systems. Many are operated at the county level, but some municipalities have incurred the expense of installing their own systems.

The siren is supposed to be reserved for confirmed weather, such as a tornado warning, but the author has witnessed their misuse from overzealous public officials who feel that they are sighting a weather emergency. This can cause confusion, as citizens are informed to seek media outlets for further information. If the media outlet is not reporting on this perceived threatening weather, it leaves the citizens to wonder why the siren was sounded. If this is repeated, the citizens will have less confidence in the system and begin to ignore its activation.

Fig. 6–5. Emergency warning siren. Having these units dispersed properly in a jurisdiction allows the warning of people who are outdoors when an emergency occurs.

FEMA News Photo.

Emergency Broadcast System

According to *Fire Officer's Guide to Disaster Control*, second edition, "The Emergency Broadcast System is the accepted means of releasing vital civil defense information to the general public during periods of emergency. Facilities and personnel of the entire nongovernmental communications industry are used on a voluntary basis to provide a nationwide functional system operated under appropriate government regulations and in a controlled manner consistent with security requirements during a national emergency."[16] Emergency response personnel at the command level must know how to activate the Emergency Broadcast System. Delays in learning the process when needed may cause undue harm to citizens.

DEVELOPMENTS IN COMPUTERS AND INFORMATION MANAGEMENT

Information management

The adage of the productive emergency manager is, "Work smarter, not harder." With advancements in planning and response software, the emergency manager is able to have needed information at the stroke of a key. Many software vendors are working diligently to provide emergency response planners and responders with the latest and most innovative programs available.

The utilization of databases has progressed greatly in the past few years. Most state fire marshals now require electronic submission. The use of these programs can deliver resource lists to the field, provide numbers for recall of off-duty firefighters, and store information pertinent to a specific facility. Technology developments allow almost any information the planners want to be retrieved from the field. Many of the databases will interact with others, such as a records management system interfacing with a computer-aided dispatching program.

The developments of the GIS as discussed earlier will speed in delivering response and recovery services to the areas needed at a much faster pace. Allowing all response personnel to utilize these systems will allow information to be available concerning the street layouts, the public utilities and their associated shut-offs, and preincident plans for various facilities.

Personnel accountability

The latest use of information technology involves the tracking of personnel. Computer programs that are linked to a firefighter's personal alert safety system (PASS) are able to track the location of the person. It also supplies information concerning the status of their supplied air and what temperatures they are being exposed to at the time. The next advancement predicted by the author is a personal AVL. This will allow command personnel to see what resources are in the affected areas. Utilizing reports received with the GIS and the personal AVL will allow command personnel to dispatch the closest responder, regardless if they are in a vehicle or on foot to the person needing assistance. It will also allow the command personnel to predict the time it will take to reach the person in need of service.

DEVELOPMENTS IN PERSONAL COMPUTERS AND INFORMATION MANAGEMENT

As discussed throughout the chapter, the developments in the personal computer, coupled with software enhancements, have given a completely new aspect to emergency response agencies. The personal computer will likely be as common in a fire truck as the SCBA within 10 years. This personal computer will also not be tied to the vehicle as more wireless and mobile technology is introduced. This computer will likely be transportable with the crew leader providing response information needed to make the best choices for his or her crew. These computers will also give the crew leader the most up-to-date information about any type of call they are responding to, thus providing the crew leader information that was initially unavailable when they responded to the call. Loading hazardous materials action plans or structural collapse marking systems on the computers will allow a crew not trained in these specialty areas enough information to make a valuable contribution to the call.

As technology gets smaller, the utilization of personal computers will likely become a personal carry item for each responder, giving them directional information and providing information about them to the command personnel. Often there are certain responders with specialty knowledge, but the command personnel may not realize this expertise or know the exact location of those responders on a large incident. In the future, the commander will be able to query the available personnel to find who has special knowledge and where they are currently at in the field of the disaster.

COMPUTER-AIDED DISPATCHING

Not long ago, dispatchers had to take a call for help, determine from the caller the address, and look up the address on a card or paper to see what emergency service units were needed based on the type of call. They then had to match this list of needed units to what was available at the time of the call. If not all of the units were available, they had to substitute units based on the additional units listed in the card for the address. This was known as the *run card system*. This entire process could take two to three minutes and was prone to many human errors, thus resulting in the wrong units being dispatched to the call. Once the emergency units responded to the call, the dispatchers had to track all of the units as to their status, thus providing more opportunities for errors. If the dispatcher was coordinating a one-station fire department and three police patrol units, this system could be relatively reliable and inexpensive. However, if the dispatcher was trying to use this system in a large metropolitan area that utilized automatic and mutual aid on a frequent basis, this task was overwhelming.

Fortunately, this situation changed dramatically with the advent of CAD. This computer software can complete all of the decisions listed previously in a nanosecond, providing the dispatcher information concerning the needed units on the screen. The computer can also track the status of all of the units, either through dispatcher's input or by receiving signals from mobile data computers located in the emergency response vehicles. The CAD system can generate calls, offer questions to determine the level of service needed, and even locate the nearest units based on their current locations. Figure 6–6 shows how extensive a large city dispatch center can become. Computerization helps lessen the burden on the dispatch personnel.

Fig. 6–6. Typical metro area dispatch center. Note the number of consoles and dispatch/call taker positions needed.

Photo courtesy of associate author Randall W. Hanifen.

The CAD system also interfaces with MDCs, giving response units a text message of the specifics of the call. It also interfaces with either GPS or GIS, which will show the streets, building locations, fire hydrants, and any other relevant data needed by responding units. Secondly, the CAD system allows storage of information based on the location address. Therefore, a list of building representatives can be stored and viewed when needed, as well as having a call history of the address to see what prior activities have occurred. From a police standpoint, it can be used to determine if the residence is a known drug location or if a violent offender resides at the address.

As discussed earlier, the utilization of GPS, GIS, and AVL allows the CAD system to have a recommended list of units for each type of call and address. In addition, it can update that list based on the current location of the response vehicles, thus delivering the needed service as quickly as possible.

Proper level of response

A part of the dispatcher's job is to pick the appropriate level of response and guide the caller with prearrival instructions. This is often the case with EMS calls. The dispatcher is often required to give first-aid instructions, such as how to stop the bleeding or to administer CPR. With the addition of ProQA emergency medical dispatching or a similar program, dispatchers have an automated advantage. According to the National Academies of Emergency Dispatch, "ProQA guides [the dispatcher] through the process of collecting the vital information from the caller, obtaining the patient's status, choosing an appropriate dispatch level, and instructing the caller with medically approved protocols until the dispatched units arrive at the scene."[17] This allows for more consistent dispatching and instructions to the caller from a variety of levels of training in EMS.

The last advantage of a CAD is the ability of the CAD program to transfer information to a records management system (RMS). By transferring the call information, times, and narratives of the call to the RMS, the RMS administrator can recognize increased accuracy on critical data such as location and response times. In addition, there will be a decrease in the amount of time spent imputing data.

IMPLICATIONS FOR DISASTER PLANNING

The communications section of a disaster plan should include any problems that could be encountered during the communications system's use during a disaster. It should also list plausible solutions to these problems. Although no planner can think of every possible failure, the plan should include the following:

- Addressing system overload and developing communications plans and SOPs to eliminate overloading the system.
- Meeting with telephone communications companies to determine how vulnerable they are to certain disaster situations and what backup plans they have in place.
- The plan should note that often the most heavily damaged areas will not be able to report their problems due to lack of phone service, thus crews should conduct rapid needs assessments (RNAs) to see the extent and span of the damage.
- The plan should cover backup communications facilities and equipment and how these backup systems will be integrated and utilized.

CONCLUSION

Communications systems are a very complex and integrated portion of the emergency response system. However, at a moment's notice, they can be rendered inoperable. This chapter has detailed the many different components of a communications and information system. With the ever-changing technology, it is imperative that emergency managers stay abreast of these changes to ensure that their response agencies have the best ability to communicate within their agency and with other responding agencies in a time of a disaster. Backup and redundant systems are a necessity; otherwise, many dollars are wasted on a system that is not functional when it is needed most. If used correctly, the advancements in communications and information management can be the emergency manager's greatest companion.

REFERENCES

1. *Dictionary.com.* 2007. Communication. Definition 2. Retrieved February 2, 2007 from: http://dictionary.reference.com/browse/communication.

2. *NIMS online.com.* 2007. Communications and information management. Retrieved February 8, 2007 from: http://www.nimsonline.com/nims_3_04/communications_and_information_management.htm.

3. Ibid.

4. InterAgency Board. 2007. About the IAB. Retrieved February 9, 2007 from: http://www.iab.gov.

5. InterAgency Board. 2006. *Standard Equipment List.* Section 6, Interoperable communications equipment. Retrieved February 10, 2007 from: http://www.iab.gov/Download/sel_section6_2006.pdf.

6. Ibid.

7. *P25.com.* What is Project 25? Accessed August 26, 2008 from: http://www.p25.com.

8. Project 25 Technology Interest Group. 2006. *What Is Project 25?* What is required for P25 compliance? August 28. Retrieved February 10, 2007 from: http://www.project25.org/modules.php?name=Content&file=viewarticle&id=2.

9. U.S. Department of Homeland Security. 2007. Tactical interoperability communication scorecards. Retrieved February 10, 2007 from: http://www.dhs.gov/xprepresp/gc_1167770109789.shtm.

10. Ibid.

11. *GIS.com.* 2007. What is GIS? Retrieved August 26, 2008 from: http://www.gis.com/whatisgis.

12. U.S. Department of Homeland Security. 2004. *National Incident Management System.* March. Chapter 2. Command and management. p. 11. Retrieved February 12, 2007 from: http://www.nimsonline.com/docs/NIMS-90-web.pdf.

13. *World Book Encyclopedia.* 2006. Amateur radio. Vol. 23. Danbury, CT: Scholastic Library Publishing. p. 157.

14. Ibid.

15. Emergency Radio.org. 2008. Katrina, the untold story. Retrieved July 15, 2008 from: http://emergency-radio.org/stories.html.

16. Kramer, W. M., and C. W. Bahme. 1992. *Fire Officer's Guide to Disaster Control.* 2nd ed. Saddle Brook, NJ: Fire Engineering Books and Videos. p. 167.

17. National Academies of Emergency Dispatch. 2007. What is ProQA? Retrieved February 12, 2007 from: http://www.emergencydispatch.org/articles/whatis.html.

Frequent Threats—Fire, Transportation, and Hazmat

3

CATASTROPHIC FIRES 7

INTRODUCTION

Fires occur each day. Fortunately, few make it to the magnitude of being considered a disaster. This chapter will not attempt to cover all aspects of firefighting. These aspects are covered in great detail by books and magazines from PennWell, including *Fire Engineering*, as well as the National Fire Protection Association (NFPA) and numerous other book publishers. This chapter will focus on fire events that result in disasters.

What defines a fire as a disaster or a catastrophe? One of the defining factors is when a large loss of life occurs, such as the Station nightclub fire. Sometimes disasters include the numerous losses of lives of fellow fire service personnel, such as the Worcester Cold Storage Warehouse fire. In this fire, multiple firefighters lost their lives fighting a fire ignited by vagrants in a vacant building. These are always the most difficult for the responders.

Another defining factor for a disaster can be the building in which the incident occurs. Many high-rise fires become disaster situations due to the sheer number of firefighters and other responding personnel required to bring these events under control.

The occupancy or type of building also can affect an incident's ability to turn into a disaster situation. Fires in occupancies such as hospitals or nursing homes can have disastrous results. Often the occupants are unable to evacuate themselves and are easily overcome by the smoke or fire. Planning for fires in these facilities can be on the scale of planning for a natural disaster in the municipality due to the number of needed responders and the number of potential victims.

Due to space limitations, builders are now combining two of the most demanding facilities to fight fire within: high-rises and nursing home facilities. Proper planning and exercise with the emergency plan is the only hope for successful operations within these facilities.

FIRE—STILL A PRIMARY FOCUS

According to statistics from the U.S. Fire Administration, about 1.6 million fires occur in the United States each year, in which 3,675 people lose their lives. These fires caused $10.7 billion in direct property loss.[1] Fortunately, the majority of these fires do not become catastrophic. However, in the first two months of 2007, there was an inordinate number of civilian fire deaths, making many of the nation's top fire and safety officials wonder if the nation is experiencing a repeat of the 1970s. During that decade, the fire problem was so severe the president established a commission to investigate the country's fire problem. Out of this commission came the report *America Burning*.

Overall, the fire frequency and deaths have trended toward a decline. Some of the reasons for this are the utilization and installation of working smoke detectors. Advancements in many building codes now require these units to be hardwired into the residential structure. In addition, some communities now require sprinklers even in single-family homes. These communities with residential sprinkler ordinances have made media headlines for their advancement of fire protection.

URBAN CONFLAGRATIONS

By definition, an *urban conflagration* is a large, destructive fire that spreads beyond natural or artificial barriers; it can be expected to result in large monetary loss and may or may not include fatalities. These fires are even greater than a *group fire*, which is a fire that destroys a group of buildings at an industrial campus or a city block. An urban conflagration moves beyond a block and destroys whole sections of a city.

One of the conflagrations that is well known in U.S. history is the destruction of Jamestown, Virginia on January 7, 1607.[2] Some of the notable urban conflagrations in more recent years occurred in Oakland, California in 1991; Brooklyn, New York in 1991; Patterson, New Jersey in 1991; Rhode Island in 1990; Chelsea, Massachusetts in 1983; Anaheim, California in 1982; Lynn, Massachusetts in 1981, and Houston, Texas in 1979. Most U.S. cities are older and contain many of the key ingredients for another urban conflagration. It is the responsibility of a city's planning team to recognize these hazards and provide prevention and response strategies to combat them when they occur.

Causes of urban conflagrations

Although fires can occur for a variety of reasons, planners should recognize prevalent past factors that have become sources of ignition in conflagrations. Many of the conflagrations have occurred as secondary events to natural disasters, such as volcanic eruptions, floods (either from torrential rains or from a hurricane), earthquakes that rupture natural gas lines, and lightning from powerful thunderstorms.

There are man-made ignition sources as well, including carelessness with open flames and smoking materials. The need for fire to warm homes and conduct many industrial processes makes it unlikely that fire will be eliminated as a part of everyday life. However, it is possible to increase the safety of its use, such as shut-off valves for the fuel powering the fire and the creation of self-extinguishing cigarettes. Prevention by way of altering human behavior has seen an increase in the last decades and has resulted in fewer events caused by problems such as smoking in bed.

Given this list of popular ignition sources, it will remain difficult to control the ignition of fires unless the studies for changing climates and weather patterns are founded. Until the technology exists to prevent natural disasters, which does not seem likely to be developed any time soon, planners must examine ways to minimize and control the conflagrations quickly.

Some of the methods to ensure that damage is limited and the fires are brought under control quickly are the following:

1. **Creating natural breaks in the buildings.** Fortunately, many people enjoy what are termed *open spaces*, which are grass areas that allow the appeal of nature within urban areas. By specifically designing the placement of these green spaces, the municipality can essentially create fire breaks.

2. **The inclusion of sprinklers in as many occupancies as feasible.** Although codes vary by area, many common codes exist in relation to requirements for sprinklering buildings. Ensuring that these codes are adhered to, and that the systems remain operational, will limit the spread of fire within any building.

3. **Ensuring adequate volumes and pressures of water from the municipal water system.** If adequate volumes and pressures of water are supplied when needed, the firefighting forces are not hampered in their ability to perform their jobs.

4. **Ensuring secondary water sources are available.** Many urban departments do not carry rigid suction hoses used to draft water from static sources. This is based on the supposition that their cities have a hydrant on every block. This viewpoint does not consider, however, how this system would work in a post-earthquake environment. How will firefighters locate and access these hydrants in deep water?

5. **Employing and ensuring that an adequate number of firefighting forces are available at all times.** The trend in many midsize cities is to reduce the amount of fire protection forces available. Reliance on just enough personnel to handle the daily volume will lead to deficiencies in firefighting operations at large-scale fires during the crucial first minutes of arrival. If adequate amounts of manpower and equipment are present early, the fire can be quickly brought under control, but it can grow exponentially if not controlled early. According to the *Fire Protection Handbook*, "Staffing levels for cities of 250,000 or more population range from 0.5 to 2.9 firefighters per thousand population, with a median of 1.1 to 1.5 per thousand . . . Fire departments operating emergency medical service transports need additional personnel to maintain basic fire company strength."[3]

Types of urban conflagrations

Studies have been conducted over the years that have classified urban conflagrations into five general types. According to the 17th edition of the *Fire Protection Handbook*, those five types are:

1. **Those originating in hazardous materials industries lacking adequate means for confining or extinguishing incipient fires.** The structures usually contain large quantities of highly flammable or explosive materials and are located in a congested district among other combustible occupancies without protection or adequate private or public water supplies or fire protection. In the Chelsea, Massachusetts conflagration, 300 buildings were involved, with a resulting loss estimated at $1.3 million. The main contributing factors in this conflagration were poor water supply, dry weather, high winds, closely stored combustible materials, narrow streets, no firebreaks, and delayed reporting.

2. **Those occurring in closely built-up residential sections containing combustible houses with wood shingle roofs.** All residential areas with combustible construction and multiple exposures belong in this classification. Even where the fire protection forces may be superior, the lack of adequate water supply, coupled with high

winds, permit fire to get out of control. Examples of this type include an apartment building conflagration in Dallas, Texas on March 21, 1983. Another is a large fire that occurred in the Westwood section of Los Angeles in late 1989, which received little publicity because it occurred on the same day that the Berlin Wall was opened.

3. **Wildland, forest, or brush fires that sweep into a town or towns over a wide front.** The problem of wildland/urban interface has taken a front seat in the problems being addressed by many fire service leaders. As is discussed in the next section, many fires have destroyed homes and left numerous homeless.

4. **Conflagrations due to explosions causing fire to spread over a wide area.** One of these explosions occurred on May 4, 1988. "A series of explosions on May 4, 1988, near the City of Henderson, Nevada, claimed two lives, injured approximately 372 people including 15 firefighters, and caused damage estimated over $100 million. The explosions affected a large portion of the metropolitan Las Vegas area and caused the activation of disaster plans by several agencies."[4]

5. **Multiple fires in one area or city as a result of an earthquake or civil unrest.** This type of urban conflagration occurred in 2001 in Cincinnati, Ohio. This started as a Cincinnati police officer was trying to apprehend a juvenile suspect, but it led to six days of rioting. This also occurred in Los Angeles in 1992 with the acquittal of the police officers who were accused of beating an unarmed African-American man.

Contributing factors in urban conflagrations

Every American has enjoyed lower home cost due to the prevalent construction material of the country, wood. In the associate author's travels, it is a personal observation that many other countries utilize exclusively block construction, which limits any fire to the unit in which it began. In the United States, especially in urban areas, the construction is of wood and fails to provide adequate firebreaks. Under normal conditions, this type of construction is faster and more economical to build, but during fire conditions, it is the primary contributor to the rapid spread of fire that will result in an urban conflagration.

One of the biggest problems in the urban/wildland interface is the use of wood shingles. Many believed these shingles would be more ecologically fitting for their homes in the woods. Unfortunately, these shingles are very combustible and allow the fire to spread from the wildland to the nearby homes. Another contributing factor beyond the construction features is the lack of an adequate water supply. Although homes are built into the wildland, it is often the case that adequate water supplies do not accompany these homes.

In terms of inadequate water supply, it seems to be the current belief that if facilities install a sprinkler system, the fire department can rely upon this system solely and will not have to ensure that supplemental water for manual fire suppression is adequate. It is the opinion of the associate author that the best advent in fire protection is the sprinkler, but too much reliance on these systems takes place among planners. Remembering the K-mart warehouse fire in Falls Township, Pennsylvania in 1982 will allow proponents of exclusive reliance on sprinkler systems to see that this is a less-than-wise opinion.

Duration of urban conflagrations

Many times despite the valiant efforts of firefighting forces, urban conflagrations will continue to burn as long as they are either fanned by winds or as long as a supply of fuel exists. The fires that start in or involve wildlands will behave as described in the latter section. As a rule, both wildland and urban fires will travel in a downwind direction. These fires will also continue to burn for a period of days. As with other disasters, these events will last for multiple operation periods and will need planning, logistical, and financial support in order to be brought under control.

Firestorms

Under certain conditions, a stationary mass fire called a *firestorm* might develop. A firestorm is not strictly a conflagration, since its behavior is quite different.

Cause of firestorms. Firestorms havc occurred because of warlike events. Many times during past war activities, bombings on cities have led to firestorms. It is feared that with the continued rising tension between countries in the Middle East and the United States, it is quite feasible that warlike strikes could occur in a major city. As well as the explosive effects, the secondary firestorm would burn much of the area. Proper control of the firestorm would limit the disaster and prevent further loss of life.

Effects of firestorms. The fires are the obvious effect of a firestorm. However, as is the case with many fire deaths, it is the toxic gases, such as carbon monoxide, that will kill people who are not directly in the fire. In the case of a firestorm, this could be up to 20 miles away from the fire, based on its size.

> Firestorms also cause severe property damage over a wide area. In Hamburg, Germany, in World War II, within twenty minutes after one of the extensive incendiary air raids, two-thirds of the buildings in a 4.5 square mile area were completely involved in fire. Two days later, when the area had cooled down sufficiently to permit the fire department to enter, it was estimated that more than 55,000 persons had perished. Before the series of attacks were over, some 30 square miles had been damaged, with a complete burnout of 12.5 square miles; 750,000 people were made homeless as the result of the total loss of about 300,000 dwelling units.[5]

This will likely be much of the same type of report that will come back from the current Iraq war. When the shock and awe phase of the United States attack occurred, it mimicked the attack on Hamburg, Germany. Fortunately, this type of attack has not occurred in the United States, but as the United States continues to engage in war with various countries, the need for preparedness for this type of disaster increases.

Duration of firestorms. Fortunately, these types of events have a history of not lasting as long as other types of disasters. Some historical events have lasted days due to the intense damage and repeated bombing of an area, but if only one attack of medium size were to occur, it was expected to last approximately three to four hours according to a 1968 report, *Fire Aspects of Civil Defense.*[6] Today's munitions and WMD products could deliver a bigger devastation, but it is likely dependent upon the fuel that feeds the fire both in terms of the WMDs and the objects that they would strike.

WILDLAND/URBAN INTERFACE FIRES

According to the International Fire Chief's Wildland Fire Policy Committee,

> The fastest growing fire problem in the United States is fire in the wildland-urban interface (WUI). This problem is growing each year as millions of people continue to build homes in high-risk interface areas and as local, state and federal government attempts to address fuel modification in

the interface continue to meet social resistance to a balanced fuels reduction plan. The direct threat to life and high-valued property in the interface (suburbs) has been demonstrated in cities, towns, counties and communities throughout the country. The direct and indirect costs of an extreme fire event can reach multi-billion dollar dimensions; impacts on the infrastructure of a community can include damage and disruption of water supply, utilities and transportation systems. These fires threaten high-valued community assets including recreational resources, cultural icons, endangered species, grazing crops, jobs, and community health & commerce.[7]

Since 1984, the term *wildland/urban interface* has received attention, as evidenced by nearly every fire service organization and trade journal dedicating a section to the problem. In 2004, *Fire Engineering* provided with its January publication a supplement, *Wildfire Report, a Supplement to Fire Engineering*. This supplement contained information on the "Montano Complex Fire/Urban Interface Fires," which was authored by Bobby Halton, chief of operations for the Albuquerque Fire Department. His article examines the fire, which occurred in the Bosque section of Albuquerque. The article contains many of the issues relevant in this discussion of wildfires and their effect on the urban interface, and thus it will be utilized here as a case study.[8]

Many people believe that only California is plagued with the wildland/urban interface problem, but this problem, as shown in the case study, occurs in many states throughout the country. CNN's report on wildfires stated, "Although the California blazes garnered intense media coverage, they represented just a few of the year's many damaging wildfires. According to the National Interagency Fire Center, more than 85,000 fires burned a total of 4,918,088 acres in the United States in 2003."[9] In addition, "The 'Webber' fire consumed more than 4,000 acres of both the Tonto and Coconino National Forests near Pine in central Arizona, about 100 miles northeast of Phoenix. The blaze was discovered on March 29, 2004, and the USDA Forest Service declared the fire fully contained on April 2, 2004."[10]

Even in Ohio, the wildfire problem exists. "In Ohio, wildfires usually burn 4,000 to 6,000 acres a year. Thus, Ohio forestry officials are launching a program to help residents here protect themselves and their homes from similar blazes. The FireWise Ohio program brings together local fire departments and citizens who live in forested areas to develop fire safety plans and teach homeowners about wildfire protection."[11]

California is, however, known for many of the costliest fires, due to its extensive wildland/urban interface. According to a CNN report, "In fall 2003, Californians faced one of the worst wildfire seasons ever, as blazes burned more than 500,000 acres and killed more than 20 people. The damage was focused in five southern counties: San Bernardino, Los Angeles, Riverside, San Diego and Ventura."[12] Furthermore, "In 2004, nearly 80,000 wildfires burned just less than 7 million acres in the United States, according to the National Interagency Fire Center (NIFC). The NIFC also calculated that federal agencies spent nearly $900 million fighting these blazes."[13] CNN also reported that "in 2005, more than 8 million acres burned in 66,753 reported wildfires across the country, according to the National Interagency Fire Center."[14]

Causes of wildland fires

Traditionally, fires that occurred in the eastern portion of the United States were attributed to careless human acts, such as leaving campfires unattended, discarding lit cigarettes, etc. In the West, lightning, industrial operations such as logging, and sparks and heat from trains traversing the wooded areas were known to start fires. However, some of the latest and most devastating fires have been attributed to careless human acts and arson.

Wildland fire behavior

It is not the intent of the authors to provide detailed training on the behavior of wildfires. However, the basics will be described here, as it is important to understand how the spread of a wildfire will occur when planning and managing an incident involving a wildland fire. For a full, detailed training, the authors suggest texts such as *Essentials of Firefighting*, 4th edition, from IFSTA, or *Wildland Firefighting Practices* by Joseph Lowe. Both texts will provide in-depth coverage of the behavior of wildfires.

A number of factors affect the behavior of wildland fires, including terrain, weather, wind, fuels, and season. It is well-known that fire travels faster uphill than downhill and is more difficult to fight on steep slopes than level ground.

When weather is damp or wet, fuels soak up the moisture, and brush or grass is difficult to burn. However, when the weather is hot and the humidity is low, wildland fires can explode with the intensity of rapid combustion, making crown fires in timber almost impossible to stop.

Even in the absence of strong winds, such as Santa Ana winds, a fast-moving fire can generate its own updrafts, particularly in canyons, causing burning brands to be carried high in the air and dropped a long distance ahead. This results in spot fires over a wide radius as the wind changes its heading. When strong, desiccating winds are prevalent, the fire moves with such rapidity that nothing short of a well-coordinated air and ground attack, coupled with wide natural barriers and clearings, can control its spread.

When high-velocity winds blow over a steep, rough, broken topography, a highly turbulent airflow is created in the canyons. Wind directions may switch 180° and back again in only a few seconds. Dense smoke clings to the ground, and burning embers are driven into unburned, dried-out fuel ahead of the fire. If such fuels happen to be finely divided and have a high oil content and a large surface-to-volume ratio, as is the case with most chaparral fuels, it becomes unstable.

Natural fuels vary greatly in the rate and intensity of burning. Fires in heavy brush and timber burn with great intensity, but they still burn more slowly than in dry grass and leaves. Dense fuels propagate fire better than sparse fuels. Due to these unique fire behavior characteristics, it is imperative that all operating personnel, especially crew supervisors, are keenly aware of and knowledgeable about wildland fire spread.

The season for forest fires varies by region. The fire season in the southwestern portion of the United States occurs from May to June, while the California season is June to December. The eastern portion of the country has two seasons. The first season is after the spring ground cover dries out, and the second is toward the end of October, when all of the dry leaves have fallen from the trees.

Forest fire protection responsibility

Responsibility for fire protection on federal lands is centered primarily in the Department of Agriculture, the Department of Interior, and the U.S. Forest Service. Each state has responsibilities, due to the majority of the wildland residing on state park property. Although the local fire departments are not directly responsible for these areas, it is likely that they will be summoned to the incident. In the western portion of the United States, the state and federal agencies are integrated well into these incident responses. However, due to the lower frequency of occurrence, the eastern portion of the United States still relies upon the local fire departments, supplemented by state and federal forces. This should sound familiar, as this is the intended chain of events that is covered and expected as part of the National Response Framework.

Community preparation

The emergency planners of small communities, especially executive fire officers of small communities with wildlands bordering their communities, should employ many of the wildland/urban interface preparations and tactics. Some of these preparation activities include the following:

- Training firefighting personnel in the behavior and management of wildfires.
- Preparing or participating in the preparation of prevention programs as they relate to forest fires. Some of the methods to accomplish this prevention are:
  - Use of local radio and television programs, such as the U.S. Forest Service's Smokey the Bear campaign.
  - Deliverance of junior fire department instruction in schools, with an emphasis on forest fire safety.
- Preparing and exercising mutual aid compacts to ensure that adequate amounts of resources are available for the long duration of these events. In addition, specialized equipment and apparatus are likely to be needed during a forest fire.
- Investing moneys in supplemental personnel in times of forest fires. If a limited budget does not provide year-round protection through employing paid personnel, the employment of seasonal firefighters may be the best preparation for a community. This is often a good summer employment opportunity for graduating high school students.
- Learning from the National Fire Academy's distance learning course "Community and the Fire Threat," as well as the disaster planning course.

Organizing and equipping for rural fire control

Because fighting wildland fires is much different in many aspects of strategy and tactics than that of structural firefighting, small departments are encouraged to employ the standards and equipment listed in NFPA 1143, *Standard for Wildland Fire Management*. By employing the suggestions of this standard, small departments can amass a solid firefighting force, with the proper equipment to battle the next wildland fire. The IAFC Policy Committee on Wildland/Urban Interface has developed the document *Advancing Wildland Fire Training for Fire Departments Implementation Plan*. This in turn addresses the 2002 congressional report *The Changing Role and Needs of Local, Rural and Volunteer Fire Departments in the Wildland Urban Interface*. This plan describes implementation actions 1a–1c from the 2002 congressional report, as well as the 2004 taskforce report *Advancing Wildland Fire Training for Fire Departments*. The goals of the implementation plan are to:

- Ensure firefighter safety is of the utmost importance…in every wildland firefighting suppression operation.
- Through enhanced training opportunities, technical and financial assistance, and interagency coordination, facilitate safe, effective and efficient wildland and wildland/urban interface (WUI) response within local areas, and thereby increase the…capacity of rural and volunteer fire departments (hereafter referred to as RFDs).
- Facilitate increased use of trained and qualified rural and volunteer fire departments on local incidents by state and federal agencies. Enhance coordination…and cooperation…between the structural and wildland disciplines of the American fire service.[15]

Further information can be found on this plan through the IAFC's Web site at http://www.iafc.org/associations/4685/files/AWFTimplementation.pdf.

Actions to be taken when a wildland fire threatens the community

As with any emergency event or disaster, there are certain actions that must be employed to ensure the proper response to, control of, and recovery from the disaster. The authors have prepared a checklist of actions that should be taken or considered by the incident commander upon the discovery of a wildland fire that threatens the community adjacent to the fire.

1. Activate the municipality's emergency operations plan.
2. Establish an incident command system.
3. Consider the enactment of the area command component of NIMS.
4. Establish a planning section that has contact with the nearest National Weather Service so that the incident commander has adequate forecasting support.
5. Establish a recon group that can gather information about the threat to the community and provide this information to the planning section chief to include in the incident action plan (IAP). Consider the use of aircraft to acquire the recon knowledge.
6. Establish a public information officer (PIO) and ensure that he or she is prepared to work in and be a part of a joint information center (JIC). The constant and direct instruction and information given to the public will be of the utmost importance.
7. Through the local EOP and the existing emergency management assistance compacts (EMACs), ensure that the needed resources (operational and support) are summoned quickly.
8. Direct the planning section to establish areas that will require evacuation and estimate the resources needed to accomplish this task.
9. Upon establishment of evacuation zones, direct the operations section chief to assign personnel to evacuate these areas. This should be coordinated with the public information officer. A place of refuge must be designated for the evacuation of the citizens.
10. Ask the operations section chief about the success, or lack thereof, of the current operations. Be prepared for assignment of the tasks of conducting direct and indirect attacks, constructing fire breaks, lighting backfires, and establishing an air operations branch.
11. Establish a communications plan that will incorporate the number of responders, the command structure, and the various radio frequencies needed, such as aircraft channels.
12. Be prepared to activate an EOC and establish a unified command structure to ensure coordination of police and public works activities that will support the control and suppression activities.

Warning to be given to the public

The public's willingness to help prevent and recognize wildfires and the conditions that perpetuate their ignition and spread will often determine the level of activity the fire crews can expect. The first step in garnering the public's support is to enable them to recognize that the problem is one that affects them. Until people recognize the harm or benefit to themselves, they are not likely to participate in any activity. This is much like raising money to fight cancer; people are more likely to donate if they have experienced it personally or have a loved one who is affected by cancer.

One way of alerting the public to the concern is to utilize news footage of prior conflagrations and then issue an action statement or instructions on how to gauge when the conditions are prime for ignition of a wildland fire. Some of the signs the public should be alerted to are as follows:

- They should be alerted when there are heavy winds, coupled with increased static electric discharges.

- They should be urged to be familiar with the topography of their immediate vicinity and how the fire will spread in this area, such as rapid movement on steep hills, etc.
- They should be alerted to clear the dry brush from around their homes.
- They should be informed of the importance of a fire escape plan, especially those living on hills and dead-end streets.
- They should be reminded of the speed at which fires can travel. For example, fires can move two to three blocks a minute when given dry fuel and proper winds.
- They should be informed of possible alerts and planning for temporary accommodations during the extreme peaks in ignitable conditions.
- They should be informed of the proper documents to evacuate with in the case of their homes being destroyed. The list of documents could include passports, driver's licenses, insurance and financial documents, and jewelry. It should be emphasized that all items should be hand cartable in case they need to walk away from their homes.
- They should be informed of the proper contents of an emergency kit, such as nonperishable foods, flashlights, and battery-operated radios. They must be prepared to live without electricity for an extended duration.
- Homeowners should consider the pretreatment of their homes with fire retardants, such as the product Barricade.

Wildland control methods

As stated previously, this text is not a strategy and tactics manual for wildland firefighting. Instead, it will focus on some of the basic control methods. This focus will allow those planning for and responding to these types of disasters to properly understand the resources (both operational and support) needed to effectively bring the incident to a successful end.

Many of the firefighting tactics and the scale of operations are different than those of a typical structure fire. The first major operation difference is the area in which the incident spans. As with many types of disasters, the square miles affected are what make the events so taxing and complicated. With a wildfire, its area affected could be expanding exponentially while operations are being established. Not only must the incident commander and the EOC personnel be accustomed to dealing with large-scale incidents, they must have the forethought to request the resources needed to outrun the fire rather than merely keep up with its spread. Figure 7–1 shows wildland firefighters in action.

Fig. 7–1. Wildland firefighting. Note the fire is very large compared to a typical structural fire. Not understanding fire spread in the wildland setting can put firefighters in great danger. Training on wildland firefighting is important for firefighters in the urban/wildland interface.

FEMA News Photo.

From an extinguishment standpoint, this discussion will begin with the possibilities for extinguishment. Based on the fire tetrahedron, which is fuel, heat, oxygen, and an uninhibited chain reaction, interrupting the uninhibited chain reaction on a large scale can be ruled out. Furthermore, it is nearly impossible to smother a fire of this magnitude. Therefore, the two remaining options are to remove the heat through cooling or to interrupt the fuel supply. Based on the two possible ways of addressing the fire tetrahedron, one can develop the direct attack, or application of water, and the indirect attack, or removing the fuel through firebreaks, backfires, and soil barriers.

Direct attack can utilize water applied from hoselines or from tankers delivering water from the air. Indirect attack may involve identifying a break that the fire crew can create through the removal of fuel sources. One of the ways that is popular in the eastern portion of the United States is the utilization of a leaf blower to blow all of the underbrush back into the fire, thus creating a ring with no fuel sources.

One of the most important resources in the direct attack is the utilization of air tankers to drop water onto the fire. Although this is a common occurrence during the fires in the western portion of the United States, in the areas not regularly affected by wildland fires, it can be an element the command staff is not regularly accustomed to utilizing. The apparatus utilized during these events can range from helicopters to a C-130 fixed-wing aircraft. The capacities of these units range up to 3,000 gallons. Proper resource typing through NIMS will ensure the command personnel receive the correct resources.

Thankfully, the resources typing portion of NIMS will alleviate confusion in certain areas of the country when it pertains to tankers. Many rural fire departments have identified their rolling water tender apparatus as *tankers*. This could lead to real confusion when an air tanker is requested for an air support operation during a wildland fire.

The use of helicopters during wildland fires has had marked improvements in the management of these incidents. Some of their uses include the reconnaissance of the area of the fire and the level of involvement in various areas through the thermal imaging cameras that can be attached to the aircraft. The helicopters can be utilized to transport and deploy firefighting forces into heavily wooded and remote areas. They can also be utilized to drop fire retardants.

History has shown that planning and operations must include the following components to be successful:

- Planning for the personnel necessary to sustain a multiple-month operation.
- Planning for the necessary logistical support for the personnel and the necessary equipment for the operation.
- Proper establishment of area command and EOCs, as well as the utilization of the unified command structure. This allows coordination of local, state, and federal resources, all of which will likely have jurisdiction over areas involved in the fire.
- Planning for and utilization of an air operations branch. This is essential in the overall coordination of an attack on a wildland fire.
- Proper planning for and enactment of evacuation procedures. Preparing reports of the spread of the fires will allow the evacuation branch officer to assign resources to the proper areas that are anticipated to become affected over the next operational period.

Predictability of fires

The conditions conducive to wildland/urban interface fires and conflagrations are well known, such as a wind velocity of 13 mph or more, humidity of 40% or less, and rainfall of 0.01 inches or less on the day of the fire and for three days preceding it. The National Weather Service releases Fire Weather Forecasts at http://www.spc.noaa.gov/products/fire_wx/, in which they show the conditions and the likelihood of wildland fires based on climatic conditions. Their scales show critical and extremely critical areas, and they give a text description of the conditions providing these warnings.

Limiting loss from wildland/ urban fires and conflagrations

Having knowledge of the leading factors that contribute to wildland/urban interface fires and conflagrations, it is not difficult to suggest actions that can prevent them. Some of the actions suggested by the authors are the following:

- Elimination of inferior construction, such as those with cedar shingles or combustible siding
- Installation of extinguishing systems and fire detection systems within combustible buildings
- Elimination of overgrowth near occupancies
- Ensuring that an adequate and reliable water distribution system is available
- Adequate availability of firefighting forces, both public and privately funded
- Current and operational EMAC agreements

Through a cooperative effort of the Communities Committee, National Association of Counties, National Association of State Foresters, Society of American Foresters, and the Western Governors' Association, the document *Preparing a Community Wildfire Protection Plan: A Handbook for Wildland-Urban Interface Communities* was created. This document offers an eight-step process for developing a wildfire protection plan that addresses the core elements of a community protection program. These steps, as outlined in the document, are as follows:

1. Convene decision makers
2. Involve federal agencies
3. Engage interested parties
4. Establish a community base map
5. Develop a community risk assessment
6. Establish community priorities and recommendations
7. Develop an action plan and assessment strategy
8. Finalize community wildfire protection program[16]

Figure 7–2 provides a checklist to implement this wildfire protection plan. Readers are encouraged to copy this figure prior to beginning the process.

Checklist for Preparing a Community Wildfire Protection Plan

☐ **1. Convene Decision Makers**
- Form a core team of represenattives from local government, local fire authority, and state agency responsible for forest management

☐ **2. Involve Federal Agencies**
- Identify and engage local representatives of the United States Forest Service (USFS) and Bureau of Land Management (BLM)
- Contact and involve other land management agencies as appropriate

☐ **3. Engage Interested Parties**
- Contact and encourage active involvement in plan development from a broad range of interested organizations and stakeholders

☐ **4. Establish a Community Base Map**
- Work with partners to create a baseline map of the community that defines the communities WUI and displays inhabited areas at risk, forested areas that contain critical human infrastructure, and forested areas at risk for large-scale fire disturbance

☐ **5. Develop a Community Risk Assessment**
- Develop community risk assessment that considers fuel hazards; risk of occurrence for homes, bussinesses, and essential infrastructure; other community values at risk, and local prepardness capability
- Rate level of each risk factor and incorporate into base map

☐ **6. Establish Community Priorities and Recommendations**
- Use base map and community risk assessment to facilitate a community discussion that leads to identification of local priorities for fuel treatment, reducing structural integrity, and other issues of interest, such as improving fire response capability
- Clearly idicate whether priority projects are directly related to protection of communities and essential infrastructure or to reducing wildfire risks to other community values

☐ **7. Develop an Action Plan and Assessment Strategy**
- Consider developing an detailed implementation strategy to accompany the CWPP, as well as a monitoring plan that will ensure its long-term success

☐ **8. Finalize Community Wildfire Protection Plan**
- Finalize the CWPP and communicate the results to community and key partners

Fig. 7–2. Wildland/urban fire planning flowchart

Source: Created from information provided by Communities Committee, National Association of Counties, National Association of State Foresters, Society of American Foresters, and the Western Governors' Association. 2004. *Preparing a Community Wildfire Protection Plan: A Handbook for Wildland-Urban Interface Communities.* March.

OCCUPIED HIGH-RISE BUILDINGS

Many senior fire officials have referred to high-rise buildings as a vertical city. This is true in many respects, especially as it relates to the number of people and the amount of infrastructure present. Furthermore, high-rise fires take the community's firefighters out of their apparatus and make them walk with all of their equipment to the fire. Because of the time this walking could take, fires that progress unchecked could become catastrophic prior to the arrival of the fire company. Finally, there are no easy ways to escape from these fires. The occupants, fire personnel, and combustion products are trapped in an enclosed environment. This often equates to loss of civilian lives and the lives of firefighters.

On February 23, 1991, the Philadelphia Fire Department responded to a 38-story office building, known as One Meridian Plaza. When they arrived at 8:40 p.m., they found that a fire was present on the 22nd floor of the building. This building was not fully sprinklered, and thus the fire burned for nearly 19 hours. Some of the factors contributing to this fire hampered firefighting efforts as well; these factors are examined in this section. Figure 7–3 provides a synopsis of the key issues related to high-rise firefighting.

Operational Considerations for Highrise Firefighting
Summary of Key Issues

Issue	Comment
Highrise SOPs	Standard Operating Procedures (SOPs) are needed for highrise buildings because of the special challenges presented and the strategies and tactics that must be employed to perform safely and effectively in a highrise environment
Highrise Operational Training	The infrequent occurrence of highrise fires relative to other types of fires in most cities necessitates special highrise training drills for firefighters
Knowledge of Fire Protection Systems	Fire departments must be familiar with the capabilities, limitations, and designs of the fire protection systems that they will rely on during a fire. Effective tactical operations may depend on the functionality of these built-in systems
Pre-fire Plans for Highrises	Pre-fire Plans are essential for highrise buildings. These plans should address, at a minimum, the types and capabilities of the mechanical fire protection systems, the layout of the building, the locations of the elevators and stairwells, and evacuation considerations
Inspection and testing of Fire Protection Systems	Fire departments should work closely with the building management to ensure that all fire protection systems are tested on a regular basis
Failure of one or more components of a Highrise Fire Protection System	Fire departments should anticipate what would happen if one or more components of a highrise's fire protection system were to fail. They should have contingency plans in place in case a system fails to work properly
Pressure Reducing Devices	Problems with pressure reducing devices have been noted in several major highrise fires in the United States. The consequences of these problems have been inadequate water pressure or volume from the standpipe systems to fight the fires
Highrise Hose and Nozzle Combinations	Many highrise standpipe systems are designed around the concept of the fire department using a specific combination of hoses and nozzles. The use of other combinations can create serious problems
Highrise Automatic Sprinklers	Very Few Departments have the capability to rapidly extinguish a large volume of fire that is above the reach of aerial streams. Properly maintained sprinkler systems have proven successful in controlling and extinguishing these fires and protecting building occupants
Compartmentation	Compartmentation in highrise buildings is important to help contain a fire, but sprinklers provide the best protection
Evacuation of Occupants	Highrise building occupants rely on the construction of the building and the functionality of the protection systems to protect them during a fire. Occupant evacuation training is essential to a workable response to an emergency. Such training should include existing drills, information on the building layout, when to stay in place, and so forth

Fig. 7–3. High-rise operational considerations—summary of key issues

Source: Information extracted from the U.S. Fire Administration. 1996. Special Report: Operational Considerations for Highrise Firefighting. USFA-TR-082. April.

High-rise SOP and ICS components

The first issue that needs to be addressed, according to *Operational Considerations for Highrise Firefighting*, is the establishment of a high-rise operations policy. Although departments are of various sizes and the hazards protected vary by community, certain issues still must be addressed during a fire incident at a high-rise building. The report lists some of the following issues as key to any policy:[17]

- **The level of response.** The policy should determine the number of resources and the type of resources that are needed on the first, second, third, and subsequent alarms. Some departments may have specialized resources, such as a high-angle rescue team or a helicopter unit.
- **The duties of the incoming units.** This will vary by the number and time of arrival of the firefighting forces dispatched to the incident. One of the first tasks that must be accomplished in a high-rise operation is determining what floor the fire is located on and the extent of the fire. High-rise fires will not be visible from the exterior of the building except in large fires or very poorly constructed buildings. Both are key issues to the safety of the responders.
- **Establishment and structure of command.** The utilization of a NIMS-compliant ICS system is important on any fire, but as with any disaster, the number of resources needed will necessitate a solid command structure. There are also divisions and groups that are specific to high-rise firefighting operations. Some of these are given in the following:
  - *Command.* This person is often on the first floor of the building so as to have access and a working space. In light of the 9/11 attacks, it can be seen that there may be some controversy on whether to establish an interior command post or to be outside the building. The author still believes that under most conditions, it is best for the commander to be located on the first floor. Some high-rise structures have fire command rooms, such as the one pictured in figures 7–4 and 7–5.

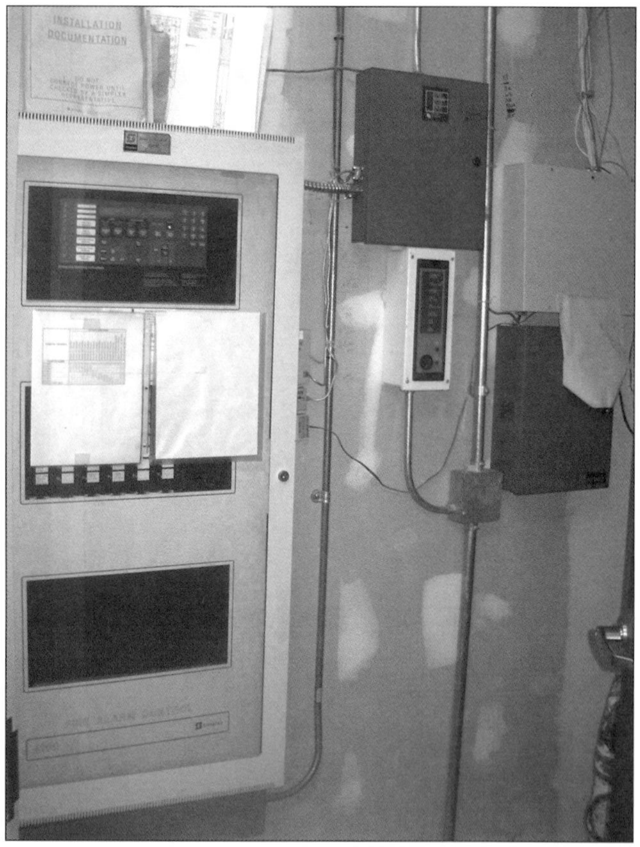

Fig. 7–4 and Fig. 7–5. Fire command room in a high-rise structure. This room is located on the first floor and contains the alarm system panel, communications equipment, and public address system.

Photos courtesy of associate author Randall W. Hanifen.

- *Lobby control.* Controlling the elevators and utilizing the alarm and information systems built into the building. Many structures are now being assembled with a command room, in which public address systems, elevator status panels, alarm panels, and built-in communications systems are controlled from a central location. This person should also assign a group to ensure that any fire pumps are working correctly. Pictured in figures 7–6, 7–7, and 7–8 are an elevator status panel, a high-rise communications system that utilizes phone handsets, and the phone handset jacks that are built into each floor of a high-rise building.

Fig. 7–6, Fig. 7–7, and Fig. 7–8. Certain high-rise buildings are constructed with the following components: an elevator status panel, a high-rise communications system that utilizes phone handsets, and phone jacks on each floor to communicate with the fire command room.

Photos courtesy of associate author Randall W. Hanifen.

- *Fire control groups.* It is important to establish fire lines on the floor of the fire, the floor above the fire, and the floor below the fire. Based on the built-in fire protection and its design, these lines could be attached to the standpipe system or may require that firelines be advanced up interior stairwells.

- *Evacuation groups.* Evacuating an entire high-rise building is complicated and chaotic at best. A systematic evacuation can be attempted, but many of the occupants who are capable of leaving will do so upon recognizing there is a problem. Ensuring evacuation occurred on the fire floor, the floor above, and the top two floors of the building will ensure that the occupants in the most danger of succumbing to fire or the products of combustion will be removed from the toxic environment. As noted in figure 7–9, the stairwells are not large enough to accommodate evacuees and fire department operations. This problem will need to be identified and addressed by the evacuation group leader.

Fig. 7–9. Stairwells in high-rise buildings are often narrow, which can hamper simultaneous evacuation and hoseline advancement.

Photo courtesy of associate author Randall W. Hanifen.

- *Interior staging group.* Because of the distance from the ground level and the energy and time needed to ascend the distance to the fire floor, an interior staging area is established in order to provide rehabilitation and fresh air to firefighters. This interior staging area is also staffed with the rehab group and is supported by the support group, which will deliver spare equipment and air cylinders to the crews.
- *Ventilation groups.* These groups will vary from the traditional firefighting ventilation, in that horizontal ventilation is nonexistent, and the vertical ventilation must use the building's structure in order to accomplish the task. The usual tasks of this group are to pressurize the stairwells and ensure that the hatches at the top of the stairwells are open. They are also to control the HVAC, which can severely intensify the fire, especially if the building has plenum construction.

- **Size-up and verbal communication.**
  - *Relay important information.* Because the commander will be blind to the incident, it is important to relay good information from the operating fire companies to the command personnel. Some of the items that need to be relayed include the following:
    - o The location of the fire and the floor(s) on which it is located.
    - o The actual number of floors in the building, noting any floors that are numbered out of order. It was once common for builders to skip floor 13 due to superstition. This could cause problems if the crew is operating in this vicinity.
    - o Presence or reports of persons trapped.
    - o Whether the building is being evacuated and which floors are evacuating. Some newer buildings have a fire alarm code that allows only the floor of the alarm and the floor above and below to have an active fire alarm. All other floors would be oblivious to a fire condition in the building.
    - o From which side of the building this information is being relayed.
    - o From which stairwell the crews are making an attack. In most high-rise buildings, especially those built today, the design includes more than one stairwell. It is important to note which

stairwell is being utilized in order to prevent opposing hose stream. In the case of the sealed environment of a high-rise, this can cause extreme burns and possible death to a firefighting crew.

- **Water supply.**
  - *Water supply issues are critical to the success of the firefighting efforts.* Water supply issues are capable of causing the biggest problems during operations. As will be shown in the "Lessons learned" subsection, water supply has been an issue on each of the fires. It will continue to be a problem until frontline personnel have an in-depth knowledge of the systems they are operating, and the fire prevention officials work to verify the system's operation and that codes will in fact deliver the water needed during an operation.
    - o Pressure reducing valves. In order to prevent the hoselines from being pressurized to the point of exploding or being unable to be handled, the system designers utilize pressure reducers to limit the pressure and volume on the floors that will be above the regulated pressure, often 175 psi. If these valves are incorrectly set, the crews will not receive the water volume and pressure needed to attack the fire.
    - o Incorrect hose and nozzle configurations. Fire officials, especially operations level personnel, must have an understanding of the water pressure available through standpipe systems. Furthermore, they must design the hose and nozzle configuration to accommodate the water pressure restrictions. Often, the pressure available is 100 psi. Starting with a standard fog nozzle that needs 100 psi at the nozzle will not work due to the friction loss in the hose. Use of smooth bore nozzles or a breakaway nozzle is best.
    - o Outrunning the sprinkler/standpipe combination. In the newest buildings that do not exceed a certain height by fire code, a combination sprinkler and standpipe system can be installed. When this occurs, the standpipe is a branch of the floor's sprinkler system, and thus the water capacity is for both the sprinklers and the standpipes. Utilization of too much water from the standpipe will reduce the effectiveness of the sprinkler system. It is important to consult local building and fire codes to determine the specifications of the local systems. Figures 7–10 and 7–11 depict clues for responders that a combination system is present. Note the combination system writing on the fire department connection (FDC), as well as the branch off of the standpipe that feeds the sprinklers.

Fig. 7–10 and Fig. 7–11. This FDC states it serves a combination system. If the FDC is not labeled, firefighters should examine the standpipe in the stairs to see if a branch line comes off of the standpipe to service the sprinklers on each floor.

Photos courtesy of associate author Randall W. Hanifen.

- Contingency plans for water supply. Many executives say, "It's not a problem, it's sprinklered." Although sprinklers would have prevented many of the large high-rise fires, one must plan for the system to have a malfunction or be disabled in some fashion. Contingency planning could involve advancing hose from the ground level to the fire or using prepiped aerial devices to connect hoselines to upper levels of the stairwell.

- Firefighters, fire officers, and fire executives must read, understand, and participate in NFPA standards. Such standards include NFPA 13, 14, 20, and 25, which are the *Standard for the Installation of Sprinkler Systems, Standard for Installation of Standpipes and Hose Systems, Standard for the Installation of Stationary Pumps for Fire Protection*, and *Standard for the Inspection, Testing, and Maintenance of Water-Based Fire Protection Systems*, respectively. A class offered through a residency at the National Fire Academy, "Fire Protection Systems for Emergency Operations," is a great class that will explain many of these standards and how to apply them to the operation.

- **Elevators.** This is often the only way to effectively deliver personnel and equipment to the levels of the fire. However, improper use or use at the wrong time can become disastrous for firefighters. SOPs should address the following:

  - *The proper procedure for utilizing the elevators*, specifically, how to utilize the firefighter controls.

  - *When to utilize the elevators.* Some departments have a cut off of when the stairs will be utilized. Others specify how many floors below the fire the elevator will be used.

  - *Who will take possession of the elevator?* Some policies dictate that a firefighter will remain with the elevator to control it at all times. Figures 7–12 and 7–13 show the firefighter's service key and the instructions that are present in buildings with a firefighter's service for the elevators.

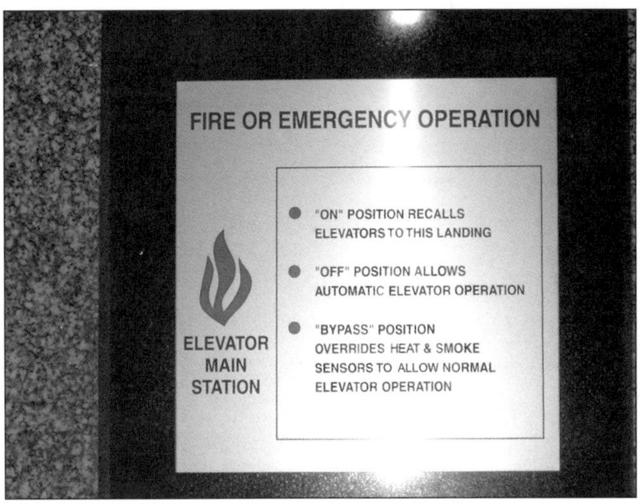

Fig. 7–12 and Fig. 7–13. These photos depict the firefighter's service for elevators. Understanding how to utilize this system will enable the fire department to utilize the elevators during their firefighting operations. Responders should consult their SOPs on the utilization of elevators during firefighting operations.

Photos courtesy of associate author Randall W. Hanifen.

- **Evacuation and stairwells.** One of the biggest problems in a high-rise fire is the simultaneous use of the stairwells for fire suppression activities and occupant evacuation. Many training materials have attempted to direct firefighters to establish one stairwell for evacuation and another for fire suppression. This does not work due to the occupants leaving via the closest exit. Another problem would be people delaying their evacuation, believing there is not a true emergency. Post-9/11, this has diminished due to people seeing the results of delaying their evacuation. Some options to alleviate this are the following:
  - *Utilize the public address system.* The public address system within the building can be used to announce who needs to evacuate and where they need to go to evacuate.
  - *Assigning a firefighting company to assist and control an evacuation.*
  - *Only evacuate three floors at a time.* SOPs such as this keep the high-rise's stairwells from becoming too crowded during an evacuation.
- **Safety.** Although the 1996 report states that few of the policies the consulting firm reviewed addressed safety, it is of the utmost importance in a large-scale incident, such as a high-rise fire. Some suggestions for providing safety oversight and issues to observe are as follows:
  - *Coordinated attack.* As stated previously, the opposing stream attack can be deadly to operating forces.
  - *Security and safety of the elevators.*
  - *Falling debris from the buildings.*
  - *Properly operating building systems.* This could include standpipes, pressurized stairwells, and fire pumps.

- *Signs of fatigue in the firefighters.* These are physically intensive fires that may have a long duration. Proper rehab and rotation of personnel is important. Also, monitoring their health will prevent overexertion.

- *Ensuring proper accountability of all operating forces.* These events can utilize a great number of firefighters, and tracking all of them is vital.

- **Helicopter operations.** Although few departments have the resources available for this operation, it is an area to assess if they are going to operate at the scene. Ensuring they are not fanning the fire is important. In addition, the pilot must be cognizant of the updrafts that can be created by the smoke to ensure that the helicopter does not crash.

As is evident, there are numerous items that need to be addressed in an SOP, but merely writing all of these issues down on paper is just the beginning of ensuring that successful high-rise firefighting will occur. There are many other issues as well, such as the preplanning of specific buildings, training firefighters to perform the actions in the high-rise SOP, and for them to have a full understanding of how each system and construction feature will perform in a building during fire conditions. If the first plan of action fails, alternatives can then be derived. Furthermore, one must know the building to ensure that all of the systems are actually operating.

Pre-incident planning for high-rise structures

By understanding many of the issues brought about in the high-rise policy, the first-due companies, in cooperation with the fire prevention bureau within the fire department, can begin to develop a preplan of the building. Many departments have drawings of the buildings, which they term *preplans*. Although this is far better than no information at all, the drawing is just one of the components of the preplan. Figure 7–14 is an example of the drawing portion of preplan.

According to Michael J. Serapiglia, SFPE, author of chapter 6 of section 7 of the NFPA *Fire Protection Handbook*, a preplan should contain several elements:

- Building Construction
  - Wall construction
    - o Compartmentalization
  - Roof Construction
  - Ceilings and Attics Specifically Plenums in High-rise buildings)
  - Floor Construction
    - o Means of Egress
    - o *Number of staircases*
    - o *Number of elevators and floors of crossover for banked elevators, known as sky lobbies*
  - Vertical Openings
    - o *Smoke venting from stairwells*
    - o *Pressurized stairwells*
  - Horizontal Openings
- Occupancy
  - Type
  - Life safety Concerns
    - o Evacuation plans established by the building management
  - Operational Hours
- Protection
  - Automatic Sprinklers
  - Water supplies
    - o Fire pumps and their control locations
  - Special Protection Systems

- *Standpipes*
    - o *Zoned or all one system*
    - o *Pressure reducing devices*
    - o *Sprinkler system division*
    - o *Location of FDC and/or standpipe connection*
    - o *Pressure to feed the connections*
- Fire Alarm System
    - o Public address capable, or a separate system

- Site Considerations
    - Access
        - o Key card entry
        - o One-way doors, reentry from stairwells
    - Security
    - Exposures
    - Utilities
        - o Elevator controls and elevator mechanical room
        - o HVAC controls and overrides
    - Environment
- Outside Assistance[18]

(*Note*: Items in italics are additional items added by the authors of this text. The authors believe that these additions are important, specifically to high-rise structures.)

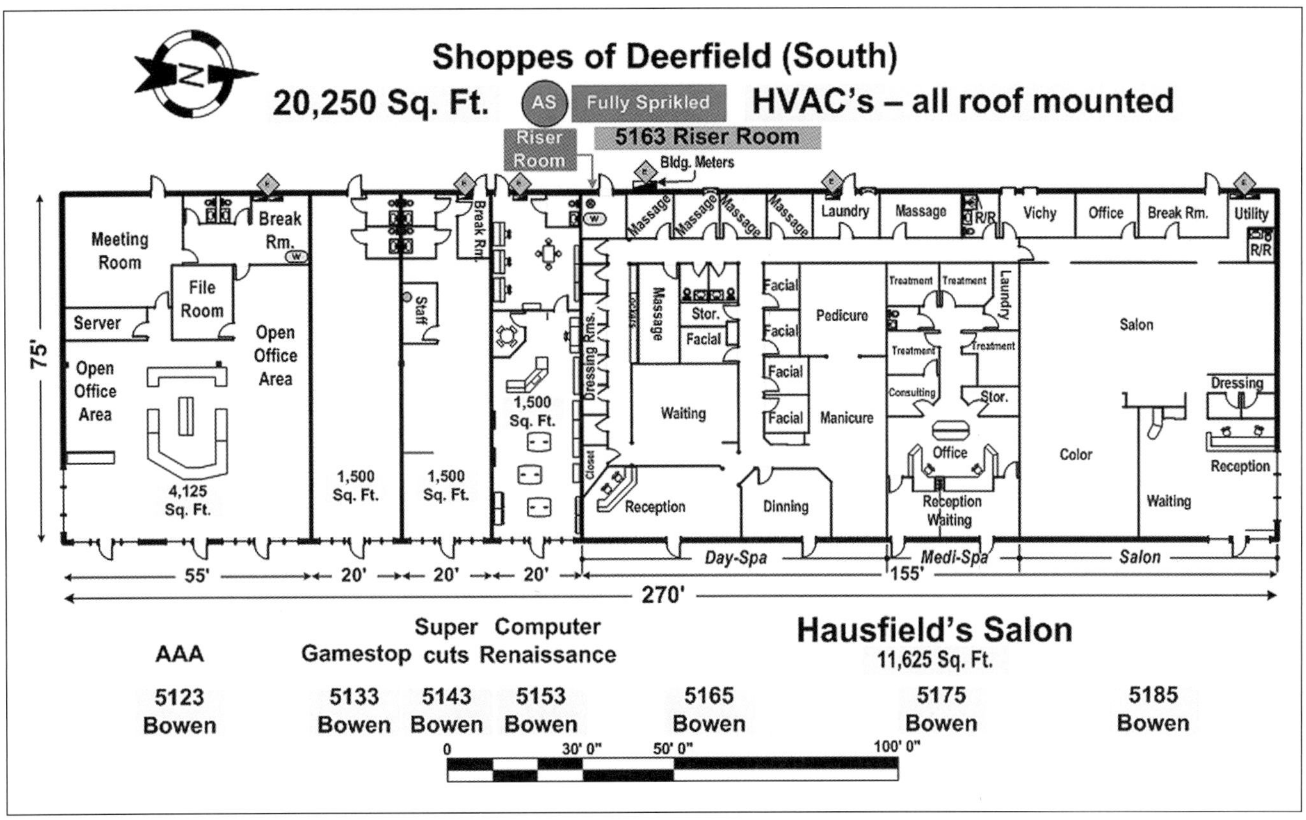

Fig. 7–14. The drawing portion of a preplan for fire department use. Because of the wide variety of buildings that responders can be expected to encounter, it is imperative that a preplan program be implemented to gather important building and response information prior to the incident.

This is a good template from which to begin the preplanning activities of a specific high-rise building. However, more detailed information can be found in NFPA 1620, *Recommended Practices for Pre-Incident Planning*, and in *High-Rise Fire and Life Safety* by John T. O'Hagan, which is published by PennWell Publications.[19,20]

Training for high-rise operations

Training will likely make the difference between a successful firefighting operation and an unsuccessful one at a high-rise building. Due to the very low frequency of actual fires within these buildings, firefighters may not be accustomed to operating within the environment and being able to adjust their fire attack strategies and tactics to the unique structure of a high-rise.

The first and most important operation that fire companies must drill on is the deployment of hose into the fire area. Because this fire attack line cannot be deployed from the fire apparatus, it must be carried into the building and attached to the standpipe system. This involves coordination among the fire company members to carry the hose to the connection and flake the hose out in a fashion that will allow advancement from the staircase into the fire floor. Companies must also be proficient in advancing hoselines up the staircases in the case of overtaxing the standpipe system. This can be a labor intensive and timely operation.

The second operation that must be covered in a drill is the operation of the elevators. As discussed, the elevators are crucial to delivering personnel and equipment to upper floors. This training should cover how to operate the elevators on the firefighter service mode, how to recognize signs of compromised safety for the elevator, and how to enact the particular policy on the utilization of the elevators.

The last issue is to ensure that all firefighters understand the specific components of the building. This includes the standpipes, pressurized stairwells, the compartmentalization, and the distance from the standpipes to the fire. It also includes how the fire codes coincide with these distances, how the zones on a standpipe system operate, and how the fire alarm, public address system, and evacuation plans assist the firefighting operation.

Examples of problems encountered at past high-rise fires

Many of the issues discussed in this section were found to affect the overall firefighting and rescue operation present at the World Trade Center on 9/11. According to the report *NIST NCSTAR 1-8: The Emergency Response Operations*, downloaded from http://wtc.nist.gov/oct05NCSTAR1-8index.htm, the following were problems encountered at the 9/11 WTC-1 and WTC-2 operation:[21]

- Overloaded command due to the sheer number of companies arriving at the incident.
- The loss of the OEM command center hampered unified command operations.
- The lack of an established unified command resulted in valuable information not being distributed to all responders.
- The collapse of the towers completely destroyed the current incident command system being utilized by FDNY and other responders.
- The interior public address system was severed when the plane crashed into the building.
- The evacuation was not systematic and overcrowded the stairwells.
- The lack of foresight in the construction of the stairwells prevented the ascent of firefighters due to the descent of building occupants.
- Loss of clear and understandable radio transmissions, which contributed to the following:
  - Loss in accountability of responders.

- – Inability to relay vital information, such as impending collapse and the crash of the second plane.

- – Lack of clear and concise orders.

- Loss of the elevators on the upper floors due to plane crash.

- The standpipe loss was not examined, partly because the time of ascent was greater than the duration of the towers standing prior to collapse. In examinations of the event by television shows on the Discovery Channel, it is theorized that these would have become inoperable due to severing during the crash, as was the sprinkler system disabled due to the crash.

Hotel disasters

Of the greatest life safety concern are fires within hotels and motels. As discussed in the previous section, high-rise buildings lend their own set of problems. Coupled with a high occupancy load that may be asleep and/or has a delay in recognizing there is currently a problem in the building, it is evident why hotel fires can become disasters. This was the case with the disaster that faced the Las Vegas Fire Department on November 21, 1980. According to the National Fire Protection Association,

> The second-largest loss-of-life hotel fire in United States history took place on November 21, 1980, at the 26-story MGM Grand Hotel and Casino in Las Vegas, Nevada. This fire, which killed 85 people and injured almost 700, provided a wake-up call for the industry to improve fire safety standards in hotels around the country...
>
> About 5,000 people were in the resort when the blaze started to burn in earnest. Many were trapped in their rooms, in the corridors, and in stairwells, and most of the victims died at the scene or in Las Vegas Valley hospitals. The investigation determined that the rapid fire spread was due to a series of installation and building design flaws. A wire at the point of fire origin that had been improperly grounded could've been discovered had the area been inspected. A compressor wasn't properly installed. A piece of copper wasn't insulated correctly. A fire alarm never sounded. A stairwell that was a crucial escape route filled with smoke. The laundry chutes failed to seal, and defects existed in the heating, ventilation, and air-conditioning systems. All of these factors contributed to the spread of smoke.[22]

This fire was followed by a second tragic fire in Las Vegas. On the night of February 11, 1981, "just 81 days after the devastating MGM Grand fire, a blaze started at the Las Vegas Hilton."[23] Ironically, this occurred while it was being retrofitted with more modern fire safety equipment. Firefighters, using the knowledge they had learned from the MGM fire, used local television networks to notify people to stay in their rooms and not go out into the halls and stairwells. In the authors' opinion, many of the lessons learned likely aided operations; thus only eight people died in this fire compared to the 85 people who died in the MGM Grand fire.

Based on these tragic fires in the early 1980s, one would think that this type of catastrophic fire would be a problem of the past. Unfortunately, on November 1, 2006, "Six people died and another was in critical condition today after a fire raced through a residential hotel in downtown Reno, Nev., late Tuesday night. Dozens of others were injured. Police arrested a woman who was a resident of the hotel today on charges of arson and murder in connection with the fire, which largely gutted the 84-year-old Mizpah Hotel. She was identified as Valerie Moore, 47."[24]

These tragedies continue to occur on a regular basis and can have disastrous effects. Planners and fire officials can have a marked improvement on the outcome or prevention of these fires by employing the following actions:

- In conjunction with state fire officials, conduct regular fire and life safety inspections of the properties.
- Insist on sprinkler ordinances that are retroactive to ensure that all hotel occupancies are fully sprinklered.
- Preplan each of these occupancies to ensure that adequate building construction features and built-in suppression will aid firefighting companies in the rapid control and extinguishment of an active fire.
- Work with hotel management to ensure their emergency procedures support the rapid detection and evacuation of the structure, in accordance with the design of the building. The associate author has witnessed hotel employees resetting the alarm upon its sounding because they did not want to wake up the guests.
- Ensure that an adequate amount of firefighting resources are summoned to extinguish the fire, ventilate the structure, rescue trapped occupants, and assist hotel guests to evacuate. Barring large cities, this cannot be accomplished by most jurisdictions' resources, so mutual and automatic aid is a necessity.

With these planning and response criteria, a jurisdiction can make attempts to prevent most fires, as well as quickly control and evacuate occupants if a fire does occur. These are often very chaotic fires due to the vast number of occupants involved, as well as the number of firefighting resources needed for the incident. Strict uses of the ICS system will aid in establishing order in the midst of the chaos.

NURSING HOME AND HOSPITAL FIRES

Although many more fires occur in nursing homes, it is fitting that these two occupancies be placed together in terms of planning for and responding to fires. The true problem in both of these occupancies is that many of the occupants are unable to rescue themselves. In addition, they have machines and medicines that are preventing their deaths under normal circumstances, much less after being exposed to fire conditions.

History has shown, as reiterated in *Fire Officer's Guide to Disaster Control,* 2nd edition, "About 2,000 fires occur in nursing home medical facilities in the United States each year. Hardly a year goes by that a fire disaster does not strike one of these occupancies that are largely populated by elderly people."[25] According to the National Fire Protection Association (NFPA), "In 2003 in Hartford Connecticut, 16 people were killed in a nursing home fire. This is nearly 11% of the 148 patients in the facility at the time of the fire."[26] Because of the constant fire and related death problems in these facilities, the NFPA has issued the following emergency procedure advice for staff at nursing homes:

It is important to realize that people living in nursing homes may not be able to evacuate because of mobility or other disabilities. Proper planning, training, and practice of all staff are essential in order to provide for the safety of residents. It is important for staff to know that patient safety is their number one priority.

Adequate planning

- Every facility should have written fire procedures that are understood and practiced by all staff. Staff should be responsible for knowing and carrying out their part of the plan. That includes doctors, nursing staff, kitchen staff, maintenance, volunteers, and others.
- Response procedures should be practiced regularly.
- There should be a clear "code word" agreed upon beforehand for the facility to alert other staff in case of fire.

Quick response

- Call out the code to alert staff.
- Activate the fire alarm.
- Evacuate everyone in immediate danger.
- Close doors to contain smoke and fire.
- Once the fire is contained to the room of origin behind closed doors, never reopen the door or reenter the room to extinguish the fire.
- Close all doors to patient rooms.
- In evacuating, make sure no patient is left behind.[27]

By having the staff at the facility trained in these procedures, coupled with ensuring that all fire and life safety codes are within compliance on a regular basis, a community can be better prepared for a fire in one of these facilities.

Factors responsible for death or injury

This chapter has examined a variety of fire emergencies that can become catastrophic and have a history of producing numerous deaths and injuries. From all of the various types of fires, the authors have compiled a list of 10 factors that have contributed to deaths or injuries in fire disasters. They are:

- Attempts to extinguish incipient fires without notifying the fire department, resulting in a delayed alarm.
- Failure of occupancy management to alert employees, customers, patrons, guests, and residents at the first sign of smoke or fire, so that planned evacuation can take place.
- Failure of customers, patrons, guests, and residents, upon seeing smoke or fire, to sound an alarm and alert others of the danger.
- Failure of customers, patrons, guests, and residents to use the designated means of egress.
- Failure of these persons to close doors behind them to help confine the fire.
- Actions taken by some persons to break windows, permitting air to enter and intensify fire.
- Lack of confidence of building occupants that the fire department will rescue them, so they jump out of windows.
- Combustible interiors (walls, ceilings, carpeting, furniture).
- Inadequate exits (inward swinging doors, open stairs, revolving doors, lack of panic bars, unlighted, insufficient number).
- Absence of automatic fire detection and fire extinguishing systems (which would prevent many of the previous factors).

Table 7.1 Fire Defense Planning by Government Division

Fire Defense Planning by Government Division

Issue	Fire Department	Municipal Admin.	Planning & Zoning	Water Department	Building Department
Adopt & Utilize NIMS	Adopt, Plan, Train, Obtain resources according to NIMS	Adopt, Train, check compliance	Train to IS 100 & 700	Train to IS 100 & 700	Train to IS 100 & 700
Planning for Extended Operations	Proper forces available, Request Mutual Aid	Passage of Disaster Proclamation to support personnel & Logistics		Calculate water capacity & identify valves to supply scene for extended periods	
Learning from the Past	Research past events and determine risk hazard present in the community	Support research and findings from FD. Pass legislation to prevent similar occurrences	Support research and findings from FD. Review current zoning code for improvements	Calculate supply vs. demand based on the research provided by FD.	Ensure codes are updated regularly to prevent subsequent tragedies. Proactive Code Writing
Improving Fire Control	Provide recommendations for improvement based on hazard analysis	Support recommendations from FD. Approve grants to offset costs		Provide personnel to scene to improve water flow	Provide personnel to scene to aid safety officer
Creation of Barriers	Supply Planning and Zoning information on fire breaks	Support land use designations that provide fire breaks	Ensure land use plans include fire break considerations		Ensure Building codes had adequte fire separations
Ensuring adequate water sources	Apply applicable standards and codes to determine water supply adequacy	Support water supply installation based on applicable codes and standards	Ensure zoning codes do not conflict with adequate water supplies and appliances	Work collaboratively with FD and building department to create an adequate supply	Ensure that buildings are constructed with adequate water supplies and construction features
Establishment of a Planning Section	Participate in a planning committee of Admin, Planning & Zoning, Water & Building Depts. To plan for fire defense	Participate in a planning committee of Admin, Planning & Zoning, Water & Building Depts. To plan for fire defense	Participate in a planning committee of Admin, Planning & Zoning, Water & Building Depts. To plan for fire defense	Participate in a planning committee of Admin, Planning & Zoning, Water & Building Depts. To plan for fire defense	Participate in a planning committee of Admin, Planning & Zoning, Water & Building Depts. To plan for fire defense

PLANNING FOR CATASTROPHIC FIRES

As stated by Dwight D. Eisenhower, "In preparing for battle, I have always found that plans are useless, but planning is indispensable." Although the need for written plans has been demonstrated throughout the text, it is also important to realize that the planning process is much more important. If responders are only reading a plan of how to enact a response to a particular disaster, it is likely that they will have little success in the operation of the plan. They will also have no idea if the plan is working or not until it is too late. Table 7–1 illustrates what planning steps are needed by each government agency for fire defense planning.

Table 7–1. Fire defense planning by government agency. Planners and executives must realize that fire defense planning is not limited to the fire department and must involve numerous key participants if a successful plan is to be created and implemented.

So what planning must be done? Pre-incident planning, especially for fires and the associated buildings or areas, is the best activity that all levels of the fire department personnel can undertake on a regular basis. As discussed previously in the chapter, preplanning is more than just a drawing of the building, but rather a full plan for emergencies at a particular site or area. A preplan will do the following:

- Reduce the command burden to manageable portions.
- Help to improve the quality of on-scene mitigation decisions.
- Communicate to all involved agencies their role within the control effort.
- Provide a lasting framework from which the mitigation effort can be tailored to a specific incident.

These are explained in further detail in the following text:

1. **Reducing the command burden.** Establishing command components that are likely to be used at an incident, as well as what tasks those positions will accomplish, will alleviate the incident commander from the need to micromanage the incident. This is, of course, coupled with training on the plan. It should be noted that the current trend is to establish policies for operations at certain types of occupancies and provide a generalized plan that identifies site-specific differences. This will prevent a "tunnel vision" approach to trying to adhere to a certain plan when adjustments are needed.
2. **Improving quality of on-scene decisions.** Having command personnel participate in the planning process and conducting exercises with the given plan will reduce the command personnel's newness to the operation. When a critical decision is needed, research shows that humans will revert to their training. This is very true in emergency operations with emergency personnel.
3. **Communicating to all agencies their role within the command system.** The use of unified command has proven to enhance operations at the scene of any incident. However, if the agencies involved are unsure of their responsibilities, efforts either may be duplicated or forgotten. Matching these responsibilities in the preplan should closely reflect the duties outlined for each organization in the municipality's emergency operation plan (EOP).
4. **Providing a mitigation framework.** Although each fire is different, there are specific components that must occur at each incident. Providing policies that dictate these operations, coupled with particulars of the site or building, will help all responding personnel understand what needs to be accomplished in order to successfully mitigate the incident.

Critical factors in planning

The following is a list of issues that have historically proven important in planning for dealing with catastrophic fires:

1. **Never underestimate the speed of the fire.** Estimating the speed at which a fire will spread is an important talent of an incident commander. If the commander is constantly sending his forces to chase the fire rather than getting ahead of the fire, it will be a long day on the fireground. This estimating process is very dependent upon the building or site of the fire. As it pertains to wildland fires, the topography will greatly affect the speed. On the other hand, structure fires are controlled by the construction of the building and the combustibility of the contents and structure.

2. **Do not assume that there will only be one fire at time.** The history of wildfires shows that these fires can multiply by flying brands and wind shifts. In the case of civil disturbances, many fires can occur simultaneously. (Civil disturbances are examined in detail in chapter 12.) In a high-rise building, fire can leap floors by lapping outside the building, producing multiple fires. If the resources available are only able to handle what is currently occurring, it is likely that the firefighting effort is already behind.

3. **Do not underestimate the time it takes to muster a massive attack.** One of the problems associated with many drills and exercises related to command incidents is their inability to utilize real-time scenarios. Often these trainings do not force the incident commanders to wait on the response time of companies, thus giving them the sense that resources are available instantaneously. When summoning additional forces, it must be considered that the response time includes the request processing time, dispatch time, turnout time, and travel time. All of these components add up to the total time it takes between the request and the actual performing of the requested action.

4. **The chief should not assume that he or she will be at the incident.** Many chiefs have a tendency to believe that they should have all of the knowledge, as they will be the ones in control of the incident. Often this will not be the case, and the information needed will not be available. Continuity of operations plans have become commonplace and should be adopted in every government organization.

5. **Do not assume the pre-incident planning will work as expected.** Although preplanning is the best nonemergency function emergency response personnel can perform, planners and command personnel should not become blinded by their plan. Inability to determine the effectiveness of the plan and adjust accordingly will result in a failing operation. Commanders must always have a backup operation plan in mind and planned for at any incident.

Improving operations

Planners and emergency response executives must use a holistic approach to improving operations. Often emergency response executives want to bolster the number of firefighters, the number of firehouses, and the amount of apparatus, but they fail to examine and improve inadequate water distribution systems. There are certain planning actions that can improve the overall ability to prevent or adequately respond to catastrophic fires. These are the following:

- **Adopting and utilizing the National Incident Management System.** Adopting and utilizing the system will allow responding personnel, regardless of their jurisdiction, to operate within the incident and have a common understanding of their responsibilities during the incident.

- **Planning for extended operations.** Many emergency response operations are of short duration, thus leading planners to think that all incidents will only last for a few hours, at most. Most catastrophic fires extend at least one day in duration. Plans must address summoning resources to handle multiple operational periods, as well as providing logistical support for the numerous personnel involved.
- **Learning from the past.** There is much to be learned from disasters of the past. Often the problems encountered at a particular incident are repeatable in another jurisdiction. Studying what went wrong or right and addressing the issues that led to the problem or to a successful response will allow planners to prepare appropriately.
- **Improving fire control.** Communities must ensure they have adequate numbers of fire protection forces available to the community. Although no community will be able to handle all emergencies, communities should have forces commensurate to the hazard present within the community.
- **Creation of barriers.** Assisting in zoning codes as well as fire codes that create natural breaks between wildland and urban areas, as well as creating breaks within a city, will aid in creating natural lines of defense against massive wildland and urban conflagrations.
- **Ensuring adequate water sources.** Water supplies are one of the most crucial aspects in fire control. Fire executives and planners must measure the water system against the hazards and ensure it is capable of handling these hazards. Many of the largest fires have progressed due to inadequate water supplies, such as the many wildland fires and many of the high-rise fires.
- **Establishment of a planning section.** This will allow secondary plans to occur concurrent to the current operation. This also will allow flexibility if a plan is not effective based on the circumstances.

CONCLUSION

This chapter has examined the basics of many types of fires and how to plan for and control these various fires. Ensuring that plans for these fires are developed, as well as conducting the drills and exercises to ensure their relevance, will help prepare a community for a large fire. Often the fire service is seen as not doing much because of a lack of fires, but this is often a sign of hard work on the part of fire service personnel toward proper prevention and planning.

REFERENCES

1. U.S. Fire Administration. 2006. Quick Stats. December 28. Retrieved May 20, 2007 from: http://www.usfa.dhs.gov/statistics/quickstats/index.shtm.
2. Kramer, W. M., and C. W. Bahme. 1992. *Fire Officer's Guide to Disaster Control.* 2nd ed. Saddle Brook, NJ: Fire Engineering Books and Videos.
3. National Fire Protection Association. 2003. *Fire Protection Handbook*. 19th ed. Quincy, MA: NFPA. p. 7-21.
4. U.S. Fire Administration, Federal Emergency Management Agency. *Fire and Explosion at Rocket Fuel Plant in Henderson, Nevada*. Retrieved July 15, 2008 from: http://www.interfire.org/res_file/pdf/Tr-021.pdf.
5. Bond, H. 1946. *Fire and the Air War.* NFPA; as quoted in Kramer, W. M., and C. W. Bahme. 1992. p. 189.

6. Office of Civil Defense. 1968. *Fire Aspects of Civil Defense, TR-25*. July. Washington, DC: U.S. Government Printing Office. p. 4.

7. International Association of Fire Chiefs. 2007. IAFC Wildland Fire Policy Committee. Retrieved February 27, 2007 from: http://www.iafc.org/displaycommon.cfm?an=1&subarticlenbr=151.

8. Halton, B. January, 2004. "Montano Complex Fire/Urban Interface Fires." *Fire Engineering* Wildfire Supplement, pp. 3–21.

9. *CNN.com.* 2006. *Wildfires.* Retrieved February 27, 2007 from: http://www.cnn.com/SPECIALS/2006/wildfires/interactive/gallery.2003.fires/content.5.html.

10. *CNN.com.* 2005. *Wildfires.* Retrieved February 27, 2007 from: http://www.cnn.com/SPECIALS/2005/wildfires/interactive/gallery.2004.wildfires/content.1.2.html.

11. *WCPO.com.* 2003. Ohio launches initiative to protect homes from wildfires. Retrieved February 27, 2007 from: http://www.wcpo.com/news/2003/local/10/31/wildfires.html.

12. *CNN.com.* 2004. Fires ravage California. *Wildfires.* Retrieved February 27, 2007 from: http://www.cnn.com/SPECIALS/2004/wildfires/interactive/gallery.2003.fires/frameset.exclude.html.

13. *CNN.com.* 2005. Introduction. *Wildfires*. Retrieved February 27, 2007 from: http://www.cnn.com/SPECIALS/2005/wildfires/interactive/gallery.2004.wildfires/content.1.1.html.

14. *CNN.com.* 2006. *Wildfires*. Retrieved February 27, 2007 from: http://www.cnn.com/SPECIALS/2006/wildfires/interactive/gallery.2005.wildfires/frameset.exclude.html.

15. International Association of Fire Chiefs. 2006. *Advancing Wildland Fire Training for Fire Departments*. p. 1. Retrieved March 1, 2007 from: http://www.iafc.org/associations/4685/files/AWFTimplementation.pdf.

16. Communities Committee, National Association of Counties, National Association of State Foresters, Society of American Foresters, and the Western Governors' Association. 2004. *Preparing a Community Wildfire Protection Plan*. March. Retrieved March 1, 2007 from: http://www.stateforesters.org/pubs/cwpphandbook.pdf.

17. U.S. Department of Homeland Security, FEMA, U.S. Fire Administration. 1996. *Special Report: Operation Considerations for Highrise Firefighting*. USFA-TR-082. April. Retrieved September 2, 2008 from: http://www.usfa.dhs.gov/downloads/pdf/publications/tr-082.pdf.

18. Serapiglia, M. J. 2003. *Fire Protection Handbook*. 18th ed. Pre-incident planning for industrial and commercial facilities. Section 7, chapter 6. Quincy, MA: National Fire Protection Association. pp. 7-85–7-92.

19. National Fire Protection Association. 2003. *Recommended Practices for Pre-Incident Planning NFPA 1620*. Quincy, MA: NFPA.

20. O'Hagan, J. T. 1977. *High-Rise Fire and Life Safety*. Tulsa: PennWell.

21. National Institute of Standards and Technology. 2006. *NIST NCSTAR 1-8: The Emergency Response Operations*. Retrieved September 3, 2008 from: http://wtc.nist.gov/oct05NCSTAR1-8index.htm .

22. National Fire Protection Association. 2001. Looking back…The 1980 MGM Grand fire in Las Vegas. March/April. Retrieved July 15, 2008 from: http://findarticles.com/p/articles/mi_qa3737/is_200103/ai_n8938965.

23. Knapp, G. "MGM Grand Fire's 22nd Anniversary." Retrieved Nov. 2 from http://www.lasvegasnow.com/Global/story.asp?S=190275.

24. *Iklimnet.com*. 2007. Hotel fire in downtown Reno kills six. *Hotel Fire New Journal-29*. Retrieved March 3, 2007 from: http://www.iklimnet.com/hotelfires/hotel_fire_news_29.html.

25. Kramer, W. M., and C. W. Bahme. 1992. p. 209.

26. National Fire Protection Association. 2003. Deadliest fires in facilities for older adults. March 3. Retrieved March 3, 2007 from: http://www.nfpa.org/itemDetail.asp?categoryID=795&itemID=20732&URL=Research%20&%20Reports/Fact%20sheets/Safety%20in%20other%20occupancies/Nursing%20homes/Deadliest%20fires%20in%20facilities%20for%20older%20adults.

27. National Fire Protection Association. 2003. Nursing home safety. March 3. Retrieved March 3, 2007 from: http://www.nfpa.org/itemDetail.asp?categoryID=809&itemID=20757&URL=Research%20&%20Reports/Fact%20sheets/Safety%20in%20other%20occupancies/Nursing%20homes/Nursing%20home%20safety#a.

TRANSPORTATION DISASTERS 8

INTRODUCTION

According to the Bureau of Transportation Statistics, in 2005, the United States was comprised of 3,995,635 miles of roads, 95,830 miles of rail line, and 26,000 navigable channels of water.[1] This amazing amount of transportation routes produced 43,443 highway deaths, 789 railway deaths, 769 water-related deaths, and 616 aviation deaths in the year 2005.[2] When these tragic events occur on a small scale, they usually are handled by only one jurisdiction and are not considered widely newsworthy. However, as was the case with the recent bus crash in Georgia involving an entire baseball team from Ohio, a transportation event can easily turn into a disaster.

Regardless of how many are killed at a time, or the form of transportation involved, an observation made by Dr. Anne Wright Phillips, M.D., as a member of the National Commission on Fire Prevention and Control, will be applicable. Dr. Phillips stated, "The public will look to the fire chief and fire fighter for knowledge and direction and perhaps to blame. Simply because one is a fire chief, one is expected to know how to deal with emergencies."[3]

Transportation accidents can be divided into three general classes: those in the air (landing and takeoff of aircraft), those on the ground (highway and railroad disasters), and those on water (rivers, lakes, and seas). These kinds of accidents may be the result of an impact or collision, the ignition of hazardous materials, or the very threatening potential of a massive fire or an explosion. In many instances, a transportation disaster simultaneously becomes a hazardous materials incident, with additional associated hazmat ramifications. These are discussed in the next chapter.

On responding to a transportation accident, the emergency responder can assume that people will be endangered and that rescue operations might be the first necessary action. Transportation accidents almost always involve human casualties, if only a driver, a few passengers, or railroad or airline crew. The second important consideration for the emergency responder is to estimate how much water, foam, or other extinguishing agents will be necessary for the task. This reality, plus the availability of firefighting and rescue crews, determines how quickly the emergency responder must radio for assisting fire, EMS, and police units to support the operation.

AVIATION DISASTERS

On January 15, 2009, the world was riveted to graphic news about how U.S. Airways Flight 1549, which left Newark, N.J., suddenly lost both engines to a bird strike and successfully ditched in the Hudson River west of Manhattan. As the plane floated, all passengers and crew were rescued, and Captain Chesley B. "Sully" Sullenberger III became a national hero. This kept intact a more than two-year record of no commercial aviation fatalities, but the good fortune was about to end. Less than a month later, on February 12, Continental Flight 3407, also from Newark, plunged into a home in Buffalo, N.Y., killing all 49 passengers and crew, along with a resident in the home.

Most people can easily recall the terrible destruction that occurred when a Pan American Boeing 747 exploded over Lockerbie, Scotland. As a result, 259 passengers and crew on board died, as well as 11 people on the ground. This tragedy, which occurred on December 21, 1988, was traced to a terrorist bomb, signifying the international aspect of disasters and indicating the widespread reach, directly or indirectly, of international terrorism.[4] On September 11, 2001, the terrorism mechanism changed from trying to down the aircraft to using aircraft as weapons. The discussion of this concept is examined in more depth in chapter 13.

Aviation disasters, however, are not new. The *Hindenburg* ended in a disastrous fire on May 6, 1937, while landing at Lakehurst, New Jersey, at the end of one of her transatlantic flights. In 32 seconds, her flimsy bag containing over 7 million cubic feet of hydrogen went up in flames, along with 32 passengers.[5]

The possibility of disaster is not limited to passenger aircraft; it also includes military and civilian cargo planes. It is not confined to fixed-wing aircraft but also includes cargo and passenger helicopters, the capacity of which is increasing.

Types of disasters

Aircraft accidents have occurred because of collisions with other aircraft in the air or on the ground or from impact with tall structures or with the ground. The aircraft may be disabled because of other reasons, such as wind shear, ice on the wings, sabotage, ingestion of birds by the jet engines, loss of a door or engine, chemical spill, bomb explosion in the cargo compartment, or by fire. Figure 8–1 illustrates the probability of survival, expressed in a percentage, based on the angle of impact.

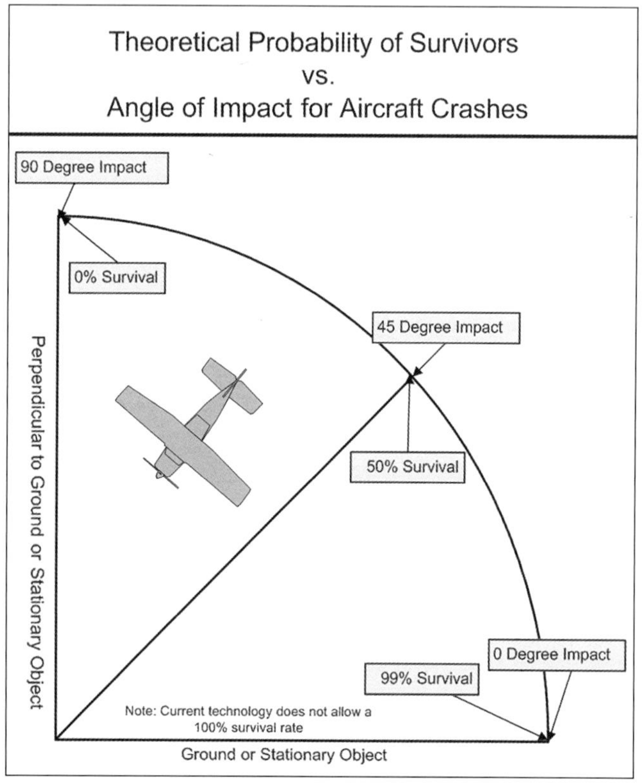

Fig. 8–1. Probability of survival based on angle of impact. Note that a 100% survival rate is not possible based on current technology.

Causes of aircraft disasters

Airport hazards have caused a number of accidents. These include inadequate clearance of operational runways, insufficient navigational aids, and runways that are too short or are obstructed by hills, bays, power lines, blast fences, dikes, or unimproved terrain. In poor visibility conditions, aircraft have struck blast fences and dikes in the undershoot areas off runways.

Overrun portions of runways create another area where accidents occur, owing to inability of aircraft to brake sufficiently to avoid hitting boundary fences or to avoid swamps, bays, or hills. On September 20, 1989, USAir Flight 5050 crashed into the river at the end of runway 31 at New York's LaGuardia Airport.[6]

In some cases, the aircraft failure is not the cause of the airline disaster. This was the case on May 11, 1996 when ValuJet Flight 592, a DC-9 aircraft, departed Miami for Atlanta. At 2:10 p.m., the plane began experiencing electrical problems and requested a return to Miami. Seconds later, shouts of "Fire, fire, fire, fire" were recorded on the plane's cockpit voice recorder coming from the cabin.[7] This crash was later determined to occur due to improperly packed oxygen generators that caught fire, thus causing the crash that killed 110 people. To make this disaster worse, many of the bodies were difficult to recover because of the location of the crash in the Florida Everglades.

Prevention of aviation disasters

While aircraft accidents are noteworthy, especially of commercial carriers involving multiple fatalities, the excellent safety record of the nation's airlines must be kept in perspective. As the number of flights continues to increase dramatically from all major airports, and as the number of passenger miles flown escalates by the millions, the accident rate per million passenger miles flown actually continues to decrease through time. In 2006, the Bureau of Transportation Statistics reported that 505.6 million passengers participated in an airline flight and that 7,066 departures were recorded.[8] These numbers have traditionally grown each year.

One reason why air transportation has achieved such a remarkable safety record is that safety on aircraft starts on the drawing board. Fire detection and extinguishment systems are normally installed to cover areas that have inherent fire potential. Flammable liquid systems are separated from ignition sources, and fire hazardous areas are compartmented from control systems and structural components of the aircraft.[9]

A disaster may be averted in an aircraft-ground impact accident if properly equipped and trained firefighters can rescue passengers before they are overcome with combustion products or are trapped in flames. The ability to evacuate large aircraft rapidly is critical in disaster prevention. This was well illustrated back in 1965 when a Boeing 727 had both main landing gears separate from the plane while landing at Salt Lake City's airport. About 37 seconds after touchdown, the ramp crews were responding to the accident site and doors and window exits were being opened. Even with the heavy, dense smoke that was pouring into the cabin, 50 of the 91 occupants successfully escaped before flames enveloped the whole interior of the aircraft.[10] According to a Department of Defense procedures manual dated December 15, 1994, the response time expectation for ARFF equipment is to make it to any portion of the runways or overruns in one minute after prepositioning or three minutes of an unannounced emergency.[11]

Examples of air disasters

The discussion of disasters in chapter 1 described regular examples of multiple fatality air disasters. In this section, a few of the more noteworthy air disasters of the past several decades are discussed.

On March 3, 1974, near Paris, France, a Turkish DC-10 jumbo jet exploded and crashed in a forest when a cargo door blew off soon after takeoff from Orly Airport. The 346 people on board were killed, making it the worst disaster in aviation history at that time. That record was broken in 1977 when two Boeing 747 planes collided at the airport in Tenerife, Canary Islands; 582 persons died as a result.[12]

As mentioned previously, on September 20, 1989, at 11:26 p.m., USAir Flight 5050, a Boeing 737/N400, crashed into the river at the end of runway 31 at New York's LaGuardia Airport.

Table 8–1. US Air 5050 response apparatus and personnel. The table illustrates the large number of response apparatus and personnel needed to handle an actual aviation disaster.

US Air Flight 5050 Response, Apparatus, and Personnel

Agency	Vehicles	Boats	Helicopter	Personnel
Port Authority of New York & New Jersey	26	1	1	150
FDNY	25	1	-	150
New York Police Department	70	28	-	700
Emergency Medical Services	60	-	-	100
U.S. Coast Guard	-	10	3	200
Department of Corrections	10	1	-	100

Immediately, numerous vehicles from various agencies responded as shown in table 8–1. This large contingent of responding persons and vehicles resulted in a successful rescue of 55 passengers and 6 crew members. There were only 2 fatalities.

On January 25, 1990, Long Island's exclusive North Shore was visited with tragedy when Avianca Flight 52 from Bogotá and Medellin, Colombia, ran out of fuel attempting a second approach to Kennedy Airport. The relatively small Oyster Bay Fire Protection District was suddenly faced with its largest ever emergency operation. What followed was the largest "mutual aid" response in the history of the Nassau County Fire Service. A heroic rescue effort resulted in 89 passengers surviving and being sent to area hospitals; 73 people died upon impact. This was the incident that became newsworthy when drugs from Colombia were found on and inside the bodies of many of the victims.[13]

The fact that more than one-half of the 161 people aboard survived was credited in part to the fact that there was no fire or explosion due to the lack of fuel. The survivals are also credited to the efforts of firefighters, police, and medical personnel from dozens of Long Island communities who assisted in the rescue effort. The most seriously injured were flown out by helicopters, while others waited for ambulances, sometimes up to two hours due to the narrow accessibility along a two-lane highway.[14]

On October 31, 1994, 68 people died as Flight 4184, which departed from Indianapolis Airport and was scheduled to land at Chicago's O'Hare Airport, crashed about 20 miles south of Gary, Indiana. The reports attributed the crash to high winds and driving rains.[15]

On September 22, 1998, a disaster occurred with Swissair Flight 111 (SR-111, SWR-111). The Swissair McDonnell Douglas MD-11 was on a scheduled airline flight from John F. Kennedy International Airport in New York City to Cointrin International Airport in Geneva, Switzerland. The flight crashed into the Atlantic Ocean southwest of Halifax International Airport. The crash site was about 5 miles from shore. All 229 people on board were killed.[16]

According to the National Transportation Safety Board,

On November 12, 2001, about 0916:15 a.m. eastern standard time, American Airlines Flight 587, an Airbus Industrie A300-605R, N14053, crashed into a residential area of Belle Harbor, New York, shortly after takeoff from John F. Kennedy International Airport, Jamaica, New York. Flight 587 was a regularly scheduled passenger flight to Las Americas International Airport, Santo Domingo, Dominican Republic, with 2 flight crewmembers, 7 flight attendants, and 251 passengers aboard the airplane...5 people on the ground were killed, and the plane was destroyed by impact forces and a postcrash fire.[17]

The A300-600, which took off just minutes after a Japan Airlines Boeing 747 on the same runway, flew into the larger jet's wake, an area of very turbulent air. The first officer attempted to keep the plane upright with the rudder. The strength of the air flowing against the moving rudder stressed the aircraft's vertical stabilizer and eventually snapped it off entirely, causing the aircraft to lose control and crash.[18]

Emergency action in air disasters

Aircraft crashes that result in a large number of fatalities are likely to involve passenger rather than cargo planes. They also could involve helicopters of large capacity that shuttle people between airports and the roofs of high-rise hotels and office buildings, or between suburban heliports and major urban air terminals. Such crashes may occur on or off airfields.

Cargo aircraft accidents may present a hazard to firefighters from leaks, spills, or fires involving radioactive materials and other dangerous chemicals. The other factor that emergency responders must consider is the area in which the aircraft has crashed. Many airports originally were located on the outer loops of urban areas. This made them convenient to large urban areas but kept the noise away from the residents. As suburban sprawl has occurred in many urban areas, these airports are now in the middle of highly populated areas.

Maintenance of good communications is essential in any disaster control operation, but it is especially important when aircraft approaching an airport have malfunction problems. Federal Aviation Administration (FAA) personnel assigned to control towers at airports are responsible for directing traffic in the immediate vicinity of the airport as well as movement of aircraft on the ground.

Ground control is one of several frequencies used by the tower. It is used mainly for controlling traffic on the ground and is shared by other agencies in addition to the tower and the aircraft, such as emergency crews, airport operations, and airline operations. To obtain clearance for fire department vehicles to cross runways, acquire information relating to emergencies, and inform the tower of conditions at the scene of an emergency, it is important that the fire department communicate directly with the tower. This can be accomplished by providing fire units with a radio having the ground control frequency or providing the tower with a radio having fire department frequencies.

Although the tower does not dispatch municipal fire apparatus, there should be a direct telephone hotline to fire stations at the airport. This way the tower can provide instantaneous information regarding crash or emergency conditions to all emergency agencies at the airport and the fire department. Upon receiving this information, the fire department dispatcher can send units to handle the emergency without having to repeat the information the tower has given to fire companies. At Cincinnati's fire station #18, which serves the Lunken commuter airfield as well as the surrounding residential area, a red phone connected directly to the airport control tower sits in the alarm room. Affectionately dubbed "The Bat Phone" by the members of this station, it has provided early alert for numerous airport emergencies over the years.

When planning for this hotline, planners could consider the use of a cellular phone kept by the first due apparatus or the incident commander. With the call volume of large metropolitan departments, no one may be present to answer the hotline phone if it is located in the station.

When the tower gives notification of an emergency, it describes the type of plane, its location, and the nature of the trouble. It may supply more information by radio while the companies are responding or are on standby. While keeping the fire department informed, the tower will divert air traffic, close runways, and relay messages from the plane as needed. Information that may be available includes exact time of arrival, wind direction and velocity, number of people on board, type and quantity of fuel remaining, change in runways, and specific facts relating to the plane's problems. Likewise, the tower may relay from the fire department to the aircraft pilot observed hazards such as blown tires, locked wheels, broken struts, fuel leaks, and fire.

Fire department and airport personnel should establish an agreement (preferably in writing) concerning duties and responsibilities among all agencies that would be involved in an aircraft incident. A useful annex to the fire department's disaster plan might be "Emergency Operating Procedure for Airport Emergencies."

Chicago air crash plan

The city of Chicago has adopted a disaster plan that incorporates an example of action to be taken if a jetliner crashes in a congested area of the city. It prefaces its advice with the observation that "in a disaster of this type, the Chief Fire Officer will have a terrific fire fighting problem on his hands that could develop into a spreading fire and exposure hazards and serious loss of life." The example starts with the question, "What action is necessary?" and then advises size-up of the situation: spreading fire–exposure hazards; life hazard–plane occupants, building occupants; and the response that might be needed. Also it advises that the fire alarm office may be requested for necessary response and be given the disaster plan number to be placed into operation (e.g., plan 1, 5–15 victims; plan 2, 15–30 victims; and plan 3, 30 or more victims).[19]

Hazards present at crash sites

The response to some aircraft disasters may involve hazards with weaponry. During the 2004 presidential race, associate author Hanifen worked a presidential detail in a northern suburb of Cincinnati. The president of the United States arrived in the presidential helicopter. This was partially a show for the crowd, but also accommodated logistics of his quick departure. During this visit, the Secret Service provided every security aspect for the president, except crash/rescue response. This aspect is left to the local community fire department. At the briefing, the crew was informed that the rescue of the president was the first priority, and that we could encounter a variety of weapons, but the secret service would not be explicit concerning the contents or locations. In this case, pre-incident planning was not an option.

This would not be any different than would be encountered in most communities that surround military airbases. Although a good portion of accidents are on the property of the military base, it is very feasible for the crash site to exist outside the boundaries, within an adjacent community. Networking and training with the military firefighters that serve the nearby military base can provide responders information on the hazards of typical aircraft that depart and arrive from the airbase. Second, the military is very adamant that their crews handle the aircraft that crashed. Consequently, a good working relationship among military and civilian response officials will aid their strategies and allow their strategy to be incorporated into the overall response plan, which could include fire and fatality problems at the site of the crash.

HIGHWAY DISASTERS

Road transportation accidents may involve autos, buses, and trucks, and they require prompt notification of emergency services so rescue and firefighting can begin at the earliest possible moment. With the prominent use of cellular phones, the summoning of assistance to an accident is faster than ever. Unfortunately, fast-acting news radio and television stations may bring the swift appearance of spectators, and most of them will hinder rather than help. Some spectators could even become casualties in the event of explosion. Some cell phone users have also caused false summons for apparent accidents or emergencies.

Causes of highway accidents

Mechanical failure, adverse weather, and driver incompetence account for a large percentage of accidents on the highway. Multiple collisions involving trains, tank trucks, buses, street cars, and other vehicles (with each other or with stationary objects alongside the road) are not uncommon. These collisions frequently end in fire and a loss of life.

Collapsing bridges, landslides, and flash floods have also resulted in multiple deaths to people traversing the highways. Figure 8–2 shows the destruction present at the I-35 bridge collapse in 2007. Explosions of dangerous chemical cargos in fires have resulted in fatalities among firefighters, spectators, and nearby building occupants.

Fig. 8–2. I-35 bridge collapse. Note the many hazards presented by this disaster.

FEMA News Photo.

Examples of motor vehicle disasters

Not all motor vehicle disasters are the result of fire, but responses by fire departments to motor vehicle fires throughout the United States accounted for 290,000 calls in 2005. Thus, it should not be surprising that a few of these result in very large monetary losses or a large loss of life. The National Fire Protection Association reports that 520 civilian deaths occurred because of those vehicle fires.[20]

Auto disasters

Dense fog that blankets a highway for several miles can be the case of a chain-reaction pileup of motor vehicles of disaster proportions. Such fogs are not uncommon on highways bordering the ocean, but they also develop in inland valleys in certain weather conditions. Without warning, motorists suddenly find themselves enveloped in a medium known as *tule fog*, which can cut off vision beyond the hoods of their vehicles.

On January 2, 1991, 70 automobiles, tractor trailers, and other vehicles were involved in a deadly chain reaction accident that killed 3 people and injured more than 19 on Interstate 215 in Salt Lake City. Like most chain reaction accidents, fog was blamed, which in this case had reduced visibility to zero.[21]

A record-setting accident occurred on April 30, 1975, when nearly 200 cars and trucks were trapped in a series of chain reaction traffic accidents on the Riverside Freeway near Corona, California. This was due to one of a series of fog blankets. These blankets of fog can plague freeway motorists in Southern California during the early morning hours, reducing visibility to almost zero. On this morning, rear-end collisions of cars and trucks caused pileups that extended several miles. The highway had to be closed in both directions for several hours while firefighters fought the resulting fires and helped trapped victims out of the wreckage. No one was killed, but 18 people had to be hospitalized, and property damage was inestimable. According to the National Safety Transportation Board, "The cars and a tremendous number of trucks were piled on top of each other. There were people walking around in panic and in shock just yelling."[22]

This area had been the site of previous mass collisions due to fog from the Santa Ana River. Highway patrol officers had been escorting some motorists minutes before the pileup started, but the heavy fog moved in suddenly, and the crashes began. The officers had posted temporary fog warning signs, and radio stations were broadcasting traffic advisories, but these measures were not enough.

In December 2004, nearly 70 vehicles were involved in an accident on the Ohio/Pennsylvania border. Although no one was killed in the event, 17 motorists were transported to local hospitals. Police stated the wreck was likely due to excessive speed. Because of this incident, the county was placed on a mass casualty alert, prompting doctors and medical personnel to respond to the local hospitals.[23]

Winter storms and crowded interstates are another recipe for large multiple-vehicle accidents. On February 28, 2007, two cities were the site of this combination. In Colorado, 18 people were taken to hospitals, and police closed a 12-mile stretch of I-25 for several hours.

In Washington on this same day, 60 vehicles were involved in collisions, producing at least four serious injuries.[24]

All of the vehicles in these examples were involved in multiple wrecks, and events like these can tax local emergency resources. This is a good example of a time to utilize an area command to delegate resources to the accidents needing them and to divert them from the ones who do not need them immediately.

Bus disasters

When a chain-reaction pileup involves a loaded bus, the loss of life is apt to be larger. Buses by their nature transport numerous potential victims and deserve special considerations and training by emergency service personnel. Buses are often used to transport sports teams, church groups, and tourists, which makes it quite possible that there will be a social bond among the victims. The desire to help loved ones may complicate the event. In addition, those unaffected by the trauma of the accident may succumb to emotional health problems, thus increasing the number of EMS resources needed.

Probably some of the worst disasters involving buses occur when they become consumed by fire, thus fatally trapping the victims. This was the case in 1988 in Carrolton, Kentucky, when a drunk driver hit a church bus, which was a converted school bus. The group was returning from an outing at Kings Island in Northern Cincinnati. The crash killed 27 and injured 34. Only 6 passengers escaped the fireball.[25]

This same scenario, with the exception of the drunk driver, was repeated in October 2003 in Jakarta, Indonesia. A chain reaction accident occurred involving a minivan that struck the rear of a bus. The bus subsequently burst into flames, trapping the passengers. As a result of the accident, 54 school-age girls were killed. In this case, as mentioned earlier, many of the distraught parents were transported to local hospitals as well for emotional-related illnesses.[26]

In a third case, seven young relatives were killed when their car was rear-ended by a tractor trailer, pushing it into a school bus stopped to drop off children. This event occurred in Gainesville, Florida on January 25, 2006. Due to the mass of the tractor-trailer, the crash scene was a mangled mess, creating a difficult rescue effort.[27]

Even emergency responders can become victims of these crashes. In August 2003, eight firefighters died in a highway crash in Oregon while on their way back from wildfire duty. The small van/bus struck a tractor-trailer head on. The crash caused the van to catch fire. Their colleagues, traveling in another van behind them, were the first to arrive at the scene.[28]

One of the most current events involves the Bluffton baseball team's bus crash in Atlanta, Georgia. According to CNN.com, "the charter bus, carrying 35 people...to Florida, had toppled off the overpass and slammed into Interstate 75 just outside downtown Atlanta...Six people were killed. The other 29 were injured, six seriously."[29] According to numerous news reports of the accident, the bus driver was traveling at interstate speeds and mistook an exit as the interstate continuing. This apparently has occurred in the past and is a result of the roadway design in the area. When the driver reached the end of the exit ramp, he struck the adjacent bridge abutment, causing the bus to flip and land approximately 20 feet below on I-75.

Truck disasters

A consideration of highway disasters must include trucks, and among the heaviest of trucks are those used by fire department personnel. One of the more dangerous aspects of the firefighting profession is the response to emergencies along highways by firefighters riding vehicles whose weight is measured in tons. They do not always arrive. In June 1990, two firefighters were killed and another two injured seriously when the brakes failed on a reserve pumper in Waterbury, Connecticut. The pumper crashed into a tree as the firefighters were responding to an automatic alarm at a commercial establishment. Most highway disasters, however, involve other vehicles, creating an array of bizarre emergencies.[30]

Consider the situation confronted by the Memphis Fire Department on Friday, December 23, 1988, when a propane truck accident killed 8 people and injured more than 12 others. Occurring on Interstate 240, just north of the Interstate 40 interchange in Memphis at approximately 10:00 a.m., a 10,000-gallon semitrailer carrying propane overturned on the northbound ramp leading from I-40 to northbound I-240. This was the third fatal incident involving a semitrailer at the interchange.

As the tractor-trailer went around an inclined ramp, it slid approximately 200 feet and struck a bridge abutment. When the front end of the tank ruptured, it released liquid propane, forming a large vapor cloud carried in a northwesterly direction. It traveled 1,500 feet up the expressway before igniting, forming a huge fireball. Vehicles in both the northbound and southbound lanes were caught underneath the cloud. Most occupants died instantly. Occupants of some other vehicles were able to escape, some burned severely, before their vehicles ignited. The incident grew in complexity as the tank shell rocketed and ricocheted from building to building adjacent to the interstate, igniting several structures at different locations.

Suddenly the Memphis Fire Department was faced with four different incidents at the accident site: three structure fires and the numerous automobiles on the highway. This required four separate command posts under an extended incident command system, with difficulties in communication as firefighters scrambled to understand which commander or which incident command post was communicating.[31] This was an excellent example of te need for an area command.

As discussed earlier in the chapter, onlookers (or in this case, thieves) can become victims of the accident themselves. In March 2007, 70 people were killed as they attempted to steal gas from an overturned gasoline tanker. The incident happened in Nigeria, when a truck carrying 8,700 gallons of gasoline overturned and passersby began to steal the leaking gasoline. During their efforts, an unknown source ignited the gasoline.[32]

Ignited gasoline tankers can cause severe problems when they occur in dense urban areas. This was the case in June 1992 in Cincinnati, Ohio. An overturned gasoline tanker burned for more than an hour just blocks from highly populated office buildings. The cleanup took nearly the remainder of a day. Due to the location of the accident, the traffic from three different interstates was affected.[33]

The catastrophic potential of trucks transporting highly hazardous materials, including explosive, oxidizing materials, flammable liquids and gases, poisonous substances, and radioactive materials, is growing. Millions of trucks are on the highways. Observations at I-70 and I-71 interchanges with I-270 through Columbus showed that 47 percent of placarded trucks were carrying flammable liquids.[34]

Another gas responsible for disasters has been ammonia. It is transported in great quantities in tank trucks. Although it is flammable and toxic, the Department of Transportation was slow to require a label or placard to indicate either of these hazards. In May 1976, a tanker loaded with 5,000 gallons of ammonia crashed through a guardrail at the busiest highway interchange in Texas, near Houston, and plunged into another lane of traffic 20 feet below, violently rupturing on impact. Clouds of ammonia gas billowed over the area, causing motorists to flee from their cars. As a result of the accident, 4 people died and 119 were treated at 14 hospitals.[35]

When a truckload of dynamite and blasting agent was rolling along the highway near Unionville, Ohio many years ago, its tarpaulin cover caught fire from its exhaust pipe, and two hours after the driver pulled off the road, it exploded. Thanks to the presence of mind of the driver, no one was killed. However, around the same time, another truck carrying dynamite and ammonium nitrate blasting agent was parked by a lumber and supply warehouse in Roseburg, Oregon. The driver had just completed a 290-mile run from the state of Washington. Having arrived too late to make deliveries at two points outside the city, he locked the truck and checked into a downtown hotel about three blocks from the warehouse. At about 1:00 a.m., fire broke out in the warehouse, and after it spread to the truck, an explosion followed. Thirteen people were killed and more than 125 were injured. The blast leveled the immediate 12-block area, tore a crater 52 feet in diameter and 20 feet deep at the site, and caused extensive building damage over a 50-block area, with glass breakage as far as 9 miles.[36] Although more commonly the proximate cause, it does not require a fire to trigger a cargo explosion. A gun can suffice.

Firefighters can be expected to recognize the danger of fighting a fire involving explosives, unless the warning placards have been removed from the truck, as was the case in Marshall's Creek, Pennsylvania, where four firefighters were killed in an explosives truck blast.[37]

Emergency action for highway disasters

A highway crash involving autos, buses, or carriers of hazardous cargos can require considerable rescue, firefighting, and evacuation operations. To determine the nature and extent of the emergency, it may be necessary to reconnoiter by helicopter, especially if there is apt to be a delay in getting through traffic during peak rush hours. An aerial survey should help to supplement information already obtained from the dispatcher. In any case, it will be necessary to alert all public service agencies as to the nature of the problem, the exact location, and the extent of the damage.

Initial size-up should disclose whether mutual aid forces will be needed. The scene of the accident may be on the outskirts of a community, and since the emergency vehicles will probably all be responding from one direction, it may be desirable to have the next community beyond the accident scene respond from the opposite direction. This is especially important if the emergency is on a divided expressway where emergency response needs to be made from both directions, owing to the infrequency of crossover ramps providing access to the other side. For example, consider the highway in figure 8–3, which depicts the Boston freeway system, known as the "Big Dig." Gathering information for this emergency may involve the use of cameras. In addition, maps of the entire tunnel system are needed to actually locate any emergency.

Fig. 8–3. An entrance to Boston's underground freeway system. During a project known as the "Big Dig," the freeways in downtown Boston were relocated below the city. Planners must ensure that projects such as this encompass a full range of emergency operation considerations.

Photo courtesy of associate author Randall W. Hanifen.

Auto and bus emergency action

Sometimes special equipment is required for accidents involving autos or buses, depending upon the type of emergency. If a busload of people has plunged into a lake, a scuba team could be brought in by helicopter. Boats, cranes, and heavy-duty tow trucks with powerful winches may be useful in retrieving the bus.

An auto or bus having its engine in the front is apt to take a nosedive, with the forward momentum carrying it to an upside down position on the bottom of the lake or reservoir. If the bottom is muddy, the vehicle will sink somewhat, making it difficult to open doors from either side. Some air probably will be trapped inside a tightly closed vehicle, permitting uninjured persons to survive a short while. Soon the pressure equalizes from inflow, but the chances that anyone will have presence of mind to wait until the condition stabilizes and then open a window and exhale while he escapes to the surface is not too great. (If the swimmer fails to let out his breath on the way up to the surface, he is likely to suffer from lung rupture and embolism.)

If a scuba team (or team of skin divers, if the water is shallow) is readily available, its job is apt to consist of attaching hoisting or tow cables to the sunken vehicle, or extricating the dead victims through windows and escape hatches. Firefighters should remember that self-contained breathing apparatus (SCBA) is able to provide air to rescuers up to a couple of feet beneath the water. Although this is not a preferred method of operation, it may suffice until specialty divers can arrive.

For the same reason as given before, a bus that falls off an elevated highway is apt to land upside down, crushing some passengers to death. There will be an immediate need for fire apparatus to safeguard against a fuel spill fire, as well as heavy rescue vehicles with hoists, hydraulic prying tools, power saws, comealongs, cutting tools, and cranes. Ambulances will also be needed, along with law enforcement personnel to provide traffic and spectator control.

Hazardous cargo vehicles–emergency action

In the next chapter, the topic of hazardous materials is covered in detail. If a highway emergency involves trucks or trailers loaded with hazardous materials, considerable caution is required, and a number of questions need to be answered, if possible. The first question, "Who is in charge?" should be resolved under the disaster plan. Does the responsibility for directing operations at a highway emergency fall upon the local police department if the location is within the city limits? Does it fall upon the state highway patrol (or troopers) if it is outside the city limits? In any case, the decision of how firefighters should be deployed will rest with the fire chief or highest-ranking fire officer in that jurisdiction.

The action that the fire chief calls for will depend upon a number of factors. These include the following:

- The type of hazard, such as a leak, spill, or fire, or a combination of these hazards
- The form of the materials, whether gas, liquid, or solid
- The terrain, which could be level or could slope into a river, flood control basin, or storm drains
- Types of exposure (inhabited or not) that may be subjected to heat, poison gas, or toxic products of combustion
- Availability of mutual aid and backup water supplies
- Availability of police assistance to stop traffic, remove spectators, and notify the occupants of exposed structures
- The probable amount of time available before an explosion is apt to occur

It can be difficult for firefighters to identify the hazardous materials that may be involved. There is no guarantee that required identification for hazardous cargos will be on the carrier or will be readily visible. The U.S. Department of Transportation continues to improve its identification system for hazardous materials. Although placarding is required on multiple sides of a vehicle, the wreckage may not make their identification prevalent.

Even if the vehicle is labeled to give firefighters and the public adequate warning of the nature of the hazard, there are situations where the sign or placard may not be visible in dense smoke. The truck driver, if not killed in the accident, may have no knowledge of the hazards involved. In addition, there are exemptions for placarding cargo based on the quantity present. Therefore, the incident commander must assume hazardous materials are present in trucks until ruled out by the shipping papers carried in the cab of the vehicle.

Table 8–2. Recent railroad disasters and their fatalities and injuries, as well as the cause. Any jurisdiction containing rail lines of any type is subject to possible disasters.

Sample List of Recent Railroad Disasters

Date	Location	Deaths/ Injuries	Cause
10/17/2000	England, UK	4/102	Faulty rail shattered due to condition known as rolling contact fatigue
4/18/2002	Crescent City, FL	4/142	Derailment due to hot weather misaligning the track, know as "sun kink"
6/23/2003	Maharashtra, India	51/not reported	Derailment of special holiday train
7/2/2003	Warangal, India	22/not reported	Brake failure, subsequent fall from bridge by train
3/11/2004	Madrid, Spain	191/not reported	Terrorist attack; detonation of bombs on four commuter trains
6/17/2004	Bombay, India	20/100	Monsoon induced land slide, which caused train to fall off of bridge
11/3/2004	Washington, DC	0/20	O.O.S. metro train rolls back into station and strikes another train
1/26/2005	Glendale, California	11/100	Double-deck commuter train collides with vehicle and derails
4/25/2005	Hyogo, Japan	107/549	Train derailment, which caused the train to crash into an occupied apartment building
7/13/2005	Ghotki, Pakistan	150/not reported	Chain reaction crash involving 3 trains due to missed signal by initial train
10/29/2005	Pradesh, India	114/numerous	Track swept away by floods, causing derailment
7/3/2006	Valencia, Spain	41/47	Train was speeding into curve at twice the speed limit and derailed
8/21/2006	Qalyoub, Egypt	35/17	Bridge collapsed onto a train
6/5/2007	Victoria, Australia	11/23	Train collided with truck at railroad crossing
8/1/2007	Benaleka, Congo	100/200 appx.	Passenger train derailment

An incident commander can take certain practical steps under emergency conditions involving hazardous materials. There are suggested procedures for handling a variety of emergency conditions that are described in many current hazardous materials texts. Such situations could include tank truck fires, accidents involving explosives, rocket propellants, cryogenic materials, pesticides, chlorine leaks, ammonia spills, flammable liquid and oil spills on the highway, and LP-gas tank truck emergencies.

RAILROAD DISASTERS

Devastating accidents have destroyed freight and passenger trains. As with highway transportation, these may be caused by human error, mechanical failure, sabotage, or extraneous factors, such as landslides, blizzards, and other storms. Collisions, derailments, and pileups, often followed by fire and explosions, are important factors in railroad disasters, especially when hazardous materials are involved. Table 8–2 depicts recent railroad disasters and their associated fatalities and injuries based on information from online sources.

Freight train disasters

For five days, from July 8 to July 12, 1986, the community of Miamisburg, nestled in the scenic Miami River valley of southwest Ohio, was in a state of emergency. A derailment caused tank cars of phosphorous to ignite, triggering an evacuation of Miamisburg and several surrounding cities. In an attempt to recover its expenses in handling the incident, the city of Miamisburg negotiated for several years with the railroad to recover more than $433,000 of expenses incurred by the city of Miamisburg and responding mutual aid fire departments.[38]

In April 1990, a freight train derailed in Worthington, Pennsylvania, just outside of Pittsburgh. This incident rapidly became a major hazmat fire as six crude oil cars, two coal cars, and one car loaded with sodium hydroxide began to burn and explode. Hundreds of firefighters converged on the accident, and over a two-day period, brought it under control. This indicates yet another dimension of the tactical complexities imposed on the modern fire service. Often it must choose between air pollution from not extinguishing a hazardous fire and water pollution from the ensuing runoff if extinguishment is attempted.[39]

Vinyl chloride escaped from a 48,000-gallon tank car and immediately ignited. Attempts to control the fire with water did not prevent the vinyl chloride from exploding in a huge fireball about 45 minutes later, causing death and injuries.[40,41]

In April 2004, a deadly blast resulted when two trains collided, one carrying oil and the other carrying liquefied petroleum gas (LPG). The site of the crash was in North Korea. Reports of up to 3,500 fatalities were reported in the April 23, 2004 edition of the *Washington Post*.[42]

The fact that such accidents happen should not be surprising in view of the vast amount of dangerous chemicals that are shipped by rail. The Southern Pacific Railroad each year moves over 100,000 carloads of hazardous material in California alone. This includes cargo in boxcars, in trailers on flatcars, and in tank cars. Probably the greatest problem area today, according to the superintendent of the hazardous materials control for the Southern Pacific Transportation Company, is with liquefied petroleum gases and flammable liquids in transit.[43]

Passenger train disasters

On January 4, 1987, one of the most memorable train disasters in recent U.S. history occurred. A Conrail locomotive operated by personnel who tested positive for drug usage skidded through a switch right into the path of a speeding Amtrak train. The wreck of train No. 94–the worst in Amtrak's history–killed 16 people and injured 184, and left a serious question regarding rail safety in America. Rescuers from throughout Baltimore County converged on the scene, including teams flown in from the renowned Shock Trauma Center 15 miles south of the accident. Thirty-seven medic units and 10 helicopters were used to treat the 184 injured passengers. Within hours, the disaster scene was compounded by the expected influx of federal railroad administration officials and numerous law enforcement agencies. Fire service personnel and first responding units operated under a command complicated by the additional agencies involved. As disaster and mitigation become more complex, it can be expected that disaster command centers will include many new agencies and require new dimensions of cooperation and coordination.[44]

Passenger train disaster planning should include subways if present in the community. Figure 8–4 shows how close passengers at the station are in relation to the train. If a derailment were to happen in the station, many people would be at risk.

Fig. 8–4. Subway rail lines operate in very close proximity to the boarding passengers and have little to no safety protection to prevent the passengers and the subway cars from coming into contact with each other. *Photo courtesy of associate author Randall W. Hanifen.*

On August 2, 1999, a train wreck in Gaisal, India claimed the live of 226 passengers, with the possibility of 200 more becoming fatalities. This disaster involved two passenger trains colliding at a high rate of speed (60 miles an hour). Many of the passengers likely never knew what happened, as most were asleep at the time of the crash. According to railway officials, one of the trains was on the wrong track. Although this death toll seems high, it was fortunate that the loss of life was not greater, as the total passenger load for the two trains was more than 2,500.[45]

Like the airlines, however, millions of passenger miles are traveled for each fatality. For example, high-speed Amtrak trains carry 60,000 people between Washington and Boston alone, on a daily basis.

Although Amtrak does have an occasional derailment or accident involving fatalities, its record is relatively safe. There are times, however, when its safety record is questioned. For example, on May 4, 1990, *USA Today* reported that Amtrak had its third accident in 10 days.

The safety record question for passenger trains arose again in 2002, in London, England. According to a *Cincinnati Enquirer* article on May 12, 2002, London newspapers called for a public inquiry after a fatal crash the previous week. In this crash, 7 people were killed and 70 more were injured when a high-speed train jumped the tracks and smashed into the station platform. This was the sixth fatal crash on the public rail system in five years.[46]

Emergency action for railroad wrecks

The initial action and size-up for a railroad accident is about the same as that for a highway accident. However, if the incident occurs in a remote location, an aerial survey by helicopter may be of greater importance than most roadway emergencies where access, though sometimes difficult, is usually obtainable by motor vehicle.

Passenger trains–emergency action

When a train wrecks, the train dispatcher should be notified immediately and be told the accident location and tracks involved. It will be necessary to watch for other trains, and it is practical to put out flag men in both directions until notification is received that the trains have been stopped. It is important that responders immediately request that all electric power be cut off and electrified rails be deenergized, as might be needed in a subway or elevated railway. Confirmation must be received that this has been done.[47] Response agencies should adopt a formal lockout-tagout procedure that mimics *29 CFR 1910.147, The Control of Hazardous Energy*.[48] Appendix A of this standard provides a procedure for complying with the standard. More information on this standard is located at: http://www.osha.gov.

Even though the fire department may share control of such emergencies under a community's disaster plan, the command post should be established nearby. The fire service probably will be requested to supply ladder equipment for gaining access to the top of overturned cars or to supply forcible entry tools, acetylene cutting torches, heavy duty jacks, hydraulic prying equipment, power saws, and lighting equipment. The fire service may also be asked to supply rescue companies for extrication of victims and engine companies to handle outbreak of fire. Planners for the fire and rescue services should consider their territory when considering equipment portability in their purchases of rescue and firefighting equipment.

Preplanning for emergencies along the railroad tracks is a good activity for fire and rescue crews. Many communities do not build homes and businesses directly adjacent to the railroad tracks due to noise. Therefore, using GIS and GPS to locate access points and nearby water sources for firefighting activities will pay dividends when the emergency arises. In the case of subways, it is important to know the location of evacuation routes and built-in fire protection. In the case of elevated railway systems, preplanning access points and places for fire department ladders to access the railways will make an incident that occurs in a difficult access location easier to mitigate.

It may be necessary to disconnect individual cars and get the train moved to perform rescues; elevating platforms, aerial ladders, and cranes may be of assistance under these circumstances. When ambulances are summoned, responders should notify the hospitals where the victims will be taken so that preparations can be made for their reception. Establishment of the transportation officer position within the ICS system will aid in ensuring that area hospitals receive an even distribution of critical and noncritical patients. Overloading a particular hospital only succeeds in moving the disaster, not mitigating it.

A triage system should be used to sort and classify the victims in accordance with the priority needed for treatment. This objective should not be defeated by permitting less seriously injured persons to be whisked off to the hospital by private citizens before the critically injured have been loaded into ambulances and transported to these hospitals. Otherwise, the nearest hospitals will be filled with noncritical cases, and the seriously wounded may have to be taken to more distant ones. Use of triage systems is discussed further in the section concerning mass casualties and mass evacuations.

The coroner may have to handle victims whose lives cannot be saved. Activation of Disaster Mortuary Operational Response Teams (DMORTs) can speed the removal and care of the fatalities. The police department may need to control the spectators and establish barriers to vehicular traffic in the vicinity.

Freight trains–emergency action

As with truck accidents, until the nature of the cargo is ascertained, a highly defensive posture should be maintained in freight train accidents. Manifests or waybills, usually carried in the locomotive (previously in cabooses when they were widely used) should indicate whether hazardous materials are being transported. Not all hazardous materials require a warning placard; not all placards are indicative of every dangerous reaction that a product can create. Furthermore, such placards are often difficult to locate in twisted wreckage and dense smoke.

If the name of the chemical can be determined, but the hazardous materials references do not list it, a responder must not assume it is harmless. The responder should contact the dispatcher to call CHEMTREC at (800) 424-9300 if there is any doubt about the material's toxicity or instability. It is important to try to obtain more information before committing firefighters to a close attack. Due to the large volumes of hazardous materials that can be carried by train, and possible vapor clouds of correspondingly large magnitude, a railroad incident provides an ideal application for a CAMEO program.[49] More information on the CAMEO program can be found at the following Web site: http://www.epa.gov/emergencies/content/cameo/index.htm.

A new program available from NOAA's Office of Response and Restoration is the chemical reactivity worksheet. This program will allow a responder to enter the names of chemicals into the program to see if a reaction could occur among them. In the case of a train derailment, this could become very important, as the wreckage may leak chemicals from various cars, thus causing reactions among chemicals that were not designed to be mixed. More information on the chemical reactivity worksheet can be found at the following Web site: http://response.restoration.noaa.gov/type_subtopic_entry.php?RECORD_KEY%28entry_subtopic_type%29=entry_id,subtopic_id,type_id&entry_id(entry_subtopic_type)=328&subtopic_id(entry_subtopic_type)=24&type_id(entry_subtopic_type)=3.

If the accident has occurred in a nonpopulated area where no buildings are exposed, emergency responders should withdraw all personnel and notify the railroad that the responders will stand by one-half mile from the site. They should remain there until given assurance by the railroad's safety personnel (not the brakeman, conductor, or engineer) that the materials involved do not present an unusual hazard to the firefighters or other responders. The lives of personnel should never be risked merely to protect cargo and railroad equipment.

If the accident occurs in a business district, such as a derailment of tank cars and boxcars piled on top of one another, and one of the ruptured tank cars is releasing a large volume of liquefied gas, responders cannot assume that it is nontoxic or nonflammable. Everyone responding to the emergency must don self-contained breathing apparatus and full protective clothing while laying lines to protect exposures and getting occupants of nearby structures evacuated to a distance of at least 2,000 feet. When such defense action is taken promptly, a disaster can be averted, as was the case at the Dow Chemical Plant in Freeport, Texas. About 10 minutes before the tank car of ethylene oxide exploded, employees saw the tank venting and immediately had the plant evacuation alarm sounded. Plant personnel were able to clear out of the area and find protection behind remote buildings and structures.[50]

It is not safe to assume that responders will have time to prepare for an explosion. The boxcar of bagged ammonium nitrate did not detonate in the Traskwood, Arkansas train derailment until 40 minutes after the accident. However, the tank car of propane at Kingman, Arizona exploded about 15 to 20 minutes after vapors escaping from its relief valve became ignited. Thirteen firefighters were within 150 feet of the tank car when it exploded. Twelve died from thermal burns–their turnout coats and whatever street clothes they had been wearing burned off their bodies. Another man died and 95 spectators who had gathered along the highway about 1,000 feet from the explosion were injured.[51]

In this instance, the fire department was attempting to keep the tank car cool with a 1-inch booster line until larger backup lines could be laid from a water supply 1,200 feet away. As demonstrated here, a small hose stream is apt to be insufficient to prevent a BLEVE (boiling liquid expanding vapor explosion). Even 500 gpm directed from portable or fixed turrets cannot prevent such an explosion unless the streams are able to bathe the area where impinging flames are causing the metal shell to heat and weaken from internal pressure of the heated product.

Firefighters cannot rely upon suddenly increased sound intensity or flare-up of the burning product to give sufficient warning to withdraw their forces prior to the explosion. Therefore, unless it is necessary to take a calculated risk to save lives, they should not attack such fires at all where flames are impinging upon the shell of any tank car. Even if the product is nontoxic and nonflammable, a violent rupture can take place from vapor expansion that relief valves cannot handle, especially if the car is inverted and the valves are in the liquid space.

If cars are piled up, and firefighters cannot see whether or not flames are impinging on tank car shells, they must assume that they are. No one but the reconnaissance team should be allowed to approach the wreck until the product is identified and it is evident that the personnel and resources are available to handle the situation. While making this determination, all fire companies and everyone else must be kept clear of the fire to a distance of at least 2,000 feet. The table of isolation distances in the *North American Emergency Response Guidebook* should be utilized, based on the types of railcars seen through a pair of binoculars. The first pages of this manual will provide guidance based on the type of railcars involved. The isolation distances should always be based on the worst of the railcars involved; this will ensure a safe distance.

If nursing homes or similar institutions are nearby that will take considerable time to evacuate, it may be possible to set up portable monitors, left unattended, with large tips to direct heavy streams on the exposed tank shells. If a good water supply is not readily available, even by relaying from distant hydrants or ponds, the railroad may be able to bring in tank cars of water, along with cranes and other emergency equipment.

If the disaster consists of exploding bombs, as was the case in the railroad yards at Antelope, California, on April 28, 1973, firefighters may have to protect homes that are not heavily involved in fire and disregard buildings that are beyond help. In this incident, almost 2 million pounds of bombs exploded over a period of several hours, leveling the town of Antelope, and destroying a Citrus Heights fire station and 205 railroad cars. About 50 people were injured, and millions of dollars in damage occurred to buildings within a wide radius. Thousands of residents in a two-mile area in western Citrus Heights and Roseville were evacuated. Some of the firefighting forces, not learning until after the emergency was over that the county sheriff was legally in charge of the disaster, disregarded orders from his deputies to stay out of the area where exploding bomb fragments were igniting homes. It took about two hours to get the emergency operations center functioning following the initial explosion at 7:52 a.m. Unexploded bombs that flew through the air, along with great quantities of flying shrapnel, landed great distances from the railroad yards. It was considered a miracle that no one was killed.[52]

MARINE DISASTERS

Disasters on the water have involved passenger and cargo vessels. Passengers are transported on ferryboats and cruise ships; ferryboats that carry autos are rather large ships of a typical hull design. Some boats on major rivers in the United States are large vessels having stern or side paddle wheels and are practically floating hotels on inland waters. For relatively short trips, such as between Copenhagen, Denmark, and Malmo, Sweden, hydrofoil ferry boats are used. In inland waters, such as the San Francisco Bay, ground effect ferry boats have been used to move passengers from one port to another.

A limited number of passengers are carried on some freighters, but the cargo vessels apt to be involved in a disaster are more likely to be those used for transporting hazardous materials.

Causes of marine disasters

Shipwrecks can be caused by a vessel going aground or colliding with another ship, a rock, or an iceberg. Other reasons for voyages ending in disaster include boiler explosions, fire, storms, and capsizing and sinking.

Fire and explosion have been leading causes of disasters originating with vessels in harbors or alongside piers and wharfs. Usually the barge or ship has a cargo of hazardous materials that becomes ignited and results in a detonation before the fire can be suppressed. Barges loaded with a highly toxic liquefied gas have broken loose in a storm and drifted miles downstream in a navigable river before finally sinking.

Passenger ships

Despite tremendous strides in passenger ship safety, disasters with large losses of life have occurred through time and will continue as long as people continue to use the waterways.

On April 10, 1991, the Associated Press reported that an Italian ferry crashed into an oil tanker causing a floating inferno that killed 139 people. That same day, 13 sailors were killed and numerous others were injured when a ship tanker fire broke out not far from the original disaster. This two-fisted calamity that hit the Italian waterways was an eerie coincidence that showed how suddenly simultaneous disasters can strike an area.[53]

On December 20, 1987, more than 4,300 people lost their lives in the worst peacetime sea tragedy in history. On this day, the ferry *Doña Paz* sank after a collision with the tanker *Vector* in the Tablas Strait. The tanker was carrying petroleum products, and thus a fireball ensued that engulfed both ships. It was noted that the ferry was only licensed to carry 1,500 passengers.[54]

On March 28, 2007, at least 60 people were reported dead after a boat capsized off the coast of Guinea in West Africa.[55]

There have been other ship disasters with great losses of life, such as the sinking of the *Titanic* in 1912 as the result of striking an iceberg, wherein 1,517 lives were lost, and the collision of the *Mont Blanc* in Halifax Harbor in 1917, where 1,635 perished. However, it was the *Morro Castle* disaster in 1934, where 125 died as a result of its burning off the New Jersey coast, that prompted the drafting of fire safety standards for passenger vessels.[56]

Cargo ships

The same factors responsible for mishaps on passenger ships, such as fire, collision, and going aground, are applicable to cargo ships. While the loss of life potential is not so great on board freighters, they may be loaded with hazardous materials that could explode while in a harbor and cause great destruction and loss of life on shore.

The most devastating explosion that ever occurred in the United States originated in a ship. When longshoremen were loading the hold of the *Grandcamp* with ammonium nitrate fertilizer about 8:00 a.m. on April 16, 1947, a fire was discovered in the no. 4 hold. When the workmen were unable to extinguish the fire with fire extinguishers, the hatches were closed, steam was tried, and the volunteer fire department of Texas City, Texas was called. While hose streams were being applied to the fire, which involved wooden dunnage and paper-bagged nitrate, a tremendous explosion disintegrated the ship and pier, killing 26 of the 27 firefighters and demolishing their four fire trucks. Four hundred sixty-eight persons were killed in the blast. More than 1,000 were injured, and 128 were missing.

Flying metal weighing many tons shattered nearby structures. Pieces of the ship penetrated oil storage tanks in a refinery a considerable distance from the blast, starting fires. Other fragments started fires in the cargo ship *High Flyer*, which was loaded with ammonium nitrate berthed at a nearby pier. No attempt was made to extinguish these fires because of the great amount of rescue work that needed to be accomplished in the shattered reinforced concrete buildings of a nearby chemical plant. A later attempt to tow the burning *High Flyer* away from the pier was abandoned about 1:00 a.m., and everyone withdrew from the operation. At 1:15 a.m., its cargo detonated. The property damage was estimated at $50 million.[57]

The date of June 11, 1990, was not a good one for the shipping industry in the United States. First, in Galveston, Texas, firefighters were struggling to keep a burning supertanker with 38 million gallons of oil–more than three times what the Exxon *Valdez* spilled in Alaska–from sinking in the Gulf of Mexico. After powerful explosions sent flaming debris rocketing into the sky, the tanker's aft deck was just five feet above the surface of the Gulf of Mexico when firefighters from Galveston and the supertanker experts–Williams, Boots & Coots–succeeded in quenching the flames. On the same day, a 90-foot tear in the bottom of a cruise ship threatened to spill 110,000 gallons of diesel fuel off the coast of Cape Cod. Cleanup workers manned booms around the 617-foot ship and succeeded in corralling the fuel.[58]

On November 13, 2002, this same type of spill occurred off the Galician coast of Spain, causing the worst environmental disaster in Spain's history. The *Prestige*, a single-hulled tanker, split in half, eventually dumping more than 69,000 tons of oil, about 80% of its load. After the incident, the ship sank and continued to dump 125 tons of oil a day, thus polluting numerous miles of coastline as the spill washed ashore.[59]

Emergency action–marine disasters

It is sometimes possible to lend assistance to vessels heading for a port with a fire on board. It is important to obtain via radio a complete description of the conditions, the type of ship, the nature of the fire, its exact location, and a description of the contents and compartments on all sides, as well as above and below. It is also important to determine the type of extinguishing system or fire equipment being used, the condition of the ship's stability, its ability to maneuver and take advantage of wind for smoke and flame deflection, and any progress being made toward containment.

If the ship is not too far away and the seas are not too high, it may be possible to dispatch a large fire boat to meet it and render assistance while it is making port. If special extinguishing agents and technical assistance are requested, a shipboard firefighting team could be dispatched by helicopter with an extinguishing agents package, such as dry chemical and aqueous film-forming foam (similar to the kits used in crash firefighting). This team could help control the fire pending the arrival of the fire boat or the ship's arrival in the harbor.

Generally, fire departments with port facilities are equipped and trained to handle ship fires, and they are aware that fighting a fire in the lower reaches of a ship is comparable to coping with a basement fire, with the additional problem of stability.

It is of utmost importance for a disaster plan to identify the chain of command in harbor emergencies in order to avoid repetition of the *Normandy* fiasco. There the burning liner capsized in the New York slip because each agency working at the fire thought that another was in charge, and no one ordered the water streams stopped when the ship developed a critical list. The loss of $53 million was only the monetary value; the real loss was the badly needed troop-carrying capacity that the ship would have provided in World War II upon completion of re-outfitting.

A harbor disaster plan should indicate explicitly the respective authority of the port warden (city officials), state harbor fire marshal, captain of the port (Coast Guard), and the commander of any military base or shipyard in which an emergency may occur. The incident commander must know in advance which agency has full authority to order a burning vessel moved away from a pier or wharf (private or military) and to order a burning vessel towed to a designated location to be anchored or scuttled. The incident commander must also know whether or not the municipal fire officer will be in charge of firefighting aboard a military vessel alongside a municipal pier or in a naval dry dock. He or she must also know whether the fire chief of a naval facility where the ship fire occurs understands that the municipal fire officer in whose jurisdiction the facility is located (unless otherwise agreed upon) has overall responsibility for suppression of fires within the city. This includes the harbor and the vessels in port. In the event of an explosion like that in Texas City, which affected a wide area, the authority of the agencies likely to be involved is not immediately relevant but should eventually be delineated.

Rescue may have to be conducted under conditions of intense heat and dense smoke issuing from the passageways and hatches, necessitating a large supply of spare oxygen or air cylinders for breathing apparatus. The cargo storage plan should be obtained from the ship's captain or duty officer to ascertain the nature of materials involved. In some situations, large quantities of explosives, poison gases, unstable chemicals, or other extremely hazardous materials are threatened by fire. In these cases, the same considerations will have to be weighed as in highway and railroad cargo fires, including the need for evacuation of inhabited structures on shore. Other considerations could include limiting reconnaissance and rescue teams to the minimum number required to handle the situation and keeping the rest of the companies 2,000–3,000 feet away from the ship to fight secondary fires caused by an explosion.

If size-up reveals that the fire is one that firefighters should attack, then hoselines that have already been laid out upon arrival can be brought aboard by helicopter, over aerial ladders, via elevating platforms, or through side cargo hatches and fixed gangways. In closing down ventilation systems and trimming ventilators against the wind, firefighters should consider the effects of premature ventilation on tightly sealed spaces where combustible products of combustion may have accumulated.

Although maximum use of the ship's extinguishing systems is desirable, the fire department's portable firefighting equipment (hose, ladders) should be used wherever possible. In using hose streams or master streams from water towers, aerial ladder pipes, or portable monitors, it is important to keep in mind the need for ship stability. Simultaneous dewatering operations may have to be carried out in flooded compartments on the listing side of the ship to prevent its capsizing or foundering. The fire should be confined to the space already involved, if possible, using no more water than necessary.

If the ship has already sounded the national fire signal (five prolonged blasts of whistle or siren), other agencies may also be responding to the scene. These could include tug boats, Navy fire boats, Coast Guard emergency personnel, police boats, and the harbor emergency force. If the signal has not been sounded and these services are needed, the dispatcher can summon any necessary assistance, including longshore crews and lighters if off-loading is required for overhauling burned cargo.

One lesson learned in the port of Charlotte Amalie, Virgin Islands, is that an entire crew should not be permitted to go ashore at one time. Here a passenger ship burned to the water at the dock.

CONCLUSION

As the U.S. society has become very mobile, and as transportation modes have increased in number and type, the number of transportation disasters has increased in direct proportion. This chapter has discussed how serious multiple fatality incidents have occurred on land, both by rail and highway, and in the air and at sea. Many of the incidents chronicled in this chapter tested the tactical capabilities and resources of the local fire service agencies to the limit, and this situation is not expected to improve.

In the 21st century, people have shown a tendency to become even more mobile, increasing the frequency with which they travel for leisure and business. Likewise, as U.S. society has become more technologically complex, the number of physical goods that must be transported continues to grow dramatically. All of this leads to an increased probability that a given fire department will be called upon to handle a major transportation disaster. These incidents are compounded where multiple fatalities or hazardous materials, or both, require maximum utilization of incident command. In the following chapter, the problem of hazardous materials is covered in greater detail.

REFERENCES

1. U.S. Department of Transportation, Bureau of Transportation Statistics. 2006. *National Transportation Statistics 2006*. Washington, DC: U.S. Government Printing Office. Retrieved March 28, 2007 from: http://www.bts.gov/publications/national_transportation_statistics/2006/pdf/entire.pdf.
2. National Transportation Safety Board. 2007. National Transportation Safety Board 2004–2005 U.S. transportation fatalities. Retrieved September 3, 2008 from: http://www.ntsb.gov/Pressrel/2006/trans_fatalities_2004-2005.htm.
3. Reiser, J. 1973. Fire department classroom. *Western Fire Journal*. July. p. 43.
4. Brown, D. A. 1989. Bomb destroys Pan Am 747 in blast over Scotland. *Aviation Week & Space Technology*. January 2. pp. 28–32.
5. Bahme, C. W. 1978. *Fire Officer's Guide to Disaster Control*. 1st ed. Boston, MA: NFPA. p. 150.
6. *Wikipedia*. 2008. USAir Flight 5050. Retrieved September 3, 2008 from: http://en.wikipedia.org/wiki/USAir_Flight_5050.
7. CNN Interactive. 1996. Transcript of final moments of ValuJet Flight 592 released. November 18. Retrieved July 15, 2008 from: http://www.cnn.com/US/9611/18/valujet.recordings.
8. Bureau of Transportation Statistics. 2007. Airline Traffic Data: Eight-Month 2007 System Traffic Up 3.4 Percent from 2006. Retrieved November 3, 2008, from http://www.bts.gov/press_releases/2007/bts052_07.html#table_01.
9. National Fire Protection Association. *Fire Protection Handbook, 17th Ed.* 1991. Boston, MA: NFPA.
10. Tryon, G. H. 1966. The Salt Lake City airport crash. *Fire Journal*. September. p. 5.
11. U.S. Department of Defense. 1994. *Instruction Number 6055.6*. December 15. p. 12.
12. Infoplease.com. 2007. "Aircraft crashes." Retrieved September 3, 2008, from http://www.infoplease.com/ipa/A0001449.html.
13. National Transportation Safety Board (n.d.) Aircraft accident report. http://amelia.db.erau.edu/reports/ntsb/aar/AAR91-04/pdf.
14. *Newsweek*. 1990. February 5. p. 22.
15. Phillips, D. 1994. 68 die in Indiana plane crash. *The Washington Post*. November 1. A1.
16. PBS. 2004. Crash of flight 111, dissection of a disaster. February. Retrieved July 15, 2008 from: http://www.pbs.org/wgbh/nova/aircrash/dissection.html.
17. National Transportation Safety Board. 2004. In-flight separation of vertical stabilizer American Airlines Flight 587. *Aircraft Accident Report*. April. Retrieved July 15, 2008 from: http://www.ntsb.gov/publictn/2004/AAR0404.htm.
18. Ibid.
19. Bahme, C. W. 1968. *Fire Officer's Guide to Dangerous Chemicals*. Boston, MA: NFPA.
20. Nfpa.org. 2007. The U.S. fire problem. Retrieved March 29, 2007 from: http://www.nfpa.org/categoryList.asp?categoryID=953&URL=Research%20&%20Reports/Fire%20statistics/The%20U.S.%20fire%20problem.
21. *USA Today*. 1991. The Salt Lake City airport crash. January 3. p. 3A.
22. National Transportation Safety Board. 1976. *Highway Accident Report. NTSB-Har 76-4*. May 5. p. 2.
23. *CNN.com*. 2004. Interstate reopens after 70-vehicle pileup. December 19. Retrieved March 29, 2007 from: http://www.cnn.com/2004/US/12/19/highway.pileup/index.html.

24. *CNN.com*. 2007. Winter storms cause major pileups in two states. February 28. Retrieved March 29, 2007 from: http://www.cnn.com/2007/weather/02/28/winter.weather.ap/index.html.

25. National Transportation Safety Board. 1988. Pickup truck/church activity bus head on collision and fire near Carrollton, Kentucky. *Highway Accident Report*. May 14. Adopted March 28, 1989. Retrieved July 15, 2008 from: http://www.ntsb.gov/publictn/1989/HAR8901.htm.

26. *CNN.com*. 2003. Schoolgirls die in bus crash. October 9. Retrieved March 29, 2007 from: http://www.cnn.com/2003/WORLD/asiapcf/southeast/10/09/indon.crash/index.

27. *CNN.com*. 2006. Fiery crash kills 7 young relatives. January 25. Retrieved March 29, 2007 from: http://www.cnn.com/2006/US/01/25/bus.crash/index.html.

28. *CNN.com*. 2003. Company devastated by firefightersí deaths. August 25. Retrieved March 29, 2007 from: http://www.cnn.com/2003/US/west/08/25/firefighters.crash/index.html.

29. *CNN.com*. 2007. Six killed as bus plunges from overpass. March 2. Retrieved March 18, 2007 from: http://www.cnn.com/2007/US/03/02/bus.accident/index.html.

30. *Fire Engineering*. 1990. July.

31. Kramer, W. 1989. *American Heat*. Video journal. January. St. Louis, MO. As cited in: Kramer, W. M., and C. W. Bahme. 1992. *Fire Officer's Guide to Disaster Control*. 2nd edition. Saddle Brook, NJ: Fire Engineering Books and Videos.

32. *Fire Engineering*. 2007. In Nigeria, 70 dead after fire sweeps people scooping fuel from tanker. March 28. Retrieved March 29, 2007 from: http://fe.pennet.com/news/display_news_story.cfm?section=wire&category=HOME.

33. Kaufman, B. L. 1992. Tanker wreck was typical, yet rare event. *Cincinnati Enquirer*. June 5. B1.

34. Federal Motor Carrier Safety Administration (n.d.) HM risk assessment continued. Retrieved November 5, 2008, from http://www.fmcsa.dot.gov/Spanish/english/HMRiskFinalReport2.htm.

35. *Los Angeles Times*. 1976. Ammonia truck crashes in Houston: 4 die, 119 hurt. May 12. Part I, p. 5.

36. Vincent, L. A. 1961. Hazardous materials on the road. *State Government*. Autumn. p. 241. As cited in: Kramer, W. M., and C. W. Bahme. 1992. *Fire Officer's Guide to Disaster Control*. 2nd edition. Saddle Brook, NJ: Fire Engineering Books and Videos.

37. *National Fire Protection Association Quarterly*. 1960. Marshall's Creek explosion. October.

38. Woodhouse, L. R. 1987. Miamisburg revisited. *Nation's Cities Weekly*. July 20. p. 3.

39. *American Heat*. 1990. Video journal. St. Louis, MO. June. Vol. 4. Prog. 11.

40. Hayes, T. 1972. Houston train derailment and subsequent hearing. *Fire Journal*. March. pp. 13–17.

41. National Transportation Safety Board. 1972. *Railroad Department Accident Report. NTSB-R AR 72-G*. December 13.

42. Johnson, A. 2004. N. Korea declares emergency, shuts border after deadly blast. *Washington Post*. April 23. A1.

43. *Western Fire Journal*. 1975. Communications, rail emergencies topics at recent California chiefs conference. July. p. 39.

44. CNN broadcast. 1987. January 4.

45. Associated Press. 1999. Train wreck claims 226. *Cincinnati Enquirer*. August 3. A2.

46. Johnson, E. 2002. Train wreck rattles British public. *Cincinnati Enquirer*. May 12. A4.

47. Bahme, C. W. 1978.

48. U.S. Department of Labor, Occupational Health and Safety Administration. 1990. *29 CFR 1910.147. The Control of Hazardous Energy (Lockout/Tagout)*. September 11.

49. National Safety Council. 1991. CAMEO demo.

50. Bahme, C. W. 1978. p. 176.

51. Ibid.

52. Bahme, C. W. 1978. p. 178.

53. Associated Press Report. 1991. April 11.

54. *Wikipedia*. 2008. MV Doña Paz. October 12. Retrieved October 21, 2008 from: http://en.wikipedia.org/wiki/Do%C3%B1a_Paz.

55. Reuters. 2007. At least 60 dead after boat capsizes off coast of Guinea. March 29. Retrieved March 29, 2007 from: http://www.cnn.com/2007/WORLD/africa/03/29/guinea.capsize.reut/index.html.

56. *The World Book Encyclopedia*. 1969. Vol. 17. Chicago, IL: Field Enterprises Educational Corporation. p. 346.

57. *American Heat*. Video journal. 1989. Kansas City fatal explosion report. St. Louis, MO. May.

58. *Cincinnati Post*. 1990. June 11. p. 12-A.

59. *New Scientist*. 2003. Prestige oil spill far worse than thought. August 27. Retrieved July 15, 2008 from: http://www.newscientist.com/article.ns?id=dn4100.

HAZARDOUS MATERIALS DISASTERS 9

INTRODUCTION

Contrary to some misconception, hazardous materials disasters are not a recent phenomenon, but have been a serious problem for public safety and industrial personnel for more than one-half of a century. Chuck Bahme, retired deputy chief of the Los Angeles Fire Department, researched the subject of explosions in 1948 in preparation for writing the Los Angeles dangerous chemicals code.[1] He listed every chemical that had produced an explosive reaction during the period of 1900 to 1948, as recorded in *Chemical Abstracts*, and found that the number exceeded 3,000.[2] He had been assigned to this task and that of heading up a new dangerous chemicals detail by his department's fire chief after his response, as an engine company commander, to a devastating electroplating plant explosion. All the buildings in the neighborhood had been leveled, and the fragments of 17 bodies were gathered up in gunnysacks by the firefighters.

It was in the early 1900s that the detonation of a huge pile of ammonium nitrate in Oppau, Germany almost wiped out a whole town, killing about 1,000 of its residents. In 1929 an explosion of X-ray film killed 125 occupants of a clinic in Cleveland, Ohio. In the more than 75 years that have elapsed since then, thousands of new chemical products have been developed annually. Possibly 1,000 or more enter into commercial production and transportation each year. Another source of these chemicals comes in the form of weapons used to kill and disable citizens and responders. These chemicals and agents are known as *weapons of mass destruction*. These are discussed in relation to their use in chapter 11, but emergency planners must understand that much of today's WMD problems are essentially a hazardous materials problem.

SCOPE OF THIS DISCUSSION

This chapter will not deal with the ordinary incidents involving chemicals and other products that emergency responders handle almost daily, such as acid carboy spills, chlorine and ammonia leaks, and fires in acetylene, oxygen, butane, and liquefied petroleum gas storage areas. The focus of this chapter will be on the hazardous materials incidents that cause, or have the potential to cause, a major disaster. Chapter 8 on transportation disasters discussed those disasters arising from shipboard fires and train derailments, as well as truck and plane crashes where hazardous materials are involved. The focus of this chapter will be directed toward those disasters occurring off the seas, channels, highways, rails, and runways. It will also briefly examine the hazardous materials portion of the WMD events.

HAZARDOUS MATERIALS DISASTERS DEFINED

Aside from the transportation element, there is still much to discuss about hazardous materials disasters. The first step is to define this subject in its broadest terms. A *hazardous materials disaster* could be described as a large-scale interruption of normal activities by an unplanned release, spill, fire, or explosion involving hazardous materials that cause, or are likely to cause, widespread injury or damage to people, property, or the environment. For the purposes of this discussion, an *explosion* is defined as the "sudden going away of things from where they have been, accompanied by a loud noise."[3] This text will not attempt to distinguish among the reactions such as deflagrations, decompositions, dissociations, detonations, polymerizations, or pressure vessel ruptures. The violent destructive results are apt to be similar. If responders learn what they need to know about these dangerous reactions, they should be able to cope with them without becoming a target for things that "suddenly go away with a loud noise."

WHAT ARE HAZARDOUS MATERIALS?

Having arrived at a broad definition of a hazardous materials disaster, this discussion can next examine some of these materials that responders are concerned with. Their principal components are often various chemicals and their compounds, as well as products in which they are used. One hazardous materials incident plan defines them as "any substance capable of posing an unnecessary risk to health, safety, property, and the environment." These include but are not limited to materials that are highly explosive or flammable, poisonous and toxic chemicals, radioactive products, corrosive materials, and biological agents.[4] The plan also defines *deleterious materials* as products that in themselves do not necessarily constitute a threat or danger to human life and property, but when allowed to enter water sources, such as sewers, streams, ditches, creeks, and rivers, may be a threat to aquatic life.

To simplify this discussion, hazardous materials will be referred to as *hazmats* in the balance of the chapter. These materials frequently turn out to be dangerous chemicals. A person does not have to be a chemist to know that just about everything in the universe of known composition is comprised of chemicals. Nor is it necessary to have a PhD degree to know that practically every chemical can be dangerous under certain conditions. Even the air in the atmosphere becomes highly explosive if liquefied and absorbed in combustible materials. The word *hazardous* may apply to a condition or circumstance in which the chemical may be placed, as well as to its properties.

Chemicals are usually classed as hazardous because they may explode, react violently, become unstable, ignite, develop heat, intensify combustion, rupture their containers, corrode, or yield toxic or flammable gases, all with possible injury to life and property. Some explosions are caused by materials that are not ordinarily considered as chemicals. Powdered wood and sugar are examples. In October 2003, "A silo explosion that killed two people at a lumber company last month was caused when air rushed into a cavity of burning wood chips, igniting gases and dust, state fire officials said."[5] Examples of some of the commonly encountered hazmats that are not usually thought of as chemicals include gasoline, natural gas, cleaning fluids, paints, plastics, pesticides, fertilizers, ammunition, and drain cleaners.

CAUSES OF HAZMAT DISASTERS

There are a multitude of reasons why leaks, spills, fires, explosions, and so on can occur, and other reasons why a minor mishap may evolve into a major disaster. In general, however, a great many may be traced to "The Four Blind *I*s": ignorance, incompetence, indifference, and inertia. People who are burdened with these handicaps where hazmats are concerned could easily end up dead. Lack of vision is not found only in the ranks of the people who encounter hazmats in their business, but also in the front lines among the emergency personnel who are called upon to mitigate these hazardous mistakes. Even with all of the hazardous materials–based training and the emphasis on WMDs, these individuals still exist, even some of whom should know better.

Ignorance

Lack of knowledge on the part of the formulators, manufacturers, packers, shippers, warehousemen, and consumers who use hazmats has resulted in disasters such as the one described earlier in the electroplating plant. At this incident, apparently the chemist who combined acetic anhydride and perchloric acid in the 200-gallon electropolishing bath was unaware that a power failure to the tank's refrigerator would result in its detonation. As one of the victims, he never survived to learn the lesson of that mistake. Even more appalling was the statement made to Chief Bahme by an executive of a large chemical company in the presence of the company's chief chemist. This executive stated that there was no need for drafting regulations for liquid oxygen, since it was "so cold that it would freeze out any fire in which it might become involved"! Chief Bahme quickly dispelled that fallacy with a burning cigarette demonstration using liquid oxygen.

Firefighters who respond to hazmat emergencies without the requisite knowledge of latent dangers and proper procedures for handling hazmat incidents have also had to pay with their lives. A costly example occurred in Texas City, Texas, when 28 firefighters battened down the hatches on the *Grandcamp*'s cargo of ammonium nitrate fertilizer as fire was spreading in its hold. The firefighters were unaware of the need for maximum ventilation, and all were killed, along with almost 500 others, when the detonation occurred.[6]

Incompetence

Ignorance may be responsible for failure to recognize the dangerous potential of some hazmats, but incompetence is often pinpointed for the failure to provide adequate safeguards that might prevent or mitigate a disaster. Close colleagues of associate author Hanifen witnessed the resulting explosion when diesel fuel was poured into a house being used for a training fire. The crew failed to recognize the ammonium nitrate fertilizer that was stored in the crawl space of the house. The resulting explosion not only demolished the building but also knocked out windows for blocks. (However, the fire was lit!) Luckily, none of the firefighters was injured in the blast.

It is easy to imagine what incompetence on the part of city councils, county supervisors, state legislators, and federal regulators might produce in the way of ordinances, laws, and regulations for safeguarding hazmats. Dangerous substances might not be required to be tested, properly classified, put in safe packages, and conspicuously and correctly labeled. They might not be stored in fire resistive structures having automatic fire detection and extinguishing systems especially designed for the particular hazards. Furthermore, dangerous substances might not be transported in properly equipped and suitable placarded vehicles. The possibilities for disaster are manifold.

This was nearly the case post-9/11, when there was a push among federal officials to remove placarding from transportation vessels for security purposes. The terrorists would not know the contents, but neither would emergency responders.

If the effect of requirements for labeling hazardous materials is to conceal the most dangerous aspects of the chemicals from emergency forces and the public, disaster can result.

If testing is inadequate to determine all of the possible dangerous reactions prior to marketing a product, disaster can result.

If there is a wrong classification of chemicals that have long been known to be dangerous, the public can be exposed to greater risks than would be the case if such chemicals were properly classified.

If a chemical known to be dangerous is omitted from a list of regulated chemicals, the absence of adequate regulation can lead to disastrous results.

If adequate regulations of hazardous materials can be circumvented by issuance of special permits that exempt chemical handlers from complying with safeguards relating to container design, for example, disaster can result.

If adequate regulations are neither well understood by persons to whom they apply, nor adequately enforced by designated agencies, disaster can result.

If the promulgation and enforcement of regulations are in a large measure dependent upon the industries being regulated, disaster can result.

If reports detailing the facts relating to chemical accidents—facts that would help substantiate the need for more stringent regulation—are withheld from regulatory agencies, disasters can result.

As long as any one of these situations exists, there is little prospect that chemical disasters can be eliminated.

On the contrary, as new products proceed from test tube to tank car quantities of dangerous chemicals, the prospect of larger chemical calamities appears to be greater than ever. One example of the danger has been seen in the expanding production, shipping, and storage of liquefied natural gas (LNG). Natural gas incidents are frequent and have historically occurred in all parts of North America. One news article from the *Los Angeles Times* in 1966 stated, "Despite industry assurances that the risk of accident from importing liquefied natural gas into southern California is extremely small, new estimates indicate that serious hazards will exist, jeopardizing more than 100,000 lives and property worth hundreds of millions of dollars."[7]

A report cited in the preceding article was issued by the Rand Corporation regarding the proposal to construct LNG off-loading facilities within two populated areas (Los Angeles Harbor and Oxnard, California). The report points out that spills could occur either on land in port facilities, at sea, or in the harbor, especially during transfer procedures. Such spills could be caused by ship collisions or groundings, sabotage, natural disaster, design faults, construction or fabrication weakness, or human error. The Socio-Economic Systems, Inc. released a report at about the same time as the Rand report, in which it estimated that an LNG tanker collision 4.5 miles offshore and a subsequent fire could kill 70,000 persons and create a total property loss of $325 million.

It is estimated that a spill of 100,000 cubic meters would create a flammable cloud ranging from 17,000 to 670,000 feet in length. One group of consultants estimated that a 25,000 cubic meter gas cloud would travel 1.9 miles, but the Coast Guard predicted that the same size spill would produce a cloud 14.4 miles long.[8]

Incompetence that leads to disaster is not just limited to legislators and administrators who fail to require and put into effect the necessary safeguards for hazmats. Too often, it is found among the ranks of the everyday handlers. As reported by the Associated Press and reprinted at Firehouse.com, "The federal government's final report on the fatal 2005 explosion at BP PLC's Texas City refinery released Tuesday criticized a key U.S. worker-safety agency [OSHA] for lax oversight and reiterated claims that organizational and safety deficiencies at the British energy company led to the blast…'The combination of cost-cutting, production pressures, and failure to invest caused a progressive deterioration of safety at the refinery,' CSB Chairwoman Carolyn W. Merritt said of the latest findings."[9]

Indifference

Indifference on the part of hazmat handlers who have the requisite knowledge and competence to avoid accidents, but who choose to take a chance for various reasons, can result in disastrous consequences. Perhaps the old adage, "Familiarity breeds contempt," led to the continued use of thin metal containers for liquid hydrocyanic acid in the county adjacent to Los Angeles County. The usage of these containers was still allowed, even though it was forbidden in Los Angeles upon finding that they would rupture if dropped from the bed of a truck. The desire to save time and money could also lead to the type of indifference shown by the truck driver who illegally took his cargo of carbon bisulfide, a highly toxic and flammable liquid, through a New York tunnel in which it caught fire, menacing the lives of hundreds of motorists. In another instance, the driver of a truckload of blasting agent hid the "Explosives" placard under the seat in his cab, and when fire ensued, drove off in the detached cab, leaving the cargo with no warning sign for the responding fire companies. This resulted in the death of several firefighters and the destruction of their fire apparatus when the cargo detonated.

Indifference on the part of emergency response teams is not generally a problem today. However, there was a time when the hardy "smoke-eaters" would show their contempt for danger by not wearing breathing apparatus to hazardous areas or by rejecting the advice of the plant management that the fire they were about to attack might be unsafe to approach. Such apparently was the case when the arriving fire officers at a Norwich, Connecticut plant were warned that the fire in the benzoyl peroxide cargo might explode. When it finally did, several firefighters were killed and their fire trucks demolished.[10]

Inertia

Lastly, in this discussion of the principal causes of hazmat disasters, there are those people who are fully aware of most of the risks involved and who genuinely desire to provide adequate safeguards but just never get around to doing anything about it. They are the victims of the "blind *I* of inertia." Some examples of this are clearly evident in the inaction on the part of federal, state, and local officials, both legislative and administrative, and also on the part of industrial and public safety managers.

This inertia not only extends to the adoption of needed uniform laws and regulations to provide improved safeguards for hazmats, but also to observing and enforcing the existing standards of care, however deficient they may be in some respects.

While it is true that some industries have been slow to establish training programs for their employees relating to hazmats, and to promulgate plans for emergency action for coping with disasters, this inertia has not been entirely absent from the fire service. The fire service still has much to accomplish in a number of areas, including providing personnel, equipment, and training for hazmat problems in their response districts and establishing a target hazard inspection program. Other areas for improvement relate to hazmat reporting and identification requirements and creating a disaster plan and training with it. Fortunately, this inertia is slowly being overcome in almost every aspect of the problem, with the result that there is a much better chance of preventing more hazmat disasters in the future. The next section examines how the odds for prevention can be improved.

PREVENTION OF HAZMAT DISASTERS

Most would agree that the best way to obviate the need for controlling, managing, and mitigating disasters is to prevent them from happening in the first place. It has long been recognized that the elimination of unnecessary risks in almost every field of endeavor can be achieved through the application of the "The Three *E*s of Safety," namely, engineering, enforcement, and education. These same three *E*s can be valuable tools in the continuing goal of preventing disasters of any kind, but especially those involving hazmats.

Since it is not within the scope of this study to provide a detailed discussion of two aspects of disaster prevention, namely, education and enforcement, this discussion only addresses them briefly.

Engineering for hazmat safety

Probably the best place to commence any safety endeavor is at the drawing board. In the field of hazmat safety, it might well start with the chemical engineer's test tube and proceed to the design of the processing, storage, and handling equipment. Engineers can do much to reduce hazmat dangers arising from exothermic reactions, spontaneous polymerization, and decomposition of unstable chemicals, etc. This can be accomplished through the provision of relief valves, soft plugs, automatic dumping or flooding devices, special walls and roofs for explosion venting, backflow and check valves, dikes, drains, and sumps. Hazards can also be reduced with high-speed detection and extinguishing systems, automatic fire doors, and venting and alarm systems. Helpful as some of these systems may be in preventing disaster-causing accidents, they do not install themselves. Furthermore, not all plant operators see the wisdom of spending the money to have them provided, even though national standard-setting organizations may recommend them. Hence, in many instances, it becomes advisable to adopt laws, ordinances, and regulations that make their installation mandatory. However, merely having laws on the books is not very helpful unless they are enforced.

Enforcement

Before commencing on a conscientious enforcement program, it is helpful to ensure that the local requirements are adequate to obtain the desirable safeguards for hazmats in the jurisdiction. If they are not, it will be important to have ones adopted that will meet the needs. Many of the latest model fire codes address these issues in depth. In any case, a sufficient number of personnel must be trained to carry out the enforcement responsibility, and this is not limited to only the local ordinances, but also involves state and federally mandated hazmat requirements.

At this point, fire service personnel may be justifiably asking why so much responsibility is imposed upon fire departments—why not the public health departments or the law enforcement agencies? The answer rests in the fact that the fire service has traditionally been the one that people look to in time of disaster. It often is the case that the incident is a personal one, such as a hand caught in a bicycle sprocket, a person stuck in a bathtub, a child locked in the bathroom, or people trapped in an elevator or a collapsed building. There are many emergencies that fire departments are called upon to handle that do not involve fire of any kind. Because the fire service responds to these disasters, why not trust them to mitigate the major disasters, including those arising from hazmats? In addition, since fire departments have always been given the duty of enforcing fire codes, why not hazmat laws also? Legislators and political leaders at every level of government have come to realize that hazmat problems can erupt anywhere, from the smallest hamlet to the largest metropolis. If there is one service agency that they all have in common, it is the fire department.

Good laws are needed

Enforcement of poorly drafted laws will not prevent hazmat disasters. There are some industrial managers who are unwilling to voluntarily adopt safe practices in the absence of regulatory coercion. There are others who might urge legislators to adopt rules that only have the appearance of establishing needed safeguards for hazmats. However, because of monetary considerations, these safeguards end up being less than satisfactory from a public safety standpoint. For example, suppose a stevedore is hired to handle a product labeled as an oxidizer. He might be paid only half the wage rate he would demand if it were more accurately labeled as an explosive. Even though the product may only be intended for use as a fertilizer, it has long been known to detonate when involved in boxcar, structure, and shipboard fires.

Chief Bahme's objections to this type of labeling were rejected in Washington, DC, when he met with regulators following the Texas City shipboard explosions on the *Grandcamp* and the *High Flyer* many years ago. (See the section on cargo ships in chapter 8 for a description of this disaster.) Lack of enforcement was not the culprit in that disaster, and neither was lack of knowledge of the hazards. Instead, the problem was the lack of useful regulations that would have required a label sufficiently descriptive of the product's latent danger to warn the firefighters and hundreds of spectators of their impending danger from explosion.

Consensus standards

Perhaps it is not always desirable to leave the drafting of regulations to the industrial organizations being regulated and then pass laws forbidding state or local governments from adopting any rules that would require more effective safeguards. Furthermore, enforcement is not of great value unless the rules being enforced are drawn up by persons of integrity who truly seek to obtain adequate safeguards for hazmat handlers, the public, and the emergency responders. Codes drafted by a consensus of committee members, many of whom represent the industries to be regulated, do not always contain the best standards that could be applied to a given problem. Instead they may be watered-down standards that a majority of the committee could agree upon. If these codes turn out to be mediocre, then a requirement that they be uniformly enforced throughout the nation merely results in the uniform adoption of mediocre rules at every level of government. It may well preclude the enforcement of really good standards at the local level. This may not be a pleasant fact to contemplate, but it is a fairly accurate one. As someone once said, "You shall know the truth, and the truth will make you flinch before it makes you free."

Some of the useful consensus standards that are often referred to within model fire codes are National Fire Protection Association's (NFPA) standards 430, 432, and 434. These are: *Code for the Storage of Liquid and Solid Oxidizers*, *Code for the Storage of Organic Peroxide Formulations*, and *Code for the Storage of Pesticides*, respectively.

To sum it up then, enforcement will not be of great benefit in preventing hazmat disasters unless the regulatory and legislative bodies adopt good regulations that can be uniformly applied within their respective jurisdictions. For example, there is little doubt that hazmat disasters are more likely to be avoided if use, storage, manufacturing, and processing of hazardous materials are required by zoning laws to be restricted to relatively uninhabited areas. Nor would anyone deny that risks might be reduced by prohibiting their transport through tunnels or congested business districts, or past schools, hospitals, and densely populated residential areas. However, basic to such solutions is the determination of which hazmats are so dangerous, and in what quantities, as to justify such restrictions. That is where the third *E* becomes applicable. This requires the education of legislators, hazmat handlers, and the public as to the need for reasonable restrictions and safeguards that can be helpful in preventing hazmat disasters.

Education

To eliminate, or at least reduce, the catastrophic consequences of a hazmat disaster, education is an indispensable tool. Obviously, the educators must acquire a thorough understanding of the potential risks that hazmats pose. This is especially important if they are called upon to assist regulators in drafting rules for safe handling or in devising a warning system to provide ready identification of such hazards for everyone exposed to them.

Learning the risks

Hazmat education must start with the people who produce them. They must learn the scope of the risks that may be generated under all of the conditions likely to be encountered from the time the hazmats are manufactured until they are packaged, transported, stored, used, and ultimately discarded. There are various ways of determining a product's hazards. One way that definitely should be discouraged is to put it into production, ship it around the country, and wait to see how many fires and explosions it will cause. A better way is to test it thoroughly according to recognized standards before it is put into commercial use. If tests show that a chemical is hazardous in one or more respects, such as fire, health, or instability, then it should be listed in a suitable classification in regulatory publications.

The laws and rules applicable to the product's packing, labeling, and handling should be rigidly enforced. Perhaps a product is tested and found to be susceptible to detonation under conditions that can be reasonably anticipated to occur during normal handling in rail shipment. Could there ever be any justification for sending tank cars of that material around the country where they might be bumped or shocked during switching operations? Was any testing done before the monomethylamine nitrate solutions were shipped in the tank car that detonated in Wenatchee, Washington, killing 2 people and injuring many others?[11] At the time of the accident, this chemical could not be found in any listing of the national organizations that promulgated safety standards or transportation regulations. Why was this possible?

In addition, chemicals should undergo continual testing. Advancements in research and technology are changing at such a rapid pace that further research may indicate what was once thought to be a benign chemical may actually produce carcinogenetic effects. This has occurred in more that one instance in the past few years. Although much has occurred in educating all persons involved in the shipment of hazmats, incidents still occur that can cause explosions, such as that reported by CNN.com on March 12, 2007. On this day, a train carrying liquefied petroleum propane derailed, setting off explosions that caused the evacuation of Oneida, New York, with a population of approximately 4,000 people.

Providing the warning

A practical system must be devised and promulgated for warning the people who manufacture, transport, store, use, and dispose of hazardous chemicals and products. This system must also warn members of the public and emergency personnel who may be on the scene of a mishap where these materials may be involved. Several systems have been proposed, but two are currently used the most extensively in the United States. One of these is the NFPA 704, which is the *Standard System for the Identification of the Hazards of Materials for Emergency Response*. The other is the placarding system of the U.S. Department of Transportation, found in the Code of Federal Regulations at 49 CFR 172.332.[12] There is also a United Nations classification system for use with the transportation of products in foreign trade, which is the basis for the DOT placarding system.

In general, placarding is used for hazmats in travel, and the NFPA 704 labeling is for fixed-facility hazmats. There has been considerable improvement made in the classification, labeling, and placarding requirements in the DOT system in recent years, but there are still some chemicals assigned to a classification that fails to disclose their most serious risks. For example, nitromethane was known to be a nonpropellant capable of detonation when the first tank car containing this flammable liquid exploded in 1958, near Niagara Falls, New York. This explosion injured 180 people and caused damage in excess of $1 million to buildings within a 3.5-mile radius. Many of the injured were children in the eight schools that were damaged.

The DOT's 1987 *Emergency Response Guidebook*, in describing fire and fire explosion hazards of this material (Guide 26), had omitted any reference to its susceptibility to detonation by shock. Fortunately, the 1990 *Guidebook* was more definitive.[13] This shows the value of checking at least two sources of chemical information prior to proceeding with any action or planning that involves a hazardous material. Omissions are possible, even with the most thorough checking.

Classifications of chemicals in the safety standards and regulatory publications should reflect their most hazardous properties. How could the people who needed to know the dangerous properties of such poisonous gases as chlorine, sulfur dioxide, nitrosyl chloride, and ammonia be alerted to their lethal potentiality if the national list of regulated hazardous commodities merely listed them as "nonflammable compressed gases?"

Hazmats should be required to be labeled under an identification system that will provide adequate warning of their most dangerous properties. Any set of regulations that does not require materials capable of deflagration, such as organic peroxides, or detonation, such as ammonium nitrate/fuel oil mixtures, under fire conditions to be labeled as explosive materials fails to impart the true hazards to the public and the emergency services. To permit explosive materials to be exempt from being labeled as explosives merely because they are not primarily intended to be used as explosives cannot be justified. Firefighters and other emergency personnel may not always be able to distinguish between products labeled as "oxidizers" that are not explosive, such as sodium nitrate, and those that are, such as ammonium perchlorate.

Likewise, any set of regulations that deliberately defines a "flammable gas" in such a manner as to exclude ammonia, a gas that has been responsible for many explosions, is not fair to the public. Permitting this gas to be labeled with the same label that was required for compressed air did nothing to prevent disasters caused by this chemical. Ammonia's capability of causing a disaster was well illustrated in Crete, Nebraska, on February 19, 1969. A freight train derailed and struck a tank car of anhydrous ammonia standing on a siding, resulting in a blinding flash and release of dense clouds of ammonia gas. This gas drifted through the town, killing 8 people and putting 11 others in the hospital.[14]

It may not be possible to devise a truly comprehensive identification system that will indicate the true hazard of every chemical, compound, or cargo. It would have to indicate whether the material was mixed or homogeneous; give its trade name, chemical name, signal word, and degree of hazard; and indicate a code number to a reference supplying information for emergency procedures in a number of languages. However, it is reasonable to require that hazmats be identified with basic warning symbols for protection of the public and emergency services.

It is also reasonable to expect that the emergency services will obtain the knowledge and training necessary to cope with hazmat emergencies even though all of the desired information is not instantly available. Industrial personnel who produce, ship, store, and use hazmats should learn their risks and become familiar with the applicable labels and placards of the identification system currently in use. First responders of the emergency services need to be able to quickly assess the risks at the scene of an incident through their knowledge of the hazard information system. This system offers information by means of labels, placards, waybills, or data cards that clearly indicate the hazards or at least give the responders clues that will enable them to seek needed information from other sources. These other sources are discussed later on in this chapter.

TYPES OF HAZARDOUS MATERIALS DISASTERS AND SOME EXAMPLES

Hazmat disasters have taken many forms. The most serious disasters precipitate extensive air and water pollution. Hazmat disasters poison the wildlife in oceans, lakes, and streams, sometimes making vast land areas unfit for human habitation. Sometimes they explosively shatter a neighborhood of businesses and dwellings, killing occupants and often the emergency response personnel who attempt to rescue them. Unlike an earthquake or a tornado, these disasters do not "just happen" in most cases, but can be traced to oversight, neglect, or downright stupidity on someone's part. It is true that hazmat disasters can grow out of earthquakes, floods, and violent storms that dislodge tanks, overturn ships, trains, and trucks, and uproot factories and warehouses. However, where there has been good engineering foresight and disaster control planning, damage mitigation can often be accomplished.

A review of some major hazmat disasters that have occurred around the world will offer a better understanding of how to prevent, control, and mitigate future disasters. The fact that some of these incidents occurred many years ago does not render the lessons less valuable. The term *experience* has been defined as "that which enables you to recognize a mistake when you make it again." Obviously, it would seem worthwhile to periodically review hazmat catastrophic experiences to avoid repeating them. This survey of environmental disasters will start with those that have resulted in air pollution. Next, disasters that have polluted water or land will be discussed.

Environmental disasters

Air pollution can arise from many sources. These include factory smokestacks, chemical spills and releases, ruptured pipelines, combustion products from hazmat fires, decomposition of buried or submerged materials that come to the surface, and radioactive emissions from nuclear accidents.

Industrial emissions

In October 1948, a thick, smoggy blanket of factory fumes and fog choked the 14,000 inhabitants of Donora, Pennsylvania for four consecutive days. "Almost 6,000 people had coughing, sore throats and breathing problems, smarting, watering eyes or nausea and vomiting. People with heart disease and bronchitis were crippled by the filth in the air; many people had to be given emergency oxygen treatment. In the four days before the wind and rain came, 20 people died."[15]

In a 1952 London disaster, industrial pollutants choked the city for five days. People were unable to keep the pollutants out of their homes and struggled to navigate through the muggy darkness. Of the many thousands who became ill, more than 4,000 died.

The smog that blanketed the city of New York in 1963 for only a brief time was directly responsible for the deaths of up to 400 people.

Fortunately, the authors cannot provide any newer dates for such events, due to the strict enforcement of the Environmental Protection Agency's (EPA) regulations. Occasionally, complaints will still occur about questionable industrial emissions, but the EPA quickly investigates them. Any findings can result in hefty fines being assessed against the company polluting the environment. This is confirmed in an article by Glenn Hess, who points out that industrial chemical releases fell by 20% in North America between 1998 and 2003.[16]

Chemical leaks, spills, and releases

On December 1, 1984, a drum of sodium hydroxide (drain cleaner) in a dumpster began to give off a corrosive vapor cloud that caused the evacuation of several thousand weekend shoppers from a mall in Hicksville, New York. The first arriving fire units sealed off the rubbish chutes and stairways leading from the area under the mall (where the dumpster was) with tape and salvage covers. They also reversed the ventilation system, creating a positive pressure that assisted police and fire units in the evacuation of the four floors of the mall building. While firefighters handled the dumpster fire, the hazmat unit set up a hot zone around the tunnel entrance. They also established a medical triage and treatment area and began construction of a decontamination site near the hazmat command post. A staging area was set up near the primary command post for arriving apparatus from nearby towns; these brought additional floodlight and air charging units, along with additional ventilation fans and tarps. Using a dumpster pickup unit, the hazmat unit brought the dumpster out in the open, where they removed the lye drum and packed it in a recovery drum. The burning material was covered with an absorbent dryer compound and then sand. After thorough ventilation, the mall was reopened in less than four hours.

This event shows how easily a person with a basic understanding of chemistry and the intent to create a mass panic and casualty event could do so with materials that exist in large quantities everywhere, every day. Fortunately, this event was an accident. Chemical, biological, radiological, nuclear, and explosive (CBRNE) incidents will be discussed in more detail in chapter 11. However, planners and responders must understand that the difference between some accidents and terrorist events is only the intent.

In 1973, about 6,000 residents fled from their homes in Auckland, New Zealand to escape the deadly vapors of 12 leaking drums of a chemical defoliant that had been unloaded in the port city. While 350 persons were being treated in various hospitals, hundreds of firefighters, police officers, ambulance operators, and troops tried to neutralize the lethal gas and the liquid in the drums. With this type of chemical, death comes from respiratory failure either by inhalation or through skin penetration.[17]

Another highly toxic chemical defoliant known as dioxin, or ICDD, escaped in a cloud of gas from a malfunctioning safety valve in an Italian factory in Seveso, on July 10, 1976, commencing a medical disaster. For more than a week, none of the 100,000 residents of the area grasped the magnitude of the disaster. When animals began dying by the thousands, trees began to wither, and doctors' offices were deluged with hundreds of patients complaining of skin infections, the health minister realized that something was wrong and had the area evacuated. To halt the spread of contamination, an army of veterinarians in protective suits began to kill all surviving animals in the Seveso area. Corn fields and vegetable crops had to be destroyed because the fine vapor had been washed down into the soil by a heavy rain. Small doses of this chemical can damage not only human kidneys, livers, and lungs, but also can cause deformed fetuses in pregnant women. As a result, therapeutic abortions were commenced on August 13 at the request of women from the most contaminated region. About 5,000 persons had to abandon their homes permanently. A government scientific commission recommended that all vegetation in the locale be destroyed in special furnaces and that all buildings, including the factory (ICMESA plant) and the evacuated homes be razed within three years.[18]

A large leak of silicon tetrachloride occurred in Chicago at 122nd Street and Stony Island on April 26, 1974, about noon. The emergency was not over until May 3, 1974. A malfunctioning valve at the bottom of a 1,250,000-gallon storage tank containing 800,000 gallons of this water-reactive chemical caused a spill of about 300,000 gallons into a diked area. Mixing with air, this chemical formed a dense white cloud of hydrogen chloride, an acid vapor that has a toxic threshold of about 5 ppm. By 6 a.m., a cloud of toxic fumes five miles long and one-half mile wide had spread over the south side of the city, causing cessation of all port activities and evacuation of a residential area known as Algeld Gardens, three miles southwest of the tank terminal. The area affected by the leak included at least 30,000 persons. After flying over it, Michigan's governor ordered mobilization of three National Guard battalions to assist in evacuation of 16,000 persons who had been ordered to leave the area.

The fire department was called at 2 p.m. to furnish all available ambulances and to assist in other operations that continued for eight days. The first evacuation headquarters was set up at the Washington Carver Hospital, but it had to be abandoned when fumes seeped in and people began to collapse. A high school was then selected as a new center, and about 2,500 persons had already gathered there by 11 p.m.[19]

The fire department transported some victims to the hospitals and administered oxygen inhalation to others. They also assisted in activities aimed at getting the leak stopped, the liquid transferred, and the spill neutralized and covered. They also provided aerial reconnaissance of the cloud path. Attempts to stop the leak included the use of transfer pumps, "hot-tap" equipment, and special quick-drying cement. Attempts to neutralize the product in the diked area included the use of lime, sand, and oil; covering it with vinyl and rubber sheeting; and various diking techniques. In trying to stop the leak in the pipe between the tank and the valve, firefighters wearing self-contained breathing apparatus and chemical handler's suits tried shredded lead wool, compounds, and shrouds, but none of them were effective. The chemical suits ripped and tore easily, and nine firefighters had to be treated for acid burns caused by the liquid and vapors. When 200 gpm chemical pumps were used in an attempt to pump out the tank, the pumps burned out. They tried to get a hot tap into the tank, and they used fireboat heavy stream appliances to throw up a water curtain to contain vapor spread.

The diked area was first covered with lime, but then it rained, causing further vapor emission. The area was then covered with vinyl sheets to keep it dry, and sand was put on top to hold them down, but chemical action caused the vapors to penetrate the vinyl. It was then paved with asphalt, and tons of lime were spread on top, but this was not a complete vapor barrier. A sight scope was used to determine the depth of the liquid in the tank and the location of the hot spots under the vinyl covers. A fast-drying cement was pumped over the pipes and the leak, but it still leaked, though not as much, and the tank was finally pumped out. The acid vapors coated the lenses of eyeglasses and damaged fire vehicles.

Technical assistance was supplied by the Army Operations Center (AOC), U.S. Army Technical Escort Center (TEC) Edgewood Arsenal, Aberdeen Proving Ground, Maryland. However, the Army chemical equipment did not arrive on the site until 11 a.m., April 27.

Chicago Fire Chief Bill Foley told Assistant Chief Chuck Bahme that if he had to do it over again, he would "flash" the vapors of the spill right at the start, and get rid of it, instead of trying for days to cover it up. Many responders are focused on trying to mitigate incidents by putting out fires and then cleaning up the incident. Currently, the trend leans towards evaluating which action will do less overall damage. Many times, as was the case here, allowing a chemical to consume itself through combustion is the better and quicker option for mitigation. Response considerations will be discussed in more detail in the later sections of this chapter.

The leaking of a poisonous gas, methyl isocyanate, from a Union Carbide Corporation pesticide plant in Bhopal, India, on December 3, 1984, created a continuing catastrophe. About 2,500 people who lived around the plant died immediately, and by 1987, 8,000 more had died and 300,000 more had suffered debilitating injuries. In 1987, the Supreme Court of India ordered Union Carbide to pay the victims $470 million, apparently based upon the Indian government's contention that the plant was poorly designed and maintained. It rejected the argument that the plant had been sabotaged by disgruntled employees.

This country managed to have a period of no large-scale incidents involving hazardous materials facilities until March 23, 2005. On this date, an explosion at the BP PLC's Texas City facility killed 15 people and injured another 170. "The explosion took place in the isomerization unit, which is the phase in gasoline production used to raise the octane level. The blaze following the blast took two hours to extinguish, thus slowing the search and rescue efforts. This explosion and fire coincidently drove gas prices to the highest level ever in the United States, although no production was stalled due to nearby backup facilities."[20]

Fire-caused pollution

Air pollution can also be caused by fires, sometimes with serious consequences. On January 14, 1976, Los Angeles city firefighters were called to a fire in an industrial building in Chatsworth, a northern suburb of the city, where containers of hydrochloric acid, cylinders of chlorine, and large quantities of a plastic derivative of polyvinyl chloride were stored. Wearing breathing apparatus, the firefighters entered and quickly knocked down the fire, unaware that there were chemicals in the building. "Soon the effects of the fumes were felt. Fire fighters dropped to their knees, vomiting, and with burning sensations in their mouths, throats, and lungs, and on exposed areas of skin. Their masks provided no protection from the fumes and fire fighters in the building, unable to breathe, managed to stumble out or be dragged to safety."[21]

Of the responders, 40 firefighters, 11 police officers, and 21 civilians were rushed to eight hospitals for medical treatment. Helicopters broadcast the alarm. More than 2,000 persons had to be evacuated from the area, including 1,400 students from two schools. The fire was brought under control in 32 minutes, but the fumes of the gases and vapors could be detected 17 miles away. Students were taken to a high school in a nearby community, and other evacuees were taken to a center established by the Red Cross in the Chatsworth High School gymnasium. More than five hours after the fire was controlled, police officers were still making a house-to-house search of the area to see whether anyone had been overcome in their homes.

Despite modern fire codes, which provide detailed regulation for chemical occupancies among thousands of businesses and warehouses where dangerous chemicals are to be found, fires occur daily. Occasionally one results in a near disaster because of the release of toxic gases from chemicals. Such was the case on June 15, 1975, when fire broke out in a swimming pool chemical firm in the San Fernando Valley, where 250,000 pounds of dry chemicals used for chlorinating water were stored in drums. The pall of poisonous chlorine gas released from these chemicals, due to heat and the application of hose streams, spread over four square miles of the valley, forcing evacuation of about 5,000 people and hospitalizing 50. After the fire was checked, firefighters stopped the application of water to reduce the chlorine emission, and let the fire burn out over a period of nine hours. At times the dense white smoke billowed as high as 500 feet above the scene, forcing police helicopters to withdraw. (Two officers from the helicopters were among those treated for fume inhalation.) But for the most part, the smoke settled close to the ground and was then spread by the wind. The Red Cross established two evacuation centers at gymnasiums of secondary schools where about 2,000 persons, including 100 elderly people and invalids from a nearby convalescent home, were provided cots and mats to spend the night.[22]

More than 1,000 persons had to be evacuated from a 20-square block area in Hawthorne, California, when a chemical reacting in a detergent manufacturing plant generated clouds of irritating sulfuric acid fumes. Eleven persons had to be given hospital treatment for the effects of vapors, which caused extreme breathing problems. Police officers cordoned off streets leading into the area. Firefighters, wearing breathing apparatus, washed down the area where sulfur trioxide had been in the process of being transferred to a storage tank at the time of the mishap. Upon hitting the air, the sulfur trioxide converted to sulfuric acid vapor, and the acid fumes billowed through the plant and out into the streets.

This same chemical put another city's disaster plan for hazmats to a severe test. On January 4, 1978, a leak developed in the discharge pipe of a Baltimore, Maryland, soap company's sulfur trioxide tank, thus producing a sulfuric acid cloud that spread over the city. More than 100 persons were treated for respiratory injuries during the 4½-hour incident.

The DOT's guidebook suggests evacuating an area two miles long and one mile wide if there is an 800-square-foot spill of this toxic and corrosive liquefied gas. However, no one could advise the chief how far he would need to evacuate the 300,000 inhabitants affected by a leak from a 4-foot by 10-foot tank of sulfur trioxide.

The cloud of vapor was tracked by Maryland State Police in a medevac helicopter as it moved into Pennsylvania. Radio and television messages warned people to get off the streets and stay indoors; firefighters assisted the staffs of three nursing homes to keep the fumes from entering the buildings and used smoke ejectors to move fumes out of buildings.

A field headquarters command was established to direct three command posts where spare air cylinders were brought for replacement of depleted ones on breathing apparatus. The contaminated air precluded recharging the empty bottles with the fire department's air compressor.

The fire chief's disaster plan contemplated localizing this type of chemical emergency and evacuating persons in danger, but it did not envision the need for evacuating hundreds of thousands of residents.[23]

Air pollution from vaporized irritating toxic liquids at least gives some warning of the hazard; such is the case with the nonflammable sulfur dioxides and nitric oxides. However, if a tank car of liquid hydrogen cyanide were to rupture and vaporize, untold numbers of people could drop dead without even suspecting that the mild almond odor was highly poisonous. Like hydrogen cyanide, styrene is both flammable and unstable, but it does give off an unpleasant odor. The following incident, taking considerable time to control, also illustrates the ignorance of some high-ranking plant officials.

On August 28, 2005, an abandoned 20,000-gallon railcar of the chemical styrene leaked in a neighborhood in the eastern portion of Cincinnati, Ohio. The leak forced the evacuation of the eastside neighborhood near Lunken Airport, as well as a shelter-in-place order for a one-half mile area around the incident. The incident continued for nearly five days and closed schools in the area. This incident was unique in the fact that the location of the leak prompted shelter-in-place orders in areas of Ohio as well as in neighboring Kentucky. This incident, although not catastrophic, is a good example of the need for jurisdictions, counties, and states to have a cooperative relationship and be able to work together for a common plan. This allows proper mitigation of the incident and prevents unnecessary harm to the citizens.[24]

On August 24, 1978, 500 people from 14 plants had to be evacuated because of a leaking tank car of styrene in the city of Industry, another community not far from the scene of the preceding incident. Firefighters had the tank car moved to isolate it from other railroad cars and sprayed it with heavy stream appliances; the addition of chemical inhibiters to the tank car did not stop its heating. Two oil company experts were flown in from Houston, Texas, and they took samples from the bottom of the tank car to help determine when the vaporizing process would end. They also removed a 1-foot-diameter lid on the car to relieve pressure caused by the vapors expanding inside the car.[25]

Figure 9–1 depicts large amounts of pollutants being discharged as a result of an oil refinery fire.

Fig. 9–1. Oil refinery fire. Note the large amount of smoke and pollutants being sent into the atmosphere as a result of the fire.

FEMA News Photo.

Water pollution

Industrial wastes are a principal cause of drinking water pollution. The Mississippi River, which serves as the source of raw water for 1.5 million people along the 258 miles from St. Francisville to Venice, Louisiana, had about 10 metal pollutants discharged into it at one time.[26] Although water pollution of this type may have disastrous effects over a long period of time, it is not the kind that fire departments are apt to be concerned with in emergency situations. Flammable liquid spills from overturned tank trucks or railroad tank cars or from leaking storage tanks that end up in rivers and harbors, whether directly or by means of storm drains and flood channels, can create potentially disastrous situations.

In one memorable incident in 1972, a tank truck apparently sheared off a pipe leading to a 30,000-gallon tank of vinyl acetate at Los Angeles Harbor, starting a fire that immediately exposed other nearby tanks of acetate solvents. Within minutes, the fire spread to other tanks, causing their rupture, blowing off their covers, and sending balls of fire into the air.

Firefighting efforts involved the use of 27 engine companies, 5 truck companies, 2 foam units, 2 paramedic units, and a large crash truck brought in especially to apply its aqueous film-forming foam (AFFF). Even so, 21 of the tanks were eventually destroyed, with damage estimated at $500,000. Two of the city's fire boats were used, and one was pumping its full 10,000-gpm capacity through several lines, including a 3½-inch line.

The chief flew to the fire by department helicopter. One of his first concerns was to prevent burning liquids from reaching the harbor waters, since dikes around the tanks were only 30 inches high and were being deluged with foam and filled with water being applied to cool exposed tanks. Three pumpers were ordered to start drafting water from the dikes, and one of the airport's crash trucks was called to apply 750 gpm of AFFF from its main turret nozzle to kill the ground fire. With the assistance of Los Angeles County fire crews, the fire was under complete control in less than four hours, and pollution of the harbor was prevented.[27]

Pollution from radioactivity

Radioactivity is a source of environmental pollution that probably has the potential for more widespread and longer lasting disasters than any hazmat known so far. The authors of a *Newsweek* article on the Three Mile Island nuclear plant episode said that "it began with a brief but terrible bump in the night and grew into the worst commercial nuclear accident in history."[28] They would have added "in the United States" if they could have foreseen that six years later on April 26, 1986, one of four reactors at Chernobyl would catch fire and explode in the USSR. Unofficial reports say that 250 died in the blast that spewed cesium, strontium, and plutonium across much of the Ukraine, Byelorussia, and Russia and sent a cloud of radioactive gas around the world.

About 116,000 people were evacuated shortly after the accident. Hundreds of thousands will suffer long-term health effects.[29] Later, 135,000 people were evacuated from Ukrainian villages where the delayed reaction has killed or maimed thousands of humans and animals. For example, on a collective farm about 40 miles south of the blast site, another calf was born without any eyes, and the children "have a lot of throat trouble and illness that nobody understands," probably caused by the unendurably high levels of radiation. Soviet leaders agreed in April 1990 that $26 billion should be spent to resettle people in the affected areas.[30] In a 1987 trial of the Chernobyl plant officials, the chief engineer and his deputy got a sentence of 10 years in a labor camp.[31]

The Three Mile Island accident near Harrisburg, Pennsylvania, occurred due to mechanical and human failures that resulted in a breakdown of a reactor's core cooling system. As a result, radioactive steam leaks developed that continued to spread out over the community for several days, to a distance of 20 miles. The spread of radiation might have been catastrophic had the technicians not succeeded in preventing a core meltdown.

Although nuclear plants do not release solid or chemical pollutants into the atmosphere, they do threaten the environment in other ways. They release small amounts of radioactive gas into the air. In water-cooled plants, a small amount of tritium remains in the water when it is returned to the river or lake. Moreover, this heated water causes thermal pollution that can harm animals and plants that live at lower temperatures. One of the largest unsolved problems to date is the disposal of radioactive wastes produced by reactors. These wastes remain radioactive for thousands of years and can cause serious radiation damage to all living things. In the United States, these wastes are now stored on the plant grounds. Most scientists believe that storage in underground salt deposits would be the best solution to the problem. These storage areas might have to be guarded for thousands of years to eliminate the risk that radioactivity might be released by someone inadvertently digging into them.

Radiological considerations

Accidents involving radioactive materials cause a great deal of concern among responders and the public. This concern is due in part to a relatively poor understanding of the physical properties and effects of radioactivity, but also because it can be detected only with special equipment such as a Geiger counter. In many cases, this high anxiety will have a much larger impact on the medical response than will the actual cases of exposure. A great many people may believe they have received large doses, despite assurances from authorities, and will seek medical assistance.

A major consideration in radiological incidents is whether or not there has been dispersion of radioactive contaminants in particulate, liquid, or gaseous form. Such conditions will require decontamination procedures similar to those required with other hazardous materials. Many radioactive materials, such as industrial radiographic cameras, employ sources sealed in welded, stainless steel capsules. Accidents with these sealed sources may still involve large-dose radiation exposures to those within a few yards of the source. Unless the capsule is damaged, however, contamination will not be a factor.

Most accidents in transportation or typical industrial settings involve just a single radioactive material. Nuclear power plant accidents, however, can release a variety of radionuclides (radioactive chemical elements) that may escape primarily in particulate or gaseous forms. This may complicate medical response strategies. Some materials may be highly radioactive but may decay to background levels in just a few days. Others may be less radioactive but may have very long half-lives and cause long-term problems by entering the food chain.

A National Radiological Response Plan, developed by the federal government in the aftermath of the Three Mile Island incident, describes how federal agencies will coordinate their activities and interact with local and state government. This plan now resides as an annex of the Nation Response Plan, as described in chapter 4. The annex is the "Nuclear/Radiological Incident Annex." According to the document, it is for providing an organized and timely, coordinated response by federal agencies to terrorist incidents involving nuclear or radioactive materials, and accidents or incidents involving such material that may or may not become an incident of national significance. Principal federal agencies include the Nuclear Regulatory Commission, Department of Defense, Department of Energy, Department of Homeland Security, Department of Defense, and Department of Justice. Fire departments that are within nuclear power plant emergency planning zones or could otherwise encounter a major radiological incident should be familiar with this plan.

Other nuclear accidents

On March 22, 1975, a fire was started accidentally in the Brown's Ferry, Alabama nuclear plant by an employee who was using a candle to test for air leaks in the combustible fire wall seal. The fire was not extinguished by the plant's brigade, though its members discharged portable carbon dioxide and dry chemical fire extinguishers on the polyurethane sealing material burning in the cable orifice of the division wall between the two units. The fixed carbon dioxide system did not extinguish the fire, which extended into the cable tray system, consuming insulating and jacket materials, though the system discharged three times. The Athens Fire Department was called to help, and after surveying the fire, the chief recommended that water be used to extinguish the blaze. However, use of portable extinguishers was continued because the plant superintendent feared that water might further degrade plant operation.

About seven hours after the start of the fire, water was used for extinguishment. As a result of the fire damage, two nuclear generating units were shut down for more than a year. Property damage was estimated at $10 million, and cost of replacement power was about $120 million. There was no release of radioactivity as a result of this fire, although products of combustion of the cable jacket and insulation materials were toxic and required use of breathing apparatus during firefighting. This incident affirmed the necessity for using water on cable fires to prevent the loss of circuits not already affected. Water would have diluted corrosive vapors and limited the extent of necessary cleanup. The fire emphasized the need for plant fire protection systems.

There is some concern that sabotage might be used against nuclear power installations. The Nuclear Regulatory Commission (NRC) reported that 99 threats of violence were directed against nuclear facilities in a two-year period, while 76 other threats were aimed at federal nuclear power generators. Since the attacks of September 11, 2001, the perception of the potential terrorist threat to nuclear installations has changed significantly. Within the nuclear industry, the immediate international response was to enhance security by augmenting the forces guarding installations, increasing physical protection by installing additional security devices, enhancing protection procedures, tightening access control, increasing standoff distances for surface vehicles, reviewing and updating emergency preparedness, and generally increasing awareness of the need for close cooperation, at all levels, between government and private sector entities concerning warning and response.[32]

The NRC is said to have accepted a probable casualty figure of about 3,300 if the worst possible meltdown should occur, though this figure would vary up or down depending upon the population density and the wind. However, Henry Kendall, former director of the Union of Concerned Scientists, once estimated that between 23,000 and 36,000 persons might die after exposure to the radiation of a meltdown. Experts today, in view of the Chernobyl accident, are revising casualty estimates upward.

On May 5, 2006, Breibart.com reported that an accidental release of I-131 gas at a nuclear power plant in Minnesota exposed approximately 100 plant workers to low-level radiation. Most workers were exposed to 10 to 20 millirems (0.1–0.2 mSv), which is equivalent to a dental X-ray. The workers were equipped with protective gear at the time, and no radiation leaked outside the plant to the surrounding area.[33]

Chemical explosions causing disasters

Chapter 8 on transportation disasters discussed how certain chemicals have been responsible for disastrous explosions in trucks, tank cars, trailers, and vessels in every conceivable location, including railroad yards, business districts, harbors, highways, rivers, and on the high seas. The largest ship ever to sink was the 207,000-ton *Marpessa*, which went down in December 1969, after an explosion ripped open its hull.[34]

There have been explosions unrelated to transportation that are memorable for their great destructiveness or large loss of lives. Most were not caused by "explosives" as defined in transportation regulations, but by chemicals never intended to be used for producing an explosion.

Cellulose nitrate has not generally been considered to be an explosive, though it is the main ingredient in smokeless gun powder. However, its ability to produce an explosion has been often demonstrated.[35]

Smokeless powder was the cause of a major, second explosion in the basement of a hardware store in Richmond, Indiana, on April 6, 1968. In this incident, 41 people were killed and more than 100 injured. The first and lesser explosion may have been of natural gas escaping from the ancient and corroded street main.[36]

Explosive magazines that contain weaponry for a country can also create an explosive incident. As reported by the Associated Press on March 24, 2007, multiple explosions and fires, as well as propelling ammunition, was the scene in a small South African community. At the time of the report, the explosive incident had killed 93 people, with more than 300 injured.[37]

Though not commonly classified as explosives, powerful oxidizing chemicals have been responsible for disasters. The explosive power of ammonium nitrate was amply demonstrated at Texas City in the two ship detonations discussed previously. It was later demonstrated ashore at Traskwood, Arkansas, in the boxcar detonation following the train derailment and fire, and in the fertilizer plant detonation at Mount Vernon, Missouri.[38]

On April 19, 1995, the explosive power of ammonium nitrate was demonstrated in the worst event of domestic terrorism known to the United States, with the bombing of the Alfred P. Murrah Federal Building in Oklahoma City. This tragic event is discussed at length in chapter 11. For this chapter, it should be noted that the investigation claimed that the explosive used was ANFO, which is comprised of ammonium nitrate and fuel oil. According to the investigation, a Ryder truck full of the mixture was parked in front of the building and detonated, thus causing great destruction to the building and loss of lives. Although there are varying opinions concerning whether or not this type of explosive could produce the damage inflicted, it is still a convincing example of the power of ammonium nitrate.

The explosions in Henderson, Nevada would have been somewhat reminiscent of a similar devastating incident that had occurred in England almost two decades earlier, except that most of the PEPCON employees had time to escape. At the English chemical plant in Flixborough, at 4:53 p.m. on Saturday, June 1, 1974, a huge vapor cloud consisting of mostly cyclohexane began to rise. When the base of the cloud reached a source of ignition, a devastating explosion occurred that killed 28 persons and injured more than 100 others. In addition to the damage caused to 2,488 surrounding homes, shops, and factories within an eight-mile radius, the plant itself was nearly totally destroyed at a loss of about $48 million. Because of the fear of further explosions and possible release of 20,000 tons of stored toxic ammonia, about 20,000 people within a five-mile radius of the plant evacuated the area. It took nearly three days for the 250 firefighters to bring the fire under control, and some units continued patrol and cooling operations for an additional six days.[39]

On January 4, 1998, an explosion was averted at the Cargill Chemical Plant in Cincinnati, Ohio. On this day a fire at the 25,000-square-foot warehouse occurred, which was in the vicinity of 420 tons of ammonium nitrate. As a precaution, nearly 3,000 people were evacuated from around the plant and railroad, and Ohio River traffic was halted for hours. A *Cincinnati Enquirer* article did mention that the Murrah Building bomb was considered as an example of what could happen, but officials stated that some subtle differences were present, such as the lack of fuel oil or containment of the ammonium nitrate. The response did, however, consist of nine fire departments and two ambulance companies from Mason County.[40]

"Combination" hazmat incidents

Some hazmat incidents do not lend themselves to neat categorization. The BASF plant explosion in Cincinnati, for example, included all the elements of a hazmat operation, a major fire, a mass casualty incident, and a large-scale evacuation.[41] Many disasters may be multifaceted, involving several types at once.

At least 155 persons were treated at three Memphis, Tennessee hospitals for respiratory problems, eye irritation, and chemical poisoning resulting from exposure to smoke issuing from a burning pesticide plant. About 40 employees were in the plant when an 8,000-gallon tank of methyl parathion exploded, forming a mushroom cloud of toxic smoke that rose over 1,000 feet in the air.

Firefighters wearing breathing apparatus dodged exploding drums of organic phosphates while directing large stream appliances on the burning structure. Police officers in helicopters covered a five-square-mile area broadcasting evacuation orders issued by civil defense officials. All boat traffic on the Mississippi River was halted 70 miles south of Memphis, and railroad cars were stopped from entering the industrial area around the plant.[42]

The usual hazmat references did advise of the health danger of methyl parathion, which is commercially found formulated in a 20% solution of xylene, a toxic flammable liquid, but they did not mention that it is also an explosive that can be detonated in a fire. However, the warning can usually be found that it "decomposes at temperatures above ambient and may develop sufficient internal pressure to cause the container to rupture violently."[43]

However, the references are now updated, likely because of this tragedy. The 2004 *Emergency Response Guidebook* (*ERG2004*), states in Guide 152, which is the appropriate guide for methyl parathion, that "containers may explode when heated."[44]

On February 22, 1980, the *Los Angeles Times* reported, "A big refrigeration compressor exploded with earth shaking force at the Knudsen Corporation dairy products plant in Central Los Angeles Thursday night, sending up a huge cloud of ammonia gas that overcame or seared scores of persons."[45]

Of those injured in the incident, 22 from the plant and nearby apartment house were taken to six hospitals in seven ambulances for treatment on February 21, 1980. Among the injured were 4 firefighters; they had suffered inhalation and corrosive injuries while trying to effect a shutoff of the ruptured high-pressure pipe in the plant's engine room. Other firefighters stripped off the injured firefighters' clothes and washed the victims with streams of water before loading them into the ambulances.

There were 11 fire companies that attempted to keep the ammonia gas from reaching a high enough concentration to enter its flammable range (16%–25%) through the application of water spray on the escaping gas. This gas is quite soluble and readily converts from anhydrous to aqueous ammonia. The compressor blew up at 6:00 p.m., and it was 8:00 p.m. before the gas flow could be stopped and the ammonia cloud began to dissipate.

Pipeline rupture disasters

Pipelines carrying flammable and toxic gases and liquids sometimes rupture because of overpressure, corroded and weakened walls, and from being dug up by excavators. Lines can be found in the business districts of large cities, such as the 2-inch high-pressure line of ammonia that ruptured in the center of Los Angeles a few years back, causing evacuation of several multistory buildings that were constructed before the era of nonoperable windows. This line supplied refrigeration equipment from a cold-storage plant on the east side of the city.

Underground transmission lines of natural gas, gasoline, LPG, coal slurry, and ammonia can be found in rural areas; some extend through several states. There are more than 300,000 miles of natural gas pipelines in the United States.

Examples of pipeline rupture

In June 1989, tragedy struck twice in San Bernardino when a gasoline pipeline ruptured and exploded just 13 days after a runaway train had careened into a group of homes at the same site. Both calamities caused fatalities, injuries, and severe damage to homes.[46]

It was in a rural area near Natchitoches, Louisiana, in March 1965, when a pipeline failure resulted from corrosion in a 24-inch natural gas pipeline. It had been installed in 1944 and operated at a pressure of 750 psi. When the rupture occurred, it took about 30 minutes to bleed down 17 miles of pipe—the distance between isolation valves. The rupture tore a huge crater in the ground, and the escaping gas ignited immediately, acting as a gigantic torch to ignite seven homes, several autos, and outbuildings. The heat was so great that all buildings were completely consumed, and the automobiles melted to the ground. As a result, 17 persons were killed by the fire and at least 9 others injured.[47]

A natural gas explosion killed 11 people and created a blast crater 85 feet long, 25 feet wide, and 20 feet deep at a campsite near Carlsbad, New Mexico. The explosion occurred on August 19, 2000. The explosion took place along the Pecos River, about three miles east of U.S. Highway 285. The resulting fire from the blast could be seen 30 miles away in Carlsbad. The fire burned 40 minutes to an hour. The pipeline, installed in 1950, was reported to have been inspected on August 2.[48]

Disaster struck the Soviet Union's Trans-Siberian Railroad on June 4, 1989, when 460 passengers were killed in the blast from liquefied petroleum gas that had escaped from a ruptured pipeline and was ignited by sparks from one of two passing passenger trains. This gas explosion caused the second-worst train wreck in world history, exceeded only by a 1981 crash that killed 800 people.

TOXIC INDUSTRIAL CHEMICALS AS WEAPONS OF MASS DESTRUCTION

Although chapter 13 extensively reviews and discusses the elements of CBRNE events, it is appropriate to realize that many instances of WMD events have involved toxic industrial chemicals. Some of the chemicals that are of choice to terrorist organizations are chlorine, phosgene, ammonia, and hydrogen cyanide, most which exist in medium to large quantities in most large industrial complexes or technology producing areas.

From an article produced in March 2007 by the BBC News, it is evident that chlorine gas attacks are popular among the insurgents of the Iraq war. The article reports three attacks in the month of February, and one of those attacks forced 350 people to seek medical attention for exposure to the chlorine gas. The article also points out that the attainment of chlorine gas is easy, as it is used for water purification and as a cleaner.[49]

The toxic industrial chemicals are used as WMDs in order to inflict health problems on the human body. These effects can range from a sore throat to instantaneous death from exposure to a minute amount of the product. These chemicals are important for all local and state planners to recognize, as the WMDs are often not able to be smuggled into the United States by terrorist groups. Consequently, these groups must steal the products from local businesses in order to conduct their event. Many times due to low security in rural areas, they choose remote plants and industrial facilities. This is much the same as the methamphetamine producer's theft of anhydrous ammonia from rural farmers and coops that have large tanks of the substance for use as fertilizer.

Obviously, during a theft of the product, transfer safety is not at the forefront of the perpetrator's concerns. Therefore, as a secondary problem to contributing to another terrorist event, a leak of the toxic industrial chemical can occur. Mitigation techniques may be hindered due to the bypassing of safety equipment in order to accomplish the theft.

CONTROL AND MITIGATION OF HAZMAT DISASTERS

After viewing the many instances in which hazmats have brought about incalculable harm to the inhabitants of many communities and their environment, it is appropriate at this point to consider what can be done about such catastrophes beyond taking every possible preventative measure. It is especially important to do this in view of the fact that fire departments are usually called upon to handle such emergencies. An incident commander who responds to the scene can almost certainly predict that not every built-in safeguard designed to control the situation will function as it was designed to do. As the old familiar Murphy's Law pointed out long ago, "If anything can go wrong, it will," despite all the precautions that may have been taken to avoid it. It would appear that Murphy was right, for instance, in the 1986 explosion of the Space Shuttle *Challenger* that killed the seven astronauts on board shortly after it was launched. Probably no vehicle ever has been constructed with more carefully designed safeguards for its solid rocket booster, and yet a single design flaw in the O-rings spelled disaster.

Therefore, if disasters cannot be prevented, it would be prudent to make the best possible plans for controlling and mitigating the effects of hazmat disasters. This raises some important questions: Who should make those plans? What laws need to be considered in their preparation? What agencies should be consulted? What data needs to be assembled for the task? What should their content be? Who has to approve them? How can they be tested? Many of these questions have already been considered in an earlier chapter on planning, so it might be well to focus on those aspects that have a particular bearing on the hazmat problem. The late Robert Obermayer has outlined these aspects in his essay, "The Intricacies of Hazardous Materials Planning," which he prepared for the Open Learning Fire Service Program.[50] The authors accepted his kind offer to allow the inclusion of some of the material here under the heading of "Planning."

PLANNING

The achievement of most desirable goals is only possible through planning.

Reasons for planning

To protect its personnel, the fire department must recognize the dangers involved in the mitigation of a hazardous materials incident and plan accordingly. Planning takes on a life and safety connotation and involves an intensive effort. This effort begins with recognition of the need to plan, extends to identification of hazards, continues with identification of needed and on-hand resources, and culminates in successful mitigation of an incident.

Reasons why the fire service must look to preincident planning as the most appropriate means of facing the hazardous materials challenge include the following:

- Magnitude of the danger
- Need for specialized training, equipment, and expertise
- Number of agencies that will become involved in an incident
- Need for compliance with federal and state legislation

Federal legislation and the hazardous materials issue

This section discusses several pieces of federal legislation that apply to hazardous materials and must be considered in the development of a preincident hazardous materials plan. Indeed, in many communities, the local planning effort may have been initiated by one or more of these federal laws. Limited space precludes a discussion of relevant legislation in each of the 50 states. However, it is important to be aware of one's own state laws concerning hazardous materials. Copies of pertinent legislation may be obtained from the local public library or from the Internet.

General federal authority

Article 1, Section B of the U.S. Constitution grants the federal government the authority to regulate interstate commerce. Although this federal authority has been used to develop regulations concerning the transportation of hazmats, the federal government's intent is to support and abet local hazmat mitigation efforts. The legislation that is discussed below exemplifies this intent.

National Response Framework (Formerly National Contingency Plan)

The National Response Framework, in its hazardous materials annex, which was formerly called the National Contingency Plan (NCP), required by Section 105 of the Comprehensive Environmental Response, Compensation, and Liability Act (CERCLA), calls for preparedness and training. One important feature of the act was the establishment of the National Response Team (NRT), which is the coordinating federal agency for emergency planning. It consists of representatives of 14 federal agencies with significant responsibilities in the areas of worker safety, transportation, the environment, public health, and emergency management.

On-scene coordinators, employed within the Coast Guard and the Environmental Protection Agency, are federal officials predesignated to coordinate and direct federal responses and hazmat removals under the National Response Framework.

Federal resources may be triggered by a report to the National Response Center (NRC), which is operated by the Coast Guard and relays the initial report to the appropriate Coast Guard or EPA on-scene coordinator, depending on the location of the incident. Provisions of several federal laws, including the Federal Water Pollution Control Act and CERCLA, require the immediate notification of the NRC of any discharge or release of hazmats. Based upon an evaluation of the initial report, the decision to send an on-scene coordinator to the scene is made. The on-scene coordinator will monitor the response of the responsible party and the state or local government and may also initiate federal action.

Regional Response Teams (RRTs) are composed of representatives of federal agencies and a representative from each state within the region. During the mitigation of a major incident, the on-scene coordinator may request a meeting of the Regional Response Team to provide advice and recommendations on specific mitigation issues.

Title III, Superfund Amendments and Reauthorization Act (SARA)

SARA contains planning requirements relative to hazmats, including the establishment of state emergency response commissions, emergency planning districts, and local emergency planning committees. Appointed by the state governor, the state emergency response commissions have the following responsibilities:

- Designation of emergency planning districts
- Appointment of local emergency planning committees for each planning district
- Review of emergency plans
- Establishment of procedures for processing public requests for information about emergency response plans, material safety data sheets, and the list of extremely hazardous substances

Emergency planning districts are intended to facilitate the preparation and implementation of emergency plans. Such districts may follow existing political subdivisions or may create multi-jurisdictional planning organizations. The local emergency planning committee for each district must include, among others, members of state and local government, emergency services, community groups, the media, and owner/operators who are subject to the provision of SARA Title III. Each planning committee was expected to have a plan in place by October 1988, with annual review of the plan. The plan must address these items:

- Facilities and transportation routes relative to specific chemicals
- Response procedures
- EMS
- Training programs
- Evacuation plans
- Means of identifying populations and areas that may be at risk
- Means of detecting a spill or release

Also known as the Emergency Planning and Right to Know Act of 1986, SARA Title III requires all manufacturing employers to provide material safety data sheets (MSDS) to state and local authorities. Listing the properties, hazards, and other necessary information about every chemical on the site, the MSDS will be an indispensable aid to the planning process.

Resource Conservation and Recovery Act

The Resource Conservation and Recovery Act established a framework for the proper management and disposal of all hazardous wastes. The Hazardous and Solid Waste Amendments of 1984 expanded the law's scope and placed increased emphasis on waste reduction, corrective action, and treatment. This act authorized the EPA to issue standards and regulations for the management of hazardous wastes by the generator and the transporter, which are implemented through permits issued by the EPA. The permit application process requires the operator wishing to store, treat, transport, or dispose of hazardous wastes to characterize the wastes to be handled, demonstrate compliance with the appropriate regulations, and have a contingency plan. A community's emergency planning process must interface with the operators within the area who are subject to the Resource Conservation and Recovery Act, and the plan should be integrated with their plans.

OSHA regulations: the Right-to-Know law

Occupational Safety and Health Administration (OSHA) regulations require employers to develop and implement response plans. Specifically, employers whose employees will be responding to hazmat incidents from their regular duty station must have an emergency response plan. The plan should include a mechanism to recognize emergencies, methods to alert employees, evacuation procedures, EMS, line of authority, on-site decontamination, site control means, and a method for the evaluation of the plan.

An important OSHA regulation is the Hazard Communication Standard issued in 1982 and known as the Right-to-Know law. Again, the material safety data sheet is the mechanism through which the employee receives needed information on the specific hazmats contained on the site.

For further information, it may be helpful to read the *Hazardous Materials Emergency Planning Guide*, which is available from the National Response Center at (800) 424-8802. The document is also located online at: http://www.nrt.org/Production/NRT/NRTWeb.nsf/AllAttachmentsByTitle/SA-27NRT1Update/$File/NRT-1%20update.pdf?OpenElement.

Additional federal regulations concerning hazmats

Under the provisions of the Hazardous Materials Transportation Act, the Department of Transportation (DOT) has the responsibility for protecting the nation against the risks to life and property that are inherent in the commercial transportation of hazmats. DOT's authority ranges from issuing placards and labeling regulations to developing specifications for highway and railway hazmat transport vehicles. Several agencies within DOT have a regulatory responsibility relative to hazmats:

- Federal Highway Administration—regulates vehicular transport (http://www.fhwa.dot.gov/)
- Federal Railroad Administration—regulates rail transportation (http://www.fra.dot.gov/)
- Federal Aviation Administration—regulates civil aviation (http://www.faa.gov/)
- U.S. Coast Guard—regulates marine transportation (http://www.uscg.mil/)
- Materials Transportation Bureau—promulgates hazardous materials regulations for other DOT agencies

National Transportation Safety Board

The National Transportation Safety Board (NTSB), online at http://www.ntsb.gov/, is a part of the DOT but reports directly to Congress. The five-member board attempts to determine the cause of major transportation accidents and recommends measures to prevent recurrences. It is authorized by the Independent Safety Board Act of 1974 to assess the effectiveness of other government agencies in preventing transportation accidents and to publish its findings for public inspection. Although this agency has proven to be an effective monitor of other regulatory agencies, it has no authority to enforce its recommendations.

DEVELOPING THE HAZARDOUS MATERIALS PLAN

Preincident planning for hazmats involves a number of important considerations for proper development of a good plan. Because of the ever-present threat of hazardous materials, this topic deserves special consideration and should definitely constitute an appendix to the disaster planning process introduced in chapter 1. Here are a few guidelines that can help ensure that disaster planning will adequately treat the hazard threat. In figure 9–2, a planning flowchart is depicted. Planners are urged to utilize a written format of planning, such as the checklist in figure 9–2, so that no areas are missed when conducting the planning process.

The planning team

As the planning authority, the planning team must respect all legitimate interests. The team must attempt to balance the needs of the emergency responders with the needs of the site and vehicle owners or operator and the community at large.

The success of the planning effort will depend on the qualities of the team members. Needed qualities include the ability, commitment, authority, and resources to accomplish the planning task. Further, team members should have a wide range of experience and expertise relative to the specific aspects of the community and the community's hazmat problem. Team members must agree on their purpose, be capable of working together, and faithfully represent all interests within the community. The following is a list of potential members of an emergency planning team.

Hazardous Materials Planning Process Flowchart

1. Assess prevalent community hazards

A. Inventory facilities utilizing hazardous materials
- -Storage containers used at facilities
- -Processes present at the facilities
- -Amounts of hazardous materials at the facility
- -Number of personnel working at the facilities

B. Examine transportation routes in community
- -What types of commodities are present on the routes
- -What type of vehicles are carrying hazardous materials
- -Do rail lines intersect the community
- -What flight paths go over the community

2. Review and improve emergency plans

A. Examine current Emergency Operations Plans
- -Community
- -County/Parish
- -State

B. Examine facilities Emergency Response Plans
- -How does the local response protocols integrate

C. Examine Community's Departmental Policies
- -Fire Department, Emergency Medical Services
- -Police Department
- -Hazmat Response Teams

3. Assess capabilities

A. Assess the Capabilities of Local Responders Utilizing Resource Typing
- -Fire/Rescue Department
- -Police Department
- -Hazmat Response Team(s)

B. Assess Capabilities of Facilities Storing/Utilizing Hazardous Materials
- -Personnel
- -Specialized equipment

C. Assess Capabilities of Mutual Aid Resources Utilizing Resource Typing

4. Exercise emergency plans

A. Determine the Key Participants
- -Fire/Rescue
- -Facilities Management
- -Emergency Management Agency
- -EMS/Hospital Representatives

B. Conduct the Following Types of Exercises
- -Orientation Session
- -Tabletop
- -Functional
- -Full-Scale

5. Establishes a continuous improvement cycle

A. Establish and review/update schedule

B. Conduct steps 1–4 for each update cycle

C. Be sure to include key players in the update process

Fig. 9–2. Hazardous materials planning flowchart. This sample hazmat planning flowchart can be utilized to ensure no areas are omitted during the planning process.

Potential members of the planning team

As stated by the *Hazardous Materials Emergency Planning Guide*,

Part A. Experience shows that the following individuals, groups, and agencies should participate if a successful plan is to be developed:

- Mayor/city management (or representative)
- County executive (or representative)/board of supervisors
- State elected officials (or representative)
- Fire department (paid and volunteer)
- Police department
- Emergency management agency
- Environmental agency (e.g., air and water pollution control agency)
- Health department…
- Hospitals, emergency medical service, veterinarians, medical community (especially occupational medicine)
- Transportation agency (e.g., DOT, port authority, transit authority, bus company, truck or rail companies)
- Industry (e.g., chemical and transportation)
- USCG [U.S. Coast Guard]/EPA representative (e.g., agency response program personnel)
- Schools or school districts
- Technical experts (e.g., chemists, engineer from a university or plant)
- Community group representative
- Public information representative (e.g., local radio, TV, press)

Part B. Other group/agencies that can be included in the planning process, depending on the community's individual priorities:

- Agriculture agency
- Indian tribes within or adjacent to the affected jurisdiction
- Planning department
- Other agencies (e.g., welfare, parks, utilities)
- Municipal/county legal counsel
- Workers in local facilities
- Labor union representatives (e.g., chemical and transportation, industrial health units)
- Local business community
- Representatives from volunteer organizations (e.g., Red Cross)
- Public interest and citizens groups, environmental organizations, and representative of affected neighborhoods
- Key representatives from bordering cities and counties
- State representatives (governor, legislator's office, state agencies)
- Federal agency representatives (e.g., FEMA, DOT/RSPA [Research and Special Programs Administration], ATSDR, OSHA) and the local branches of Federal land management agencies[51]

Tasks of the planning team

During the planning effort, four tasks will confront the fire department's hazmat planning team:

- Review existing plan
- Review risk management plan information
- Perform a hazard analysis
- Assess available capabilities and resources

The plan review function may encompass a review of hazmat response plans of other agencies as well as the local community's. In addition to providing the planning team with new ideas, the review will help prevent overlap and inconsistency, while fostering interagency cooperation.

The hazard analysis will yield the necessary documentation for the support of the hazmat planning effort and the factual basis for the establishment of planning priorities. It allows planners to accomplish the following:

- Examine the local community's potential for an incident
- Estimate the potential magnitude of the incident
- Identify the most probable location of the incident
- Identify the material most likely to be involved

The hazard analysis may be simple or complex. It may focus on the most common hazard, such as the highway transportation of gasoline, or the most potentially destructive hazmat that is stored, utilized, manufactured, or transported through the community. In any case, it is the task of the planning team to determine the appropriate level of detail for the hazard analysis in view of local resources.

Hazard analysis employs several terms that should be mastered:

- A *hazard* is a situation that has the potential for causing damage to life, property, and the environment.
- *Vulnerability* is the susceptibility of damage to life, property, and the environment if a hazard reaches its full potential.
- *Risk* is the probability that an injury or damage will occur.

To illustrate these terms, an example can be considered with an elementary school at the bottom of a steep hill on a highway that is used for the transportation of hazardous materials. How does this situation represent a hazard? Where is the concept of vulnerability? Is risk involved? Here, the potential for damage exists simply because a hazmat accident may occur in close proximity to the school. Why does the potential for an accident exist? Due to its proximity to the highway, the school is vulnerable. Risk is the chance that an accident will occur. What factors may affect the level of risk that is present? Risk can be moderated or exacerbated by such factors as regulations that govern the transport of hazmats and road conditions.

Hazard analysis: three steps

As explained in chapter 1, the first step in the performance of a hazard analysis is hazard identification. Here planners seek to identify the facility and transportation situations that have the potential for causing injury to life, property, and environment. Factors to be considered include the types and quantities of hazmats, their location and routes of transport, and the nature of the hazard that is most likely to accompany an incident. A useful Web site in which MSDS sheets can be obtained for use in identifying hazards related to releases is http://www.hazard.com.

The second step, the vulnerability analysis, it to identify what in the community is susceptible to damage should an incident occur. The analysis includes the extent of the vulnerable zone, the size and characteristics of the population in the affected area, and the property that might be affected. It also includes the environmental issues of damage to sensitive areas and endangered species.

Risk analysis, the third step, assesses the probability of damage occurring due to a hazmat release. Planners may assume a worst-case scenario, a best-case scenario, or take a middle-of-the-road approach. The hazard analysis summarizes all three steps and forms the basis for a hazmat plan.

Capability assessment

After the hazard analysis, the planning team must determine the amount and composition of the resources that will be required in the mitigation effort. Following resource identification, the team must assess the availability of the required resources. Paramount considerations are the resources that are available from the facility or transporter management, other local private sector sources, and the public sector, including local, state, and federal governments.

A community base plan

Planning is certainly a local responsibility, because each community is the best judge of the local hazards, the conditions of mitigation, and the availability of resources. The community, including the local fire department, has the duty to determine what level of response commitment is appropriate. The *Hazardous Materials Emergency Planning Guide* has summed up this sentiment in the following terms, "Each community must choose the level of planning that is appropriate for it, based upon the types of hazards found in the community."[52]

As the planning process unfolds, the fire department may take the multihazard or the single hazard approach.

- The multihazard approach seeks a common planning ground for all emergencies, such as initial response, command and control needs, and communications. Following the construction of the planning framework, the multihazard approach then looks at each hazard type individually via a dedicated plan annex or appendix that contains specific information and action provisions. In addition to providing the fire department or the community with an integrated emergency management system, the multihazard approach also avoids unnecessary confusion and redundancy.
- On the other hand, the single hazard approach, requiring a less-intensive planning effort, is useful where the fire department has limited time, resources, and personnel. The single hazard planning approach assumes that any plan is better than no plan at all. This planning option may be acceptable where the hazard analysis has indicated that the community may face a small number of quite specific hazards. However, the single hazard planning approach becomes counterproductive when the number of hazards escalates so that instead of having a general disaster plan with several appendices, the fire department has many disaster plans, each addressing a specific hazard.

Specific hazmat planning issues

A number of special issues must be addressed in the hazmat planning process—training, interagency cooperation, unique health and environmental concerns of hazmats, and monetary issues.

Training. Training should acquaint department members with their role within the plan, train them in the skills and equipment that their role requires, and keep them informed about plan changes and alterations.

- First, the training effort must convey to all plan participants the social value of the role they undertake. To provide motivation, plan participants should see themselves as important in the solution of a significant community problem.

- Second, the participant should have an appreciation of the importance of each role within the plan, and the effect that each individual can have on the implementation of an effective emergency response.
- Third, the training program must include classroom work, hands-on skills, and equipment familiarization sessions that should be presented to the plan participants immediately after the adoption of the plan and on a continuing basis.

Training assistance is available from the EPA, FEMA, OSHA, DOT's Research and Special Programs Administration (DOT/RSPA), and the U.S. Coast Guard. SARA Title III, Section 305, mandated FEMA to provide $5 million for training grants within fiscal years 1987, 1988, 1989, and 1990. The FEMA training grants were designed to improve local emergency planning, preparedness, mitigation, response, and recovery capabilities. Funding continues today to provide first responder training. For more information, the reader can contact the local emergency management agency office or the local emergency planning committee.

Liaison with other agencies. Hazmat incidents, because of their potentially destructive effects, require a concerted effort on the part of many agencies for a successful mitigation. The plan must recognize the reality that the local fire department will not possess all of the required equipment or expertise; therefore, the resources and cooperation of other agencies must be sought. The fire department that attempts to address the issue of hazmats solely with in-house resources is doing a disservice to its community as well as its members.

Health and environmental issues. The planning document must look beyond the immediate mitigation of the incident and prepare for the health and environmental impacts that may accompany any release of hazmats or products of their decomposition. A review of the November 1, 1986 fire at the Sandoz Chemical Plant in Schweizerhalle, Switzerland highlights this concern.

The blaze lasted 1½ hours and occurred in a warehouse that contained insecticides, various herbicides, and more than four tons of flammable solvents. The smoke emitted by the blaze contained compounds of sulphur dioxide, nitric oxides, and sulphuric hydrocarbons. Area residents were forced to remain indoors and avoid opening their windows and doors. Further, the water runoff from the fire reached the Rhine River, killing more than 500,000 fish and rendering 11 miles of the river biologically dead, incapable of supporting fish and other forms of wildlife. Domestic water supplies were also disrupted for scores of communities along the river.

This incident began as a fire. However, a plan that required only fire control resources would have been insufficient. Although fire was involved, the incident was a hazmat release, and its impact went beyond the immediate destruction of the structure and its contents. Consequently, planners should not limit themselves to the problem of fire extinguishment.

Incident management and the underlying planning effort must consider the effects of the incident beyond the immediate damage that is evident at the scene. In environmental terms as well as in the health of the community and the operating forces, a "poorly chosen response to a hazardous materials incident can have effects that last for years."[53] Therefore, certain planning team activities are important, including the compilation of the results of basic chemical tests of major waterways (for example, the pH and dissolved oxygen concentration levels). Providing the incident commander with baseline environmental data greatly facilitates the mitigation of environmental damage.

Monetary issues. As with other types of disasters, the hazardous materials annex to the plan must consider monetary issues such as the responsibility for extraordinary expenses. These could include overtime for personnel, meals and lodging; large quantities of special extinguishing agents such as dry power and AFFF;

and large quantities of consumable supplies, such as sodium bicarbonate or dry ice. Further, the value of the material released, the container from which it escaped, and the potential for damage to surrounding valuable property make it imperative that the planning process, which involves representatives of the private sector, adopts a policy to deal with such monetary concerns. These are litigious times, and expenses resulting from the incident, such as loss of business and damage to property, will often result in litigation. Having a plan does not exempt a department from being sued, but it will simplify the assessment of possible liability and free the incident commander from this concern.

By guaranteeing vendor's payments for supplies and services rendered during an emergency, the plan can help speed delivery of needed items.

Another issue is compensation. Who will pay for damaged and lost equipment, lost business, and consumable supplies? It is far better to answer these questions in the planning process than in the courts after the event. Litigation can be costly and uncertain; a fire department may be unable to recover all of its expenditures.

The planning process may be viewed as a consensus, since it involves representatives from many interests in the community. In the area of monetary issues, this consensus process enables members of both the government and the business community to consult and agree on a monetary policy that can be incorporated into the planning document.

Legal liability issues. Another important subject that should be covered in a plan annex is the possible legal considerations related to liability of a hazmat instigator. (*Note:* The possible civil and criminal liability of the fire department and its officers is discussed in Lawrence T. Bennett's book, *Fire Service Law.*) It is not the possible negligence of response teams that will be mentioned in the annex, but that of the persons or corporations responsible for the hazmat incident. Under the laws of most states, a number of different agencies have the authority to initiate legal sanctions against the responsible party. Prosecution, if appropriate, would be coordinated between such agencies to avoid duplication of the process and avoid unnecessary hardship on the defendants.

When a hazmat spill or release occurs, the responsible person or persons (e.g., driver, carrier, processor) may be criminally liable, civilly liable, or both; in some cases there may be joint liability.

Abatement of toxic spills or releases can result in tremendous costs to public entities. The incident commander or unified command staff personnel should be familiar with civil and criminal statutes applicable to such incidents to expedite necessary cost recovery procedures. Questions regarding liability of either kind should be directed to the community's district attorney, county counsel, city attorney, or other available legal advisor.

The question of civil penalties is usually dependent upon the actual or potential damage to roadways, water resources, agricultural land, domestic livestock, air quality, wildlife, and other environmental resources. Criminal liability for misdemeanor or felony violations of the various codes (e.g., vehicle, penal, fish and game, and health and safety) can arise out of hazmat spills. Any questions regarding the applicable laws in this area should be directed to the district attorney's office, specifically the deputy in charge of toxics.

It might be well to include in the plan's annex a list of the various section numbers of each state code that can result in the imposition of civil and criminal liability arising out of hazmat accidents.

Enforcement of toxic release laws

In situations where the city or county team directs the on-scene operations, the district attorney's office should be requested immediately to determine whether any criminal or civil action is appropriate on its part. If the response teams are from federal or state agencies, then their leaders are usually responsible for enforcement of their respective laws and regulations.

Recovery of damages

Recovery of damages to public property and expenses incurred by a city or county department (or other agency involved) is usually the responsibility of that department or agency. It has to institute an action against the spiller through its respective government attorney office.

Extensive documentation of agency costs, as well as those incurred by private contractors, is required if reimbursement is to be considered from state or federal agencies.

If a state or federal agency coordinator has directed response operations, then any claim for reimbursement of costs incurred by a local jurisdiction should be submitted to that agency. Such agencies usually will only approve such expenses as have been incurred after their arrival on the scene.

With the increased awareness of operations at hazardous materials incidents by legal counsels, on-scene commanders and planners must be prepared to justify their actions and expenditures for a given incident in order to ensure full repayment.

MEDIA RELATIONS

The ever-increasing number of hazmat incidents has caused the press and television media to consider them newsworthy. It is not uncommon for their representatives to monitor the radio channels of the public safety agencies and respond to the scene of hazmat spills in a matter of minutes. They may look for the incident commander in the desire for detailed information that they can televise, transmit via radio, or write for their newspapers.

At the command post, there should be an information officer who can perform the very important function of providing factual and timely information. To appear disinterested or provide inaccurate data to media representatives would be a disservice and also would be counterproductive. The right of these people to reasonable access to the scene should be respected. Unless it would jeopardize their safety or unduly impede the handling of the emergency, every reasonable attempt should be made to help them perform their tasks.

Only the incident commander, or persons designated by him, such as the incident's public information officer (PIO), or the joint information center (JIC), should be permitted to release news concerning the overall operations. Often hazardous materials incidents will involve numerous response agencies and other agencies that will have jurisdiction in the incident. Therefore, establishing a JIC early will ensure that information from all agencies is disseminated in one nonconflicting report, thus providing the public solid, correct information in a timely manner. State laws frequently require that properly identified members of the news media be permitted to enter disaster scenes, including the scenes of hazmat spills. These same laws also require that they be warned before they enter that it may be hazardous to their safety and that they be given a recommendation to stay at a safe distance to avoid injury. If they disregard the warning and enter a contaminated area, then their equipment and personnel will be deemed contaminated and will be subsequently subjected to decontamination measures.

SUPPORTING SYSTEMS AND RESOURCES

Communications

Chapter 6 on communications described the frequent use of cellular telephones in emergency vehicles in metropolitan areas. The radio, however, is the primary means of communication for emergency services whether city, county, or fire district. Special hazmat vans are often equipped with computers with printout capability and equipment to receive data in various forms. Many of the computerized information programs are able to connect to the Internet in order to retrieve up-to-the minute information. Equipping response personnel with the appropriate communications and information equipment is vital to providing the necessary information in order to choose the correct mitigation strategies and tactics.

Volunteers

In a great many fire departments, hazmat responses are augmented by volunteer firefighters, not all of whom have received special training in coping with the risks involved. In California and some other states, volunteers who are duly enrolled and assigned duties are considered employees of the state for workers' compensation purposes under its labor code. Therefore, command personnel and planners must understand the limitations of the various levels of competency as spelled out in *NFPA 472: Standard for Professional Competence of Responders to Hazardous Materials Incidents*. This forms the basis for planning for what would be feasible for specific personnel to accomplish. Failure to do so not only jeopardizes the safety of these supporting personnel but will often result in criminal and civil lawsuits.

Resources

No hazmat disaster plan would be complete without an annex showing the fire service resources, and another annex listing those of the public works departments and private contractors. Additionally, the hazardous materials response apparatus should have a current list of vendors that could deliver additional containment and control equipment. Often the hazardous materials response vehicles are only equipped to handle an emergency for one operational period. It is difficult to stock a vehicle with an abundance of equipment for every type of incident. Therefore, planners must recognize the equipment limitations of the responding resources and make provisions to provide proper logistical support.

PLAN REVIEW AND UPDATE

As in all disaster plans, one for hazmat control should have a requirement that it be submitted for review to the appropriate agencies not less than annually to make sure that its procedures and data are up-to-date. Agencies responsible for hazmat incident response or its support should advise the designated agency, such as the county health department, at any time there is a contemplated change in procedures or notification telephone numbers.

ANNEXES

Typical annexes that are commonly included in a hazardous materials incidents response plan are illustrated in the plan of Lake County, California. Lake County has a population of about 40,000, with two small incorporated cities and a dozen villages served by mostly volunteer fire departments. Among the annexes listed are the following:

1. Incident Classification by Type and Location
2. Extracts of Pertinent Legislation
3. Notification Sequence/Responsibility
4. Hazmat Incident Notification
5. Hazmat Incident Report
6. Role Definitions (for the 23 agencies listed)
7. On-Scene Manager Checklist
8. Technical Reference Sources and Emergency Telephone Numbers
9. Activation

PREINCIDENT RESPONSE PLANS

Commencing in the 1990s, state legislatures began to question the effectiveness of hazmat response on the part of the various emergency agencies that would be called upon to handle chemical leaks and spills wherever they occurred. For example, Chief Bahme served on a committee of the California State Firemen's Association. The committee had as its main objectives the education of fire service members throughout the state and the adoption of legislation that would establish training facilities to teach and equip responding emergency forces to handle hazmat incidents in the safest possible manner. Fire departments everywhere in the nation should be involved in preparing such preincident planning efforts.

The ensuing discussion resumes with an outline of some of the materials that the late Robert Obermayer prepared for use in the National Fire Academy's Open Learning Fire Service Program.[54] Robert Obermayer stresses the importance of replacing quick action with a deliberate, thoughtful decision-making system based on a risk-benefit analysis. He asserts that contemporary fire departments are faced with at least two important responsibilities relative to hazardous materials:

- Planning—participation in the development of effective preincident response plans for hazmat incidents
- Mitigation—service at the scene of the incident

The planning initiatives of SARA Title III were a relatively new development at the beginning of the 1990s. Thus fire officers were not initially very well versed in the preparation of comprehensive hazmat response plans, though they were quite familiar with prefire planning for target hazards. However, over the past two decades, many training classes and educational programs, such as the National Fire Academy's "Managerial Issues in Hazardous Materials," have focused on how to implement the planning and response requirements of SARA Title III. In addition, the development of consensus standards, such as NFPA 471, 472, and 473, has aided fire officers in having the proper knowledge to plan and implement a hazmat response based on the hazards of the community and the resources available.

RESPONSE DIFFERENCES—HAZMAT VS. TYPICAL

The general comparison of emergency incidents with various types of disasters, as outlined in the first chapter, could also be applied to major hazmat incidents. However, with respect to the nature of the destructive impact, the decision-making system utilized, and the services required, the hazmat response differs from the typical emergency services response in most cases.

Destructive impact

Hazmat response planning can be designed as a separate annex to a comprehensive major disaster plan or can be undertaken independently. Either way, community planners assigned to the hazmat planning process must keep in mind the destructive impact that the uncontrolled release of a hazmat may exert on the community's property, inhabitants, domestic animals, and the environment. It is important to remember that it is the curtailment or mitigation of the harmful forces unleashed in a hazmat mishap that justifies the existence of the plan. If the plan does not recognize the potential danger inherent in hazmat incidents, not only will it be ineffective, but it may actually increase the risk of personal injury and property damage.

The planning team must be aware that it is not planning for a fire control operation but for an incident potentially far more destructive. In the hazmat incident, the destructive potential includes immediate, long-term, and recurring risks of damage and injury to the community and the environment, as well as to the operating forces.

Deliberate decision making

Planners must recognize that the traditional incident command decision-making process, usually calling for quick, aggressive action, will not always be appropriate for hazmat incidents. It must often be modified to suit the circumstances. Some fire officers may find it difficult at first to adopt a more deliberate approach. However, after reviewing the records of several catastrophic hazmat disasters, some of which have been mentioned previously, they may become convinced of the need for a more restrained approach in the decision-making process.

In this text, the references to these disasters will not dwell on the monetary losses, though often tremendous, but on the impact to the affected people and their environment. This will be a primary consideration.

These incidents could have occurred in many communities. Knowledge and planning are imperative on the part of all emergency services personnel, and especially on the part of fire department personnel. The incident in Switzerland was precipitated by the failure to control water runoff from a fire control operation. Sometimes, in the effort to meet the competing demands on the incident commander to heed the concerns of control strategy, tactical alternatives, and monitoring suppression and containment progress, environmental problems can be overlooked.

As mentioned earlier, the Bhopal plant in India has been considered the site of the "world's worst disaster."[55] However, the full effects of the 1986 Chernobyl meltdown may never be calculated, and thus it may surpass Bhopal. In any case, the design of the Union Carbide plant at Bhopal may not be too dissimilar from pesticide chemical production plants elsewhere. As a matter of fact, the U.S. chairman of Union Carbide, Warren M. Anderson, described the plant as exhibiting the "same design, same equipment, same everything" as operations in the United States.[56]

Services required

Another important difference between a hazmat response and the typical emergency incident is in the type of services usually required for successful mitigation. The average response requires fire suppression services or medical care. This has traditionally been provided by the fire departments with varying levels of proficiency, depending upon the amount of training and experience they have had, or by the response of one or two police patrol cars. In contrast, a hazmat incident may demand the provision of any number of technically complex and distinctly different services. Such services could include the plugging and diking of leaking tanks, monitoring of the water and air supplies, and efficient systems for determining the chemical properties and potential risks of the hazmat involved.

Though still unfamiliar to some fire departments, such technically complex services have been supplied by some fire departments for a great many years. The Los Angeles Fire Department, for example, started to respond to hazmat incidents in 1950, following the creation of its dangerous chemicals detail of five members with a captain in charge (Chief Bahme). Today, such response units in large departments may be several in number and be part of a separate command headed by a chief officer. FDNY has upwards of 11 hazardous materials units, which are under the command of the deputy assistant chief of special operations command.[57]

ORGANIZING THE HAZMAT RESPONSE

During the planning process, the fire department must determine how it will respond to the hazmat challenge. This involves the consideration of at least two issues: the first response and the hazmat response team.

First response

The objective of a disaster plan annex or hazmat plan is to organize the hazmat incident mitigation process. This begins with the initial response of the local fire department. The plan must organize the first response by assigning to it a mitigation role, addressing the problems that will affect the actions of the first response and developing operational priorities to facilitate a good size-up of the situation.

Role of the first response

A vital function of the first response commander is to recognize a hazmat incident and to identify the hazmat involved. Early recognition and identification will allow the fire department to determine the magnitude of danger and the resources and precautions that will be required. Toward this end, the plan must train all unit commanders and other department personnel in the following:

- Potential location of hazmat incidents
- Chemicals that may be involved
- Hazards of such products
- Potential for human, monetary, and environmental damage
- Equipment and supplies that will be required to prevent or moderate such damage
- Resources available from within the department to mitigate the incident

Source of hazmat data

In addition to the information provided by the hazard analysis, the plan should catalogue the following sources of information:

- The Department of Transportation Hazmat Identification System
- The National Fire Protection Association 704 Marking System
- CHEMTREC
- Shipping papers
- Consultation with the facility operator or transporter's personnel

This information is used to derive a level of personal protection equipment, as well as a mitigation strategy. (Note in fig. 9–3 the level of protection worn by the workers.)

Fig. 9–3. Hazmat mitigation activities. Having information for the chemicals involved in a hazardous materials incident enhances the decision of the proper level of personal protective equipment.

FEMA News Photo.

First response issues

When assigning tasks to the first response units, it is important to recognize that they will probably operate under the following constraints:

- They may lack specialized hazmat training.
- They may not have access to all relevant information on the situation.
- They may be unable to obtain the full department inventory of hazmat control equipment.

- To save life or property, they may be required to make split-second decisions that have life-or-death consequences, with additional potential for catastrophic monetary losses.
- They may commit or omit actions that affect the outcome of the event.

The plan must also recognize that the first response unit may experience pressure from the people who summoned it. Civilians on the scene are probably unaware of the danger of hazmats and will expect fast action, not understanding the need for deliberation. For example, consider a first arriving firefighter in command at a flammable liquids fire. The firefighter realizes that he has insufficient quantities of foam available to effect extinguishment. He has summoned an additional supply and is awaiting its arrival. Meanwhile, the transporter or fixed facility operator (realizing that each gallon of liquid consumed by the fire increases the financial loss) may attempt to induce the firefighter to start extinguishment as soon as possible. The firefighter's team members may also want to take quick action. The firefighter in charge must have the presence of mind to resist an action that his training tells him is inappropriate.

Team issues

If the plan calls for the creation of a hazmat team, the planning and management responsibilities will increase. The following issues must now be decided:

- **Identification of the team's mission.** Will the team attempt to prevent, as well as contain, leaks and spills? Will they try to prevent chemical fires and explosions? Will the team engage in cleanup activities once the incident has been controlled? In view of the fact that hazmat incidents may require costly cleanup measures for streets, buildings, streams, vehicles, etc., private sector interests may urge the planning team to make this residual cleanup a responsibility of response agencies.
- **Personnel issues.** How many members will be assigned to the team? What qualifications or level of training will they have? Will the unit be staffed with on-duty or off-duty personnel, or both?

Table 9–1. EPA PPE protection levels. Utilization of the proper PPE level will prevent injuries to responders.

Environmental Protection Agency
Levels of Protection & Usage Guidelines

Level	Type of Clothing	When Used
A	Totally encapsulating, air light chemical resistant PPE. SCBA or Supplied Air Respirator with escape pack Chemical Resistant Gloves and Boots	Severe eye skin or respiratory hazard
B	Chemical resistant, splash suit. SCBA or Supplied Air Respirator Gloves and boots	Severe respiratory of moderate skin hazard
C	Chemical Resistant Splash Suit Air-purifying Respirator Gloves and Boots	Moderate skin and respiratory hazard
D	Coveralls, safety boots, hard hat and safely glasses	Mild skin hazard, No respiratory hazard

- **Management and resource issues.** Where will the unit be located within the community? How will it be alerted? How will the plan provide for periods of personnel shortages? What level of equipment and supply inventory should be maintained?

- **Health issues.** What provisions will be made for the continual assessment of the team's health? Since many hazardous substances have long-term effects, the physical condition of team members must be carefully monitored.

- **Team response issues.** What units will respond with the team? How will the first response units be integrated into the overall hazmat response plan? What are the means through which field commanders will be apprised of the capabilities of the team?

- **Level of protection.** Which of the EPA levels of protection is appropriate? (See table 9–1.) It is important to first check the results of the hazard analysis, which will reveal just what is required to handle the hazmat problems.

Decontamination planning

Since hazmats often have a destructive effect upon the environment, unlike ordinary fire debris, they must usually be removed from the scene of the incident after the damage control measures have been taken. In order to ensure the chemicals are not prematurely moved from the site of the incident, a comprehensive decontamination plan needs to be established prior to any offensive actions by response personnel or the removal of victims. To ensure that these measures will be carried out with minimum risk to operating forces, the planning team must recognize the need for protective clothing and establish procedures to do the following:

- Procure decontamination equipment
- Identify the hazmats for which decontamination will be required
- Establish the safe level of decontamination
- Determine the best means for carrying it out

Failure to solve the decontamination problem could result in long-term health and environmental dangers arising from the released materials. Figure 9–4 shows a typical decontamination line schematic. Having a predetermined setup for decontamination can save valuable time.

OPERATIONAL GUIDELINES

The plan should provide departmental officers with operational guidelines that will ensure the achievement of personnel safety and incident control objectives outlined in the planning document. To attain these objectives, the incident commander must make a number of determinations.

Determine if the incident is an emergency services response problem

Upon arrival at the scene, the commanding officer must determine whether or not the incident is actually an emergency services problem requiring emergency action. It could be a cleanup problem that should be handled by employees of the plant or transport carrier, or by a professional hazmat disposal contractor. In some instances, the situation may call for drastic measures that no emergency services department could be expected to provide. For example, when a chemical firm in England was contaminated by dioxin in 1968, the only solution for getting rid of the toxic chemical was to level the factory and burn the wreckage in an abandoned mine shaft.[58]

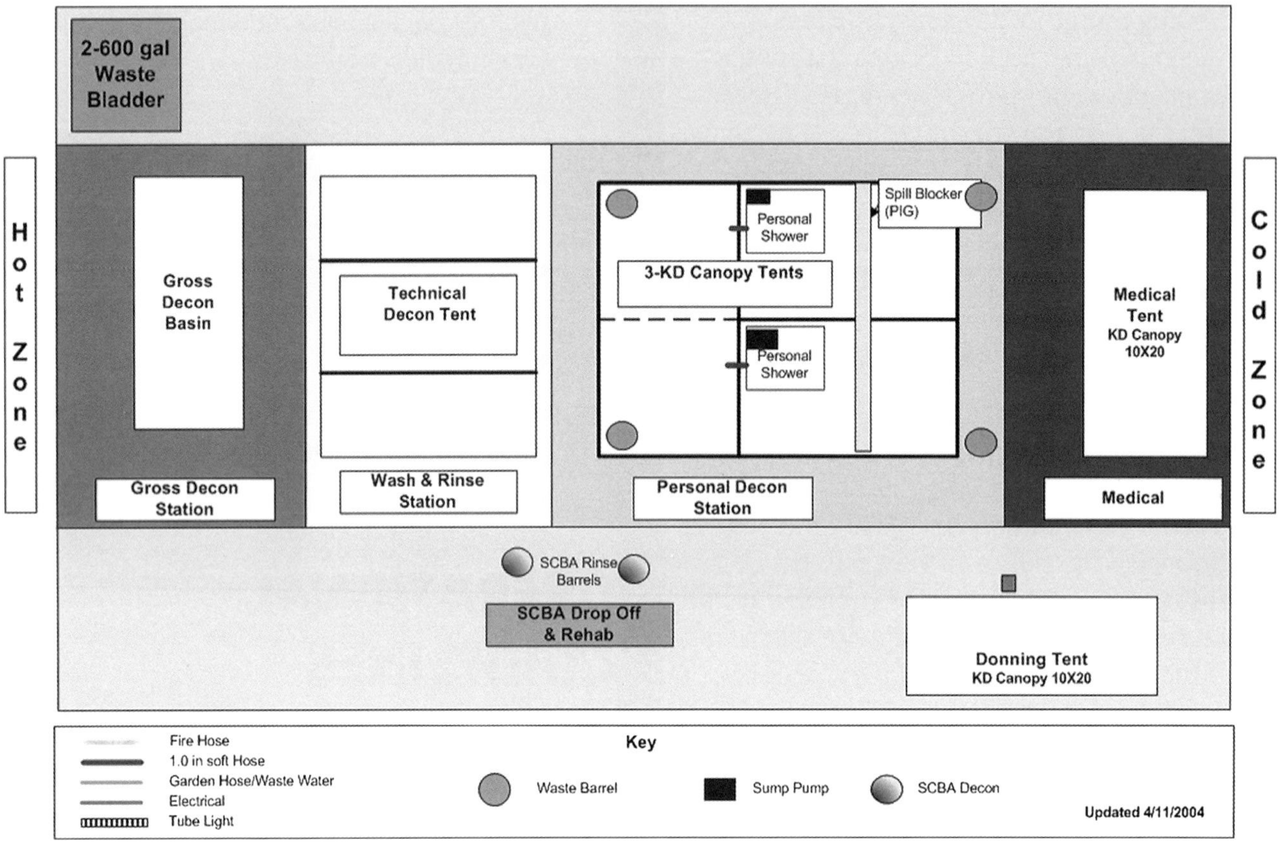

Fig. 9–4. Sample decontamination schematic. Note that the hot and cold zones are separated by the decontamination line.

Emergency services personnel should not be committed to performing cleanup operations unless there is an imminent threat to life and property. The safety of the personnel should never be jeopardized in the effort to save product, property, or the environment. Incident commanders must be aware that huge sums of money could be at stake, and transport or facility operators may seek to minimize their costs at the expense of responders.

Determine if it is a fire service problem

It is important to consider where the responsibility rests to handle an incident—is it the fire service's responsibility, another jurisdiction's responsibility, or is it a private sector problem? Often, the person reporting the incident, or a first responder not from the fire department, may not be familiar with jurisdictional boundaries, the hazard present, or other operational considerations. He may unknowingly request the response of the hazmat team. The planning document should not assume that just because the resources of the fire service are summoned to the scene of the incident, they must be used. There are valid situations where no action is the best alternative. Using the risk-benefit analysis, if no benefit is to be derived, then why commit resources and undertake a level of risk? Also, it is important not assume that others at the scene of a hazmat incident automatically want the help of the fire department's team.

Determine who is in command of the incident

Sometimes incident command is disputed among rival agencies. Each agency may feel that its role is indispensable and that its leader should be in overall command. However, an incident commander from the fire service, for instance, would be understandably reluctant to commit the welfare of his personnel to the chaotic command structure that results from such disputes. Incident command should be determined during the planning effort in order to alleviate conflicts.

Evaluate the situation

Effective mitigation requires that the incident commander identify the type of situation that exists, the extent of the incident, and the relevant hazards before determining objectives, identifying resource needs, and embarking upon a specific plan.

Various identification systems and procedures will furnish the basic identity of the hazmat and an elementary understanding of its properties. However, during the mitigation effort, additional up-to-date, detailed information will be required. Consequently, the planning effort must establish a means by which such information is readily available to operating forces.

Until recently, responders were restricted to stagnant sources of information. Unfortunately, the contemporary proliferation of hazmats tends to make some reference sources become out of date quickly. However, new technologies, including portable computers with wireless Internet access and fax machines, have dramatically increased the team's ability to quickly obtain fairly accurate data. The planning document should address the problem of proper hazmat identification through use of computerized databases.

CAMEO (Computer-Assisted Management of Emergency Operations) is an excellent planning and response computer program that was developed in 1986 through the joint efforts of the Seattle Fire Department and the National Oceanic and Atmospheric Administration (NOAA). In addition to containing information on more than 2,600 chemicals, CAMEO can predict vapor cloud paths and, through its Marplot capability, relate such information to a map of the jurisdiction in question. It does this with its ALOHA component (aerial location of hazardous atmospheres). Using this program in conjunction with the TerraServer mapping system will allow planners and responders to view exact locations and structures that will become affected by a release.

Operational priorities

To give guidance to the first response commander, the plan should include the following operational priorities:

1. Preservation of human life, both civilian and operating forces.
2. Containment of the hazmat release to the area of origin.
3. Preservation of property. (Note that this is the third priority; the safety of civilians and operating personnel should not be jeopardized for the preservation of the environment or property.)

Hazmat response team

Because hazmat incidents require special skills, knowledge, attitudes, and equipment, many fire departments have recognized the value of a hazmat response team. The hazmat response team is defined as a group of individuals whose abilities, training, and equipment inventory represent an asset in handling hazmat incidents. Some questions arise that should be answered in the organizing process.

Is there a need?

The first question for the planning body is, Does the department really need a hazmat response team? Considering the cost of training, equipment, personnel, and the potential number of incidents, perhaps it may be advisable to rely on the services of an adjacent fire department or to participate in a regional team. If the department opts to join a regional team, or to utilize an adjacent team, its planning document must address joint operational procedures and resource commitment issues.

HAZMAT APPARATUS AND EQUIPMENT

An increasing number of fire departments are obtaining apparatus for carrying a variety of gear. This includes apparatus for carrying not only special equipment useful in handling chemical leaks, spills, toxic gas releases, etc., but also for transporting communications gear, including computers and fax machines. Other items to transport could include lab equipment, reference books, files, report forms, checklists, notification and resource lists, and other items contained in the annexes of the disaster plan. Some have a desk for the hazmat incident commander, with duplicating equipment for reproducing maps, material safety data sheets, and instruction guides to give to members of his command and on-scene managers of other agencies.

Using the hazard analysis, the planning team will identify any special apparatus or equipment needed and will establish how the equipment is brought to the incident. Workable ideas include departmental purchase of some equipment, donations from facilities that contain hazmats, and possibly the establishment of multijurisdictional equipment pools.

Hazmat apparatus

Some fire departments, such as Boston, have converted beverage delivery trucks into hazmat response units to suit their needs. Other fire departments have rigs that more closely resemble motor homes, complete with toilet and shower, coffee makers, etc. Others have come up with a unique design that makes for easy access to the various types of equipment it carries. Such is the case with the New Zealand fire service.

In an excellent *Fire Engineering* article from 1990, the western division commander of the city of Auckland describes the design of its hazmat unit. He presents several photographs depicting its horizontal-sliding "concertina" doors that run the full length of the vehicle, permitting easy access to either end or both sides. It was built on a standard commercial nine-ton chassis with a stepped floor design into a low-loading deck. In loading the rig, the heavy and important items are stored near the deck, and the less frequently used items higher up. In the two years that it had been in service, its good load-carrying capability and ease of access to all equipment was working out well at incidents.[59]

The amount of equipment that hazmat teams are carrying has increased with the addition of WMD detection equipment, as well as mass decontamination units. Thus, hazmat teams would be wise to examine tractor-trailer apparatus, which are customized by many fire apparatus manufacturers to function as a hazardous materials unit. The important part of selecting an apparatus is to ensure it will meet the needs of the team and community not only today, but for a given time into the future.

Hazmat equipment

High technology equipment for communications, data storage and retrieval, spectrometers for chemical analysis, combustible gas analyzers, radiation detectors, and similar devices are being employed in hazmat emergencies more than ever before.

Equipment commonly carried in a hazmat van includes protective clothing, such as encapsulation suits, self-contained breathing apparatus, absorbents, neutralizing materials, liquid transfer pumps, and monitoring equipment for gases and radiation. It also includes inflatable shelters, showers, fans, portable generators, hand tools, tarps, placards, barricade items, decontamination showers, soaps, detergents, temporary clothing, and full recording and monitoring items for cleanup action on 100 or more fire service personnel or other victims.

Equipment pools can be a convenient procurement alternative to the purchase of expensive specialized items such as backhoes, skip loaders, and bulldozers. Where such pools can be arranged by a regional planning team, each participating agency could submit a list of the equipment that it could afford to provide by a mutually agreeable date. The regional planning committee would also establish the rules and policies under which the pool would operate.

Consumable supplies

Many hazmat incidents require the use of special supplies and extinguishing agents that are easily obtainable any time of the day or night every day of the week. Moreover, when used in large quantities in control operations, the expense could be considerable. Unlike ordinary fire incidents where the fire department consumes mostly water and air for breathing apparatus, commodities that are not difficult to replenish, a hazmat incident may require the following:

- Large quantities of various foaming agents, and dry chemical extinguishing agents for both flammable liquids and metals
- Dry ice, sodium bicarbonate, limestone, soda ash, and other agents to retard or neutralize reactions
- Sand, sawdust, absorbent materials, etc., for diking and absorbing spills

The planning committee should see that the plan provides for the procurement of large quantities of such materials on short notice. The preplan analysis of the principal hazards in the department's response area should help to determine the need for the type and amount of such supplies that may be required for the community's particular conditions. For example, in some communities, oil storage and refineries are of major concern. Thus, arrangements could be made with the local petroleum companies and neighboring fire departments to set up a bank for various foam agents, diking materials, containment drums, absorbent compounds, etc. that could be readily transported to a flammable liquids incident. It may also be possible to make similar arrangements with local industries and commercial suppliers of such commodities, whose day and night telephone numbers would be noted in the plan's resource annex.

Action/no action decision

The plan should recognize that operating forces may arrive on the scene of a hazmat incident just before an explosion or other form of material release occurs, with resources that are woefully insufficient to reverse the course of events. Since the primary objective is to save lives, the plan must give the incident commander the authority to defer any action if an unacceptably high level of risk is present. Viable tactical alternatives include decisions to take no immediate action or to withdraw, and incident commanders should be urged to err on the side of caution in such decisions.

To encourage the exercise of this tactical option, the planning document should include the criteria upon which to base the action/no action decision. Because emergency personnel are accustomed to aggressive tactics, the plan should contain a list of specific conditions that would indicate the need for immediate withdrawal or the decision not to commit department personnel. These might include the following:

- Flame impingement on a pressure vessel or tank and inability to deliver the required volumes of water at the site of the incident
- Inability to obtain special extinguishing agents
- Lack of protective equipment that is commensurate with the hazard
- Fire involving an explosive material
- Inability to adequately identify the product or its properties

The action option of the action/no action decision must not be made before completion of a risk-benefit analysis. An incident commander should first consider the level of risk to operating forces and then compare it with the expected mitigation benefit. Of course, this rule should be followed in all fire department tactical decisions, whether involving hazmats or not. However, a review of many firefighter fatalities reveals that it has often been ignored. Sadly, there have been hazmat incidents that resulted in firefighter fatalities, most notably the death of 26 Texas City, Texas firefighters in 1947 and the death of 13 Kingman, Arizona firefighters in 1973.

If an explosion or widespread release of toxic gases has already occurred prior to the department's arrival on the scene, then a decision to not take any action is not a viable option. An evaluation, sometimes with the aid of binoculars, may result in activating the evacuation plan for persons in the endangered area exposed to gas clouds. Alternately, an evaluation could result in the decision to call for the specialized equipment required for rescuing persons from collapsed structures, suppressing incipient fires, closing valves on ruptured pipes, and instituting other damage control and medical measures as needed. Actually, the fire service department may be fortunate if the hazmat explosion takes place while they are responding to the emergency. This offers two advantages. One is the realization of the responders that a subsequent explosion could occur. Secondly, it will heighten the awareness of the responder and prompt the staging of responders at a distance further away from the incident than likely without an explosion.

Regardless of the circumstances, the plan should contemplate the possibility that the fire department's first-in officer, whether a company commander or a chief officer, may be the first to arrive on the scene of the incident. He or she will be expected to notify the dispatcher where the interim command post will be located pending the arrival of the person who assumes the responsibility of incident commander. If the first response commander even suspects that the hazmat situation might become a disaster (if it is not already), there should be no hesitation to request that the hazmat disaster plan be activated immediately. This will allow the dispatcher to make the required notifications and get the hazmat unit on the scene as soon as possible. If the situation fails to become as dangerous as the rapid size-up anticipated, and the special team is not essential, there is no harm done. It can be considered an unplanned drill.

If a decision to go is made, incident commanders must make provisions to decontaminate the entry personnel and must have this set up prior to allowing responders to enter. Notice in figure 9–5 that this decontamination process can involve simple items such as plastic pools and brushes.

Fig. 9–5. Simple decontamination. Note that a simple child's pool and brushes can be used to quickly set up a decontamination line prior to entry of response personnel.

Postoperations planning

Among the more famous aphorisms, Yogi Berra's "It ain't over 'til it's over" is a classic. It certainly applies to hazmat incidents. They are not over when the leak is plugged, the spill removed or neutralized, the fire extinguished, the exposed inhabitants evacuated, the decontamination of personnel and equipment accomplished, the injured cared for, and the dead removed. There may still be much to do that the plan should take into consideration. For instance, the plan should consider the demobilization of the deployed units, including the mutual aid responders. The plan must also consider the gathering of data for the preparation of required reports. Necessary data could include information on the exposure of personnel to toxic substances for their medical records, the amount of supplies used, expenses incurred, agencies that rendered assistance, injuries to personnel, and damage to equipment.

The need for postoperation consideration of legal liabilities on the part of the disaster instigator for injuries, deaths, and property damage, and the fire department's possible liability for negligent conduct, has already been discussed. There could be a situation where the city could be fined by OSHA, as was Waterbury, Connecticut, for failure to inform the firefighters that hazmats were stored at the fire scene, resulting in cyanide poisoning of 28 of them. "While Waterbury fire fighters fought the chemical blaze, a 38-page hazardous materials report the company had filed with the city in 1984 was locked in the fire marshal's office."[60]

Recovery

When the disaster has been brought under control, there may be homeless people and animals to be fed and housed in temporary shelters. Priorities will have to be established for the resumption of services. Arrangements must be made for salvage and cleanup operations, restoration of critical services, e.g., water, electricity and sewer, and eventually, the eviction of the out-of-town volunteers from the shelters where they have been fed and housed (sometimes for weeks) ever since their arrival.

Postincident analysis

The disaster plan for hazmats should call for some "Monday-morning quarterbacking," or a postincident critique in which all the major participants sit down and analyze what went right and what went wrong. Assuming that it could happen again, how could the operations be improved the next time around?

To lessen the psychological impact of disasters, there should be some planning for the emotional support of rescuers and victims. Man-made spills of oils, chemicals, and similar hazmats could possibly undermine public confidence, especially if serious injuries or deaths have resulted from the incident. Thus, the plan should consider how to restore the credibility of the fire service and other departments and the public's confidence in the responders to provide effective emergency services.

Hazmat incident action

After the planning process is completed and the hazmat plan has been drawn up and approved, it might be well to set forth a brief outline of the actions that the incident commander will usually need to take upon approaching and arriving at the scene of a major hazmat problem. The preceding chapter suggested detailed steps for evaluation, control, containment, mitigation, and disaster management for transportation disasters. Therefore, at this point, disaster management options will only be outlined in general. Even though hundreds of tasks may have to be carried out by numerous agencies—fire, police, medical, utility, etc.—the primary job of the incident commander is to set priorities. The incident commander must decide what should be done first and by whom.

Incident commanders who are prepared due to thorough hazmat study and planning stand a good chance of making most decisions right the first time. At least that should be their goal, for if they do not make the right decision, there may never be a next time! It must be the aim of all incident commanders to get all the knowledge and training for hazmat responses that will be needed to cope with any kind of hazmat disaster that may confront them. Knowledge overcomes most fear of making mistakes, so incident commanders must aim to get it. He who aims at nothing is almost sure to hit it!

Hazmat incident response

The following outline may be helpful in responding to hazardous materials incidents.

I. Approaching the Scene

When rolling up to the scene, if there is a cloud of yellow gas or smoke in the distance, responders should ask the dispatcher if any further information has been called in about the nature of the incident, e.g., fire in fertilizer or pesticide storage, ruptured acid drum, etc.

A. *Size up carefully.*

1. Use binoculars if indicated and do not get too close.
2. With the information available, give responding units a minimum safe distance to advance.
3. Give appropriate dispatcher the size-up report, and update as needed.

B. *Take initial action to effect control.* This could include the following steps:

1. Call for activation of hazmat disaster plan.
2. Deploy arriving forces.
3. Call for additional help as may be required, etc.

II. Arrival on Scene

A. *Determine safe location to park.*

Park clear of magazines, flammable liquid spills, and buildings that may blow up from gas ignitions, ammunition, and gun powder, etc.

B. *Establish needed controls.*

1. Locate command post where easily accessible and recognizable.
2. Designate staging areas for incoming special-called units.
3. Designate "hot zones" if needed, and mark off their peripheries with barrier tape, rope, signs, etc. Prohibit anyone from entry into contaminated areas unless the following apply:
   a. Personnel are properly protected.
   b. Decontamination equipment and procedures in a designated area are being provided.

III. Evaluating the Hazards

A. *Identify the product.*

1. Check the warning signs; acrid odors, colored smoke, clouds of vapors, pools of sizzling liquids, etc.
2. Check the container type and shape.
3. NFPA identifying system labels.
4. DOT identifying systems label, placards.
5. Markings—chemical names on containers.
6. Shipping papers.

B. *Assess the risks.*

1. Access available information sources to determine risks to health, property, and environment.
2. Coordinate the evaluation process with appropriate agencies, e.g., health department, CHEMTREC, EPA, etc.

IV. Securing the Scene

Security and access control are very important in the disaster zone. It may become necessary to implement and enforce an emergency security plan in at least two respects:

A. *Access to the general area.* To keep out sight-seers and permit legitimate media and business persons to enter.

B. *Access to "hot" areas.* To permit emergency forces to enter who are properly equipped, and to keep all others out.

V. Managing Emergency Medical Services

Integrate the EMS system into the disaster control operations for the following:

A. *Triaging victims.*

B. *Providing emergency medical services.* Services provided to emergency forces and to victims.

C. *Removing victims.* Victims are removed for transport to designated medical centers.

VI. Monitoring and Decontamination

A. *General guidelines for personnel*

1. Do not commence decon operations until the product has been identified, if possible.
2. Decon of personnel should preferably take place inside the "contamination reduction zone."
3. Proper levels of protection should be used by decon personnel.
4. Try to contain the materials used in decon procedures.
5. The decon team should be properly decontaminated.
6. When possible, exposed personnel should be monitored and checked by appropriate health agencies before being released from the area.
7. Follow-up checks by such agencies should be performed as necessary.

B. *General guidelines for equipment decontamination*

1. Decon of equipment should await product identification.
2. Decon of equipment should take place inside "contamination reduction zone" only.
3. Decon teams should wear proper gear for the hazard level.
4. Decon materials should be contained as far as possible.
5. Decon teams should be properly decontaminated.
6. Equipment should be checked by appropriate health agencies before release from the area.

C. *Coordination of decontamination*

1. The incident commander should coordinate any other than primary emergency decontamination.
2. The appropriate health agency should ensure that proper decontamination occurs.
3. Additional support units, e.g., fire, hazmat, public works, etc., may be requested to assist in decon operations.
4. Normally, decon should be attempted only after technical expertise on the material has been obtained and proper safety precautions have been taken.

VII. Servicing the Forces

A. *Logistical support*

1. For hazmat unit—replenish supplies of foam, neutralizers, absorbents, compressed air or oxygen for SCBA, fuel for generators, clothing to replace torn or contaminated clothing (that cannot be decontaminated).
2. For supporting agencies assisting in the operations.

B. *Personnel needs*

1. Medical attention for the injured and exhausted.
2. Portable toilets for extended operations.
3. Food and beverages for extended operations.

4. Relief teams to permit rotation for sleep and feeding.

C. *Accounting needs*

1. At the staging area to record arrival and departure of assisting departments' apparatus and personnel.

2. At the command post to record time and services of private contractors, off-duty reporting personnel, volunteers' overtime, persons hospitalized, equipment damaged, etc.

VIII. Leveraging Resources

Review the human, equipment, and information sources that are available to assist in control and recovery actions. (Resource list should be in the hazmat plan annex.)

IX. Caring for Evacuees

A. *Determine extent of help from agencies.* Determine what help agencies such as the Red Cross can provide (e.g., food, shelter, blankets, cots, etc.) if evacuees will not be permitted to return to their homes for an extended period.

B. *Arrange for communication for separated family members.* Arrange a means of notifying separated family members (parents, spouses, or children) of the others' condition and location.

C. *Show some concern for the care of pets of evacuees.*

D. *Arrange to check homes or businesses for safety of occupancy.* Advise evacuees when it is considered safe for them to return.

E. *Advise evacuees of property security measures.* Advise them of whatever security measures will be provided for their property pending their return to it.

X. Restoring Critical Services

Assist public utility companies in determining when their services (water, power, sewer, etc.) can be safely restored.

XI. Managing Cleanup and Salvage Operations

Determine to what extent cleanup will be done by the plant management, property owner, or persons or corporations responsible for causing the hazmat incident, or other agency, and to what extent the fire department's services will be feasible and affordable.

XII. Assessing the Damages

Seek professional assistance in determining the monetary and property loss sustained in the disaster (insurance adjusters may be of assistance in this). Request the medical services management to provide a report of the deaths and injuries sustained by people and livestock at the incident.

XIII. Demobilizing the Forces

In major hazmat disasters, it may be several days before the emergency forces can all be returned to their quarters. It is the hazmat incident commander's responsibility to advise the overall incident commander when the situation has been sufficiently controlled or decontaminated to permit removal of the security barriers and to permit security forces to be released. Management's hiring of private security guards to keep the area secured over an extended period of time may obviate the need for department personnel to stay behind on standby watch, thereby speeding up total forces demobilization.

XIV. Postdisaster Operations

These matters are not part of the hazmat disaster plan, for the situation has been handled, the troops have left the scene, and the media have begun to sing the praises of the responders. The mayor is so pleased with reports that he has been receiving of the fire department's efficient management of the whole incident that he may be about to ask if there is anything the department could use to make their hazmat plan function even better (e.g., more personnel, more equipment, more budget appropriations for training programs, etc). To be fully prepared to justify an affirmative answer, it might be well to lay the groundwork.

A. *Preparing the reports*

1. Get the facts, figures, and photos on paper showing what the department accomplished with the resources now at its disposal, and what more might have been done, in a shorter time, if the resources had been augmented in certain respects. This report could support an argument for upgrading the department's response capability.
2. Gather data showing how injuries occurred and how they might be more easily avoided with improved equipment or training. This should be part of the report.

B. *Critiquing the* response *performance*

1. Assembling the major participants in the operation for a postincident analysis should help to pinpoint the weaknesses in the plan or in the manner in which it was carried out. It is not unusual to find that a plan is workable in most respects, but the communications between the participants and the coordination of their activities at the scene could be improved. Discuss ways to correct these problems.
2. If the plan needs to be changed in any way, submit the modifications for approval to the appropriate authorities and get them adopted right away.
3. Try out the modified plan in the next joint-agency drill, and expect to find that these activities function more smoothly.

CONCLUSION

It is not contended that the suggestions offered in this chapter represent the final solutions to the many complex problems in a hazmat disaster. At best they offer only partial guidance to emergency responders facing perilous situations. Notwithstanding years of study and training, a firefighter who is the first to respond to the scene of a widespread hazmat incident may face difficult choices of courses to pursue and directions to take. It brings to mind Lucy of the Peanuts comic strip and her reluctance to play on Charlie Brown's ball team. Standing out there in tall grass over her head, she exclaimed, "I don't mind playing left field if they would only tell me which way to face!" Well, emergency responders do not mind taking the risks they have to face in handling hazmat disasters if only they would be given the right directions to take. This chapter will not give responders all the right directions, but perhaps it can offer some signposts that will point toward the right course of action.

REFERENCES

1. Los Angeles Fire Department, Bureau of Fire Prevention. 1951. *Dangerous Chemicals Code.* Los Angeles, CA: Parker and Co.
2. American Chemical Society. *Chemical Abstracts*. Baltimore, MD: ACS.
3. Davis, T. L. 1991. *Chemistry of Powder and Explosives.* New York: Wiley and Sons.
4. Lake County, California. 1986. *Hazardous Materials Incident Response Plan*.
5. Hannah, J. 2003. Gases and dust caused Ohio silo blast that claimed two firefighters. *Firehouse.com*. November 20. Retrieved March 4, 2007 from: http://cms.firehouse.com/content/article/article.jsp?id=sectionId=46&id=22175.

6. Bahme, C. W. 1978. *Fire Officerís Guide to Dangerous Chemicals.* 2nd ed. Boston, MA: NFPA.

7. Pryor, L. 1966. Liquefied natural gas accident risks looming larger. *Los Angeles Times*. August 16. Part II. p. 1.

8. Bahme, C. W. 1978. p. 114.

9. Porretto, J. 2007. Probe of fatal Houston blast cites lax oversight. *Firehouse.com*. Retrieved March 20, 2007 from: http://cms.firehouse.com/content/article/printer.jsp?id=53918.

10. Bahme, C. W. 1978. p. 114.

11. *Fire Command*. 1975. Devastating blast of railroad cargo. April. pp. 32–34.

12. GPO access. 2007. *Electronic Code of Federal Regulations*. Retrieved March 20, 2007 from: http://ecfr.gpoaccess.gov/cgi/t/text/text-idx?c=ecfr&sid=585c275ee19254ba07625d8c92fe925f&rgn=div8&view=text&node=49:2.1.1.3.7.4.25.23&idno=49.

13. U.S. Department of Transportation. 1990. *Emergency Response Guidebook*. Washington, DC: U.S. Department of Transportation.

14. Bahme, C. W. 1978. p. 215.

15. U.S. Department of Defense, Defense Civil Preparedness Agency. 1972. *Your Chance to Live*. SM 3-12. p. 71. Available at: http://eric.ed.gov/ERICDocs/data/ericdocs2sql/content_storage_01/0000019b/80/35/b6/ba.pdf.

16. Hess, G. 2006. Chemical emissions are down in North America. *Chemical & Engineering News*. July 28. Retrieved March 18, 2007 from: http://pubs.acs.org/cen/news/84/i31/8431emissions.html.

17. *Los Angeles Times*. 1973. Killer gas peril hits Auckland. March 1. Part II. p. 1.

18. *Los Angeles Times*. 1976. Abortions for Italy poisoning begin today. August 13. Part II. p. 1.

19. *Los Angeles Times*. 1974. Chemical mist envelops Chicago's far South Side; 16,000 told to flee. April 27. Part I. p. 8.

20. *CNN.com*. 2005. Oil refinery explosion kills at least 14. March 23. Retrieved March 22, 2007 from: http://www.cnn.com/2005/US/03/23/plant.blast/index.html.

21. *Los Angeles Firefighter.* 1976. Phosgene fumes hospitalize 40 fire fighters. February/March. Vol. 14, no. 7. p. 1: *Los Angeles Times.* 1976. May 16. Part I. p. 2.

22. Belcher, J., and K. Roderick. 1976. Chemical fire burns 9 hours; 50 hospitalized, 5000 flee. *Los Angeles Times*. June 17. Part II. p. 1.

23. Peige, J. D. 1978. Sulfuric acid over Baltimore. *Fire Engineering*. May. pp. 22–23.

24. *Redorbit.com*. 2005. Styrene tanker finally gone. September 5. Retrieved March 22, 2007 from: http://www.redorbit.com/modules/news/tools.php?tool=reprint&id=230349.

25. Oliver, M. 1978. 500 evacuate 14 plants near leaking fumes. *Los Angeles Times.* August 25. Part II. p. 5.

26. U.S. Department of Defense, Defense Civil Preparedness Agency. 1972. p. 73.

27. Dektar, C. 1972. L.A. crash truck with AFFF halts chemical storage fire. *Fire Engineering*. October. p. 40.

28. Matthews, T., S. Agrest, and G. Borger. 1979. Nuclear accident. *Newsweek.* April. p. 24.

29. *U.S. News and World Report*. 1987. July 20. p. 14.

30. Kirk, D. 1990. In the shadow of Chernobyl, fear and deformity. *USA Today*. April 26. p. 4A.

31. *Newsweek*. 1987. August 10. p. 47.

32. International Atomic Energy Agency. 2007. Engineering Safety Aspects of the Protection of Nuclear Power Plants against Sabotage. Retrieved November 6, 2008, from http://www.

pub.iaea.org/MTCD/publications/PDF/Pub1271_web.pdf.

33. *Breitbart.com*. 2006. Minn nuclear workers exposed to radiation. May 9. Retrieved July 15, 2008 from: http://www.breitbart.com/article.php?id=D8HGDVE02&show_article=1.

34. *World Book Encyclopedia, Year Book 1969*. 1969, 1975. Chicago, IL: Field Enterprises Educational Corporation. p. 300.

35. Bahme, C. W. 1978. p. 149.

36. Ibid. p. 173.

37. Camillo, E. 2007. Deadly depot blast racks South African community. *Firehouse.com*. March 24. Associated Press report. Retrieved March 24, 2007 from: http://cms.firehouse.com/content/article/printer.jsp?id=53975.

38. Bahme, C. W. 1978. pp. 68–69.

39. Di Neo, M. 1975. Explosion at Flixborough. *Fire Journal*. November. pp. 58–67.

40. Hillenmeyer, K. 1998. Fire disaster averted. January 5. *Cincinnati Enquirer*. A1.

41. Uebelacker, C. 1990. Explosion & fire in Cincinnati kills two, injures 83. *Firehouse*. October. pp. 43–51.

42. *Los Angeles Times*. 1979. Burning cloud from burning pesticide routs 2,000 at Memphis. July 6. Part I. p. 4.

43. National Fire Protection Association. 1975. *NFPA Hazardous Chemicals Data*. p. 202.

44. *Emergency Response Guidebook*. 2004. Guide 152. p. 252. Retrieved March 24, 2007 from: http://hazmat.dot.gov/pubs/erg/g152.pdf.

45. West, R., and S. Goldfarb. 1980. Dairy plant releases ammonia cloud: 22 hurt. *Los Angeles Times*. February 22. Part I. p. 3.

46. *American Heat*. Video journal. 1989. Heat sheet. Training guide. Vol. 4. Prog. 1. (July).

47. *Fire Journal*. 1969. Bimonthly fire record. Vol. 59, no. 4 (July). p. 49.

48. *CNN.com*. 2000. Death toll from New Mexico blast rises to 11. August 21. Retrieved March 24, 2007 from: http://archives.cnn.com/2000/US/08/21/explosion.deaths.02/index.html.

49. BBC News. 2007. Iraqis killed by chlorine bombs. March 17. Retrieved March 17, 2007 from: http://news.bbc.co.uk/go/pr/fr/-/2/hi/middle_east/6461757.stm.

50. Prepared by Robert Obermayer for the National Fire Academy's Open Fire Service Program, and partially based on the writings of Chief Bahme for the same program.

51. National Response Team, National Oil and Hazardous Substances Pollution Contingency Plan. *Hazardous Materials Emergency Planning Guide*. Washington, DC: Government Printing Office. pp. 17–18. Accessible at: http://www.mass.gov/Eeops/docs/mema/emd_advisory_committee/appendix_b/state_emergency_response_commission/NRT-1%20Guidelines.pdf.

52. Ibid. p. 37.

53. Orvos, D., and J. Cairns. 1987. Hazmat and habit. *Fire Engineering*. July.

54. Prepared by Robert Obermayer for the National Fire Academyís Open Fire Service Program, and partially based on the writings of Chief Bahme for the same program.

55. Wikipedia. 2008. Bhopal disaster. October 22. Retrieved October 22, 2008 from: http://en.wikipedia.org/wiki/Bhopal_Disaster#cite_ref-Eckerman2001_0-54.

56. *Time*. 1984. India's night of death. December 17. Retrieved October 22, 2008 from: http://www.time.com/time/magazine/article/0,9171,923797-3,00.html.

57. New York City. 2008. New York City fire department. Organizational chart. Retrieved

July 15, 2008 from: http://www.nyc.gov/html/fdny/pdf/fdny_org_chart.pdf.

58. Carrol, R., and L. Jenkins. 1976. Italy 'our own Hiroshima.' *Newsweek*. August 16. p. 49.

59. O'Sullivan, K. G. 1990. A soft-sided HAZ-MAT unit. *Fire Engineering*. October. pp. 55–57.

60. *Fire Command*. 1986. City fined for failure to provide hazmat information. March. p. 10.

4

Growing Threats, Global Concerns

HOMELAND SECURITY AND CIVIL DISORDER 10

Author's note: The sections "Homeland Security—Local Threats," "Homeland Security—Imported Threats," and "Preparations for Local Security" were expertly authored by Walt Clements, associate director of global security for the Proctor and Gamble Company, and Federal Bureau of Investigations, retired.

INTRODUCTION

The Department of Homeland Security was officially created in 2002, but efforts to protect the United States from unwanted threats, both natural and man-made, are not new to this country. Whether it is termed *civil defense* or *homeland security*, America's citizens have always worked toward having peace within their borders. In this chapter, we examine threats that occur both from within the country and abroad. We discuss a common warlike activity, rioting, and present an examination of two case studies. Finally, we examine potential problems that occur during operations at events of national significance, particularly bombing incidents.

HOMELAND SECURITY–LOCAL THREATS

Although the term *homeland security* was coined after, and primarily in response to, the September 11, 2001 attack on the United States by foreign terrorists, the nation has always been vulnerable to threats and has always had a need for security. The United States is a nation where there are numerous sites with large concentrations of people, riches beyond what much of the world can comprehend, and freedoms that are envied by many but despised by some. There are thousands of structures, monuments, and historical sites that attest both to America's success and to its pride in succeeding. While Americans see these attributes as the emblems of a proud and blessed country, some see only symbols of oppression, or more simply, targets. There are two other factors that present additional challenges to the mission of keeping the U.S. homeland secure. Any serious threat to the United States, or worse, any successful attack by terrorists from without or within its borders, brings instant and worldwide media coverage, a benefit to terrorists seeking to have their message heard. The United States also has large, diverse communities that are a testament to the reality of the words of the Statue of Liberty beckoning the tired, poor, and huddled masses. However, at the same time, this diversity allows international attackers to blend easily into society both before and after an attack.

When many people think of homeland security, they imagine the continuing threat of the United States coming under attack by foreign terrorists bent on waging jihad within its borders and against its people, and the efforts of the government to prevent such an attack. While this is certainly a viable and considerable threat, it is by no means the only threat to the security of the United States. Within the jurisdiction of the U.S. Department of Homeland Security is the Federal Emergency Management Agency, or FEMA, with responsibility for responding to and managing natural disasters in the United States, among other duties.

The devastation caused by a significant natural disaster occurrence in the United States can definitely impact the overall security of the nation in a number of ways. The occurrence of a disaster at the level of Hurricane Katrina can either devastate local emergency services or leave them so overwhelmed they cannot meet the need. The utilization of law enforcement and other emergency services from surrounding locales or even distant communities can weaken resources in the donating community. If National Guard troops are activated to assist at the scene of the disaster, a vacuum may be created elsewhere. The personal suffering and loss that accompany a natural disaster are immense. However, additional threats can arise from contaminated food and water, spread of disease, and widespread crime and looting as criminals prey on a community without adequate security.

Another local threat faced by the United States that impacts homeland security is the escalation in illegal immigration across the rather porous U.S. borders. One impact of this massive abuse is felt in the economy of the country and the tax burden on its legal citizens. However, the most serious threat to U.S. homeland security is in the number of criminals and potential terrorists that illegally enter the country. According to Mark Landsbaum, the "U.S. Border Patrol has arrested tens of thousands of illegal immigrants with criminal records, ranging from suspected murderers to child molesters, since it installed a fingerprinting system last year [2004]. About 30,000 of the 680,000 illegal migrants arrested from May through December were shown to have criminal records, compared to only 2,600 identified with criminal records during the same period in 2002."[1] Interestingly, a Harris poll conducted in 2006 asked people about possible threats to the United States. Several of the scenarios that respondents listed as extremely likely or very likely, and the percentage who deemed them so, are given in the following. The poll asked,

> There are a number of possible threats that the U.S. might face. How likely do you think the following will be a major threat to the U.S. in the next five years?

(Extremely/ Very likely listed)

- A large number of illegal immigrants come into the country 55%...
- A significant natural disaster destroys large areas of a major city 43%...
- Terrorists launch a number of attacks against airplanes 26%...
- The country is attacked with biological weapons 24%...
- Major riots by groups within this country 20%...
- A city within the country is attacked with a nuclear weapon 14%
- A large scale avian flu epidemic 11%[2]

Perhaps one of the most concerning aspects arising from the threat of increasing illegal immigration is the enhanced potential for foreign terrorists to enter the United States in this manner. Testifying before the Senate Select Committee on Intelligence on February 16, 2005, Deputy Secretary of Homeland Security Admiral James Loy stated, "Several al-Qaeda leaders believe operatives can pay their way into the country through Mexico and also believe illegal entry is more advantageous than legal entry for operational security reasons."[3] Even with the more stringent security protocols in place at the nation's airports today, areas of weakness are still being discovered that must be shored up in order to truly address the threats facing the United States. Thomas Frank reported in the March 27, 2007 issue of *USA Today* that the U.S. Department of Homeland Security's Transportation Security Administration was increasing screening of airport workers. This was in response to learning that nearly 1 million employees had access to airplanes and other restricted areas without going through security.[4]

The potential for a significant natural disaster and its impact on the people, economy, and in many ways, the security of the United States is a threat that must be considered and prepared for carefully. The flood of illegal immigrants and its potential effect on U.S. security, particularly in regard to the very real potential of foreign terrorists infiltrating this way, is also a threat that must be seriously addressed.

A third significant threat to U.S. security that comes from within its borders is domestic terrorism. Domestic terrorism is not new. Domestic terror groups such as the Weather Underground, the Symbionese Liberation Army, the Puerto Rican nationalist group FALN, and other similar groups in the 1960s and 1970s were idealistic groups using criminal and terrorist actions to try to achieve their goals. Though people were killed as a result of many of their actions, generally they were more interested in social and political change and winning converts to their cause than they were in killing innocent people. Many of the bombings committed by these groups were done at night on unoccupied buildings. To much of society, they were viewed more as criminal groups than terror groups.

America's view of domestic terrorism changed, however, in 1995. In that year, Timothy McVeigh parked a rented truck laden with ammonium nitrate and nitromethane in front of the Alfred P. Murrah Federal Building in Oklahoma City and detonated it, killing 168 of his fellow Americans. Suddenly greater attention was paid to the "terrorists among us," and the threat to the U.S. homeland from domestic terrorists became more real.

Domestic terror groups can include militias, the most openly militant groups of the conservative right wing that dress in camouflage and conduct paramilitary-style training with semiautomatic weapons. In many cases, militias profess a strong suspicion of government, with some advocating a violent overthrow of federal government. These include white supremacy groups such as Aryan Nations. In 2002, federal agents were investigating a New Jersey man, and a Texas man and common-law wife, with ties to white supremacy and a group called the New Jersey Militia. The federal agents raided a residence in Noonday, Texas and recovered "nearly 500,000 rounds of ammunition, 65 pipe bombs and briefcases that could be detonated by remote control…[and] 800 grams of almost pure sodium cyanide."[5] Clearly, these groups and the threat they pose need to be taken seriously.

Radical antiabortion groups may present a threat as well. Incidents of the assassination of doctors performing abortions and bombings of abortion clinics have been recorded. Some believe this can be traced to the release of a 1993 "cookbook" created by the Army of God. This book "instructs would be pro-life terrorists on how to use bombs, threats, and other violence to close down abortion clinics."[6]

In 2005, however, the Federal Bureau of Investigation told a Senate hearing that environmental and animal rights extremists who have turned to arson and explosives are the nation's top domestic terrorism threat, not the militant antiabortion groups, the militias, or the white supremacists. Of concern are groups like the Animal Liberation Front, the Earth Liberation Front, and the British-based group SHAC (Stop Hunting and Animal Cruelty), which is active in the United States. Such extremists are far surpassing other domestic terror groups in the amount of damage caused and the number of crimes committed.[7] These domestic groups have espoused goals of saving animals from cruelty and saving the planet from pollution, but they have employed arson, bombings, theft, vandalism, harassment, and letters rigged with razor blades to promote their agendas.

Local threats to domestic U.S. security need not always be associated with a particular group. In a free and democratic society such as that of the United States, the ability for those who desire harm to the United States to blend in easily and live among the general populace is great. Sympathizers with radical Islamic teachings, indeed actual or self-proclaimed members of al Qaeda, have arisen among the U.S. citizenry. This includes Americans such as John Walker Lindh, who was discovered fighting for the Taliban against the United States in Afghanistan. He had been converted to radical Islam and elected to see his former homeland as the enemy. Another is Adam Gahahn, a young American man who converted to radical Islam and now represents al Qaeda as "Azzam the American." Gahahn was charged in 2007 with treason in the United States. He began with an interest in the Islamic faith and eventually was recruited into the jihad. Raised on a goat farm in Riverside County, California, Gahahn allegedly turned to Islam in 1995 and moved to Pakistan in 1998. He has appeared in five al-Qaeda videos referring to the 9/11 attacks as "the blessed raids on New York and Washington."[8] Referring to previous attacks in Europe, he stated, "Yesterday, London and Madrid. Tomorrow, Los Angeles and Melbourne, Allah willing. And this time, don't count on us demonstrating restraint or compassion."[9]

Radical Islamic prison gangs such as Jamiyyat Ul Islam Is Saheeh, or JIS, have also been known to spread outside the penal system and have been linked to possible terror plots against military installations and synagogues in the Southern California area. Another American who became an Islamic terrorist is Jose Padilla. Padilla, a Chicago gang member and ex-convict, converted to Islam and then disappeared in 1998 after visiting Afghanistan. Several years later, he was in Chicago on a reconnaissance mission for an alleged al Qaeda plot to detonate a conventional explosive containing radiological material (i.e., a "dirty bomb"). Fortunately, his mission failed, and he was arrested.

To truly address domestic U.S. security, it must be recognized that the threats come both from abroad and from within. Terrorism is no longer a threat that comes from "over there," but is also a threat that must be dealt with on U.S. soil, and it sometimes involves U.S. citizens.

HOMELAND SECURITY—IMPORTED THREATS

Ask most people to name the foreign terrorist organization posing the greatest threat to the security of the United States, and chances are they will name al Qaeda. This response is understandable, given the significant high-profile attacks on the U.S. embassies in Tanzania and Kenya in 1998 and the attacks on Washington, DC and New York City in 2001, as well as the continual media references to al Qaeda in the war on terrorism. The threat posed by al Qaeda is not to be minimized, of course. It is important to realize, however, that al Qaeda is far from the only foreign terrorist organization posing a threat to domestic U.S. security.

The U.S. Department of State's report on global terrorism in 2005 listed 40 designated foreign terrorist organizations (FTOs), al Qaeda being only one of these. Each of these 40 terror organizations was either responsible for the death or kidnapping of an American citizen, financed by state sponsors of terrorism, or otherwise deemed worthy of inclusion in this annual report to Congress. In addition to the 40 groups designated as FTOs, the U.S. State Department also lists 40 additional selected terrorist organizations deemed of relevance in the global war on terrorism. Some of the groups designated by the secretary of state as FTOs were the Armed Islamic Group (GIA), Hizballah, Abu Sayyaf, and Hamas.[10]

The Armed Islamic Group is an Islamic extremist group with aims to overthrow the Algerian government and replace it with a fundamentalist state. Beginning in 1992, the GIA has been responsible for the killing of tens of thousands of Algerians and hundreds of Europeans. Hizballah is a Lebanon-based radical Shia group that takes its ideological inspiration from the Iranian revolution and was formed in 1982 in response to the Israeli invasion of Lebanon. Hizballah is known to be responsible for numerous attacks on U.S. and Israeli targets, including the devastating suicide truck bombings perpetrated in 1983 against the U.S. embassy and U.S. marine barracks in Lebanon. Abu Sayyaf is a small, violent Muslim terrorist organization operating primarily in the southern Philippines. Engaging in kidnappings for ransom, bombings, beheadings, and assassinations, Abu Sayyaf kidnapped 17 persons, including 3 Americans, in 2001 from a resort in the Philippines. One American was murdered and a second killed during a rescue attempt. Hamas, formed in 1987 as an outgrowth of the Palestinian branch of the Muslim Brotherhood, has employed both political and violent means of establishing an Islamic Palestinian state in Israel. Hamas has used suicide bombing as a terror tactic, and while they have not directly targeted U.S. interests, they make little or no effort to avoid targets frequented by foreigners.[11]

What do these four designated FTOs, out of a list of 40, have in common? In addition to al Qaeda, each of these four terrorist organizations is currently operating, or has recently operated, within the United States.[12] These operations may have been fund-raising, recruitment of operatives, surveillance of prospective targets, or more. They have operated in areas of the United States where one might least expect them: Hizballah in Charlotte, North Carolina; Hamas in Kansas City, Missouri; and the Algerian Armed Islamic Group in Seattle, Washington.[13] In a statement before the Senate Select Committee on Intelligence, January 11, 2007, Federal Bureau of Investigation Director Robert S. Mueller III stated,

> Throughout 2006, al Qaeda made efforts to align itself with established regional terrorist groups, such as the Salafist Group for Preaching and Combat, or GSPC, that may expand the scope of the threat to the homeland. In addition, al Qaeda is also finding it easy to attract individual members of these groups who align closer to Bin Laden's ideology and crave a more global agenda. This strategy has been particularly successful in recruiting individuals from Pakistani and Kashmiri militant groups operating overseas, as was evident in the recently disrupted al Qaeda–related airline plot out of the United Kingdom...
>
> While al Qaeda, its affiliates, and independent Islamic jihadist groups inspired by the global jihad remain the primary threat to the U.S. homeland, other groups, such as the Iranian-supported Lebanese Hizballah, warrant attention due to their ongoing fundraising, recruitment, procurement, and capability to launch terrorist attacks inside the U.S.[14]

The threat of terror is not the only threat these terrorist organizations have brought into the United States. The FBI has determined that Hizballah associates and sympathizers engage in a wide range of fund-raising activities in the United States. These include criminal activities such as money laundering, credit card fraud, bank fraud, and narcotics trafficking, to name a few. Palestinian terror groups such as Hamas engage in fund-raising, promoting propaganda for the Palestinian cause, and proselytizing in the United States as well.

Clearly, the threat posed to the security of the United States from abroad is not only great, but it comes from a number of sources. Such sources include foreign terrorist organizations acting independently, those acting in concert with al Qaeda, and al Qaeda itself. What are some of the primary threats posed by these foreign terror groups? In his January 2007 speech to the Senate Select Committee on Intelligence, FBI Director Mueller presented the FBI's assessment of al Qaeda's strategy for conducting an attack inside the United States. Director Mueller states,

For example, we believe:

- Al Qaeda is still seeking to infiltrate operatives into the U.S. from overseas who have no known nexus to terrorism using both legal and possibly illegal methods of entry.
- We also believe, if it can, Al Qaeda will obtain and use some form of chemical, biological, radiological, or nuclear material.
- Al Qaeda's choice of targets and attack methods will most likely continue to focus on economic targets, such as aviation, the energy sector, and mass transit; soft targets such as large public gatherings; and symbolic targets, such as monuments and government buildings.[15]

In recent years, concerted antiterrorist efforts by the United States and coalition forces have degraded al Qaeda's central command infrastructure, decreasing its ability to plan and carry out massive attacks. Some analysts, however, are seeing this as leading to a new phase in the global war on terrorism. They expect that al Qaeda will continue to spread its anti-Western ideology to other groups and geographical areas, and affiliated or independent terror groups will begin to present the greatest threat to the United States from abroad.

The FBI reported that in 2006, al Qaeda and its sympathizers continued their attempts to make global jihad accessible to English-speaking Western Muslims. Large amounts of violent Islamic extremist propaganda were disseminated in English via media outlets and the Internet. Al Qaeda's official media component, known as Al-Sahab, produced 48 propaganda videos in 2006, the most ever released by al Qaeda in a single year.

Certainly the smuggling of a chemical, biological, radiological, or nuclear device into the United States and the dispersing or detonation of the weapon present the greatest imported threat to U.S. security. While the actual detonation of a nuclear device is unlikely due to the required technical expertise and challenges associated with doing so, terrorist groups do have access to simple chemical and biological agent recipes obtained via the Internet or other training materials. Radiological material to create a conventional bomb that disperses radiological material (a dirty bomb) is available in many industrial locations in the United States. On March 22, 2007, police in Philadelphia recovered a stolen industrial gauge that contained small amounts of radioactive material. Another threat potentially considered, and possibly planned by foreign terrorist groups, is the poisoning of water and food supplies that would create panic among the population. Furthermore, cyber attacks could affect U.S. national security, military operations, and emergency services and would pose a significant threat to the U.S. economy.

Perhaps one of the most concerning potential threats that may be imported to the United States by a foreign terrorist organization is the tactic of suicide attacks, particularly against mass transportation. Though the only suicide attack by foreign terrorist organizations in the United States to date was the 9/11 attack, the use of suicide attacks against U.S. interests has occurred in the past. In 1983, there was an attack on the U.S. marine barracks in Beirut, and there were suicide attacks against U.S. embassies in 1998. Most disturbing, however, is the relative "success" and worldwide media attention achieved by terrorists through the use of suicide bombings on mass transit in Madrid and Moscow in 2004, and in London in 2005. Suicide attacks on public buses and trains in Israel have occurred with alarming frequency. It can no longer be assumed that this is a tactic practiced by Islamic extremist terrorists in other parts of the Western world and could not possibly be employed in the United States. Indeed, it may be more a question of *when* rather than *if*.

The threat posed to the security of the U.S. homeland by foreign terrorist groups is broad, diverse, and perhaps imminent. Foreign terrorists are not deterred by periodic setbacks that merely delay achieving their intended goals. They are not interested in merely converting Americans to their way of thinking, nor are they interested in a negotiated settlement. Islamic extremists have only used negotiation as a tactic to give them more time to prepare for attacks. Indeed, Islamic jihad terrorists do not see time in the same way as the Western world perceives it, and they will wait years to execute an attack with the intention of drawing the enemy into a state of complacency. Americans have no choice but to remain vigilant, prepared, and engaged in the struggle against terrorism that will present a threat to their homeland in the foreseeable future.

CIVIL DISTURBANCES: A SOCIAL PHENOMENON

"'The social model is one of the reasons for the riots,' said Johnny Munkhammar, a director of the Swedish research organization Timbro in Stockholm."[16] This statement by Mr. Munkhammar sums up the root cause of many civil disturbances. Although some are related to celebrating a sports team's victory or to college parties, many of the events that can truly be termed *rioting* are related to a group of society who feels oppressed.

It is not the intent of this section to provide a sociology education on rioting and the reasons surrounding its presence in the world. However, in order to properly plan for a civil disturbance event, planners must understand the issues that lead to rioting and compare them to the current affairs of their jurisdiction. In many cases, city leaders have played Monday morning quarterback and realized that telltale signs of an impending riot were blatantly obvious. Although a study of riots will often show that a precipitating event sparked the violence, the atmosphere present in the town can tell emergency service leaders when precautions should start and when to look for a riot to begin.

Additionally, planners should understand the issues that are surrounding the current riot situation, as their workforce might become the target of the violence.

There are a number of explanations for the causes behind rioting:

> Prior to the "spark" or precipitating incident, riot events are foreshadowed by a gradual escalation of resentments or grievances held by people who live in the riot area. These feelings, correlated with the structural characteristics of riot communities, represent the underlying conditions of riot genesis—the fuel that feeds the fire when the appropriate spark is provided. There are three general sets of structural explanations for the origins of riot violence: economic, political, and demographic. The first of these economic explanations suggests that people riot in response to conditions of abject poverty. A second and related economic explanation is that people riot when they see themselves as deprived relative to members of higher economic strata. Relative deprivation may spark rioting among those whose economic fortunes are improving but not fast enough to fulfill their rising expectations. By contrast, riots may also develop as a response to political disfranchisement, breaking out in areas where particular groups are politically underrepresented and angry about their lack of access to institutional power. Similarly, riots may represent a reaction to racial/ethnic segregation, which often is combined with economic deprivation and political exclusion. By this measure, places with the highest levels of minority racial/

ethnic population will be more riot-prone. Finally, the origins of rioting might also be discovered in processes of demographic change that alter the racial/ethnic composition of communities, affecting cultural control of institutions, and the psychic well-being of longtime residents who fear such changes. Thus rioting may be related to a general process of ethnic succession and competition.[17]

Just watching the news every day can lead a planner to understand that these structural explanations are present. The media has done a great job of hyping various citizen groups and making them believe that social injustice has occurred. In the Cincinnati area, which is a site for one of the case studies, the associate author has noticed that many times the news media seems to focus on the racism and social injustice portion of any news story, rather than reporting only the facts of the incident.

Weather is another factor that leads to social disturbance. Most, if not all, social rioting has occurred in a warm climate. As noted in some research, riots evolve when a large group of individuals converge and engage in violence. During cold months, individuals do not spend time on the street conversing in groups. Instead, they are left to remain indoors to view news coverage of events that lead to a perceived disenfranchisement. Thus, shortly after weather permits their socializing in groups, they meet with a preconceived agenda, which can lead to shared attitudes among the group. Upon a precipitating event, which rekindles their hatred and discontent, they begin to retaliate and cause violence, which can lead to a riot.

CINCINNATI: A CASE STUDY

On April 9, 2001, the city of Cincinnati, Ohio experienced the first of numerous nights of rioting in its downtown and urban residential areas. As noted in the social phenomenon section, this riot's root problem did not start on April 9. During the previous six years, the Cincinnati police had altercations with black males that resulted in the deaths of 15 black males. During the same period, no white males were killed by police.

Numerous events of high media attention occurred in the months prior to April, including the mistreatment of a black businessman during a routine traffic stop. Bomani Tyehimba filed a lawsuit in 1999 against the city of Cincinnati, claiming police illegally ordered him out of his car, handcuffed him, and roughed him up, all at gunpoint, during what should have been a routine traffic stop.[18] Shortly after this incident, another black male, Roger Owensby, died because of a struggle with Cincinnati police. Owensby died, according to the coroner's report, of asphyxiation, which was attributed to a chokehold applied by police. On the morning of the media reports of Owensby's death, another black male was shot after stealing a police officer's gun and shooting another officer.

As these incidents piled up, and kept occurring on a more frequent basis, tensions rose between the African-American community and the Cincinnati Police Department (CPD). As community leaders pointed out their being targeted by police, the police union pointed out statistics of high crime rates by black males, thus showing how at odds the two groups were at the time.

On the evening of April 8, 2001, an unarmed black male, with known warrants, was running from police. He stopped, and according to police statements, appeared to be reaching for a weapon when he was shot and killed by Officer Steven Roach of the CPD. This was the final incident that united a large group of frustrated citizens. On April

9, 2001, a large group barged into a city council meeting demanding answers to the shooting on the previous evening. When they were turned away and were told the CPD did not have the investigation completed, they then marched to the police station and stood outside demanding answers. When the crowd would not disperse, police began firing bean bag rounds at the crowd. Beginning at nightfall, formal rioting began in the urban residential neighborhoods surrounding the downtown area and moved into the downtown area. Police moved in with force and pushed the rioters back out of the downtown area. For the next three nights, rioting occurred in the residential neighborhoods surrounding the downtown area. On the fourth night, a curfew was imposed and kept in place for the downtown and adjacent residential neighborhoods. This, coupled with a weather shift to rain, quelled the violence.[19]

During the rioting, many downtown stores were looted and set on fire. Emergency services were deployed in a task force style for response. Because rioters were shooting at police and firefighters, large complements of fire personnel and police acting as force protection were sent to every call.

According to a Cincinnati fire district chief, the fire companies from all of the stations within a given fire district were corralled into the district's central firehouse and deployed from this location in task force complements. This included engine companies, ladder companies, and ambulances, coupled with a number of police patrol cars. Figure 10–1 depicts this task force complement and an associated ICS configuration to support the configuration. The district chief of that fire district established an area command within the district and triaged and dispatched companies to the simultaneous calls for service based on the priority of the call. Many smaller fires had to go unchecked due to stores and residential buildings burning simultaneously with these smaller events.

The city continues to face a violence and racism problem even six years later. The CPD was placed on oversight by the Department of Justice, and the department focused on police-community relations after the event. The county police department that protects the area outside of the city limits was brought in to assist the CPD in curtailing the problem. Changes have been made, and the social problems reduced in Cincinnati.

LOS ANGELES: A CASE STUDY

On April 29, 1992, the city of Los Angeles experienced numerous days of rioting. This riot had many similarities to the Cincinnati riot. This riot was related to race relations and relations with the police department as well. According to research into the causes, the dissention between Hispanics and African-Americans began about 10 years prior to the incident, when Hispanics began to heavily populate the south side of Los Angeles, and immigrants bought many of the stores in the local neighborhoods. Due to these two trends, African-Americans saw their jobs being taken away and reissued to lower-paid Hispanics. This was compounded by a recession and the perception that the Los Angeles Police Department (LAPD) used racial profiling in their daily operations.

In March 1992, an ex-con ran from police. When he stopped his car, he began a confrontation with the four LAPD officers who were pursing him during the chase. The police retaliated with tasers and nightsticks. A person who came out of his apartment caught this confrontation on amateur video. This video became a media event, resulting in the charging of the officers for excessive force. Coincidently, the officers were Caucasian. The ex-con was Rodney King, and thus the 1992 riot is often referred to as the Rodney King Riot.

Later that month, a young black girl was gunned down by a storeowner, who was Korean, over a bottle of soda that he felt she was stealing. This event was caught on video as well. This had certain citizens groups outraged, as the storeowner only received a parole and community service sentence from the judge. The jury had suggested a lengthy prison sentence.

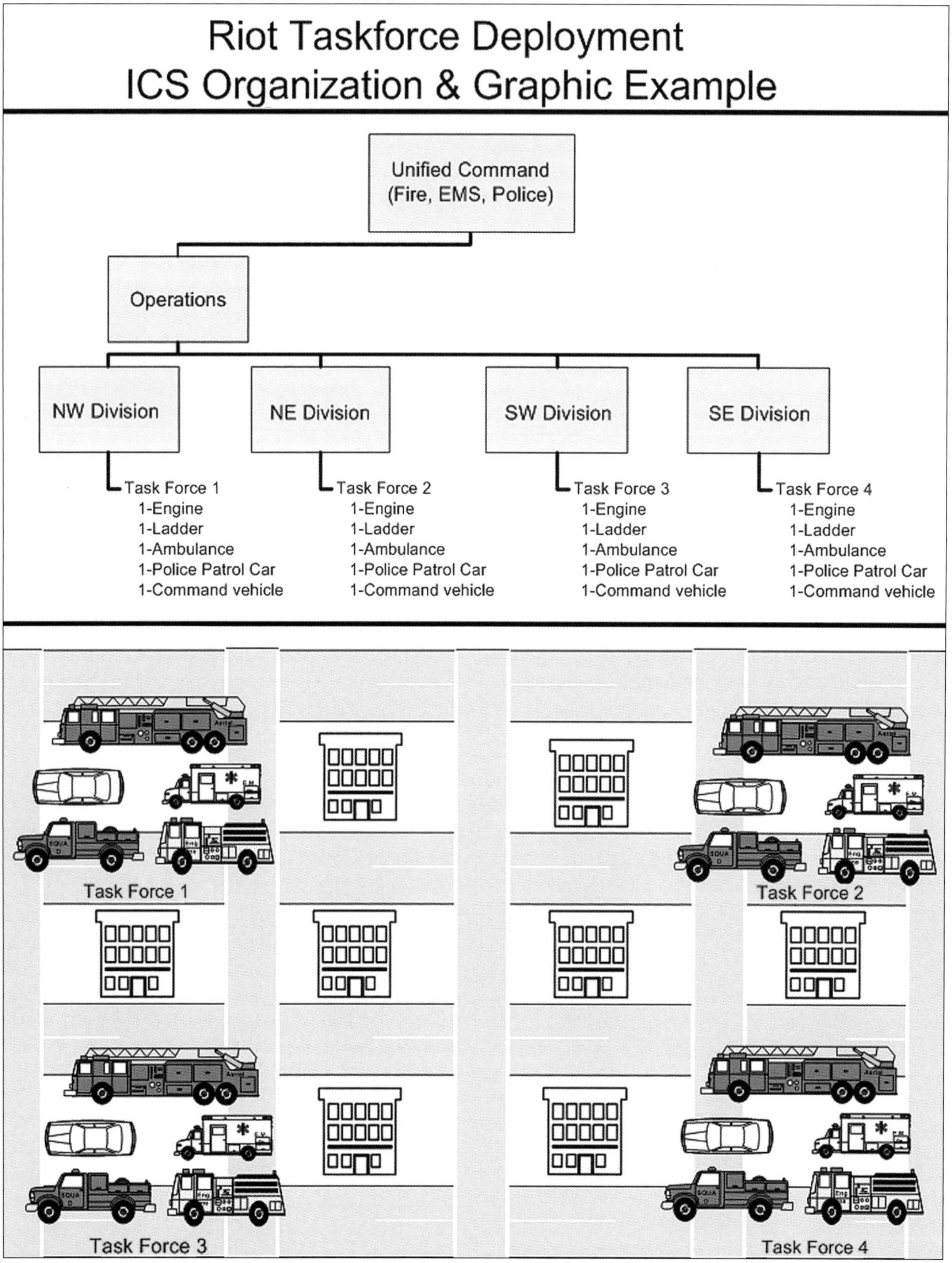

Fig. 10–1. Sample task force deployment of forces and distribution of forces for a community dealing with a civil disturbance.

The final high-profile event that led to days of rioting was the trial of the officers accused of excessive force in the beating of Rodney King. In a trial over what appeared to be an open-and-shut case of police brutality, the officers were acquitted of all but minor charges. This shocking verdict, based on the information available to the public, began rioting that lasted for six days.

The riots began in the same fashion as those in Cincinnati, with a group of demonstrators showing up at the courthouse to voice their displeasure with the verdict. In addition, a crowd arrived at the police headquarters. The rioters outnumbered the police, causing a retreat of police from the intersection where they encountered the rioters.

Helicopters videoed the rioting, catching many violent events. One of the biggest media events was the live footage of a white truck driver pulled from his vehicle and beaten repeatedly by rioters. The police did not show up to help the truck driver. Just after this instance, another person was pulled from a vehicle and beaten in the same fashion. The appearance to the public was that the LAPD was not responding to the rioters.

The increased rioting led to numerous vandalism and arson events, as rioters looted stores and threw Molotov cocktails into businesses. The fire department was answering one call a minute for fires.[20] During their responses, they were fired upon by rioters, thus making standard operations at fires nearly impossible.

As the rioting peaked, the LAPD needed assistance from the National Guard. Establishment of a curfew took place within the city, which did help quell the violence. A backup force of numerous military personnel with tanks and large firearms finally put an end to the rioting. Although some rioting took place in a sporadic fashion after the arrival of military forces, the riot essentially ended after four days. The show of force is credited with stopping the riots. Since this show of force stopped the widespread rioting, one wonders if a large showing by the LAPD in the initial phases might have stopped the violence that marked the commencement of the riots. The answer may never be known or proven, but it may suggest how planners and commanders could respond to future threats. The use of mutual aid forces and joint training for riot situations by city police and the neighboring departments could prove to halt any rioting in the early phases of a civil disturbance.

PREPARATION FOR LOCAL SECURITY

Whether the threat to U.S. security comes from an attack by domestic or international terrorism, or as a result of the chaos that can follow a significant natural disaster or civil disturbance, it should be clear that every citizen has a responsibility to be a part of the solution. This may mean active service performed by highly trained professionals in law enforcement, the fire service, emergency medical services, and even National Guard troops. Or it could mean simply relying on the "eyes and ears" of alert, responsible citizens to report suspicious activities that could prove crucial to preventing the attack or quelling the chaos of the aftermath.

In the days and months immediately after the 9/11 attack by international terrorists on the United States, security was at the forefront of nearly everyone's thoughts. A new U.S. government office was created, the Department of Homeland Security. Businesses rushed to purchase and install the "latest and greatest" security countermeasures. They peppered their access/egress points, mailrooms, shipping docks, and nearly every area except restrooms with closed circuit television, metal detectors, and sophisticated alarm systems. Americans received instructions from the government to report anything suspicious. They were given advice on using the infamous "plastic and duct tape" solution to the threat of a biological or chemical attack.

The scene, while understandable when considering the shock the nation was experiencing, was strangely reminiscent of the governmental warnings of the 1950s and 1960s. Schoolchildren of that era were taught to "duck and cover," as if such action would have proved lifesaving in a true nuclear attack. Airports instantly became heavily patrolled by armed police officers, and in many cases, National Guard troops. The contract security guards who had previously scanned passenger luggage were replaced by Transportation Security Administration security personnel en masse, charged with preventing even the smallest nail file from getting on the plane.

There were two pieces missing from the well-intentioned actions of businesses, citizens, and the government in the wake of 9/11, in their zeal to enhance homeland security. First, all of the enhanced security devices in the world are rendered nearly impotent unless there are trained personnel monitoring the data provided by the devices. Installing metal detectors without knowing what a bomb or other weapon looks like is silly. Even when security personnel think they know what they are looking for (at airports, this can range from the ominous knitting needle to the common chainsaw), their "target" can expand the moment a terrorist tries to ignite a shoe or intelligence suggests explosive materials can be smuggled in a bottle of shampoo. Second, merely asking citizens to report suspicious behavior, without at least defining examples of what suspicious behavior may look like, leaves the decision solely in the hands of well-meaning but uninformed citizens.

Security is proactive. The object of security is to prevent the undesired activity from occurring in the first place. This is true whether it is the work of the TSA agent screening baggage, the police officer on patrol, or the fire service personnel alert to signs of bomb construction as they perform fire inspections. It is true for the security guard at the mall or the average but alert citizen going about his or her daily business. Homeland security in regard to protecting the country and its people from a terrorist attack, the aftermath of natural disasters, or significant civil disturbances can be viewed from both the preincident mode and the postincident mode. Effective security preparation centers on security awareness, security assessment, and security enhancement. Postincident security covers much of the same ground but may have to be implemented much more quickly.

Preincident security preparation

Security awareness. The most basic tenet of security is awareness. Some people see the world as they would like it to be instead of how it is. As such, they tend to disregard any information they learn that does not fit their model of the world. They go about their day blissfully unaware of the dangers that surround them, believing that if they do not see the dangers, they do not exist. Criminals know this, and purposely select victims who are not paying attention to their surroundings or any impending threat. Like the antelope contentedly eating grass as the lion approaches stealthily, they become victims without ever knowing what hit them. Even citizens who do recognize that the threat of terrorism or other criminal activity may be in their midst may not really understand what "suspicious behavior" is. If they do, they may hesitate to report it to authorities out of fear that they may accuse an innocent person or be criticized for overreacting.

To develop successful security preparation, however, requires the development of greater security awareness on the part of emergency personnel and the civilian population. What might have happened if the shopkeeper who sold Timothy McVeigh the large quantity of ammonium nitrate had reported his suspicious feelings when McVeigh refused discounts he was due on the purchase? What would have happened if the flight training suspicions raised by the 9/11 terrorists were reported, and taken seriously, sooner? Security awareness does not mean viewing the surrounding world with suspicion, or worse, paranoia. It does mean viewing the world in the context of understanding what is normal and noting what is not. An unusual vehicle parked on the street may be normal. The same vehicle with people sitting in it at 3:00 a.m. is not. A stranger asking for directions may be normal. A stranger videotaping security at a building or drawing diagrams of a building is not. Fire department personnel noting protective clothing, chemicals, and laboratory equipment as they conduct an inspection of a university chemistry department building is normal. The same scene in the basement of a home discovered in response to a minor fire is not.

Israeli terrorism experts have recently been training U.S. airport security personnel to "look for the bomber, not the bomb." It has been recognized from years of Israeli experience, and more recently from lessons learned in Iraq, that improvised explosive devices can look like anything. Thus, awareness efforts now focus more on the unusually nervous passenger with a one-way ticket and no luggage, who avoids eye contact and paces nervously while waiting to pass security. Awareness is being alert to what does not belong, recognizing it may represent a threat, and reporting it rather than chalking it up to coincidence.

Security assessment. The second key piece to successful security preparation is assessment. Simply put, security assessment first involves determining the vulnerability to attack or security compromise by determining the attractiveness of the target, whether to terrorists, criminals, or looters following an instance of natural disaster or civil disorder. Next it involves the likelihood that the target may be attacked or compromised. What kind of things should warrant a security assessment? Certainly, security assessments are important for critical infrastructure components such as public and commercial transportation systems, water and food supplies, schools, nuclear and other power generation sites, and military facilities. Security assessments are also necessary at sites where chemical, biological, nuclear, or explosive compounds are housed or transported. Other sites that should be assessed for security include symbolic sites, any location where large numbers of people may be concentrated (i.e., sporting events, malls, etc.), and commercial sources of weapons. Figure 10–2 is a photo of one of the security measures at a historic battleship in the harbor of Boston.

Fig. 10–2. Security measures at a historic battleship site. Note the heavy traffic barrier to prevent vehicles from being driven into the exhibit that could carry explosives.

Photo by Randall W. Hanifen.

Security assessment should focus on the overall security precautions in place. These could include positive identification checks, closed-circuit television that is actively monitored, and metal detectors if site appropriate. A security assessment should also consider general access control. Do only those authorized to be there get in? In addition, the level of security response to violations of the general access controls or permitted presence monitoring must also be evaluated. The security assessment may determine that there are weak controls that would allow the security of the site to be compromised, either passively or in the midst of an emergency situation such as a disaster, riot, or following a terrorist attack. In this case, security enhancements should be promptly made.

Security enhancement. To have effective security preparation for any anticipated or unanticipated homeland security event, there must be an awareness of the potential for a threat and an assessment of the risk of the threat, resulting in the implementation of security enhancement. A good example is the public transportation system in some major U.S. cities. Even after 9/11 when significant enhancements were made to the security screening of airline passengers, there were few security enhancements made to the public transportation system. Israel had experienced bombings of public buses for years, and threats had been made against public transportation systems on more than one occasion. It was not, however, until the terrorist train bombings in Madrid in 2004 or the London train bombings in 2005 that New York City residents began to see heavily armed police at subway stations checking packages and visually screening passengers. Still, security enhancements must be meaningful.

As noted earlier, installing metal detectors and closed-circuit television cameras is virtually meaningless without someone monitoring the equipment who knows what to look for and recognizes what presents a threat to security. Security enhancement may be as simple as training security personnel to recognize surveillance-related activities and report them to law enforcement. It could involve ensuring that fire stations that were previously open to the public 24 hours a day are now fully secured from unauthorized entry as the terrorist threat of stealing fire apparatus or uniforms is realized. In short, security enhancement is accomplished by recognizing through assessment what the security weaknesses or vulnerabilities are and taking proactive steps to correct the weaknesses.

Postincident security preparation

If an attack by terrorists or criminals, or the chaos that may accompany a natural disaster or significant civil disturbance, is unanticipated or otherwise not sufficiently planned for, the enhancement of security may have to occur much more rapidly. Emergency responders who normally experience victims welcoming their arrival may find themselves drawn into an ambush and fired upon. Fire units responding to a "routine" medical call may find themselves in the midst of a chemical or biological attack. Schools, businesses, hospitals, or other critical infrastructure sites such as public transportation or even police stations that normally maintain an adequate level of security may find their structures damaged by natural disasters to the point that security systems are inoperable or even under attack by looters.

While preincident security preparations strive to "expect the unexpected," in reality, it is not feasible to effectively secure against all possibilities of attack. Commercial businesses could have security enhanced to prevent looting in the event of a cataclysmic disaster by installing machine gun posts at the door, but no one would shop there any more. In spite of the fact that it is not feasible to make adequate security preparations for every possible contingency, it is possible to plan ahead for those types of additional security resources that may be needed in the event of a rapidly evolving scenario of significant magnitude. This may include planning staging locations for National Guard troops or developing mutual aid arrangements with law enforcement agencies from neighboring states or municipalities. It could also involve the creation of auxiliary or reserve forces to help secure facilities or structures.

Preparations must be made in advance to heighten the security of the nation's critical infrastructure and other sites vulnerable to attack by criminals or terrorists, or those subject to compromise during the course of a natural disaster or significant civil disturbance. This is certainly easier than trying to secure sites in the midst of an emergency situation. When considering security preparations, however, it should be recognized that in business, security is generally not considered revenue generating. As such, it may receive a lower priority than it truly demands.

When threat levels are high, as in the wake of an actual terrorist attack or very credible government warnings, security is typically enhanced. When the visible threat or the "clear and present danger" appears to subside, so does security awareness. This fact is not missed by terrorists during their planning stage. They would certainly rather attack a target with less security preparation than one prepared for every likely contingency. Thus, security or law enforcement planners should keep in mind that a determined adversary with enough time and planning can have a reasonable chance of success attacking any target. Figure 10–3 shows some of the active security enhancements that are in place at significant sites.

POTENTIAL PROBLEMS AND FIRE DEPARTMENT OPERATIONS AT CIVIL DISTURBANCES

Fire department as targets

As the previous parts of this chapter demonstrate, the fire department is a target. Although the fire service is thought by many to be the "good guys," it is seen that certain groups of people and certain events will have the fire department trying to defend its own safety. Many would proclaim that the firefighters should stage outside of the areas of violence and await a safe scene. After all, this is what the curriculum for both firefighters and paramedics teaches the responders. However, this may not be feasible. If fires are allowed to go unchecked for long periods, they can progress to conflagrations in large cities, thus killing and injuring thousands.

Operating in hostile environments

Operations, however, are not able to resume as normal. The first problem that units will encounter is the lack of full alarm assignments. In large cities, fire companies have a specific function, such as advancing hoselines or truck operations. These one-track operations will likely not work during these periods, and all companies will have to perform numerous firefighting functions, within the limitations of their equipment. Planners should take into account high call volume periods when they are specifying apparatus and equipment. In addition, firefighters must only deploy the tools and equipment needed to complete the task at hand and then quickly resecure the equipment. They must plan for equipment theft or tampering during the operation. The turning off of fire hydrants is common. Securing the wrenches or leaving a person at the hydrant is a way to deter the shutting off of water. This person should be prepared to encounter violence in the event rioters want to disrupt fire department operations. In figure 10–4, it is evident that upon civil unrest, the

Fig. 10–3. Active security measures in place at key sites throughout America as a result of enhanced security in the post-9/11 environment.

Photo by Randall W. Hanifen.

citizens begin a type of marshal law that involves hurting or killing others to preserve themselves.

Fig. 10–4. Civil disturbance measures by citizens. Often average citizens will take the law into their own hands when normal law enforcement forces have failed. This can present a danger to responders.

FEMA News Photo.

Upon arrival, the incident commander and/or the company officer should dictate a rally point in case a hostile event ensues. Firefighters are often not equipped with PPE or firearms to defend themselves; therefore, fleeing the area is often safest option. If the fire department is operating in conjunction with the police department, the police department should be advised of the exit strategy and rally point, so they can provide coverage for this escape.

Assigning units in a task force configuration will help many of the problems discussed in this section. Ideally, a task force would be comprised of at least the following units:

- Engine company
- Truck/ladder company
- Ambulance
- Police units

Depending on the call volume, the size of the riot and the ability to summon mutual aid and off-duty forces will dictate how large each task force is able to be during the operation. Many times, the city experiencing the riot is reluctant to summon mutual aid due to the high-risk nature of operating in such an environment.

Managing the incidents

The number of calls will quickly exceed the resources on hand. Therefore, using companies wisely will be the only hope for answering the important calls and mitigating the incident. In order to accomplish this task, an EOC and area commands should be established. The EOC will allow for proper coordination between police, fire, public works, and elected officials. Decisions about state and federal assistance are better made when all involved parties are present and can provide relevant information.

Establishment of area commands based on districts or neighborhoods will keep the units in that particular district working the most important calls at that time. Additionally, the area commander can redeploy forces to aid other operating forces, thus mitigating the incidents quickly and safely. By keeping the units assigned to that district in the district, time and resources are not wasted by having units drive from one side of the city to the other to answer the calls in the order they are received.

Ideally, the communications center will have a liaison in the EOC that will provide the EOC staff the list of calls. They can then process them to the appropriate area commander. This allows the EOC to see the big picture but does not overload them with assigning every call that is received. The EOC is also able to organize planning and logistical issues at the overall incident level.

Once received by the area commander, it can be added to the list for the area, and the calls can be prioritized. Area commanders can also advise the EOC if they have exceeded their resources.

Preparing the vehicles for safety

In the past, fire companies have taken a number of actions to prepare their vehicles for response to an area of rioting. The first step is to remove equipment from the exterior of the vehicle. Often, fire companies will mount tools and equipment to the exterior of the vehicle for easy access. Any tools or equipment not permanently secured are subject to theft.

Fire companies may encounter gunfire when responding to calls and at calls. Many fire companies place duct tape in the windows of the apparatus in order to prevent the glass from shattering on the crew in the event they are hit by gunfire.

In the 1960s, rioting took place on a more frequent basis, thus prompting some fire departments to install firearms on the vehicle. This again became something considered during the rioting that took place post–Hurricane Katrina. This is certainly a decision that should have forethought when considered. A safer alternative could involve assigning a SWAT team member of the police department to ride on each of the fire apparatus during civil disturbances, as members of the fire company need to be busy fighting fires rather than holding rioters at bay with firearms. Having a loaded firearm on the vehicle also could result in the weapon being used against the fire crew by rioters.

FIRE DEPARTMENT OPERATIONS AT BOMBING INCIDENTS

The most common terrorist operation is a bombing. The authors will cover the specifics of the various weapons used by terrorists in the next chapter, but this section will discuss what operations need to occur when a fire department responds to a bombing incident. In addition, it is important to consider that operations are centered on the explosive effects of a bombing incident. As such, explosions for natural or accidental events, such as natural gas explosions, will mimic the operations of bombing incidents to a degree.

In the past, the associate author has overheard statements from fire executives about how much validity and necessity are really present in the additional training and method of operations that has ensued post-9/11. To this, planners and response executives must be reminded that Oklahoma City is not one of the largest cities in America, and the idea behind terrorism is to scare citizens and make them change their level of personal security. What better way than to bring the next bombing incident to the local suburban shopping mall? This event may not kill and injure the large numbers as 9/11 did, but would this not make citizens think about their personal safety each day, regardless of where they live? In addition, who will respond to these incidents when (not if) they happen?

As in every large-scale terrorist event, the local fire and police are dispatched to deal with the event. There also are many WMD response teams, structural collapse teams, and evidence collection teams that are available from state and federal departments. However, it will continue to be America's first responders who encounter and mitigate the initial operational period of any event.

Planning for bombing events

Planning for bombing events takes a culmination of the right people with the correct training, the right equipment, and the correct plan being employed. These three related aspects of response need to be commensurate with each other. Having a great deal of equipment, with no plan or personnel to operate it, will not produce a positive outcome. Neither will a detailed plan that does not have the properly trained personnel and equipment to support its strategy and objectives.

The first step planners must undertake is to determine the level of response their jurisdiction plans to provide for a given incident. Due to the vulnerability analysis, the jurisdiction may recognize very few hard and soft targets and recognize that they do not have a large force of responders and cannot support a highly technical cache. Providing an awareness or operations level response to a bombing incident is a good start.

However, the planners first must identify where and how to obtain technician level responders and equipment and determine the length of time the responders will take to arrive at the incident.

Components of a bombing incident

Bombing incidents usually involve two aspects: building collapse and mass casualties. In order to respond effectively to a bombing incident, plans, personnel, and equipment must be present to deal with these aspects. Because terrorists want to have a high-exposure media event that leaves an impression in the minds of all citizens, they will often ensure that both aspects are present. Bombing an empty building would possibly destroy it, but if no one is hurt, the feeling of security is still present. If the bomber detonates an explosive in a crowd in an open venue, some feeling of security is gone, but as soon as the victims are removed, the visible signs of terrorism are gone. In order to make a lasting impression, bombers want to hurt people and have a picture that will last for a decent period of time. Thus, a collapsed building, such as the Alfred P. Murrah Federal Building or the remains of the 9/11 attack, will forever be embedded in the minds of Americans. The 1993 bombing of the World Trade Center is not as imprinted in the minds of Americans because most of the damage was underground on the inside of the building. News footage was dark, and the real destruction was not seen through the media outlets.

First responder actions

The first arriving fire company will be awestruck at the number of tasks that need to be completed in a minimal time, as well as the size and complexity of the incident. Even the most seasoned fire officers will find themselves overwhelmed. The key to these incidents is to enter into a command mode. Reconnaissance missions must be assigned to determine the extent of the damage and to begin to estimate the injury and fatality totals. Staging areas must be created in order to systematically assign companies to tasks, preventing even more chaos than is present from the incident.

First arriving command officers must begin to activate their emergency operations plans and begin to involve local and state officials as soon as possible to acquire the proper resources to deal with the incident. Hospitals will begin to become overwhelmed, as those who are mobile will begin to seek treatment.

Control of utilities will make preparations to save lives. Many times a fire will occur secondary to a building collapse. A crucial decision must be made as to putting more water in the building, thus drowning trapped victims, or letting the fire burn, thus burning trapped victims. Often, sprinklers should be shut off, and firefighters should advance hoselines only to areas currently on fire. This will ensure a limited amount of water is dispensed on the building, preventing the subsequent drowning of victims. Controlling electricity and natural gas flows to the building will increase responder and victim safety.

It is important for a commander to establish triage and treatment areas. Although corralling numerous victims is nearly impossible, it will allow responders to remove victims to the proper location. If victims approach responders seeking care, they can be directed to the proper location.

Depending on the extent of damage, collapse rescue teams need to be summoned early. Depending on the team, a structural engineer from the jurisdiction should be summoned to establish safe areas for responders. Although firefighters will want to rush in to help, they must understand that secondary collapse is a great danger at bombings. Most victims able to leave will do so if directed. Having initial arriving companies use loudspeakers to direct victims out of the wreckage and to a triage area will eliminate extra victims that could become entrapped.

The summoning of a mass casualty unit is an early need. Hospitals will become overrun by the walking wounded. Thus, temporary care facilities will need to be erected near the site of the bombing in order to care for the wounded until enough transport resources are available to transport the wounded to facilities that are not overloaded.

Dangers and coordination of responders

As mentioned earlier, emergency responders, specifically firefighters, are important to the public and are held in high regard. This fact is often used by terrorists. The events of 9/11 would have been very sad even without the 343 firefighters who perished, but their deaths added a level of fear and anger on behalf of the public. Terrorists know this and pride themselves in killing responders. This is often accomplished with a secondary device. Planting another explosive in the building or in the vicinity to hurt and kill responders is not out of the question. Responders must keep this in mind. Deployment of bomb-sniffing dogs is a good tactic for the police services in order to possibly find any secondary devices. Dogs are often able to maneuver through the collapsed structure more quickly and easily than humans, thus they can go into all areas in search of a secondary device.

Coordination among responders is important. Firefighters will focus on fire and rescue problems, the EMS will focus on caring for the injured, and the police will want to evacuate survivors and preserve evidence. Training all entities to understand the objectives of the others will help each entity accomplish its goals. This is often best accomplished through the use of a unified command. By having all entities represented, the response efforts can be coordinated, increasing safety and efficiency.

Utilization of state and federal assets

Declaring a disaster early and officially will enable state and federal resources to be deployed to the incident. Due to the necessary reaction time, this declaration should be made as soon as possible. Local leaders must remember that by design of the Nation Response System, the federal responders are not coming to take over their incident, but rather to aid them. Some of the federal-level assets that are useful at a large-scale bombing of a building are the following:

- Incident Management Teams (IMATs)
- Disaster Medical Assistance Teams (DMATs)
- Urban Search and Rescue (US&R) Teams
- Disaster Mortuary Operational Response Teams (DMORTs)
- Civil Support Teams (CSTs)
- FBI

All of these response teams have a cache of equipment and specialized training that is often not available to local responders. Events of national significance should utilize their services in order to properly mitigate the event effectively and safely.

CONCLUSION

Homeland security planning and response are not new to America's emergency services. Americans have entered a new era in which all levels of government are now committed, and in some cases legislated, to plan for and respond to these events. Understanding the enemy and its mission is at the core of proper planning and response. If planners know what to expect, they can plan for and respond to the incident. Often America is the last place for events to occur, and therefore planners should keep an eye abroad for upcoming trends in homeland security.

Rioting is not new, nor are the reasons people begin riots. Planners must know the impending signs of a riot and prepare responders for response to these hostile events. Learning from recent events will enable responders to continue to respond as safely as possible to riots.

Finally, those trying to make a statement will continue to use explosives to bomb buildings and other venues. Proper response to these incidents will ensure that responders can save as many lives as possible, while maintaining their level of safety. These events are large in scale and often require expertise above the local level. Providing a coordinated response with local, state, and federal responders will offer the best assistance to the victims of the incident.

REFERENCES

1. Landsbaum, M. 2005. Al-Qaeda's illegal immigration threat. March 7. *FrontPage Magazine.com*. Retrieved March 22, 2007 from http://frontpagemag.com/Articles/Printable.asp?ID=17216.
2. *Angus Reid Global Monitor.* 2007. Illegal immigration is biggest threat in U.S. February 21. Retrieved March 22, 2007 from http://www.angus-reid.com/polls/index.cfm/fuseaction/viewItem/ItemID/14810.
3. Landsbaum, M. 2005.
4. Frank, T. 2007. *USA Today*. March 27.
5. Gold, S. 2004. Case yields chilling signs of domestic terror plot. *Los Angeles Times.* January 7. Retrieved March 22, 2007 from http://www.commondreams.org/headlines04/0107-03.htm.
6. *MILNET.* 2006. Domestic terror: what it holds for the new millennium. Retrieved March 22, 2007 from http://www.milnet.com/domestic/Dom-Terror.htm.
7. Associated Press. 2005. FBI surprise on top domestic terror threat: it's not abortion foes or Klan, but animal and eco-extremists. May 19. Retrieved March 22, 2007 from http://www.msnbc.msn.com/id/7908466.
8. *CNN.com.* 2006. $1 million for arrest of American al Qaeda charged with treason. Retrieved March 26, 2007 from http://www.cnn.com/2006/LAW/10/11/gadahn/index.html.
9. Ibid.
10. U.S. Department of State, Office of the Coordinator for Counterterrorism. 2005. *Country Reports on Terrorism 2004.* April. Department of State Publication 11248. Retrieved March 14, 2007 from http://www.state.gov/documents/organization/45313.pdf.
11. Ibid.
12. Emerson, S. 2002. *American Jihad: The Terrorists Living Among Us.* New York: Simon & Schuster.
13. Ibid.
14. U.S. Department of Justice, Federal Bureau of Investigation. 2007. *Robert S. Mueller, Director, Federal Bureau of Investigation: Statement Before the Senate Select Committee on Intelligence, January 11, 2007.* January 11. Retrieved March 26, 2007 from http://www.fbi.gov/congress/congress07/mueller011107.htm.
15. Ibid.
16. Lynn, M. 2006. France's rioters reveal social, economic sloth. *Bloomberg.com*. Retrieved April 3, 2007 from http://quote.bloomberg.com/apps/news?pid=10000039&cid=lynn&sid=aJkyKmBAEmWg.
17. Law Library. 2007. Riots: behavioral aspects—precipitating events and underlying conditions. Retrieved April 3, 2007 from http://law.jrank.org/pages/1969/Riots-Behavioral-Aspects-Precipating-incidents-underlying-conditions.html.
18. Horn, D. 2001. Cincinnati: 2001 year of unrest. *Cincinnati Enquirer.* December 30. Retrieved July 17, 2008 from http://www.enquirer.com/unrest2001.
19. Ibid.
20. National Geographic Channel. 2006. The final report: LA riots. October 4.

TERRORISM 11

INTRODUCTION

Terrorism is the buzzword of the emergency services in the present day. Grants are awarded, training sessions are offered, and piles of equipment are procured in order to conduct the war on terrorism. The terrorists have been identified as the enemy, but how much is known about the enemy? There are large caches of equipment amassed for a response to a terrorist incident, but what will an incident look like? How will responders act at an incident as a result of the enhanced training? Why should those who are not planners or responders in America's largest cities concern themselves with terrorism?

The authors examine terrorist enemies, both foreign and domestic. A description of hard and soft targets will allow planners and responders to identify the key differences. Next, the discussion addresses what the terrorists might use to inflict damage. A discussion on the basics of chemical, biological, radiological, nuclear, and explosive (CBRNE) terrorism is presented. Finally, defensive and offensive actions at actual events are discussed, and the command and coordination efforts that must take place at an event of national significance are described. The authors' aim is to deliver the latest information on terrorist groups and their tactics. Nevertheless, the reader is warned that terrorists change their tactics constantly in order to make the next attack a surprise and to inflict fear in citizens and responders. To not change tactics would be to remove the shock factor of the terrorist incident. Thus, readers are encouraged to continually stay abreast of the latest intelligence findings and investigate how those findings affect their planning and response to these incidents.

SEPTEMBER 11, 2001

September 11, 2001 has forever changed the way emergency responders view their work. It is not the intent of this section to examine all of the events surrounding the 9/11 attacks, as this could fill many books. Interested readers are encouraged to research further the events surrounding the attacks and the response that took place after the attacks. The authors do wish for everyone who reads this section to pause for a moment and pray for the civilians and responders who lost their lives in this event, as well as for their families. We pay special tribute to the police officers and firefighters who paid the ultimate sacrifice. Figure 11–1 shows the timeline of the activities leading up to the event, the crashes and subsequent building collapses, and the response and aftermath.

The attacks

On the morning of September 11, 2001, the skies were clear and the day was calm. At 8:46 a.m., this tranquility was destroyed as American Airlines Flight 11 crashed into the North Tower of the World Trade Center (WTC 1). The FDNY crew attending to a gas leak in the area witnessed this event and began a response to the site. (Although it is unusual for a plane to strike a high-rise building in New York, it is not something new and has occurred since 9/11.) The crew initiated a high-rise firefighting and rescue activity.

At 9:02 a.m., United Airlines Flight 175 struck the South Tower of the World Trade Center (WTC 2). News media had covered the original plane crash and reported the fire. It was not until the second plane struck the WTC 2 that Americans realized they were under attack. The response continued and was upgraded by both the FDNY and the New York police, in conjunction with Port Authority police. This response involved the ascending of numerous floors of both high-rise buildings and the evacuation of numerous people from the buildings.

At 9:37 a.m., a third plane, American Airlines Flight 77, slammed into the Pentagon, crashing through many rings of the building. This crash was caught on surveillance cameras, and the plane's speed led to belief that a missile hit the building. Because of the crash, a large fire ensued in the building.

At 10:03 a.m., the fourth plane in the event crashed after passengers took over the plane from the hijackers. The plane crashed in a field near Shanksville, Pennsylvania. The fourth plane is reported to have been en route to the White House.

Results of the attacks, building collapse, and fires

Despite numerous other buildings being struck by planes, the WTC towers were of a newer generation of buildings built of a steel frame. This steel frame was covered in a protective fire-resistive material, but due to the impact and raging fire in the building, the steel was heated above the temperature at which it would lose its tensile strength. This, coupled with the weight of the structure above the fire, caused the buildings to collapse in a pancake fashion. At 9:59 a.m., 56 minutes after being struck by the plane, the WTC 2 tower collapsed. At 10:28 a.m., 102 minutes after being struck by a plane, the north tower collapsed in the same fashion.

The two buildings collapsed in a pancake fashion, but the debris and fires that resulted from the collapse caused all of the other buildings on the WTC campus, including the building housing the New York EOC, to either collapse or burn from hostile fires. This amount of debris falling caused destruction to a large portion of the FDNY apparatus. It also created a large accountability and control problem, as the original command posts were established in the lobbies of the towers.

Problems encountered

Although the FDNY is considered one of the most prepared departments, the resulting investigations into the attacks and the response recognized unexpected problems encountered by emergency response personnel. These have been the basis for much-needed change in the profession of emergency response. It helps to ease the pain of the loss of fellow fire service personnel to know that improvements are being made to prevent subsequent deaths in future attacks or other disasters.

Communications and interoperability are at the forefront of the problems. The firefighters reported not being able to transmit and receive within the building due to the repeaters on site either being inoperable or turned off. The lack of ability for

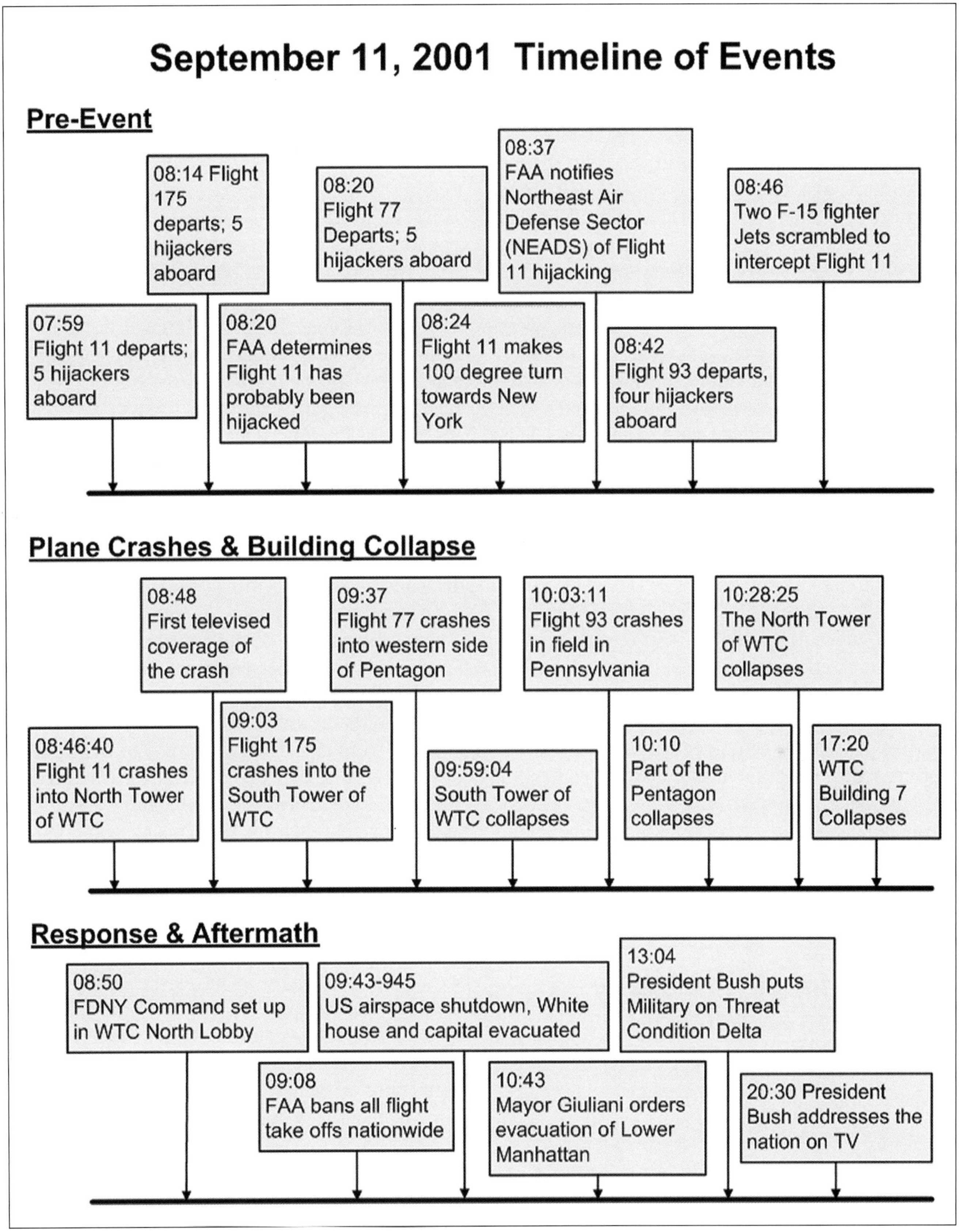

Fig. 11–1. Timeline of pre-event, plane crashes and building collapses, and response and aftermath activities.

the Port Authority and New York responders to communicate with each other hampered warnings of the buildings collapsing.

Problems with the design of the buildings hampered firefighters from ascending the building. Stairwells were not a priority in the construction of the buildings due to their having elevators. Elevators are often used in response to fires and alarms in tall buildings due to the time and energy needed to ascend stairs. (Readers are referred to chapter 7 for further information about high-rise firefighting tactics.) When the elevators were severed due to the impact, all operations, including simultaneous evacuation and the response of fire crews, occurred in a stairwell that was only wide enough for one firefighter with equipment. This problem, as well as the construction methods of high-rise buildings, is under review.

The common practice in the response to high-rise fires is to establish a command post within the fire command room or the lobby of the building. This provides information on the building systems, allows the use of built-in communication systems, and offers a large working area for these large-scale incidents. For the majority of these incidents, seeing the building is not important, as the fires are usually contained within the building. However, this can cause numerous problems when the buildings collapse. In the case of 9/11, many of the command staff were killed in the collapse, thus causing complete destruction of the current strategy and the accountability set up by the command personnel.

Changes for all emergency responders and planners

Responders quickly realized that they could become targets in their response to disasters. The thought that terrorist activities only occurred abroad was proven false that day. Planners realized that planning for events that have a high probability of occurring, rather than for what could occur, was likely not the correct approach in large-scale exercises. Many emergency service executives lived under the misconception that "it won't happen here." That mentality was shattered, and many reexamined their preparedness levels.

Planners began to recognize the idea that disasters take a collaborative response and planning effort. Each organization could not operate on its own. Responders began to recognize that beyond the ability to communicate on the same radio channels, many incident command systems were in use, terminology differed, and many organizations failed to understand how to work within an ICS system. Born out of the attacks was the National Incident Management System, described in chapter 3. This system leveled the playing field among all of the organizations responsible for emergency response. Due to its ties to government funding, it was virtually mandated for every state and local jurisdiction. Requirements for planning, training, certifying, and responding to disasters now had a common document.

A reemphasis on the funding of emergency responders ensued after the attacks. Many remember the president of the United States was in the process of cutting the Assistance to Firefighters Grant Program just months prior to the attacks. The attacks spurred numerous grant dollars to equip and train emergency responders in WMD training. The Assistance to Firefighters Grant Program was funded. Furthermore, appropriations were increased significantly due to the realization that the fire department would be the first to arrive and begin mitigation techniques at any type of disaster, including an attack on the country.

A reemphasis on funding key federal programs, such as the national disaster response components, occurred in the post-9/11 environment. Due to the lack of national disasters, and the unrealized potential of a terrorist attack, these programs were only minimally supported, thus putting many of them in a state of being nearly unable to be deployed.

A better understanding resulted that America's first responders are the ones who will deal with a disaster until the arrival of state and federal resources. With this understanding came an increase in preparing all first responders with the right equipment to respond to the potential environments. Prior to the recurrence of worries about emergency personnel responding to CBRNE attacks, responders were forced to either scramble to find the large sums needed to upgrade equipment or to respond with inadequate equipment.

INTERNATIONAL TERRORISM

International terrorism has a long history and has been used by a broad array of political organizations to further their objectives. Such

Table 11–1. A sample list of terrorist organizations in operation.

Sample List of Terrorist Groups Known to Operate in the Past 5 Years

Note: Not an all-inclusive list

Terrorist Group	Domestic	International	Mission / Cause
Al Qaeda		X	Establish a pan-Islamic Caliphate throughout the world
Armed Islamic Group		X	Overthrow the Secular Algerian regime, replace with Islamic state
Al-Jihad		X	Overthrow Egyptian Government, replace with Islamic state
Palestine Liberation Front (PLF)		X	Overthrow Egyptian Government, replace with Islamic state
Hizbollah		X	Increasing its political power in Lebanon, opposing Israel, and peace negotiations in Middle East
Al-Aqsa Martyrs Brigade		X	Drive the Israel Army and people from West Bank, Gaza Strip, and Jerusalem, and Establish a Palestine state
Hamas (Islamic, Resistance Movement)		X	Establishing an Islamic Palestine State in place of Israel
Jamaat ul-Fuqra	X		Purify Islam through Violence
Aryan Nation	X		Neo-Nazis
Ku Klux Klan	X		White supremacist
Animal Liberation Front (ALF)	X		Property damage and animal release
Earth Liberation Front (ELF)	X		Property desruction related to environmental issues

organizations include right-wing and left-wing political parties, religious groups, revolutionaries, and ruling governments. For many years, Americans viewed this crime as something that happened in other countries. If American citizens were victims of terrorist attacks, they must have been traveling or living overseas or were passengers on international flights or cruises that operated in dangerous locations. The attacks in Oklahoma City and at the World Trade Center and Pentagon destroyed that belief forever. There are volumes written on the subject of international terrorism, covering subjects from its roots and history to terrorist tactics and future predictions. For the purposes of this text, the discussion considers what international terrorism is, and what emergency responders need to know about it.

The FBI defines international terrorism as "the unlawful use of force and violence against persons or property to intimidate or coerce a government, the civilian population, or any segment thereof, in furtherance of political or social objectives." The FBI further states that "international terrorism involves violent acts, or acts dangerous to human life, that are a violation of the criminal laws of the United States or any state, or that would be a criminal violation if committed within the jurisdiction of the United States...Acts of international terrorism transcend national boundaries in terms of the means by which they are accomplished, the intended persons they appear to intimidate, or the locale in which the perpetrators operate."[1]

Terrorism is a criminal act that influences an audience far beyond the immediate victims. The strategies of international terrorists include committing acts of violence that draw the attention of the local population, the government, and the world to their cause. Terrorists plan their attacks to generate the greatest possible publicity and choose targets that symbolize what they oppose. Table 11–1 is a list of some local and international terrorist organizations known to be operating currently.

The true effectiveness of the terrorist act lies not in the act of terror itself, but in the reaction of the public and the government to that act. Think back for a moment on that fateful September day. As tragic as the tremendous loss of life was in New York, Washington, DC, and Pennsylvania, what effect did those attacks have on the majority of Americans who were not directly injured or suffered family members killed in the attack? Many experienced real fear for the first time, an almost instant deflation of an attitude of invincibility. Hundreds of thousands refused to fly on commercial aircraft, sending numerous airlines to the brink of bankruptcy from which some still have not fully recovered. Feelings of fear and retaliation resulted in acts of ethnic discrimination and worse. And a man named Osama and a group called al Qaeda were no longer just threats that resided overseas.

It is exactly this desire to generate fear that leads international terrorists to target and strike "soft" targets such as shopping centers, office buildings, and houses of worship.

International terrorist groups can generally be divided into three categories: loosely affiliated extremists, formal terrorist organizations, and state-sponsored terrorists.

- **Loosely affiliated extremists.** These terrorists are motivated by political or religious beliefs and may pose the most urgent threat to the United States. Sunni Islamic extremists such as Osama bin Laden and individuals associated with his al Qaeda organization have already demonstrated the desire and capability of carrying out large attacks in this country. This has resulted in "large-scale casualties and destruction against U.S. citizens, facilities, and interests, as demonstrated by the August 1998 bombings of U.S. Embassies in East Africa" and the horrific attacks of September 11, 2001.[2] The threat from al Qaeda, however, is only part of the overall threat from international terrorism in the United States. The radical international jihad movement encompasses individuals from varying nationalities, ethnic groups, tribes, races, and terrorist group members who work together in support of extremist Sunni goals...The single common element among these diverse individuals is their commitment to the radical international jihad movement, which...promot[es] the use of violence against the "enemies of Islam" in order to overthrow all governments which are not ruled by Sharia, or conservative Islamic law. A primary tactical objective of this movement is the planning and carrying out of large-scale, high-profile, high-casualty terrorist attacks against U.S. interests and citizens.[3]

- **Formal terrorist organizations.** Another category of international terrorist groups refers to autonomous and usually transnational organizations with their own infrastructures, personnel, financing, and training. These organizations are able to create and implement terrorist actions on a worldwide basis.[4] A number of these organizations maintain operations and support networks in the United States.[5] "For example, extremist groups such as the Palestinian Hamas, the Irish Republican Army, the Egyptian Al-Gama Al-Islamiyya, and the Lebanese Hizballah have a presence in the United States whose members are primarily engaged in fund-raising, recruiting, and low-level intelligence gathering."[6] Prior to the 9/11 attacks, Hizballah was responsible for more American deaths than any other terrorist organization, including al Qaeda, as a result of their bombings of the U.S. embassy and the marine barracks in Lebanon in 1983, the attack on the U.S. embassy in Beirut in 1984, and the hijacking of TWA Flight 847 in 1985. While Hizballah has not to date carried out a terrorist attack in the United States, in 2000 the FBI arrested 23 suspected supporters of Hizballah in such locations as Charlotte and Lexington, North Carolina.

- **State sponsors of terrorism.** The last type of international terrorism centers on countries that view terrorism as a means to shape foreign policy. Some of the countries identified by the U.S. State Department as state sponsors of terrorism include Sudan, Libya, Cuba, Iran, Iraq, Syria, and North Korea. While Syria has not directly been involved in conducting recent terrorist activities, "the country still provides safe haven to international terrorist groups and rogue extremists."[7]

General characteristics of international terrorist groups

- Terrorist groups seek to intimidate by promoting fear.

- Some international terror groups have advanced weaponry, but they are generally militarily weaker than the governments they fight.

- They employ unconventional warfare tactics, and their training can include physical and mental preparation, weapons and explosives, political and religious indoctrination, and psychological warfare and communications.

- International terrorists do not equate tactical success with mission success. A specific terrorist attack may not achieve its desired results, but the terrorists may still view the act as successful if it publicizes their cause.

- These groups are usually urban-based and highly mobile. If they are urban-based, they have access to mass transportation, and they likely have access to forged passports and safe havens in many countries.

- International terrorists generally organize and operate clandestinely in cells of three to five members. A cell may only have contact with another cell or the next higher command level. As such, the capture of one or more terrorists rarely results in the compromised identity of the entire terrorist organization. Terrorist organizations are well versed in covert surveillance techniques and intelligence gathering.[8]

Common methods of attack of international terrorism

Bombings are the most common method of terrorist attack. Typically, improvised explosive devices are inexpensive and easily constructed, and generally small and easy to conceal. Bombs can be made from materials readily available to the average consumer, as in Oklahoma City in 1995, or can be massive flying bombs made from fully fueled commercial jetliners, as occurred on September 11, 2001. Today, detonation is high tech and wireless, and from anywhere in the world a terrorist can detonate the device with a call to a bomb's cellular phone trigger. Terrorist bombs are usually accompanied by shrapnel, making them antipersonnel. While acts of individual suicide bombings have rarely occurred within the United States to date, the success measured by the fear generated in other countries using this tactic makes it a viable concern for the future. The use of CBRNE devices remains a concern as well, though use of these agents may or may not be associated with an explosive dispersal.[9]

Another favorite method used by international terrorists involves attacks using firearms, generally employing a concealable semiautomatic rifle, possibly modified to fire full auto. There are three categories of shootings: those where escape is intended, those where escape is unlikely, and those where escape is not even contemplated. In the latter instance, responding emergency personnel should always consider the likelihood that the terrorist is wearing an explosive garment. An example of international terrorists using firearms as a means of attack on U.S. soil can be seen in the 1993 murder of two CIA employees by a Pakistani immigrant in front of CIA Headquarters in Langley, Virginia.

Hijacking of aircraft or ground transportation can create a hostage situation that terrorists may utilize to bargain for demands. It might also create the threat of the aircraft or vehicle being purposely flown or driven into a selected target for maximum devastation. This was demonstrated on 9/11, when hijacked aircraft were crashed into ground targets. Other scenarios could involve international terrorists hijacking a school bus full of children, driving a tanker truck containing chemical, biological, or nuclear waste into a chemical manufacturing factory, or commandeering a passenger train.

Considerations for emergency personnel

America cannot be lulled into a false sense of security because there have been no significant acts of international terrorism on American soil since September 11, 2001. The "War on Terror" is not, and should not be expected to be, a short-term operation. There is little doubt that there will be future attacks in the United States by international terrorists, and emergency personnel will be expected to respond. Police and fire services should understand that international terrorists are well prepared to unleash savage, merciless, and sustained campaigns of terror and will use tactics that are the result of more than 50 years of evolution.[10] The fight against modern international terrorism is not an ideological fight. They cannot be convinced to live in a democratic society, and they will not be appeased without their religious goals being achieved. As such, emergency personnel need to be prepared for a prolonged engagement with terror. Figure 11–2 serves as a reminder of the massive destruction that can occur as a result of a terrorist attack.

Fig. 11–2. The aftermath of the 9/11 attacks. This photo is a reminder of the large-scale destruction and loss of life associated with a terrorist attack.

Emergency personnel should understand that international terrorists are not merely criminals. They think and operate in a different way. International terrorists are not as interested in escaping arrest or death as they are in achieving their goal. In many cases they have shown a propensity to not only go to their deaths while inflicting as much death and devastation as possible, but also to take responding emergency personnel with them. Secondary explosive devices set to detonate after fire and EMS personnel arrive, attacks with automatic weapons on the crowd of onlookers that gathers at the scene of an attack, and booby traps at the scene of a disaster are all possible. The possibilities are limited only by the imagination of the terrorist.

The best proactive, preventative tactic to reduce not only the potential for an international terrorist attack, but also to enhance the degree of safety for emergency personnel, is a heightened sense of awareness and attention to potential preincident indicators of terrorism. Emergency personnel should maintain a high level of security with regard to emergency response apparatus, uniforms and badges, self-contained breathing apparatus, and procedure manuals. It is important that personnel remain alert to indicators of surveillance or other intelligence-gathering operations and false alarms that could have been engineered to calculate response times. They must be aware of the potential for secondary explosive devices or secondary attacks while on the scene of an emergency.[11]

DOMESTIC TERRORISM

The FBI views domestic terrorism as the unlawful use, or threatened use, of violence by a group or individual that is based and operating entirely in the United States or its territories without foreign direction, and which is committed against persons or property with the intent of intimidating or coercing a government or civilian population in furtherance of political or social objectives.[12]

The country was shocked when Timothy McVeigh used a truck bomb to destroy the Alfred P. Murrah Federal Building in Oklahoma City on April 19, 1995. Many believed the country was under attack again by foreign terrorists, since the first attack on the World Trade Center had occurred a mere two years earlier. Americans were certain they would see news footage of some Middle Eastern person arrested at a major airport trying to flee the country. Perhaps the only thing that shocked Americans more than the horrific devastation of that April day was learning that the deed had been perpetrated by an American. Attention was focused for a while on the internal threat from domestic terrorism. However, that attention seemed to fade as foreign terrorists struck U.S. embassies overseas, U.S. warships, and eventually perpetrated the unthinkable attack of 9/11. As the nation declared war on terror, many people interpreted that war as targeting international terrorists on foreign lands, and memories of Oklahoma City began to fade in the minds of most Americans.

To focus solely on international terrorism as the only threat of terror to the United States and to minimize the continued threat posed by domestic terrorism would be foolish indeed. Those within the United States bent on achieving their twisted domestic-related goals through violent acts of terrorism have not disappeared, nor have they lost focus on their intent to succeed. U.S. Marshals Service Chief Inspector Geoff Shank said, "Not a lot of attention is being paid to this [domestic terrorism], because everybody is concerned about the guy in a turban. But there are still plenty of angry, Midwestern white guys out there."[13] As examples:

- In May 2004, William Krar, 63, of Noonday, Texas was sentenced to "11 years in prison after he stockpiled enough sodium cyanide to kill everyone inside a 30,000-square-foot building. Krar, described by federal prosecutors as a white supremacist, also had nine machine guns, 67 sticks of explosives and more than 100,000 rounds of ammunition."[14]
- In May 1995, Larry Wayne Harris, a member of the neo-Nazi organization Aryan Nations, was arrested in Ohio after allegedly misrepresenting himself when ordering three vials of *Yersinia pestis*, the bacteria that cause bubonic plague.
- In March 1995, four members of the Minnesota Patriots Council, a right-wing militia group advocating the violent overthrow of the U.S. government, were convicted of conspiracy for planning to use ricin, a lethal biological toxin.
- In the 1980s, a little-known survivalist group called The Covenant, the Sword, and the Arm of the Lord (CSA) acquired a large drum of cyanide with the intent of poisoning the water supplies of major cities.

Domestic terrorist groups represent interests that span the full spectrum of political and economic causes, social issues, and concerns. The people and groups range from the white supremacist antigovernment types and militia members to ecoterrorists and people who hate big corporations. These include violent antiabortionists and nationalists who envision a separate state for minorities or ethnic groups.[15]

Generally, domestic terror groups can be broken down into three categories:

- **Right-wing extremist groups.** Right-wing extremist groups often adhere to the principles of racial supremacy and embrace antigovernment and antiregulatory beliefs. Right-wing extremist groups continue to present a serious domestic terror threat in the United States as they seek to do away with a liberal, democratic government and create a fascist state in its place. Included in this category are groups such as the Ku Klux Klan, Aryan Nations, and the World Church of the Creator. Such national-level, formal right-wing groups have been known to espouse extremely violent tactics for promoting their cause, including the broad use of explosives and numerous recorded attempts to possess chemical and biological weapons. The actions of right-wing extremists have also been called *vigilante terrorism* by some scholars.[16] Racist prison gangs such as the Aryan Brotherhood are included in this grouping, as well as the gangs known as Public Enemy Number 1 and the European Kindred.[17]
- **Left-wing extremist groups.** Left-wing groups "generally profess a revolutionary socialist doctrine and view themselves as protectors of the people against the 'dehumanizing effects' of capitalism and

imperialism. They aim to bring about change in the United States through revolution rather than through the established political process."[18] With the dismantling of communism in the 1980s, the left-wing movement lost some of its momentum, and law enforcement made significant strides in dismantling much of the left-wing domestic terrorism infrastructure. This is not to say, however, that there are no remaining left-wing domestic terrorist groups operating in the United States. Among these are the extremist Puerto Rican separatist group known as Los Macheteros, and anarchist and extremist socialist groups such as the Workers' World Party, Reclaim the Streets, and Carnival Against Capitalism. Anarchists, working individually and in groups, were responsible for much of the damage in Seattle during the World Trade Organization meeting in 1999.[19]

- **Special interest extremists.** "Special interest terrorism differs from traditional right-wing and left-wing terrorism in that extremist special interest groups seek to resolve specific issues, rather than effect more widespread political change. Special interest extremists continue to conduct acts of politically motivated violence to force segments of society...to change attitudes about issues considered important to their causes."[20] Special interest extremist groups cover a broad range, including the fringes of the animal rights, antiabortion, antinuclear, environmental, and other special interests. The FBI reports that during the 1990s, right-wing domestic terrorism surpassed left-wing domestic terrorism. However, in the early 21st century, special interest extremism has emerged as one of the most serious domestic terrorist threats.[21]

Among special interest extremists, authorities have seen decreasing incidents attributed to antiabortion or antinuclear activists. However, they have noted an increase in the domestic terror incidents attributed to animal rights and environmental extremists, particularly the Animal Liberation Front (ALF) and the Earth Liberation Front (ELF). Two separate special interest extremist groups, ALF and ELF, have joined forces on at least some occasions and have jointly claimed credit for acts of domestic terror. The most destructive practice of both ALF and ELF is arson using improvised incendiary devices equipped with crude timing devices. Instructions for constructing these devices can be found on ALF and ELF Web sites, and their attacks are usually preceded by significant intelligence gathering, substantial surveillance, and considerable planning. Both groups have been connected to dozens of acts of vandalism and arson in the United States since 1996. These include an arson fire at a lumber company causing $1.5 million in damages (where ELF was spray-painted at the site) and an arson fire at Brigham Young University's Ellsworth Farm, where animal experiments are conducted. Crimes also included destruction of animal research laboratories at universities, threats to individuals, and sabotage against equipment.[22]

One of the most serious domestic terror series of incidents to date involving ecoterrorists were those associated with a group known as "The Family." More than a dozen suspects were indicted in 2006 for a series of firebombing incidents targeting U.S. Forest Service ranger stations, meat processing companies, lumber companies, and a high-tension power line. Three other members were arrested with bomb-making materials and were plotting to destroy federal property, cell phone towers, and power generation facilities. "The Family" has also been linked to the 1998 arson of a ski resort in Vail, Colorado causing $12 million in damages.[23]

Emergency responders should take care to avoid underestimating the resolve or lethality of domestic terror groups. There are numerous domestic terrorist groups still active in the United States, and many are heavily armed and have at least considered the possible acquisition of chemical or biological agents as weapons. The ability to create a disaster scenario on U.S. soil is not reserved solely for foreign-born terrorists, as Timothy McVeigh so chillingly proved in 1995. To illustrate this point, one need look no further than

Demetrius Crocker, 39, of McKenzie, Tennessee. In October 2004, Mr. Crocker was arrested by FBI agents after he tried to acquire chemical weapons and explosives to blow up a government building. According to the FBI affidavit, Crocker had inquired where he could obtain nuclear waste or nuclear materials, and professed an "absolute hatred" for the government. Crocker later told an undercover FBI agent, "It would be a good thing if somebody could detonate some sort of weapon of mass destruction in Washington, DC," and Crocker later bought what he thought was Sarin nerve gas and C-4 explosive from the undercover agent.[24] The "Crockers" and "McVeighs" of the United States continue to pose a threat to its security, and they continue to have the attention of law enforcement. Recognizing that the potential threat to U.S. security from domestic terrorism can be every bit as dangerous as the threat from outside the country will be the key to effectively preparing for and responding to domestic terror acts.

SOFT VS. HARD TARGETS

Terrorists do not choose their targets at random. In some cases, terrorists have been known to choose their targets years in advance of the actual attack. In the 1960s and 1970s, the traditional politically motivated terrorist groups tended to choose their targets where they could gain the most media attention with little loss of life, and where the terrorists could survive or escape after carrying out the attack. Political terrorists strove for little loss of life in order to avoid alienating the populace that they saw themselves fighting for against the "oppressive government." There was no real desire to become a martyr, so escape was of paramount consideration. As religious fundamentalist terrorism has grown, and in most cases overshadowed many political terror groups of past decades, so has the target desire of these groups. As 9/11 so adequately demonstrated, massive loss of life with no concern for (or perhaps a desire for) martyrdom is now a focus of target selection by these groups. The apparent willingness to die during the commission of a terrorist attack is what makes the suicide attack so frightening, and so difficult to stop.

Whether the goal of the terrorist is to strike a high-profile target with little loss of life and a good chance of escape, or strike a populated civilian target resulting in high casualties and virtual suicide of the attacker, there are commonalities in their decision process for selecting a target. In fact, the strategies employed by terrorists to select a target are much the same as those employed by the common street criminal in selecting the next victim. First, there is at least some degree of surveillance as part of a risk analysis. That is to say, the terrorist, like the street thug, does not want to work harder than necessary to achieve the maximum effect desired. The terrorist will need to assess the cost in resources necessary to achieve this goal, which could include the financial costs of learning to fly a passenger jet or the technical difficulty in obtaining highly enriched uranium to construct a nuclear weapon. The terrorist must also assess the likelihood of success. To better understand the strategic process of terrorist target selection, one must understand the differences between *soft* and *hard* targets.

- **Soft target.** A soft target is a target that is relatively unguarded (military protection or other forms of security), has ineffective or inefficient security (determined by surveillance or "testing"), or that is difficult to protect from terrorists. Examples include subways, schools, malls, parks, entertainment venues, storage facilities, and potentially some pharmaceutical or chemical manufacturing facilities.

- **Hard target.** A hard target is one that has a significant (and generally obvious) security presence in order to deter a terrorist attack. The mere fact that security is present and visible, however, may not necessarily ensure the target is viewed by terrorists as a hard target. Obvious security does not always translate into effective security, and it should not be assumed that terrorists

> will not test the viability of the security present to determine its effectiveness. Some examples of hard targets include military installations, nuclear power plants, and government agencies.

With the appropriate amount of effective security, of course, a soft target can be transformed into a "harder" target, but not always into a hard target that would dissuade terrorists from attacking. For example, most sports stadiums would be considered a soft target, since the security provided, even during events, is minimal and generally designed to reduce the occurrence of petty crime or to assist fans. When the sporting event is the Olympics, however, the considerable security afforded such an event makes the target harder. Similarly, airports, which were generally considered soft targets prior to 9/11, have been made harder targets with the additional screening procedures and enhanced visibility of security. One would be hard pressed, however, to consider either the Olympic Games or airports as hard targets despite the millions of dollars and man-hours devoted to security. These sites are simply too difficult to protect adequately. Both have significant numbers of civilian visitors and substantial numbers of access and egress points. These factors give rise to opportunity for terrorists. Coupled with the potential media coverage and likely mass casualties such an attack would generate, they remain very attractive targets.[25]

Suzanne MacNevin, writing in the Lilith Gallery, presented "The Top 100 Terrorist Targets in the United States." Among the groupings listed were political targets, economic targets, ecological targets, and educational targets. MacNevin notes that ecological targets (i.e., forests, dams, National Parks, etc.) are extremely soft targets because they are unguarded. She further states that pyroterrorism (deployed by means of massive forest or grass fires) can destroy American agricultural and manufactured goods, homes, and neighborhoods, with damages in millions of dollars.[26] Most political targets, such as U.S. embassies, the Statue of Liberty, etc., are generally well guarded. In contrast, economic targets, like gas stations, oil pipelines, or bridges, and educational targets, like schools, school buses, or summer camps, are generally very soft targets.

Perhaps the potential terrorist target that has attracted the most attention of authorities, yet generally remains a soft target because of the relative difficulty to protect it from a terror attack, is mass transit. One need look no further than the al Qaeda attacks on the Madrid (March 2004) and London (July 2005) trains or the multiple suicide attacks on buses in Israel to understand that this potential target has not escaped the radar of terrorists. In a statement issued in October 2002, the Federal Bureau of Investigation warned that debriefings of al Qaeda detainees indicates the group had considered directly targeting U.S. passenger trains, and that "captured al-Qaida photographs of U.S. railroad engines, cars and crossings have increased concern."[27] Despite the increased presence of heavily armed police officers in New York City subway stations, there remains no viable means of detecting or preventing a terrorist from boarding a mass transit system vehicle with enough explosives or chemical, biological, or radiological material to produce mass casualties. The tightly crammed passengers and limited means of access and egress only add to the softness of rail mass transit.[28]

Mass transit systems, of course, do not present the only soft target for terrorists in the U.S. transportation system. Each day hundreds of thousands of trucks travel U.S. highways and pass through (or deliver in) U.S. cities loaded with flammable, corrosive, or explosive materials. Most vehicles transporting nuclear materials or military grade chemical/biological agents are heavily guarded. However, a gasoline tanker purposely driven into a school or hospital (both very soft targets) would cause significant psychological and economic damage to the United States. Should the stolen tanker be equipped with additional explosives and transformed into a vehicle-borne improvised explosive device (VBIED), the impact could be even more devastating.

Terrorist target selection is not random. It is preceded by a risk analysis to determine the cost to achieve the goal and the likelihood of success and will likely focus on soft versus hard targets.

Given this information, what can be done to reduce the likelihood of a terrorist attack? In addition to presenting an obvious display of effective security, there are six key signs of potential terrorist targeting that should be understood by law enforcement, emergency personnel, and facility security, as well as the general citizenry:

- **Surveillance.** During the planning phase of a terrorist operation, surveillance will be conducted on potential targets to determine how soft or hard the targets are. They will determine the strengths, weaknesses, and number of emergency personnel that would respond to an incident (this can include sending false fire alarms), as well as routes to and from the site. It is important to note anyone recording or monitoring activities, drawing diagrams, using vision-enhancing devices, taking unusual photos, or having floor plans or blueprints in his or her possession without a legitimate reason.

- **Elicitation.** Elicitation is attempting to gain information about a place, person, or operation. This can include asking multiple and unusually specific questions about a critical infrastructure or manufacturing facility, maritime port, power plant, or other potential target. Terrorists may also attempt to monitor and gauge bridge and tunnel usage rates, make unusual inquiries about shipments, or ask questions about how a military base operates. Terrorists may also place people in key roles in sensitive work locations.

- **Tests of security.** Terrorists will also use tests of security to gather data as part of their early planning and risk analysis. This has also been called *red teaming* and is usually conducted by driving by the target, moving into sensitive areas, and observing both the speed and the effectiveness of the response by security or law enforcement. They would be interested in knowing the routes by which emergency personnel would respond (to potentially detonate a secondary device to incapacitate arriving emergency personnel). The terrorists may also try to penetrate physical security barriers or procedures in order to assess the strengths and weaknesses of the barriers. Alternately, they may take a quieter approach by gaining employment in key locations to monitor day-to-day activities and record both the normal routine and the response to emergency situations. Figure 11–3 shows typical physical barriers that are installed for security enhancements.

- **Acquiring supplies.** There should be a heightened awareness to the acquiring of supplies that could potentially be used to construct a weapon. This could be someone purchasing or stealing explosives, weapons, or ammunition, or it could involve someone purchasing or storing harmful chemicals without a legitimate reason. Terrorists would also find it extremely useful to have in their possession equipment, identification, and uniforms associated with law enforcement or emergency responders, military uniforms, airport security passes, or even police, fire, or EMS vehicles. How many security guards at a chemical plant or biological laboratory would consider refusing entry to an arriving fire engine or ambulance with lights and siren activated?

- **Suspicious people who do not belong.** This refers to profiling behaviors, not profiling people. This includes suspicious border crossings, stowaways on a ship, or people working or lingering in an area who just do not fit in because of their demeanor, language usage, or unusual questions they may be asking.

- **Dry runs.** Before executing the final attack plan, and even during the determination of whether the target to be potentially selected is soft or hard, the terrorists may conduct a dry run. Examples of a dry run being conducted include someone monitoring a police or fire radio frequency and recording emergency run times, or mapping out routes and determining the timing of traffic lights and flow. Multiple dry runs are normally conducted at or near the planned target area. [29]

Fig. 11–3. Physical barriers installed to prevent terrorist attacks. These measures are meant to prevent car bombs, which have become prevalent internationally.

It is important to note that the process of selecting a terrorist target, the risk analysis, and the determination of the softness or hardness of the target, as well as other preincident indicators noted above, may take place over a period of months or years. Even with the most diligent attempts to enhance security, some venues will never reach the "hard target" level. As such, in order to prevent an attack, diligent and continued attention must be paid to the signs indicating terrorist target selection is taking place.

WEAPONS OF MASS DESTRUCTION: CHEMICAL, BIOLOGICAL, RADIOLOGICAL, AND NUCLEAR (CBRNE) THREATS

When one pictures battle scenes from World War I, the most common depiction in old war movies or history books is of the costly and seemingly stagnant employment of trench warfare. This was a horrible way to fight, with troops on both sides alternately confined to muddy, cold trenches in the ground or charging headlong into the machine gun bullets of the enemy waiting just outside the trench. Neither side seemed to gain any advantage or make any progress. Then, on the sunny spring day of April 22, 1915 in the quiet Belgian town of Ypres, a new and horrible weapon was introduced to the world. The Allied soldiers began to hear a slight hissing sound as the winds traveled in their direction from the German trenches. French Colonial soldiers saw a green-yellow fog moving toward them, about head high. The cloud enveloped thousands of French Colonial troops, and they began retching and dying, with thousands incapacitated. The Germans had released a lethal cloud of chlorine gas. After the war, the memory of the terrible injuries and high death rate caused by the use of gas led most countries of the world to sign a treaty banning the use of gas in future wars. Nearly 80 years later, on March 20, 1995, the Japanese cult/terrorist group Aum Shinrikyo unleashed sarin gas in the Tokyo subway system in an effort to kill people who worked in Tokyo Police headquarters or nearby government buildings. Because of the low purity of the sarin used and the air filtration system of the subway system, there were only 12 deaths among the more than 5,000 victims. There was a new weapon of mass destruction in the hands of terrorists.[30]

It is not surprising that possible use of CBRNE weapons by terrorists strikes fear in the hearts of the civilian population and first responders alike, just as the use of poison gas created great fear among World War I troops. The very word *terror* comes from the Latin word *terrere*, which means "to frighten." Terrorism is a form of warfare that uses fear to achieve its goals. It is frightening to consider a terrorist detonating a chemical or biological weapon in a U.S. city, poisoning the drinking water or food supply, delivering a radiological dispersion device (RDD, or a dirty bomb), or in some other way secretly and quietly attacking U.S. citizens. The possibilities can bring nightmares to even the most hardened veteran, let alone the civilian population.

Firefighters and police officers are accustomed to dealing with threats they can touch, see, and hear. A terrorist bomb is, of course, tangible. As terrible as the effects of its detonation may be, they can be seen and felt, and emergency first responders can manage the aftermath. When the threat is invisible, however, this perception changes, and fear increases. How do responders know what chemical or biological agent they are battling? How do they stop it? What will its effects be on citizens or on the responders? Responders may wonder if they have already breathed in the invisible substance that will kill them, and whether it will prove fatal the next day or 10 years later.

Certainly, the use of such ominous weapons by terrorists generates nightmares. But how likely is the use of these weapons by terrorists against the United States? Testifying before the Senate Select Committee on Intelligence on February 16, 2005, then-CIA Director Porter Goss said, "It may be only a matter of time before al-Qa'ida or another group attempts to use chemical, biological, radiological, and nuclear weapons."[31] Indeed, Osama bin Laden was quoted in the December 1998 issue of *Time* (referring to CBRNE weapons) as stating, "Acquiring these weapons for the defense of Muslims is a religious duty. If I have indeed acquired these weapons, then I thank God for enabling me to do so. And if I seek to acquire these weapons, I am carrying out a duty. It would be a sin for Muslims not to try and possess the weapons that would prevent the infidels from inflicting harm on Muslims."[32] What were once considered to be weapons only in the domain of military forces can now be manufactured and utilized by almost any determined group or person.

There are three fundamental steps that are involved in any attack that would involve a chemical, biological, radiological, or nuclear weapon. First, the lethal materials needed to construct such a weapon must be obtained. As U.S. border security is increasingly enhanced, including greater scrutiny of the raw materials and finished products entering into the country, the smuggling of weapon components is becoming increasingly difficult. Greater border security forces terrorists to utilize materials already found in the country to construct weapons, requiring increased emphasis on the security of domestic chemical plants, research laboratories, and nuclear power plants. Second, the lethal materials must be made operational, or constructed into a bomb or some other form of delivery device. Such delivery devices can be as simple as the plastic bags of sarin punctured by an umbrella by a member of the Aum Shinrikyo group on Tokyo subways, or as dramatic as a planned explosion of a train carrying chlorine as it rolls through a sleepy Midwest town. Depending on the complexity of the plot, constructing the device may require people with specific technical knowledge or specialized equipment. Third, the weapon or device must be delivered or placed in the area targeted for attack. The individual actions required for each of these fundamental steps varies based on the lethal material being utilized, so each of the four categories of weapon should be considered separately.[33]

Chemical agents

While chemical and biological agents are sometimes discussed together because of their similarities, there are distinct differences that should be noted. Chemical weapons are generally sensitive to environmental conditions, and therefore can usually be expected to dissipate quickly, which dictates the selection of an enclosed target area for maximum effect. In addition, chemical weapons can be manufactured from products available in this current economy, and they are generally easier to obtain. Chemical weapons have been more widely used in terrorist attacks to date. In April 2004, members of an al Qaeda cell were arrested in Jordan with enough chemical weapons and poisonous gas to kill more than 20,000 people. Their plan was to attack a Jordanian military intelligence facility and the U.S. embassy in Amman, among other sites. In addition, police raids on al Qaeda cells in Britain, France, Spain, Russia and other countries have found detailed instructions for the manufacture of ricin, a poison extracted from castor beans that is relatively easy to manufacture.

Most chemical agents are found in liquid form and can produce a variety of effects depending on the strength of the compound, volatility, and resistance to evaporation. Factors of temperature, wind speed, and surface composition may also be a factor in determining the eventual effect of the weapon. Chemical agents may be classified as follows:

- **Nerve agents.** These are described as agents that "disrupt the mechanism by which nerves communicate with the organs they stimulate."[34] Nerve agents cause the body to receive the wrong signals from the nerves, thus disrupting the muscle function of the body. When a large dose is received, death can occur in a short time, usually preceded by tightness in the chest, nausea, convulsions, and blurred vision. Nerve agents are generally clear and colorless and are heavier than water, thus affecting lower terrains. Included in this grouping are sarin, VX, and tabun. Nerve agents are not commercially available and require significant chemical expertise to produce.

- **Blistering agents.** As the name implies, blistering agents will cause blistering on exposed skin and on internal organs if breathed in or ingested. Death will generally occur through respiratory failure, preceded by blindness. Common blistering agents are mustard gas, phosgene oxime, and lewisite. Mustard agents are not commercially available, but they do not require significant expertise to produce.

- **Choking agents.** Phosgene gas is one of the deadliest examples of a choking agent. Phosgene gas damages the respiratory system. It causes the victim's lungs to fill with water, resulting in suffocation. Chlorine gas also falls in the choking agents category.

- **Cyanides.** Terrorists have considered using a number of toxic cyanide compounds, including sodium and potassium cyanide (white to pale-yellow salts that can be used to poison food or can be mixed with chemicals to enhance skin penetration). Other such compounds are hydrogen cyanide and cyanogen cyanide, which are colorless to pale-yellow liquids that turn to gas at room temperature. Exposure to cyanide can cause nausea, vomiting, confusion, vertigo, coma, and death.

Biological agents

While attacks using chemical weapons have been one of the most common tactics used by terrorists, it is clear that the use of biological agents has been considered, and indeed, planned. Mohammed Atta, one of the 9/11 terrorists, had explored obtaining a crop duster plane and had actually requested a loan from the U.S. Department of Agriculture to purchase a passenger plane and outfit it with a chemical tank.[35] Ahmed Alhaznawi, another 9/11 terrorist, had sought medical attention in 2001 for an infection consistent with cutaneous anthrax. In 1984, members of a cult in Oregon, in an attempt to prevent voters from reaching the polls, used a spray bottle to contaminate restaurant salad bars with *Salmonella*, resulting in more than 1,000 victims becoming sick.

Materials to construct crude biological weapons as simple as common food poisons that can be delivered by a spray bottle are readily available at clinical microbiological laboratories. Other biological agents can be obtained at government facilities, medical schools, graduate schools of microbiology, graduate programs in virology, pharmaceutical research laboratories, and even (as has been successfully tried) ordered via the mail or Internet. With low costs, a widespread knowledge of microbiology, and equipment available in many countries, terrorists can produce these deadly biological agents with relative ease.

Biological agents can be disseminated in liquid or powder form. While producing liquid agents can be relatively simple, disseminating it in aerosol format is more challenging. Powders can be disseminated more easily, though high-quality powders are more difficult to make. Some examples of dissemination of biological agents include the following:

- Using a crop duster aircraft
- Using a water spray bottle to infect the food supply
- Using the postal system, as in the anthrax cases
- Filling light bulbs and dropping them in crowded areas

Some examples of biological agents include the following:

- **Anthrax.** A single-cell organism that attacks the skin, lungs, and gastrointestinal tract, produces flu-like symptoms in one to seven days, and progresses to cause fever, difficulty breathing, and acute respiratory distress. It is estimated that 1 gram of anthrax is capable of killing millions of people.[36]
- **Botulinum toxin.** A protein made by the *Clostridium botulinum* bacterium, this is one of the most toxic compounds known. Exposure affects the nerve cells, resulting in palsy, spasms, and paralysis. The amount of exposure determines how quickly the victim will die.
- **Plague.** The "disease of choice" because of its effectiveness and deadly characteristics. It is generally spread by respiratory transmission, and if untreated, will result in death within 24 hours. Symptoms range from general aches and pneumonia to hemorraghic pneumonia. While attractive to terrorists because of its potential for a large body count, it is difficult to employ as a weapon because of its relative instability.[37]
- **Ricin.** A plant toxin that is 30 times more potent than the nerve agent VX by weight and is readily obtainable by extraction from common castor beans. It has no strong odor, is off-white in color, and there is no treatment for ricin poisoning once it has entered the bloodstream.[38]

Some researchers argue that the technical and organizational obstacles terrorists would face in order to effectively disseminate a biological agent on a large scale would be considerable. However, most agree that the threat of a biological agent being utilized by terrorists on a smaller, low-technology scale, such as contaminating food or dissemination in an enclosed space, is becoming more likely.

Radiological/nuclear

Assessments on whether terrorists could actually detonate a nuclear weapon in the United States vary. The U.S. nuclear arsenal and domestic supply of the necessary components to construct a nuclear weapon (i.e., highly enriched uranium) are very well protected. Unfortunately, it is not possible to be as confident that the necessary ingredients are as well protected in other regions of the world where they exist. This leaves open the potential for terrorists to obtain a weapon or the necessary components elsewhere and smuggle them into the country. Great efforts are made to protect the U.S. borders. However, the size of a nuclear weapon (or the necessary components to construct one), when compared with the vast number of containers and shipments of products arriving at

U.S. ports each day, demonstrates that smuggling is entirely possible.

Recognizing that a nuclear attack can happen on U.S. soil is the first step to preparation, and state and local governments should be prepared to detect radiation and ensure the public is adequately educated about the dangers. While the difficulties associated with obtaining and deploying an actual nuclear device present a challenge for terrorists, the technical difficulties associated with constructing and deploying a radiological dispersion device (RDD, or dirty bomb) are much less of a challenge. Essentially, an RDD consists of conventional explosives packed into radioactive material. While the detonation of such a device does not cause a nuclear explosion, it can be effective in causing the dispersion of radioactive material over a large area, thus resulting in widespread contamination. The radioactive material needed to construct a dirty bomb can be found in many places, from hospitals and dental offices to nuclear power plants and research facilities.

The most common method of dispersing radioactive material available to terrorists is through combining it with conventional explosives. Other methods include the following:

- Concealing the radiological material in a high-traffic enclosed area (known as a *passive RDD*)
- Converting the material to aerosol format and dispensing it in an enclosed area
- Spreading it via an unwitting public on air currents (known as *atmospheric RDD*) [39]

Most experts agree that it would be difficult to construct an RDD with massive destructive power, but the goal of terrorists is not always a high body count. Use of even a small, localized device would undoubtedly result in broad-scale panic, disrupting normal life both in the affected area and areas where the populace fears the next attack. It could potentially result in significant economic harm due to evacuations and decontamination efforts.

The possibility also exists that terrorists will obtain the required fissile material—highly enriched uranium or plutonium—to construct an improvised nuclear device capable of causing a yield-producing nuclear explosion. While less likely to occur than the detonation or use of an RDD or dirty bomb, it is feasible.[40]

CBRNE prevention strategies

The common theme presented here is that in order for terrorists to successfully deploy a CBRNE weapon of mass destruction, they must first obtain the necessary materials. The terrorists then must construct a device capable of delivering the agent. Therefore, a significant key to prevention of such an attack is to prevent the terrorists from getting their hands on these critical materials. Diligent regulatory efforts are critical to successful prevention. This must be coupled with the use of detection devices where appropriate, securing both sales and transportation methods of the materials, and ensuring rapid response capabilities are in place. Other tactical methods include the following:

- Establish working relationships between local and state representatives with security representatives at both chemical industry sites and sites that use and/or store radiological materials.
- Remain alert to reports by hospitals of unusual spikes in illnesses that could indicate exposure to chemical or biological agents.
- Equip law enforcement and emergency responders with radiation detection devices and advertise the fact that they are so equipped.
- Harden the security of facilities that use or store the materials needed to construct a CBRNE device.

- Educate the public to reduce the potential for widespread panic through an understanding that casualties associated with known attacks in the past have been relatively minimal.

- Enlist thc public in identifying and rapidly reporting unusual circumstances or activity that could indicate interest by terrorists in either the materials necessary to launch a CBRNE attack or in sites most susceptible to such attacks. These sites could include enclosed, crowded areas such as sports arenas or mass transit.[41]

SOURCES OF DANGER

On September 11, 2001, citizens of the United States were horrified and shaken by the reality that a major terrorist attack could, and did, happen on U.S. soil. Figure 11–4 shows the destruction and look of disbelief on the face of an emergency responder. The danger that the United States would be attacked again did not diminish after the first attack on the World Trade Center in 1993; indeed, it increased. Americans saw that the 1993 WTC terrorists were tried and convicted in a court of law, the World Trade Center reopened for business, and several other terrorist plots were successfully dismantled. Terrorists, on the other hand, saw an unfinished task, a vulnerable country, and "weapons of opportunity" at every airport.

Fig. 11–4. Massive destruction causes disbelief in the face of a responder in the surreal aftermath of the 9/11 attacks.

FEMA News Photo.

Why did no one raise the alarm when numerous Middle Eastern males were expressing interest in learning to fly an aircraft but had no interest in learning how to take off or land one? If alarms were raised, why were they not taken seriously? When Timothy McVeigh bought large quantities of fertilizer from one source as he initiated his plan to use a vehicle bomb at the federal building in Oklahoma City, he paid cash and refused discounts he was entitled to for the purchase. The seller thought that was strange but did not recognize it as a possible threat. Much has been learned from America's head-on crash with reality on 9/11, and should either of those scenarios occur today, the reaction of witnesses would likely be much different. As enlightened citizens, Americans have recognized more sources of danger.

Bombs are the weapon of choice for terrorists. While it is much more likely that conventional explosives will be used as opposed to true nuclear bombs or other chemical, biological, or radiological dispersal devices, the possibility of the latter cannot be ignored. It is also important to recognize that the materials or components terrorists would use to construct conventional or dirty bombs, as well as to deploy a chemical or biological agent, need not be smuggled into the United States from some third-world country. Instead, they may be found largely within the borders of the United States.

ANFO (ammonium nitrate/fuel oil)

The combination of ammonium nitrate (AN) and with fuel oil (generally diesel fuel or kerosene) to form an explosive material was used by domestic terrorists to destroy the Alfred P. Murrah Federal Building in Oklahoma City. In that attack, Timothy McVeigh combined AN with nitromethane, a more sophisticated variant of ANFO. ANFO was also used by foreign terrorists to bomb the World Trade Center in 1993. In both cases, the explosive material was concealed in a vehicle. The use of ANFO concealed in a vehicle has been a popular way to deliver destruction and death by terrorists since 1970, when it was first improvised by student protesters in Wisconsin. Since then, it has been adopted by the Provisional Irish Republican Army, FARC and ETA rebels in Colombia, and various Palestinian extremists and other Islamic extremists, as well as by domestic terrorists.

Though improvised bombs made with agricultural grade AN are less sensitive and less efficient than the explosive grade type of AN, both are capable of great devastation. ANFO is popular primarily because of its low cost and high stability. Furthermore, in most locations, AN does not have to be classified as an explosive for transport purposes, since it is considered to be an oxidizer. ANFO has an explosive efficiency of approximately 80% of TNT, and it decomposes through deflagration with high velocity. Even unmixed AN can decompose explosively, as evidenced by the disaster in Texas City, Texas in 1947, in which a ship loaded with AN caught fire and exploded, with 567 deaths resulting.

Clearly the power of ANFO as an explosive weapon has been demonstrated, along with the propensity of terrorists to use ANFO in vehicle-borne improvised explosive devices (VBIEDs). The next step is to examine how easy it is to obtain these components. Ammonium nitrate is widely used in the United States as a fertilizer in the agricultural industry, and diesel fuel is available at any truck stop and at most gasoline stations. The combination of ammonium nitrate with fuel oil as an explosive is the "most widely used explosive in coal mining, quarrying, metal mining, and civil construction."[42] ANFO usage makes up about 80% of the total North American annual usage of 6 billion pounds of explosives. In fact, most mines prepare the ANFO explosive compound on site using the same No. 2 diesel fuel that powers their trucks.[43]

Given the relative stability and explosive power of ANFO, along with the availability of its components throughout North America, the next concern to address is its ease of preparation. How difficult is it to correctly mix the components to make the bomb? Wikipedia, the online encyclopedia, states that "in an ideal stoichiometrically balanced reaction, ANFO is composed of approximately 94.3% AN and 5.7% FO by weight...In practice, a slight excess of fuel oil is added as underdosing results in reduced performance while overdosing merely results in more post-blast flames."[44]

Chemical/biological agents

Certainly ANFO is one of the most widely available (through open purchase or theft) conventional explosive materials that terrorists could get their hands on in the United States. While it is more likely such conventional explosives will be used before more sophisticated weapons of mass destruction, it is still important to recognize the sources of chemical or biological components that could be used for dispersal by terrorists.

The most common sources for the dangerous components needed by terrorists to effectively deploy or detonate a chemical or biological weapon are the home laboratory, purchase of the material from a biological supply business, or stealing the components. If the terrorists use a home laboratory to develop the substances, it can mean either impure, low concentrations or fairly potent agents, depending on the process used to purify and produce them. In some instances, this process can be as simple as improperly processed canned food (botulinum toxin) or a very complex process (nerve gas). Documents found in Afghanistan were reported by the Central Intelligence Agency to indicate al Qaeda's interest in the production of more effective chemical agents such as mustard gas, sarin, and VX.[45] Ricin, for example, can be produced from the seeds of the common ornamental plant that produces castor beans, and al Qaeda affiliates have reportedly produced a how-to manual for the ricin extraction process.

Many terrorists, however, tend to rely on a genuine lab to initially produce the agent, then try to purchase it on the open or black market or steal it. Generally, chemical or biological agents are stolen from industries that use poisonous chemicals in bulk or that ship or store the chemicals in relatively accessible areas. Theft offers the easiest method to obtain these chemicals, particularly chlorine and phosgene, which are both commercially available. Such chemical components are generally transported over the highways, and terrorists would not need to expend much time conducting surveillance to determine when and where shipments would be made. They could then either hijack the truck to obtain the chemicals or consider the truck to be a ready-made chemical bomb and crash it into a target.

Medical and university research facilities are also possible sources of small quantities of hard-to-find chemical or biological agents, and these sites are generally poorly protected.[46] These more specifically include the graduate schools of microbiology and virology, and medical schools in general. Animal diagnostic laboratories, biotechnology research laboratories, and human pathology laboratories all contain the lethal building materials for a crude biological weapon. Production instructions for CBRNE agents found in actual al Qaeda and other jihadi literature and manuals include instructions for cyanide, hydrogen sulfide gas, mustard gas, botulinium toxin, ricin, plague, and cesium-137. These instructions were assessed by Salama and Hansell. They determined that while the production instructions from mustard gas as contained in the manual are not sufficient for production of the agent, the instructions for the manufacture of the other agents (though considered crude) are valid and might work.[47]

Radiological/nuclear

Most experts agree that it is highly unlikely terrorists will secure a true, yield-producing nuclear weapon that can be detonated in the United States. For the most part, nuclear plants and military facilities in the United States are well guarded, and it is unlikely terrorists will obtain the components needed for a nuclear weapon from these sources. Unfortunately, the security afforded nuclear weapons in some other regions of the world is not always as strong, and the opportunity may arise for stolen nuclear material to be smuggled into the country. The International Atomic Energy Agency (IAEA) reports that there has been a sixfold increase in nuclear material since 1970. More troubling, the IAEA reported in 2001 that since 1993, there had been 175 known incidents of illicit trafficking in nuclear material and 201 incidents of illicit trafficking in other (medical/industrial) radioactive sources worldwide.[48]

While the delivery and detonation of a nuclear bomb in America is not considered likely due to difficulties terrorists would have in obtaining the weapon, such confidence is not as strong when considering the deployment of an RDD. An RDD is merely the combination of conventional explosives and radioactive material, which is dispersed into the air when the detonation occurs. Some have referred to this threat as a "weapon of mass disruption," rather than of destruction, because the fear generated by such a device is likely more than the actual casualty-causing effects. Nevertheless, it is a source of danger to be considered. RDDs can be constructed using radioactive sources commonly used in everyday life. The radioactive sources used in radiotherapy alone number nearly 10,000 and are found in medical radiation locations such as hospitals and medical laboratories. Many more are used in industry to check for welding problems in pipes, cracks in buildings, or in the preservation of food.

One of the most common radioisotopes used in industry is cesium-137. Thousands of devices use cesium-137, including moisture density gauges used in the construction industry and leveling gauges used to detect liquid flow in pipes. Other devices include thickness gauges used to detect the thickness of sheet metal and other products, and well-logging devices used in the drilling industry. Cesium-137 is also used in medical therapy to treat cancer. Exposure to cesium-137 cannot initially be felt, tasted, or smelled, and although it remains in the body a short time, high exposures can result in serious burns and death.[49]

There are a multitude of industrial and medical uses for cesium-137 in the United States, and as such, there are multiplied opportunities for terrorists to obtain it for purposes of making it a component of a dirty bomb. As an example of what could occur in the United States, consider events in Goiania, a major city in Brazil. In 1987, scrap scavengers broke into an abandoned radiological clinic, stole a highly radioactive cesium-137 source, and moved it to a junkyard to sell for scrap. They broke the encasement and cut the 20-gram capsule of cesium-137 into pieces. As a result, 14 people were overexposed, 249 were contaminated, and 4 died. More than 110,000 people had to be continuously monitored, and at least 85 houses had to be destroyed.[50]

It is evident that communities are surrounded by sources of danger—components that can be used by terrorists to construct and detonate a conventional bomb or to disperse a chemical, biological, or radiological agent. Everyone must work together to keep these components out of the wrong hands.

DEFENSE POPULATION PROTECTION

The fight against terrorism is a war, and in order to fight it successfully, there must be coordination of all the elements involved in homeland security. Preventing terrorism is primarily the responsibility of federal and national agencies. These agencies have the capability to conduct preemptive attacks on terrorist organizations and to dismantle the terrorist infrastructure through infiltration, arrests, and capture. Emergency services at the state and local level are primarily responsible for the initial response to a terrorist incident, with duties including containment of the incident, saving lives, and preventing additional casualties and property damage. A response to a terrorist incident that is coordinated, simultaneous, rapid, and professional is critical to success.

Reacting to a terrorist attack is one of the most complex and challenging operations emergency responders will ever be exposed to. The difficulty in reacting appropriately to such an event is primarily due to the surprise factor, which is inherent in a terrorist event, and indeed, is planned that way by the attackers. Responding appropriately to a scene that may encompass dangers never before faced by the responder, such as chemical, biological, or radiological hazards or secondary explosive devices carried by suicide bombers, presents significant challenges. Responders must be prepared for anything. In addition to rescuing victims and extinguishing fires, responders to a terror attack must also recognize that the incident site is a crime scene. They must take care not to destroy evidence. The response to a terror attack will undoubtedly drain local resources, particularly if chemical, biological, or radiological weapons have been deployed. Civilians miles from the scene may be in need of evacuation or rescue from the dispersal of the attack weapon, and on top of all these challenges, the scene will attract media from around the world.

The most pressing issue emergency responders will encounter when responding to a terrorist attack, and more specifically a CBRNE attack, is to secure the area and ascertain the severity of the threat. In recent years incidents have been recorded in which a secondary device was employed with the intent of killing emergency responders, and in some countries, a secondary assault was perpetrated by the terrorists to kill or harm rescuers responding to the scene. Extra caution should be employed if the explosion seemed to do little damage. A small explosive device may have been utilized by the attackers for the purpose of dispersing a chemical, biological, or radiological agent. It might also be used for the purpose of "drawing in" rescuers to become victims of a much larger secondary blast. Even if conventional explosives were utilized without the dispersal of a CBRNE agent, there will be a primary inhalation threat to responders from concrete dust and the likelihood of airborne asbestos from building components.

According to Clark Staten, ERRI, "In most cases, both a primary and secondary perimeter must be established. A thorough search of these perimeters must be a priority at the onset of the incident."[51] If chemical, biological, or radiological agent dispersal is even slightly suspected, a large downwind area may need to be established to rapidly evacuate civilians to minimize casualties. If the evacuation of the civilian population is deemed warranted, care must be taken to avoid panic and hysteria to the extent possible. Responders must be properly equipped, including the utilization of hazmat protective gear, and decontamination stations must be established for both responders and civilians directly affected by the attack. Nevertheless, they must recognize the psychological effect both the attack and the vision of decontamination stations or responders in full protective gear will have on the population. The responders' desire to rescue and protect the civilian population may be hampered by the emotional response of the victims. The depictions of the aftermath of CBRNE attacks by Hollywood may cause people to assume that the situation is much worse than it is. In extreme cases, the psychological reaction of the population to a genuine or suspected release of a chemical, biological, or radiological agent could evolve into incidents of mass hysteria, rioting, and opportunistic crime. Thus, extensive attempts must be made to protect the public from themselves.

It is anticipated that early intervention and media statements from technical experts and political leaders can help alleviate public feelings of confusion and fear and lead citizens toward more appropriate behaviors.[52] The natural reaction of the civilian population is to flee as far as possible from the area of the attack, yet responders must remember that proper decontamination of victims is necessary before they leave the area and spread the contamination. All chemical weapons, for example, have the potential for secondary contamination of ambulances, hospitals, homes, and other people encountered by victims as they leave the scene. Ideally, decontamination stations should be set up at all mass transportation sites serving the area in the event of a chemical attack.

The next critical issue faced by responders as they endeavor to protect the population from further endangerment in a CBRNE attack is the identification of the substances involved with the attack. As with common industrial hazmat incidents, a priority in managing the incident is to determine the identity and physical properties of the substance that has been released. Only after the dispersed agent has been identified can an effective outer perimeter be established, neutralization plans formulated, decontamination processes initiated, and emergency medical treatment plans engaged.[53]

Greater detail is provided elsewhere in this chapter on the types of CBRNE agents that may be deployed via conventional explosives or nonexplosive dispersal methods. Regardless of the type of attack encountered, and especially before identification of any dispersed agent can be established, there are general safety guidelines to be followed at any terrorist incident:

- Suspect the worst at all terrorist incidents until the air is tested.
- Use barrier protection (anything that can be placed between the responder and the suspected agent to reduce the time/degree

of exposure and provide distance from the hazard) until the area is tested and determined to be safe.

- Establish decontamination stations as soon as possible.
- Contain all runoff from decontamination activities.
- Be alert for structural collapse.
- Be alert for secondary explosive devices.
- Responders must remember that they are part of a crime scene and should preserve all evidence when possible.[54]

Responders must always remember that they are of little help to the population they strive to protect if they become victims themselves through insufficient or ineffective self-protection. Whenever a CBRNE incident is suspected, responders should immediately put on protective equipment. They must remember the time-distance-shielding rule for protection—responders must spend as little time as possible in the hazard zone, keep their distance from the area, and shield themselves from the hazard at all times, decontaminating as soon and as thoroughly as possible. In the real world, and in the midst of a terrorist attack, the appropriate equipment needed to detect and identify the type and physical properties of a chemical, biological, or radiological agent may be nonexistent, or at least not readily available. As such, responders should be familiar with the early indicators of a possible CBRNE attack.

Early indicators of a CBRNE attack

- Unusual numbers of people dying in an area from strange causes
- Unusual numbers of sick or dying animals, birds, or fish
- Lack of insect life where it should be seen
- Unusual numbers of people in an area complaining of blisters, rashes, nausea, disorientation, difficulty in breathing, convulsions, localized sweating, reddening of the eyes or skin, or any irregular symptoms
- Strange-colored smoke coming from the area of a detonation
- Explosions that seem to do very little damage, release an unusual amount of smoke, leave droplets of liquid in the area, or leave fragments covered with liquid or droplets
- Unusual appearance of liquid droplets, particularly where there should be none
- Abandoned aerosol sprayers in the area of sick people
- People reporting unusual odors or tastes
- Unexplained mists or hazes in urban areas
- Sudden or unexplained appearance of low-lying clouds
- Unidentified, low-flying aircraft (particularly crop dusters) over a populated area[55]

In addition to ensuring the safety of the responders, rescuing/evacuating victims, and securing the area of the attack, responders will face the challenge of establishing a safe command post for the incident. Ideally, this command post and staging areas should be located well outside the most pessimistic danger area. However, sometimes responders or command officers find themselves suddenly or unexpectedly in an area from which they cannot escape or find that there is no choice but to establish a command post close to the incident site. In these instances, certain defensive measures should be taken to help reduce (though not eliminate) the risk.

On-scene emergency defensive measures

- Locate the command post upwind from the hazard hot zone.
- Locate the emergency refuge or incident command post inside a building rather than in a vehicle. Buildings provide better physical protection and are easier to seal.
- Select the most interior room available. Since many hazardous fumes are heavier than air, an interior room on a middle floor should be considered.
- Shut off central fans and air-handling equipment.
- Establish a decontamination station at the entrance to the facility.
- Windows and doors should be covered with plastic and sealed with duct tape.
- If possible, a HEPA filter should be placed in the temporary field station and a positive room pressure generated.
- Keep the room sealed until the chemical or biological plume has completely passed the area.
- Keep escape vehicles in a nearby garage or other protected area in the event emergency evacuation is needed. Cover them with blankets if left outdoors.
- Have a site evacuation and rescue plan.[56]

Protecting the civilian population from the imminent threats presented by a terrorist attack, and particularly an attack employing the use of CBRNE weapons, will undoubtedly be one of the most challenging and dangerous scenarios a responder will encounter. Training and preplanning to ensure coordination of effort and maximum possible safety of the responders will help make the response by emergency services as effective as possible. It will allow the defense of the civilian population to occur with maximum success.

OFFENSIVE ACTIONS

The best possible terrorist attack scenario is the one that never happens. While obviously desirable, it is unlikely the terrorist attacks of September 11 will be the last time the United States is attacked, domestically or abroad. Every citizen must recognize this threat, and emergency responders should prepare for it. However, it is important to determine what can be done from a proactive stance to reduce the threat as much as possible.

Some of the offensive actions taken by the United States to thwart terrorism are visible and obvious. The War on Terror comes alive in living rooms across America as U.S. troops seek out and capture insurgents in Iraq and Afghanistan. Sanctions are placed on countries that support or encourage terrorism, and financial resources of suspected terrorist organizations are seized. Other offensive actions against terrorism occur daily, though few if any of these are readily apparent. The United States continues to face threats from Islamic extremists who are prepared to sacrifice their lives to carry out their attacks and from domestic terrorists who easily blend in with society. Furthermore, other assorted groups plan to bring harm to U.S. citizens to further their agenda or retaliate against the American way of life.

Perhaps the most effective yet generally unnoticed offensive action against terrorism is the development of intelligence. Terrorist organizations operate by design, and from necessity, in secrecy. Though in hindsight some of the activities carried out by the 9/11 terrorists now seem obviously suspicious, they were not viewed as suspicious at the time (at least to the point they were taken seriously). Similarly, Timothy McVeigh's purchase of a huge amount of ammonium nitrate while refusing a discount caused no concern at the time he did it. Today, terrorists know that they must operate even more secretly to accomplish the tasks necessary for them to attack America again. Surveillance gathering, recruitment of group members, fund-raising, procurement of explosives or CBRNE material, and the deployment of weapons must all be accomplished under a cloak of secrecy.

The employment of aggressive intelligence gathering tactics, coupled with a coordinated process for sharing and analyzing the intelligence, is a key offensive action to prevent terror attacks from occurring. Intelligence gathering takes place primarily in two ways. The first is the gathering and analyzing of intelligence that comes from the average citizen who employs a heightened sense of awareness, noting and reporting suspicious activity or behavior. The second means is the targeted gathering of human and technical intelligence by government agents infiltrating terrorist organizations. General Meir Dagan, head of the Israeli Bureau for Counterterrorism, said, "Investments in intelligence are invisible, whereas increased security is visible but often wasteful. The first priority must be placed on intelligence, then on counterterrorism operations, and finally on defense and protection."[57] The gathering of human intelligence through the use of infiltration techniques or informants, as compared to technical intelligence gathered using satellites or electronic eavesdropping, tends to yield greater, actionable information. A satellite image can reveal where a terrorist camp is located, but it cannot reveal what the operatives are thinking and planning. Despite the much publicized and highly criticized failures of American law enforcement and intelligence agencies to adequately share intelligence prior to 9/11, significant strides have been made. These agencies are now able to more quickly recognize, analyze, understand, and share intelligence gathered both from the field and from average citizens.

In addition to gathering, analyzing, sharing, and utilizing intelligence, effective counterterrorist operations are an essential offensive weapon in the U.S. security arsenal. Targeted counterterrorism operations may be carried out by law enforcement at the federal, state, or even local level as a response to credible intelligence indicating a clear and present danger or as the result of an extensive investigation. The FBI maintains primary jurisdiction in matters of terrorism. In the United States, much of the counterterrorism effort is carried out by Joint Terrorism Task Forces (JTTFs) consisting of an FBI-led, multiagency law enforcement team dedicated to counterterrorism investigations. Counterterrorism operations may also be carried out on foreign soil by covert operatives against a terrorist training camp, recruitment location, or hiding place. The Central Intelligence Agency (CIA), in addition to having primary responsibility for gathering intelligence abroad, maintains teams of covert, paramilitary operatives that pursue terrorists outside of the United States. The CIA occasionally works in concert with American military operations. Counterterrorism operations aimed at disrupting the broader infrastructure of terror groups in the areas of recruiting, indoctrination, training, and logistical support of terror operatives is vital both within the borders of the United States and abroad.

Other offensive actions aimed at preventing terrorist attacks can be political in nature. Ensuring a cooperative liaison exists with allied countries enhances the ability to gather intelligence and pursue investigative leads across international boundaries. Political efforts to reduce nuclear stockpiles, and in particular to negotiate treaties that help enhance the security afforded to CBNRE weapons, can be vital to reducing the opportunities afforded terrorists trying to obtain such devices. Nuclear plants and military weapons stockpiles in the United States are generally provided the appropriate level of security. However, this is of little consequence if terrorists manage to secure, and smuggle into the country, weapons of mass destruction obtained from a third-world country that can afford to build the weapon but not to protect it. The protection of U.S. ports and airports depends not only on keeping out those persons who intend harm, but preventing the smuggling of weapons or weapon-grade material in any of the millions of pieces of cargo that arrive each day in the United States. The U.S. Department of Homeland Security, and in particular U.S. Customs and Border Protection, initiated a program shortly after 9/11 known as the Customs-Trade Partnership Against Terrorism, or C-TPAT. C-TPAT enlists the aid of the major manufacturers and shippers in America in guaranteeing sufficient security controls are in place to prevent the use of private companies' supply chains as a means of smuggling in weapons of mass destruction.

Offensive actions—response

After the 9/11 attacks, responder training emphasized how to respond to incidents of a CBRNE nature. Firefighter training mimicked the response to a hazardous materials incident, as any CBRNE event is really a hazmat incident of a special type. Unfortunately, the hurry-up-and-wait mentality that is often employed at hazardous materials incidents is not applicable at WMD incidents due to the number of people involved. Preplanning the initial response to various types of incidents is the best method of operation. Each incident will dictate what strategy and tactics are employed, but a general understanding of the tactics by all responders will lessen the time needed to make crucial, time-sensitive decisions.

Knowing the agents and their categories is the first educational task of responders. Because terrorists fail to abide by the placarding rules set in place by the DOT, responders must look for the signs and symptoms of the victims as well as other clues upon their arrival. No matter how great the detection equipment issued to responders, it is the action taking place upon arrival that will clue responders to a terrorist attack. A knowledge of the signs and symptoms that occur as a result of choking agents, blister agents, and VX gas could save the lives of the responders.

The next educational point is to recognize the ability and limitations of the PPE available. Areas involving any of the chemical weapons can be entered using a level A hazmat suit, but unless one's department responds to every incident, including a sick person call, in this level of PPE, it is not realistic to think it will be donned prior to a terrorist incident. The second consideration is how long a firefighter can suppress fire and rescue people in one of these suits. Preplanning related to what PPE is available to first responding companies, such as engine companies and ambulances, will help in planning offensive action. Studies have indicated that fire gear can be used to enter a chemical environment for a limited duration. Further research is available from the Army's research branch. The authors will defer all recommendations to this source, as their units have conducted the proper research. Figure 11–5 demonstrates some of the response assets that have been obtained as a result of increased funding to prepare responders to deal with WMD incidents.

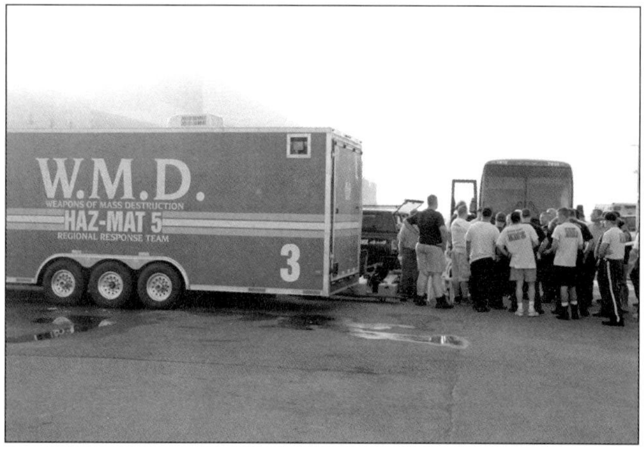

Fig. 11–5. Response equipment utilized to respond to a weapons of mass destruction incident. Increased grant funding has made the purchase of this equipment commonplace throughout the country.

FEMA News Photo.

Planning how to accomplish mass decontamination is a vital planning issue. Many sales representatives sell elaborate decontamination shelters. This can be of use for decontamination of injured and dead prior to removal from the site, but planning for how to utilize equipment used on a daily basis for decontamination is important for the first responder. Utilization of master streams, building sprinkler systems, and named hose streams should be considered.

From a planner's perspective, built-in decontamination equipment is necessary for sports areas, coliseums, and other large gathering arenas. Having people decontaminated prior to their exit of the structure will help contain the event and ensure that all victims receive the decontamination. Consideration should be given to victim's dignity when planning for decontamination. People will forego or refuse decontamination if they have to strip their clothes around other people.

Planning for specialized hazmat and WMD response teams needs to be arranged prior to the event. A jurisdiction certainly does not need to carry every piece of PPE, detection, and decontamination equipment for every possible WMD event. However, planners and response executives must know how to obtain the equipment and how long it will take the equipment to arrive.

Obtaining bomb detection equipment is important early in the incident. Even a chemical attack may be the precursor for a massive explosion. It is always important to think ahead and plan for the worst at these incidents.

Radiological equipment was popular during the Cold War period and was often issued to fire departments for their apparatus. This has occurred again in the wake of 9/11, but many responders do not know how to adequately use the equipment provided. Often the equipment is low-cost equipment that will only recognize the presence of high-powered gamma radiation. Training responders in the limitations of their detection equipment related to WMDs is important. Otherwise, they might have a false sense of security with this equipment and fail to realize they could be unknowingly exposed to many CBRNE products for which they are not protected.

Successful mitigation relies upon working with all response agencies and successfully identifying and mitigating the hazardous materials incident portion of the event, while simultaneously working to deal with the mass casualty event that is taking place.

COMMAND POST AND EMERGENCY OPERATIONS CENTER

Successful command of a terrorist event relies on a few key points:

- A current plan for operating at a terrorist event
- Training with the plan to ensure it functions as designed
- Utilizing a unified command in order to take advantage of all resources responding to the incident

The fire department within each community is often the department that is trained and well versed in the response to terrorist events. Often when a response occurs, other response agencies will look to the fire department for guidance. The amount of resources needed will frequently tax the ability of the initial incident commander. Utilizing the various levels of the ICS structure will help keep the incident commander from continuously operating in crisis mode. Second, establishing tactical radio networks will help those who need to hear critical traffic to hear the right units, rather than listening to all of the incident's radio traffic.

Selecting a proper location for a command post is important. As creatures of habit, most commanders establish a command post from the rear of their cars, directly in front of the incident. In an incident that has the potential to escalate, the commander should resist establishing a command post that will soon be in the operating zone of the incident. In addition, secluding oneself from the incident allows for a clearer mind to think through the incident. Looking at a bombed building with victims lying around will heighten the emotions of the commander and will force a tunnel vision. By not seeing the incident, the commander can work with the planning section chief and the logistics section chief first to develop a plan that addresses the current and future issues and second to provide the correct logistical support for the plan.

Establishment of a staging area and securing the perimeter of the incident will help provide order to the incident. Due to their good intentions, responders often will go to work doing what they feel is important. Past incidents have seen self-deployment to the incidents by firefighters, police officers, and medical personnel. Help by many is needed to mitigate these incidents, but having workers who are not accountable to an incident commander and who could possibly become the next victims only compounds the event.

Command post needs

Many regions and municipalities have purchased command vehicles through the homeland security funding. Many dealers have pointed out what they feel is a necessity for equipping one of the vehicles, but the required equipment will depend on the jurisdiction's hazards, communication equipment, and budget. Chapter 6 provides information on the design and needs of a mobile command center. Adequate space is needed to write down and conduct meetings among the command staff. The utilization of computers to transfer information is helpful, but old-fashioned pencil and paper can enable command of an incident through at least the first few operational periods.

Emergency operations center

Because of the complexity of these incidents, it is imperative that an EOC be established early in the incident. Aside from the immediate mitigation of the incident, emergency executives must plan for relocating citizens, as well as how to feed, equip, and support all of the resources that have been summoned. Use of state and federal resources will take place at any incident of significance; therefore, coordination of their integration must occur. The establishment of an EOC will take a large burden off the command post personnel and allow them to mitigate the incident at hand. Good communications between the EOC and the command post is important. The EOC cannot make decisions for the command post, and the incident commander cannot assume that the EOC is aware of a problem and/or has coordinated resources to handle a problem.

The EOC will need to coordinate its efforts with the state and federal EOCs. Failure to properly coordinate efforts and make the proper requests could have results such as the events surrounding Hurricane Katrina, in which federal aid was not available for an extended period of time. Properly working with the EOCs will ensure the proper deployment of response and recovery forces. (For further information on EOCs, refer to chapter 3.)

CHANGING TACTICS

No longer can emergency responders expect to respond to an incident, go to work blindly, and come home with a feeling of satisfaction for helping a fellow citizen. The mere fact that emergency personnel have become targets of violence changes how responders work at an incident. This is not the only tactic that has changed in the last two decades. Responders are also influenced by the fact that a department could be called upon to deal with a terrorist event that involves substances designed to kill everyone in the area.

When the associate author joined a federal collapse rescue team, the team focused on shoring and accessing collapsed structures. The PPE involved some BDU pants, a t-shirt, and a bandana tied around one's face to cut down on the concrete dust. Because of the emphasis on responder safety due to many illnesses related to past responses, this PPE has changed to a more scientific approach that considers respiratory hazards and skin contact of substances. Although responders will never be able to eliminate all hazards related to their response, a serious overhaul of the methodology of responder safety has occurred. Currently, a whole array of PPE options is available for the collapse responders to prevent unnecessary exposure to contaminants.

The second changing tactic is the integration of other response agencies. Through NIMS and the NRP, a true emphasis is placed on integrating fire, EMS, police, and public works responders into a coordinated response. The old days of the fire chief showing up, deeming himself in charge, and leading his men based on his objectives solely are over. Today, responders must work as a unified group to meet the overall needs of the incident. This may involve fire suppression, collapse rescue, treatment of patients, and evidence preservation.

The last changing tactic involves preparation. The emergency response community has realized that each response agency is likely to respond to the same type of incidents, and the mitigation strategies and tactics remain the same regardless of jurisdiction. Thus, the community has began to develop professional standards that need to be met prior to deploying their personnel. The National Emergency Responder Credentialing System is the first successful attempt to create training standards that are consistent across the country. Prior attempts from the IFSAC, Pro-Board, and NFPA have been steps in the right direction, but these credentials were more optional than required. The national credentialing system will prevent unnecessary responders from showing up at large disasters and attempting to get to the scene of the disaster. Self-deployment has become a problem at disasters, and this credentialing and ID system seems to be a good answer for the problem.

As events occur, emergency responders will continue to refine their tactics, just as they have in the past. One aspect will remain the same: if you call, we will come.

CONCLUSION

Although large-scale terrorist events are new to America, intelligence professionals predict that these will not be the last large-scale terrorist events. Planners must understand problems that have occurred in past responses. They must know and understand the intentions of international and domestic terrorist groups and the vehicles of destruction they will use to deliver their message.

Responders must know how to respond safely to these events. They must know what tactics offer lifesaving effects on victims, while preventing their own demise from the substance deployed. In addition, command and control functions must occur according to tested plans in order to bring about a successful mitigation of a large-scale incident. Failure to develop plans, exercise regularly, and equip and train responders with the proper knowledge and equipment will only yield failure in a time of true need.

REFERENCES

1. U.S. Department of Justice, Federal Bureau of Investigation. 2001. *Threat of Terrorism to the United States*. Testimony of Louis J. Freeh, Director, FBI, Before the United States Senate, Committees on Appropriations, Armed Services, and Select Committee on Intelligence. May 10. Retrieved March 12, 2007, from: http://www.fbi.gov/congress/congress01/freeh051001.htm.

2. Ibid.

3. Ibid.

4. Ibid.

5. Emerson, S. 2002. *American Jihad: The Terrorists Living Among Us*. New York: Simon & Schuster.

6. U.S. Department of Justice, Federal Bureau of Investigation. 2001.

7. Ibid.

8. United States Marine Corps. 2002. *The Individual's Guide for Understanding and Surviving Terrorism*. MCRP 3-02E. March.

9. Linett, H. 2006. The harsh reality of terrorism: what you don't know will get you killed. *American Cop*. 2 (8). November/December.

10. Ibid.

11. Ibid.

12. U.S. Department of Justice, Federal Bureau of Investigation. 2001.

13. Copeland, L. 2004. Domestic terrorism: new trouble at home. *USA Today*. November 14. Retrieved March 12, 2007 from: http://www.usatoday.com/news/nation/2004-11-14-domestic-terrorism_x.htm.

14. Ibid.

15. Ibid.

16. White, J. R. 2003. *Terrorism: An Introduction*. 4th ed. Belmont, CA: Wadsworth/Thompson.

17. Anti-Defamation League. 2007. Extremism at home: recent trends in domestic extremist groups. March 6. Retrieved March 14, 2007 from: http://www.adl.org/learn/extremism_in_the_news/Other_Extremism/extremism_trends.htm?Multi_page_sections=sHeading_7.

18. U.S. Department of Justice, Federal Bureau of Investigation. 2001.

19. Ibid.

20. Ibid.

21. Ibid.

22. White, J. R. 2003.

23. Anti-Defamation League. 2006. 11 indicted in five-state ecoterror campaign. *Extremism in America: Update*. January 24. Retrieved September 15, 2008 from: http://www.adl.org/learn/extremism_in_america_updates/movements/ecoterrorism/ecoterror_11_12406.htm.

24. Copeland, L. 2004.

25. Dugdale-Pointon, TDP. 2005. Terrorist targets. May 26. Retrieved October 24, 2008 from: http://www.historyofwar.org/articles/concepts_terror-targets.html.

26. MacNevin, S. 2004. The top 100 terrorist targets in the United States. Retrieved March 12, 2007, from *Lilithgallery.com*: http://www.lilithgallery.com/articles/terrorist_targets.html.

27. Associated Press. 2002. Al-Qaida: U.S. railroads to be next terrorist target. October 25. Retrieved March 12, 2007 from: http://media.www.dailylobo.com/media/storage/paper344/news/2002/10/25/News/AlQaida.U.s.Railroads.To.Be.Next.Terrorist.Target-307626.shtml.

28. Ibid.

29. New York State Metropolitan Transportation Authority Police Department. August 22, 2004. Field Information: Seven Signs of Terrorist Activity. (MTAPD Circular #: 04-107). Albany, NY.

30. Bolz, F., K. J. Dudonis, and D. P. Schulz. 2002. *The Counterterrorism Handbook: Tactics, Procedures, and Techniques*. Washington, DC: CRC Press.

31. Connors, T. P., and G. Pellegrini, eds. 2005. *Hard Won Lessons: Policing Terrorism in the United States*. July. New York: The Manhattan Institute.

32. Dowle, J. 2007. The CBRN scenario. *Law & Order*. 55. February. pp. 19–21.

33. Connors, T. P., and G. Pellegrini, eds. 2005.

34. Bolz, F., K. J. Dudonis, and D. P. Schulz. 2002.

35. Connors, T. P., and G. Pellegrini, eds. 2005.

36. Bolz, F., K. J. Dudonis, and D. P. Schulz. 2002.

37. Ibid.

38. Central Intelligence Agency. 2003. *Terrorist CBRN: Materials and Effects*. CTC Publication No. 2003–40058. May. Washington, DC.

39. Central Intelligence Agency. May 2003. Terrorist CBRN: Materials and Effects (U). (CTC Publication No. 2003–40058). Washington, D.C.

40. Ibid.

41. Connors, T. P., and G. Pellegrini, eds. 2005.

42. *Wikipedia*. ANFO. 2007. Retrieved March 15, 2007 from: http://en.wikipedia.org/wiki/ANFO.

43. Ibid.

44. Ibid.

45. Central Intelligence Agency. 2003.

46. New Jersey State Police. 2004. *New Jersey Hazmat Emergency Response Course: CBRNE Awareness*. 4th ed. Awareness Training Course #06085.

47. Hansell, L., and S. Salama. 2005. Does intent equal capability? Al-Qaeda and weapons of mass destruction. *Nonproliferation Review*. 12 (3). November. The Monterey Institute of International Studies, Center for Nonproliferation Studies. Retrieved September 13, 2008 from: http://cns.miis.edu/pubs/npr/vol12/123/123salama.pdf.

48. International Atomic Energy Agency. 2001. *Calculating the New Global Nuclear Terrorism Threat*. November 1. Retrieved March 15, 2007 from: http://www.iaea.org/NewsCenter/PressReleases/2001/nt_pressrelease.shtml.

49. U.S. Environmental Protection Agency. 2006. Cesium. October 6. Retrieved March 13, 2007 from: http://www.epa.gov/radiation/radionuclides/cesium.html.

50. International Atomic Energy Agency. 2001.

51. Staten, C. 1997. *Emergency Response to Chemical/Biological Terrorist Incidents.* Emergency Response & Research Institute. August 7. Retrieved March 20, 2007 from: http://www.emergency.com/cbwlesn1.htm.

52. Ibid.

53. Ibid.

54. Heyer, R. 2006. *Introduction to CBRNE Terrorism: An Awareness Primer and Preparedness Guide for Emergency Responders*. No. 20. January 10. Longmont, CO: The Disaster Preparedness and Emergency Response Association.

55. Ibid.

56. Ibid.

57. Tucker, J. 2003. *Strategies for Countering Terrorism: Lessons from the Israeli Experience*. March. Retrieved August 25, 2005 from: http://www.homelandsecurity.org/journal/Articles/tucker-israel.html.

MASS CASUALTIES AND MASS EVACUATION 12

INTRODUCTION

Disasters usually have a serious impact on a large number of persons. As seen in the preceding chapter on terrorism, human losses that are often staggering in ordinary disasters become incomprehensible in events such as a major earthquake or nuclear attack.

This chapter discusses the role of the fire service in dealing with large numbers of persons, both the injured, or *mass casualties*, and the uninjured, or *mass evacuees*.

In many disaster situations, the fire service is confronted with numerous fatalities and injuries and numerous people who must be moved from harm's way. A simultaneous need is created to treat the mass casualties and to effect orderly removal of the uninjured. Chapter 13, which discusses weather-related disasters, explains how Texas had to evacuate major cities in preparation for the landfall of Hurricane Rita.

MASS CASUALTIES

The first part of this chapter deals with mass casualties. The fire service assumed a key role in managing mass casualty incidents in the decade of the 1990s for two major reasons. First, it moved from simply providing first responder rescue services to providing transportation and advanced life support or paramedic services. Second, it developed a comprehensive incident command system that can be effectively applied to managing large-scale emergencies involving multiple casualties and fatalities. The fire service has continued to lead the area of response to mass casualties, as evident in the Oklahoma City and 9/11 responses, as well as in response to natural disasters such as Hurricane Katrina.

Causes of mass casualty disasters

Before discussing planning and response considerations in multiple casualty incidents, it is helpful to review the causes and effects of disasters that can generate large numbers of injuries or fatalities.

Serious students of mass casualty incidents at the National Fire Academy are quite familiar with W. Steve Collier, adjunct academy instructor and EMS director for Indianapolis, Indiana. Steve has conducted extensive research regarding mass casualties and has documented his findings in the National Fire Academy course, *Disaster and Fire Defense Planning.*[1] He has graciously agreed to provide much of the information in this portion of the chapter.

Mass casualty disasters can be natural, technological, and human-caused. The emergencies in each category will be reviewed, some of which have been covered in preceding chapters. In addition, the nature of the medical problems they cause and some common response problems are discussed.

Natural disasters

Natural disasters include all of those covered in chapters 13 and 14 and encompass those disasters caused by one or more natural phenomena. This could include earthquakes, tornadoes, windstorms, dust storms, floods, hurricanes, blizzards, prolonged cold, prolonged heat, drought, ice storms, hail storms, and insect infestation. Other causes could be volcanic eruptions and lava flows, subsidence (sink holes), mud slides, avalanches, tsunamis, and tidal waves. There have even been rare occurrences of "exploding lakes." In Cameroon in 1986, large volumes of carbon dioxide that had formed from natural causes at the bottom of a volcanic crater lake were suddenly and violently released from the water. The resulting cloud of carbon dioxide dispersed through a valley, asphyxiating some 1,700 persons living in the area.[2]

With the exception of this sort of rare phenomenon, natural disasters are fairly well understood and throughout history have had the potential for massive life loss. The following is a sampling of major natural disasters in the last 50 years:

- 1925—multiple tornadoes kill 689 in the U.S. Midwest
- 1931—flooding of the Huang He (Yellow) River kills almost 4 million in China
- 1970—a cyclone kills 300,000 in Bangladesh
- 1976—an earthquake kills 800,000 people in China
- 1986—an earthquake in Mexico City kills more than 9,000
- 1990—a cyclone again strikes Bangladesh, killing 140,000 more
- 2004—Asian tsunami kills more than 200,000, including many missing persons
- 2005—Hurricane Katrina kills more than 2,000

Chapter 13 contains a detailed discussion of the health effects and medical response problems posed by weather-related natural hazards prevalent in the United States—storms, hurricanes, tornadoes, and floods.

Chapter 14 contains information on terrain-based natural disasters—earthquakes, landslides, and volcanoes. Most natural disasters pose the same general sorts of problems, including interruption of utility services and communications, damage to key response resources and facilities, and regionwide effects that limit outside assistance. One all-too-familiar common denominator is the large number of persons affected.

Transportation disasters

Chapter 8 covered transportation disasters, which can involve large numbers of persons, especially in aircraft and train accidents. Growing population density and technological advancements that increase the number of people who can be transported at one time raise the possibility for a very large mass casualty incident virtually anywhere in the country. Most transportation accidents involve violent trauma from either impact or fire.

Major boat or ferry incidents are rare in this country but have occurred recently in the Philippines. Few can forget the ferry accident in Sibuyan in the central Philippines, when the *MV Princess* ran aground and killed 845.[3]

Highway transportation accidents normally result in fewer casualties than other modes, although chain-reaction accidents in fog can require major extrication resources.

HAZARDOUS MATERIALS INCIDENTS

As discussed in chapter 9, hazardous materials incidents pose unique problems and compound mass casualty response needs. Specific problems are listed here:

- The exact name of involved chemicals may not be known for quite some time if shipping papers or fixed facility operators are not available.
- The effects of chemical mixtures, even when the compounds are known, may not be predictable.
- Known chemicals can mix, creating new unknown hazards.
- The biological effects of chemicals are usually toxicological rather than traumatic. Thus, large-scale incidents result in numerous sick people rather than injured people. Resources may not be immediately available to deal with these patients.
- Patients and responders can become contaminated and thus continue to cause problems until they can be decontaminated. Decontamination can be difficult both in the field and in hospitals because of a lack of special facilities.
- Expert technical advice on involved chemicals, including symptoms, personal protective measures, and medical treatment, may be difficult and time-consuming to locate.

The NFPA provides excellent guidelines for dividing personnel and resources among varying danger zones during a mass casualty or mass evacuation incident. These materials include (former) *NFPA Standard 471: Recommended Practice for Responding to Hazardous Materials Incidents*; *NFPA 472: Standard for Competence of Responders to Hazardous Materials/Weapons of Mass Destruction Incidents*, and *NFPA 473: Standard for Competencies for EMS Personnel Responding to Hazardous Materials/Weapons of Mass Destruction Incidents.*

Civil disturbances

Episodes of civil unrest occur periodically in the United States. Chapter 10 discussed the widespread violence that occurred in the late 1990s and in the years following that was associated with race relations. Since that time, isolated events, such as those in Washington, DC over the Iraq War protests, erupt periodically, causing problems for fire and law enforcement forces.

The factors that facilitate either isolated or widespread civil disturbances are numerous and interdependent, making it difficult to predict future episodes. However, incidents tend to occur in highly populated urban areas that have heterogeneous populations in areas of relatively rapid social change in the distribution of power and wealth. They also occur among disadvantaged groups experiencing stress where large groups of persons are already congregated. In both situations, the potential for a mass casualty incident increases.

The specific causal factors for civil unrest may involve the response to a triggering event by government forces, particularly the police, but sometimes by fire personnel as well. Policies on the role and responsibilities of the fire department in civil disturbances should be worked out in advance with police and senior government officials.

Medical care in mass casualty situations arising during civil disturbances will be similar to other incidents with the following exceptions:

- Field medical activities may be limited by difficulties in gaining access to victims until the violent activity can be controlled.
- Hostile actions may be directed at fire and emergency medical services (EMS) personnel as part of the authority structure. This problem is worsened when fire department hoselines and personnel are used to control criminal activity.
- Fire stations and government facilities themselves may be the targets of hostile activity.

Terrorism

As shown by the Oklahoma City bombing and by the 9/11 attacks, terrorism will create large mass casualty incidents. Every jurisdiction should have a mass casualty plan in place for the event of another terrorist attack. The attacks of 9/11 saw the destruction of the response system that was initially in place, thus making it nearly impossible to care for the wounded.

The walking wounded can be a problem in these types of incidents due to their large scale and the inability to instantly provide enough resources to deal with all of the victims. Planners must account for two mass casualty incidents. The first is at the scene of the incident. The second is at the area hospitals and clinics. Plans must be in place at both locations. In addition, the manpower must be able to respond to each of those locations. Once thought to only be a problem for the emergency medical services, these events have proven that the medical community comprised of doctors and nurses needs to be prepared at their facilities for these walking wounded. Second to treating them upon arrival is the decontamination of the victims, so as not to produce an environment within the emergency room at the hospital that will condemn the facility. These recent events have caused many to consider how to deal with these events at emergency rooms. The associate author provides some suggestions based on evaluation of past drills and events:

- Decon operations must not be manpower intensive and must be located on the exterior of the building, or within the entrance. Having a self-serve decon may not decontaminate all victims completely but can minimize the contaminants being tracked into the emergency room.
- Entrances must be able to provide this decontamination. Expecting people to volunteer to delay their care to decontaminate will not happen and could cause a civil disturbance on top of the original event.
- In order to prevent problems with people volunteering to decontaminate, direction, via an authoritative system, must instruct people that decontamination will help their current situation.
- Decon setup must be a passive rather than an active event. The expectation that staff will erect a tent or decon line prior to arrival is not practical. The construction of a corridor in the building to accommodate the process and having it built into the building is the best option.
- Plans must include assembly and operation of a temporary emergency room to accommodate the large volume of people or the shutting down of the original emergency room from contaminants, or both.
- Plans must be regularly tested due to the nursing and doctor workforce's nomadic habits caused by the shortage of medical personnel.

Some acts of terrorism, such as hostage-taking, do not involve a large number of people. These incidents will not result in mass casualties but may be highly charged because of diplomatic and political implications and intense news media coverage.

Aircraft hijackings and bombings are predicted to increase and may involve a much larger number of people. The possibility obviously exists for utilizing explosives or hazardous materials to take a very large number of people hostage, such as at a sporting event. Chapter 11 addressed the real concern about the feasibility of fabricating a crude nuclear weapon. The detonation of such a device could be catastrophic even if an actual nuclear explosion did not occur, because highly radioactive materials would be dispersed.

Medical response implications during terrorist incidents include the following:

- Federal agencies, such as the FBI, may be heavily involved in reacting to such incidents. Fire and EMS responders are still learning to coordinate their efforts with these agencies.
- Terrorist incidents may last for days and take unexpected turns.
- Medical responders or facilities may become targets if it is felt that such actions will place additional pressure on authorities to meet terrorist demands.
- Bombing incidents have been known to involve delayed secondary bombs designed to detonate after responders arrive. Responders should closely coordinate with law enforcement commanders and watch for the presence of suspicious devices.

MASS CASUALTY PRIORITIES

Normally, two immediate needs face emergency responders at a mass casualty incident:

- Search and rescue
- Incident stabilization

If sufficient resources are available, both may be accomplished at the same time. Forces may be able to extinguish a fire or control a fuel spill, while conducting search and rescue (SAR) operations. Search and rescue is a multifaceted task at a mass casualty incident. Alternatives not only involve whether or not to begin search and rescue efforts, but also involve where to start.

After victims are rescued, many will need medical care. This is a high priority in all situations, but when large numbers are involved, a system should include the "Three *T*s"—triage, treatment, and transportation.

Doctors and nurses can be a valuable asset or a serious liability at the scene of a medical emergency. When they are properly identified and operating within the command system, they are extremely useful. When they do not understand their part in the plan and try to freelance, or try to take command, they become a serious liability.

Regardless of the planning effort, some emergency workers at a mass casualty incident will not be accustomed to working under a structured command system (doctors, nurses, private ambulance personnel). This makes it imperative that they be supervised by someone who understands the system. The "Three *T*s"—triage, treatment, and transportation—will be discussed first, followed by organizational considerations and incident command procedures that can facilitate these actions.

Triage

Triage is a French word meaning "to sort." Sorting victims as to their need is an essential element of a multiple victim scene. Most firefighters and EMS personnel are adept at performing this rapid evaluation on a small scale. At a multiple victim scene, it is not unusual to delay treatment of minor injuries, while turning all attention to a critically injured patient. At a scene with more victims than can be treated, triage sorts victims into the following categories:

1. Those requiring immediate care (life-threatening injuries, but survival is probable with medical assistance).

2. Severely injured victims who can withstand a short delay in medical care without immediate risk to life.

3. Victims with minor injuries (walking wounded).

4. Dead or so severely injured that death is highly probable regardless of medical attention.

Triage tags identify victims as to the level of treatment required.

Three important points need to be made regarding triage teams:

1. They must be supervised so all victims are assessed and separated.

2. The most experienced medical personnel at the scene should perform triage.

3. Triage teams should not treat or transport victims until triage is complete. At the scene of a large disaster, it may be best to have triage personnel reevaluate patients, rather than involving them in actual treatment.

The triage activity is the first medical priority in a medical disaster. The first arriving EMS unit should not leave the scene until relieved of triage responsibilities. Emergency medical teams are accustomed to treating one or two individuals, then transporting. Categorizing without treatment runs contrary to their normal role and must be overcome by rigid enforcement of the triage concept.

Planners and response executives should evaluate the possibility of having firefighting crews assist with triage and treatment. Current trends center around cross-training fire and EMS personnel, but planners and incident commanders must evaluate the situation to determine if assigning fire companies to help with triage is beneficial to the overall strategy of the incident.

Another problem arises as EMS personnel are directed to pass by victims who are so severely injured that they are categorized with the known fatalities. When manpower permits, it is best to physically separate the dead and dying from other patients. The walking wounded should be directed to another location for future treatment.[4]

Treatment

Once the triage teams have sorted victims, treatment teams will follow. Treatment teams first treat those needing immediate care. Later, delayed treatment and minor injury treatment teams will be formed. The most highly qualified medical personnel are on the triage teams. The next highest level is the immediate treatment teams. The order of arrival also dictates who is on what team. Later arriving personnel may be assigned to delayed treatment teams, although they may be more qualified than others assigned to higher-level activities.[5]

Treatment area labeling can aid sorting the patients in treatment areas. Often organizations have purchased color-coded tarps. A yellow tarp could indicate delayed care patients. A red tarp could indicate immediate patients, and a green tarp could indicate the walking wounded. Although these tarps require people to set them up, they can add some order to a chaotic portion of the incident.

The treatment group officer is a position assigned when the need arises to oversee the treatment operation. This person is responsible for coordinating the on-scene treatment area, which can include the following:

- Establish areas in which immediate, delayed, and walking wounded can be corralled during their treatment.

- Ensure that the properly certified personnel assigned to the treatment group are tending to the proper patients. As discussed earlier, higher certified personnel should tend to the patients with more serious medical needs.

- Ensure that triage tags are completed and information about the treatment of the patient is recorded on the triage tag.

- Provide oversight to the in processing of persons from the triage area and the proper discharge to the transportation group.

Transportation

Providing access to victims is an important role of the transportation group officer. A "drive through" arrangement for ambulances will keep traffic lanes open. Helicopter landing zones need to be located in safe areas, far enough away from treatment and triage areas so as not to interfere with those activities. Staging and categorizing transportation capabilities are essential in providing proper transportation.

It may be possible to transport several delayed care victims in the same vehicle. Immediate care victims often require advanced life support (ALS) personnel onboard the transport vehicle, and only one can be transported at a time.

Minor care victims usually are staged for later treatment near the site. If minor care victims need transportation, it can be provided using vehicles other than ambulances. The only way to manage transportation vehicles is to utilize a staging area or areas. All incoming vehicles must be directed to a staging area.

Communications with hospital facilities to find out how many patients can be handled is necessary; hospitals determine limits based on personnel and space. Limits established during normal conditions should not prevail during an emergency, but there is a limit. Patients should be directed to hospitals and trauma centers based on facility capacities. The transportation group officer needs to document how many patients have either been sent or are en route to all hospitals. The hospitals must be willing to exceed their norms but also recognize when they have reached their expanded limits. The mass casualty incident may require that only immediate care patients be transported. It may be necessary to set up field hospitals for delayed care patients. Even with transportation available, it is generally not a good idea to flood hospitals with the walking wounded.[6]

As discussed earlier, the transportation officer must consider how many of the victims have taken themselves to nearby hospitals, thus decreasing the capacity of the hospital.

ORGANIZING AND MANAGING MASS CASUALTIES

Operational procedures in mass casualty incidents differ considerably from routine EMS operations and require special groups and branches that interface well with the incident command system. This section of the chapter discusses the ICS management structure both in the field (on-site) and in off-site settings. Figure 12–1 depicts a common EMS branch organization at the scene of a disaster. Readers are encouraged to utilize a similar ICS organization, but must remember that each incident is unique and will require certain groups and divisions based on the strategy and tactics utilized at the scene.

Field medical operation

A number of organizational structures for managing multiple casualty situations have arisen around the country that can be considered when implementing an ICS structure based on the National Incident Management System. The authors recommend a review of chapter 3 for complete information on the NIMS, and specifically the ICS portion of the system.

Medical branch function and structure

Field medical operations generally focus on the following sequence of activities:

1. Identifying and triaging (sorting) casualties to establish priorities for treatment and transport

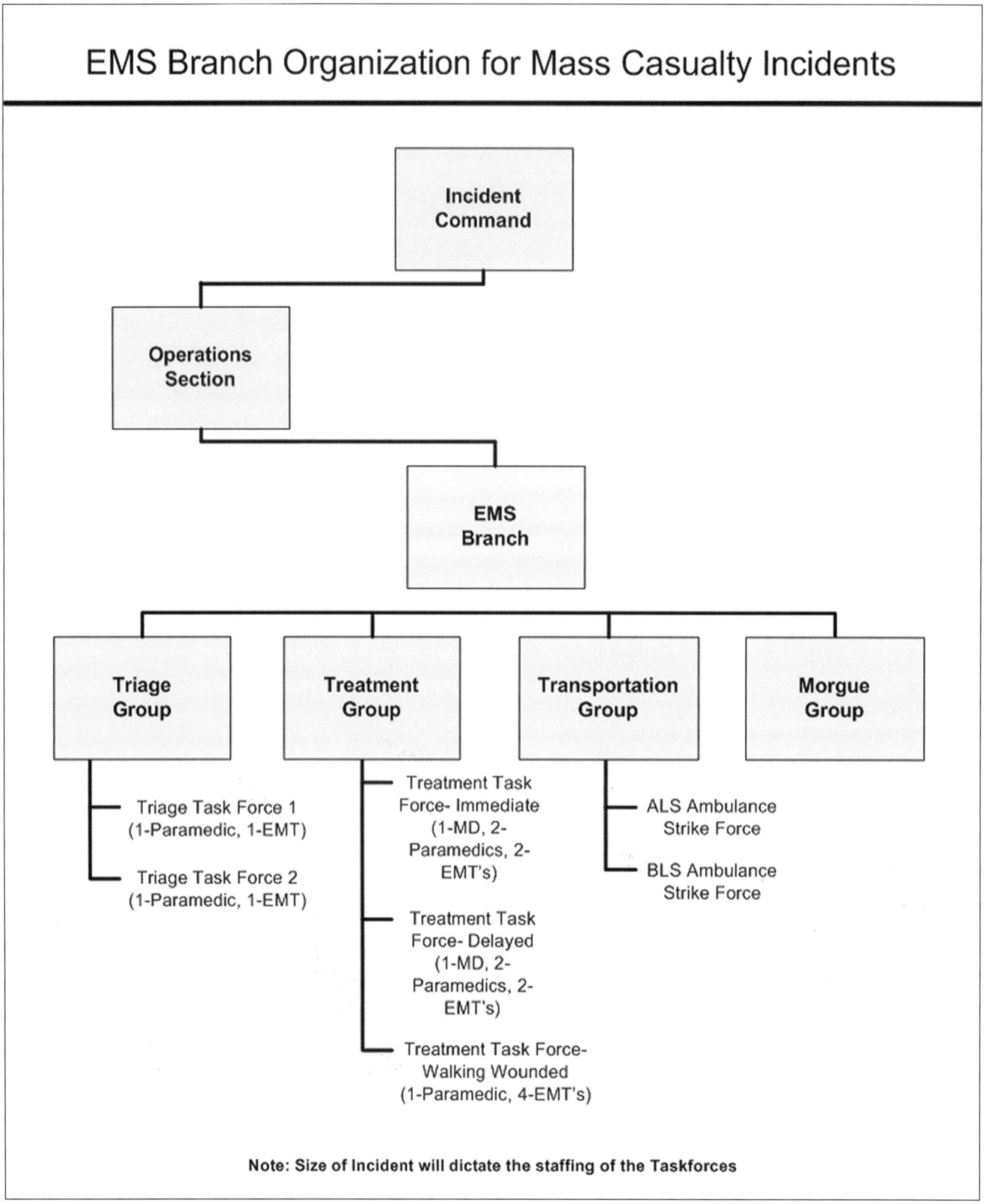

Fig. 12–1. Emergency medical services branch organization for mass casualty incidents. Commanders must properly divide resources to deal with numerous aspects of a mass casualty.

2. Treatment of casualties in order of priority
3. Loading of patients into paramedic or basic life-support ambulances
4. Distribution of patients to appropriate hospitals

The medical branch consists of a triage group supervisor, treatment group supervisor, and a transportation group supervisor. Due to the large number of personnel working in these groups, organizing the resources in strike teams and task forces will help keep the span of control within the recommended range of three to seven.

Medical (EMS) branch officer. The medical (EMS) branch officer is responsible for developing and managing the triage, treatment, and transportation groups, as well as relaying needs to either the operations section chief or the incident commander based on the organizational chart. Group leaders, such as triage, treatment, and transportation, report to the medical (EMS) branch officer. The medical (EMS) branch officer should usually be a field EMS supervisory person rather than a physician or EMS management.

Triage group supervisor. The triage group supervisor is responsible for identifying and deploying triage resources, such as strike teams, usually consisting of five two-person emergency medical technician (EMT) teams or paramedics. If fewer personnel are needed, these resources can be reduced to a single resource consisting of two EMTs or paramedics. Triage teams rapidly assess each victim and attach a tag or marker indicating a priority for treatment or transportation to the hospital based on the guidelines discussed previously in this chapter. The triage group supervisor coordinates with the treatment group supervisor in obtaining field medical care for high priority patients or moving them to one or more treatment areas. The triage group supervisor may be a physician with field medical experience or a paramedic or EMT. The latter is preferred if he or she has considerable field experience dealing with large-scale incidents in a variety of settings.

Treatment group supervisor. The treatment group supervisor is responsible for assembling treatment teams and establishing central treatment areas. In the treatment process, victims receive a secondary triage to identify changes in their condition, appropriate field treatment, and preparation for loading into ambulances. In special cases, such as those involving extensive extrication, physicians or hospital medical teams may be utilized. The treatment group supervisor is charged with coordinating the activities of these teams.

Transportation group supervisor. The transportation group supervisor has the following responsibilities:

- Requesting ambulances and other transport vehicles from staging
- Establishing a loading zone where victims are loaded onto stretchers and prepared for transport
- Organizing vehicles adjacent to the loading zone
- Coordinating with the command post or EMS branch officer to determine destination hospitals for each victim and keeping records of victim distribution

Medical supply officer. Working for the supply unit within the logistics section, the medical supply officer is responsible for providing the supplies needed to operate the mass casualty incident. Depending on the time victims are in treatment areas, the supplies required to continue an operation can be numerous. Close coordination between the EMS branch officer and the supply unit will ensure that supplies are adequate for the intended operation. Planners should preplan the location of mass casualty units. These units are often equipped with enough supplies to triage and treat 100 or more patients. Often large airports will house these units due to the potential for mass casualties during a plane crash. The emergency services are accustomed to operating on a push-style logistics system, rather than the pull-style, which will occur during extended operations. Keeping ahead of the needs will prevent collapse of a functional operation.

Medical staging officer. Medical staging may be part of a general fire or incident staging area, or it may be a separate area. A general staging officer would report to the operations section chief, while a medical staging officer would report to the medical branch officer through the operations section chief. The staging officer is responsible for keeping a designated number of ALS (advanced life support) and BLS (basic life support) ambulances and other medical equipment and transport vehicles on standby for deployment to the incident scene. Staging areas are generally several blocks from the incident site and in a location that permits uncongested ingress and egress.

Medical communications

Medical operations may use one or both of two national radio communications bands. The Hospital Emergency Administrative Radio (HEAR) network was established in the 1960s, primarily to permit administrative coordination among medical facilities during large disasters. The system consists of two VHF simplex radio frequencies. The first, 155.34 MHz, was designated as a local frequency for communications among local hospitals, one of which was designated a regional coordinating facility. This frequency has also come to be used for communications between ambulances and hospitals. The second frequency, 155.28 MHz, was designated as a regional channel for communications among regional coordination hospitals.

The advent of advanced life-support programs vastly increased communications needs between paramedics in the field and hospital-based physicians. In the 1970s, a set of 10 frequency pairs in the 450 MHz UHF band was designated by the Federal Communications Commission (FCC) for use throughout the United States. Eight of these channels are for paramedic-physician communications, and two are for ambulance dispatching. Certain local plans restrict the use of repeaters (devices that permit mobile-to-mobile communications) on some channels. However, the FCC does permit the use of repeaters on all 10 channels. Agency planners should explore the possibility of obtaining local authorization to use repeaters on additional channels for major medical operations.

The widespread use of this 10-channel UHF system permits a variety of field and off-site communications schemes to be developed that can be used by the majority of EMS agencies. A communications officer should be established to coordinate on-site communications. This officer can report to either the medical group officer or the transport control officer.

Although these two systems still exist throughout the country, advancements such as cellular phones and Internet-based information systems have replaced these systems. In the Cincinnati area, medical managers can log into a Web-based program to obtain real-time information on the capacity and status of all of the area hospitals. Planners and command personnel should have knowledge of how the medical disaster communications network works and how to obtain information about the capacity of local hospitals.

The establishment of tactical radio networks that include the hospital staff will allow a coordinated effort among field and hospital staff. This will prevent moving the disaster to the local hospital.

Extrication operations

Major extrication activities may be necessary in such incidents as transportation accidents, building collapses, earthquakes, and tornadoes. These activities are usually organized as a separate ICS branch, often called the *rescue branch*, but they require very close coordination with the medical group. The rescue branch officer is the leader of this branch and should focus on managing and planning to accomplish the following:

- Coordinate the extrications, based on priorities.

- Ensure that the correct equipment cache and rescue-trained personnel are present and are operating within the overall strategy of the incident.
- Based on recommendations of the EMS branch, freeing trapped victims. This may include the amputation of victims by physicians on scene.
- Ensure that the wreckage or building is properly shored to allow personnel to operate safely.
- Coordinate with the technical information specialists, such as structural engineers and transportation safety experts, to determine areas that are unsafe for personnel to operate.

Physician or hospital-based medical teams may be very helpful in these situations and should be involved in key extrication decisions.

Hospital coordination

The goal of mass casualty plans is to save the most lives by bringing appropriate field and definitive medical care to seriously injured victims in the shortest possible time. This does not necessarily occur by simply removing victims from the site as soon as possible. If activities such as distributing patients from the scene, transferring patients between facilities, and maximizing use of medical manpower are not coordinated, the disaster may simply be shifted from the site to hospital corridors.

Medical disaster plans should include identification of one individual, possibly a hospital administrator or physician, to coordinate with hospitals and other medical facilities concerning the following:

- Determination of the ability of surrounding hospitals to accept general trauma patients, chemical or radiological injuries, burns, spinal cord injuries, hand injuries, partial amputations, and other special cases
- Coordination of hospital patient transfers
- Coordination of medical manpower including physicians, nurses, technicians, paramedics, EMTs, and other allied health professionals
- Coordination of medical supplies for use at hospitals
- Establishment of auxiliary medical facilities in the event of very large numbers of casualties or damage to one or more primary medical facilities by fire, flooding, collapse, etc.

Medical Assistance Teams

In some jurisdictions, the medical community is organized in advance to provide invaluable lifesaving skills at a mass casualty incident. One example is Hamilton County in southern Ohio. Here the Medical Assistance Team (MAT) is comprised of physicians, nurses, and paramedics who have identification cards and will respond upon request.

The MAT is sponsored by the Cincinnati Academy of Medicine and the Hamilton County Disaster Council. All team members are linked by portable radio, which is integrated into the established disaster communications system for all the hospital emergency departments in the area. The MAT, in cooperation with the Greater Cincinnati Hospital Council, has established seven sets of "disaster boxes." All of the disaster sets can be brought to the scene by emergency vehicle in the event of a major mass casualty incident. This provides medical equipment for definitive critical care for 35 to 50 victims.

Physicians on the MAT are identified by green helmets with *MD* in reflective letters. Registered nurses wear red jumpsuits with Red Cross insignia and a white helmet marked *Nurse*. Physicians and nurses have photo-identification cards. All MAT members wear patches and/or arm bands from the Cincinnati Academy of Medicine.

On notification of a possible disaster, the medical commander of the MAT is notified immediately through dispatch. The MAT commander then decides the extent of the MAT response by evaluating the number of potential casualties, the type, size, and location of the disaster. A backup notification system for the MAT is conducted through the Hamilton County disaster net radios, located in the emergency departments of hospitals. Hamilton County dispatch can also notify the MAT.[7]

A publication entitled *Disaster Medical Assistance Team Organization Guide* is available from the Department of Health and Human Services.[8] Figure 12–2 shows a DMAT in action. Note that these teams respond with enough equipment, supplies, and personnel to create a hospital in the field. Although their response is not immediate, their services are invaluable for extended operations.

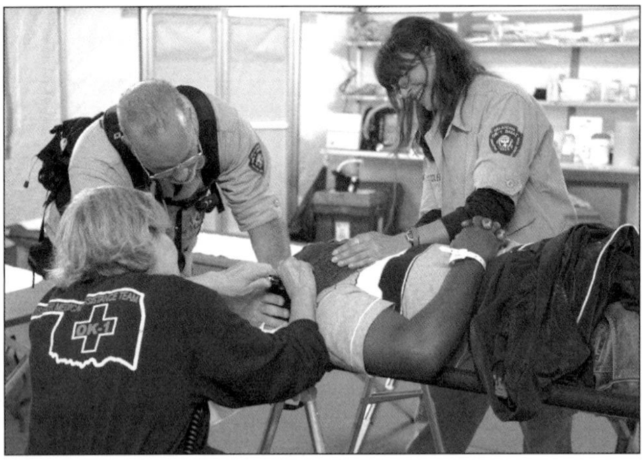

Fig. 12–2. DMAT in operation. DMATs are a great asset for long-term mass casualties. If local or regional MATs are in place, their services can be invaluable in short- and medium-term operations involving mass casualties. Commanders should also consider their use for overwhelmed hospitals.

Evacuation by helicopter

Occasionally evacuation by helicopter provides the difference between life and death to the most critically injured victims in a disaster. Even when the distance to a hospital is relatively short, a helicopter often provides a much smoother ride than one over roadways, thus reducing the amount of additional induced trauma, increasing the odds for survival. When a helicopter is to be employed, however, certain precautions are in order. The following is a list provided by the Training Bureau from the Cincinnati Fire Division.[9]

1. Landing site must be clear of overhead obstructions. Notify the pilot of the location and height of any overhead obstructions.
2. Clear the site of debris and loose objects, e.g., clothing, sheets, or blankets. Expect a wind of 60 to 70 mph.
3. Wear eye protection.
4. Never move toward the helicopter until signaled by the flight crew. If a responder can see the pilot's eyes, the pilot can see the responder.
5. Only flight crew members may open and close aircraft doors.
6. Always approach and leave the helicopter from the front.
7. When flammable or poisonous vapors are involved, select a landing site at least one mile upwind.
8. At night:
   - Do not point lights at the helicopter while it is airborne.
   - Mark the corners of the landing site and the center of the windward edge with flares.
9. Do not smoke near the helicopter.
10. Do not run near the helicopter.
11. Responders should not carry IVs or other objects over their heads. They should carry long objects parallel to the ground.
12. Be especially cautious in the tail rotor area.

Barrier tape should not be used to mark a landing zone, because the heavy winds from the helicopter will cause it to be blown about severely in unpredictable ways, perhaps adding to the calamity.

Drills

The potential for mass casualty is present in every community, and many have conducted mock disaster drills to improve existing procedures so that the jurisdiction will be better prepared to deal with the real thing. According to a *Fire Chief* article,

> A command officer arrives at an accident scene where a bus has overturned and there are numerous casualties. Ironically, the transit authority and other agencies recently participated in a mass-casualty session with the officer's organization only a few months ago.
>
> The commander rapidly makes a series of decisions. To the observer, his decision-making process appears very natural. A week later, in a post-incident critique session involving principals from the responding agencies, the command officer says he didn't consciously reflect on the lessons learned in the mass-casualty drill while making his decisions, though without a doubt it helped him prepare for the real thing.[10]

Fire and EMS departments throughout the country conduct mass casualty drills for just this reason. When time is of the essence and quick decisions need to be executed, the mind will revert to the training that has been received. Training to respond to and command these types of incidents will allow swift, correct decisions.

NATIONAL DISASTER MEDICAL SYSTEM

Most local disaster plans and resources are capable of managing casualties ranging in number from up to 100 in smaller communities to 300 or more in larger cities. Even the largest municipalities would be severely stressed attempting to deal with 2,000 to 3,000 casualties. Yet a natural disaster, such as a predicted earthquake in California or a major toxic chemical release in a large metropolitan area, could generate hundreds of thousands of casualties. Recognizing that such catastrophes are plausible, President Ronald Reagan established an Emergency Mobilization Preparedness Board in 1981 to develop a national medical response system.

The board is composed of representatives from four principal participating federal agencies:

- Federal Emergency Management Agency
- Department of Defense
- Veterans Administration
- Public Health Service of the Department of Health and Human Services

According to the Department of Health and Human Services,

> The National Disaster Medical System (NDMS) is a federally coordinated system that augments the Nation's medical response capability. The overall purpose of the NDMS is to supplement an integrated National medical response capability for assisting State and local authorities in dealing with the medical impacts of major peacetime disasters and to provide support to the military and the Department of

Veterans Affairs medical systems in caring for casualties evacuated back to the U.S. from overseas armed conventional conflicts.

The National Response Framework utilizes the National Disaster Medical System (NDMS), as part of the Department of Health and Human Services, Office of Preparedness and Response, under Emergency Support Function #8 (ESF #8), Health and Medical Services, to support Federal agencies in the management and coordination of the Federal medical response to major emergencies and federally declared disasters including:

- Natural Disasters
- Major Transportation Accidents
- Technological Disasters
- Acts of Terrorism including Weapons of Mass Destruction Events[11]

For further information about the National Disaster Medical Response System, the reader is encouraged to visit the Web site http://www.hhs.fema.gov/index.html.

Elements of the National Disaster Medical System

NDMS consists of three principal elements:

- Disaster Medical Assistance Teams recruited from around the country that can be sent to an emergency site to assist local authorities in triaging, sorting, and clearing casualties.
- A large-scale airlift system, coordinated by the military and primarily utilizing Department of Defense aircraft, to transfer victims to any of the 71 NDMS operational areas around the country.
- A network of civilian hospitals, originally established under a Civilian-Military Contingency Hospital System (CMCHS), to receive these casualties.

Types of NDMS teams

Based on the needs of the disaster, the NDMS program can send a variety of teams to assist. The types of teams are:

- Disaster Medical Assistance Team (DMAT)
- Disaster Mortuary Operational Response Team (DMORT)
- Veterinary Medical Assistance Team (VMAT)
- National Nurse Response Team (NNRT)
- National Pharmacy Response Team (NPRT)
- Disaster Portable Morgue Units (DPMU) Team

Disaster Medical Assistance Teams. The Disaster Medical Assistance Teams (DMATs) are generally sponsored by local hospitals and consist of 23 people, including physicians, nurses, technicians, and clerical staff. According to the Department of Health and Human Services,

DMATs are designed to be a rapid-response element to supplement local medical care until other Federal or contract resources can be mobilized, or the situation is resolved. DMATs deploy to disaster sites with sufficient supplies and equipment to sustain themselves for a period of 72 hours while providing medical care at a fixed or temporary medical care site. The personnel are activated for a period of two weeks.

In mass casualty incidents, their responsibilities may include triaging patients, providing high-quality medical care despite the adverse and austere environment often found at a disaster site, patient reception at staging facilities and preparing patients for evacuation.[12]

Casualty transportation and distribution

Once casualties are cleared from the disaster locale, they are transferred to military aircraft and flown to various operational areas having available hospital beds. The U.S. Air Force is responsible for tracking all patients transported by NDMS. Upon arrival at local airports, a DMAT will meet arriving patients and process them for ground transport to area hospitals.

In many cases, operational areas may need to activate local medical disaster plans to coordinate ground transportation from the airport to area hospitals, as well as hospital transfers and related activities.

Federal NDMS activation

Federal NDMS activation may occur in one of three ways:

- Upon declaration of a "major disaster" by the president at the request of a governor
- By the Secretary of Defense in response to a military conflict
- By the Secretary of Health and Human Services in response to a peacetime disaster

MULTIPLE FATALITY INCIDENTS

Incidents that result in multiple fatalities involve unique features that affect fire and medical operations.

Legal authority in multiple fatality incidents

In most states, the coroner or medical examiner is considered a high-ranking law enforcement officer. The coroner or medical examiner may have the authority to do the following:

- Determine the biological cause of death
- Investigate deaths to determine the possibility of criminal acts
- Identify fatal victims
- Preserve the personal effects of victims at the scene of the death until investigations are complete

Within the ICS organization, a morgue group officer would operate under the medical (EMS) branch officer. This individual may come from the coroner's office, EMS, the fire department, or a police agency and is responsible for coordinating initial temporary storage of remains until longer-term arrangements can be made by the coroner. In most states, the coroner or medical examiner has clearly specified legal authority and generally will not fall under the direct authority of the incident commander. However, with the states adopting NIMS, this can vary from state to state. Planners should understand the legal authority and the use of the coroner or medical examiner at the scene of a disaster prior to needing their assistance.

Investigation and identification of victims

Investigation of multiple fatality incidents can be extraordinarily difficult because remains are often badly burned or mutilated. Local laws or practices concerning movement or removal of remains by emergency personnel vary greatly, but the following general guidelines should apply in most multiple fatality incidents:

- The bodies of apparent fatalities should not be moved without approval of the coroner or medical examiner except to assess whether the victim is alive or viable for resuscitation or to gain access to live victims.
- The coroner or medical examiner should be consulted, if possible, before moving remains that may be threatened by fire, rising water, or other conditions.
- The immediate personal effects found near victims may be valuable in identifying them. Items such as purses and briefcases should be packaged, labeled, and moved with the remains.
- Jewelry and clothing should not be removed.
- Surrounding items, such as furniture, should be moved as little as possible. If the remains must be moved, some sort of mark using chalk or masking tape should be made to indicate the location and position of the remains.
- If possible, EMS or fire personnel should keep careful records, including drawings or photographs, describing where victims were found and what articles were nearby.

In severe cases where numerous victims cannot be identified, a special FBI victim identification team may be requested to assist.

Notification of next-of-kin is usually handled by law enforcement agencies, although in some areas the Red Cross has that responsibility. In cases of multiple unidentified fatalities, notifications may be delayed until all victims are positively identified. This will cause a great deal of emotional stress for relatives present at the scene. Fire or police chaplains may be helpful in such situations. According to the Department of Health and Human Services,

> The National Response Framework utilizes the National Disaster Medical System (NDMS), as part of the Department of Health & Human Services, Assistant Secretary of Preparedness and Response (ASPR), Office of Preparedness and Operations (OPEO), under Emergency Support Function #8 (ESF #8), Health and Medical Care, to provide victim identification and mortuary services. These responsibilities include:

- temporary morgue facilities
- victim identification
- forensic dental pathology
- forensic anthropology methods
- processing
- preparation
- disposition of remains[13]

Managing the remains

In large multiple fatality incidents, local resources for storing and processing remains may be inadequate. The initial problem may be insufficient body bags for removing remains from the site. Local funeral directors or state or federal military departments may be able to assist.

Local hospital morgues may have insufficient storage space. Temporary facilities such as refrigerated semi-trailers or local ice skating rinks may be used.

Major incidents may involve multiple fatalities and multiple casualties or they may result almost entirely in fatalities. In either case, medical response personnel will be heavily involved. In dealing with remains, they must be sensitive to the emotional needs of relatives and bystanders by protecting bodies from view (especially by unauthorized photographers) and by handling them with respect. They must also be aware of the importance of preserving all items found both on the remains and in the immediate area, so that positive victim identification can be made as quickly as possible.

MASS EVACUATION

In order to prevent mass casualties, the emergency services are often involved in mass evacuations of civilians from dangerous or potentially dangerous zones. The remainder of this chapter will deal with "potential victims"—the evacuees rather than the mass casualties. While police and occasionally military personnel are involved in a primary manner with evacuations, it is the fire department that usually provides information on suggested zones.

Evacuation is an important protective action for hurricanes, floods, tsunamis, volcanoes, hazardous materials accidents, nuclear power plant accidents, and crisis situations such as tornadoes and earthquakes. Another form of protective action is sheltering, and most forms of disasters will involve some combination of sheltering and evacuation.

Fires in occupied buildings frequently require removal of the occupants. Firefighters have always been involved in such evacuations, as well as evacuations of people from whole blocks.

For example, on March 28, 2007, an entire six-story office building in Houston needed to be evacuated when a fire broke out on the top two floors. Many of the evacuated were removed with the fire department aerial apparatus.[14] Estimates were at least 100 people were evacuated. In recent years as technological mishaps have dramatically increased, such as hazardous materials incidents, so has fire department involvement in mass evacuation.

Natural events

Evacuation is an action taken as a protective measure necessitated by a dangerous situation. It sometimes stems from large, but fairly infrequent, natural events such as hurricanes and floods. The impact of potential disasters is mitigated by removing the population from the dangerous area, and the evacuation may be very large.

In 1961, Hurricane Carla caused the relocation of an estimated 750,000 people in the United States, and in 1985, Elena necessitated the relocation of 300,000 people. In 2005, evacuations occurred for Hurricane Katrina and Hurricane Rita. Both had some difficulties, but the Rita evacuation had a positive outcome according to the media, and the Katrina evacuation was botched. Planning for the logistical portion of the evacuation is what proved to make the 2005 evacuations successful. Many other evacuations affecting tens of thousands of people occur numerous times each year.

After these numerous occasions, planners learned that evacuating by zip code is best. As stated in the Key Considerations for Large Scale Disasters section of this chapter, there are only so many people and vehicles that can use evacuations routes, thus, orderly, zone-based evacuations are best. As noted in a September 11, 2008, article by CNN, "About 15,000 residents were leaving Galveston's Brazoria County Wednesday after a mandatory evacuation order was issued at 10 a.m. CT (11 a.m. ET) governing one ZIP code—77541—and residents throughout the county with special needs"[15] As disaster planners and practitioners continue to to learn from these events, we will continue to see refinements that will streamline the process.

Hurricanes and floods usually occur at certain times of the year, usually allow some warning time, and are mostly confined to certain areas of the country. (The Gulf Coast and East Coast experience some effect from just about every hurricane.) Earthquakes and tornadoes usually give so little warning that evacuation is not a prime consideration. In these situations, subsequent sheltering is the dominant form of protection. Figure 12–3 is example of the massive evacuations that have taken place as a result of an impending hurricane landfall.

Technological events

A second general type of occurrence that often results in evacuation is technological or human caused events. These are principally hazardous materials leaks, spills, or fires that more directly and immediately involve the fire department. They can be divided into two types:

- On-site, as at chemical plants, factories, or storage facilities
- Transportation accidents

Thousands of new chemicals are developed each year, and citizens and officials are concerned about accidents happening in their communities. Of the 3,100 localities completing a FEMA questionnaire in 1985, 93% identified one or more hazardous materials risks as a significant threat to the community.

Because of the risk of hazardous materials incidents and because local governments will be completely on their own in the first stages of almost any incident, committees need to maintain continuing preparedness. Specific evidence of being prepared is an emergency plan. A FEMA report on 242 evacuations of all types and sizes shows that communities rely upon evacuations as a primary means of protecting their citizens. However, the majority of evacuations cited were implemented on an ad hoc basis, without a viable emergency operations plan (EOP).[16] Each jurisdiction should not

Fig. 12–3. Massive evacuation. Depicts the massive evacuation efforts that have resulted from an impending hurricane landfall.

FEMA News Photo.

only have an all-hazard EOP, but also a complete and sound evacuation annex.

Joseph Weber is a retired deputy fire chief for the city of Los Angeles and one of the most respected instructors at the National Fire Academy. Joe has graciously agreed to allow the incorporation of these guidelines, with some modification by the authors, into this discussion on mass evacuations. Much of the information that follows originated with Joe and his research in producing the National Fire Academy's open learning course, *Disaster and Fire Defense Planning*.[17]

It is strongly recommended that a function annex to a disaster plan be drafted. In response to Title III, the National Response Team (NRT) produced the *Hazardous Materials Emergency Planning Guide*, a product of the cooperative efforts of the 14 federal agencies that constitute the NRT.[18] Below is information from the section of the NRT guide that corresponds to a functional annex on evacuation. It is not a model but a guide to the topics that should be covered.

EVACUATION PROCEDURES

According to the *Hazardous Materials Emergency Planning Guide*, the following should be addressed in preparing evacuation plans:

- Title of person and alternate(s) who can order/recommend an evacuation
- Vulnerable zones where evacuation could be necessary and a method for notifying these places
- Provisions for a precautionary evacuation
- Methods for controlling traffic flow and providing alternate traffic routes
- Locations of shelters and other provisions for evacuations (e.g., special assistance for hospitals)
- Agreements with nearby jurisdictions to receive evacuees
- Agreements with hospitals outside the local jurisdictions
- Protective shelter for relocated populations
- Reception and care of evacuees
- Re-entry procedures[19]

Evacuation is the most sweeping response to an accidental release. The plan should clearly identify under what circumstances evacuation would be appropriate and necessary. DOT's *North American Emergency Response Guidebook* provides suggested distances for evacuating unprotected people from the scene of an incident during the initial phase. This is known as the Table of Isolation Distances, which is found in the rear of the text, in the green-bordered pages. It is important to distinguish between general evacuation of the entire area and selective evacuation of a part of the risk zone. In either case, the plan should identify how people will be moved (e.g., city buses, police cars, private vehicles). Provision must be made for quickly moving traffic out of the risk zone and also for preventing outside traffic from entering the risk zone. If schools are located in the risk zone, the plan must identify the location to which students will be moved in an evacuation and how parents will be notified of this location. Special attention must also be paid to evacuating hospitals, nursing homes, and homes for the physically or mentally disabled, and those in low-income areas who cannot provide a means for their own evacuation.

Maps (drawn to the same scale) with evacuation routes and alternatives clearly identified should be prepared for each risk zone in the area. Maps should indicate precise routes to another location where special populations (e.g., schools, hospitals, nursing homes, homes for the physically or mentally disabled) can be taken during an emergency evacuation. They should also show the methods of transportation during the evacuation. In areas frequented by evacuations, signs, such as the one in figure 12–4, can distinguish the evacuation route.

Fig. 12–4. Evacuation signs inform evacuees of the proper evacuation route. Areas frequented or that have the possibility to be evacuated should have these sign posted, as well as maps of where these routes are located.

Photo courtesy of associate author Randall W. Hanifen.

Consideration of when and how evacuees will return to their homes should be part of this section. It should also include a description of how other agencies will coordinate with the medical community.

Copies of evacuation procedures should be provided to all appropriate agencies and organizations (e.g., Salvation Army, churches, schools, hospitals) and could periodically be published in the local newspapers.

KEY CONSIDERATIONS IN LARGE-SCALE EVACUATIONS

Items to be weighed when planning, or when actually evacuating, will surface by determining the answers to the questions *why*, *who*, *how*, *how long*, *where to*, and *how much*. Answers to these questions will determine what goes into the plan and what is important in the on-scene decision-making process.

Why evacuate?

Listed below are factors to be weighed in making an evacuation decision:

- The accident expansion potential. How much material is left to explode or spread as a gas, etc.?
- What are the odds for expansion?
- What are the possibilities of in-place shelter?
- How can conditions such as weather affect the outcome?
- What are the estimates of warning time if the condition worsens?
- What are the effects of a worst-case situation compared with the adverse effects of an evacuation?
- What is the estimated time to evacuate compared with the time available in a worst-case development?

Citizens evacuate when four conditions are met:

1. They have accounted for the safety of their immediate household.
2. They have been given by authorities, or have personally developed, a plan for protective action.
3. They believe that a threat does exist in the environment.
4. They perceive that upon impact, this threat could result in some level of damage to their person or property.[20]

Figure 12–5 lists steps that citizens can take when being evacuated. These forms should be distributed when citizens are preparing to evacuate. Placing the forms on the Internet will allow people to immediately access the forms.

FEMA officials have determined that separating pets from families compounds evacuation problems, so, when possible, pets should be evacuated with the family.

It is interesting to note that these conditions were met for the latest large-scale hurricanes, but many citizens remained in the thought they could ride them out, even with news media predicting total devastation.

FEMA's Citizen Evacuation Guidelines

Always:	**If Time Permits**
Keep a full tank of gas in your car if an evacuation seems likely. Gas stations may be closed during emergencies and unable to pump gas during power outages. Plan to take one car per family to reduce congestion and delay.	Gather your Disaster supplies kit.
Make transportation arrangements with friends or local government if you do not own a car.	Wear sturdy shoes and clothing that provides some protection, such as long pants, long-sleeved shirts, and a cap.
Listen to a battery-powered radio and follow local evacuation instructions.	Secure your home: Close and lock doors and windows Unplug electrical equipment, such as radios and televisions, and small appliances, such as toasters and microwaves. Leave freezers and refrigerators plugged in unless there is a risk of flooding.
Gather your family and go if you are instructed to evacuate immediately.	Let others know where you are going.
Leave early enough to avoid being trapped by severe weather.	
Follow recommended evacuation routes. Do not take shortcuts; they may be blocked.	
Be alert for washed out roads and bridges. Do not drive through flooded areas.	
Stay away from downed power lines.	

Fig. 12–5. FEMA evacuation guidelines

Who is evacuated?

Details of an evacuation are difficult to include in a hazardous materials plan, since the portion of the population at risk cannot be accurately predicted. In hazmat incidents, the evacuation area depends upon the size and location of the incident, the material involved, geographical considerations, and weather conditions. The area to be evacuated may include the area of potential danger due to drifting gases, radioactivity, or subsequent explosions, as well as the area of actual damage. Large-scale hazmat evacuations have occurred because of gas plumes. Chapter 9 introduced the system called Computer-Aided Management of Emergency Operations (CAMEO) for predicting the path and size of chemical gas dispersions.

If people can make their homes reasonably airtight, are they safer at home from poisonous gases or radiation than they are with the risks of evacuation? Immediately following 9/11, officials from the Department of Homeland Security began telling citizens to protect themselves with plastic and duct tape for the windows. These two items were a necessity in a person's disaster kit.

If evacuation is decided upon, the area must be exactly defined. In a disaster exercise in Mississauga, near Toronto, Ontario, a grid was outlined by street names so that the public could easily understand the limits of evacuation. Additional zones for expanded evacuation were mapped out by street names and natural boundaries. This was in contrast to designated areas with a radius of 5, 10, or 20 miles at the Three Mile Island (TMI) nuclear power plant emergency in Pennsylvania in 1979.[21]

At TMI, a selected portion of the population (pregnant women and children) was advised to evacuate, but this would probably never be the case in a chemical hazmat incident.

People will evacuate if they perceive the threat to be real.[22] In September 2005, this statement proved correct in New Orleans. A CNN article prior to the landfall of Hurricane Rita, just after the deadly results of Hurricane Katrina, stated, "'Rita scares us to death, at least it scares me to death,' Louisiana Emergency Response Medical Director Dr. Louis Cataldie told CNN. Cataldie is in charge of the makeshift morgue in St. Gabriel, about 70 miles from New Orleans. On Wednesday, 628 of Katrina's victims were at the morgue, spokesman Bob Johannsen said. About 100 have been identified."[23]

When an area is designated and evacuation is ordered, everyone must leave, and security must be established. The knowledge that their property and pets are safe helps to persuade people to leave.

Who orders evacuation?

This is a matter of planning. In small-scale emergencies, the first responder, first on-site incident commander, or the individual specified by the department SOP should have authority to evacuate exposed buildings or the immediate area. The plan must designate the person or persons with authority to order large-scale evacuations.

How?

Sometimes people leave by themselves without an order. Although only an evacuation advisory for pregnant women and young children was given at TMI, about 144,000 people, or 39% of the total population within 15 miles of the reactor, evacuated.[24] This may have been because the public perceived a unique danger that distance would minimize, but spontaneous evacuation is an uncertainty in emergency operations planning.

When an evacuation decision is made, the public must be informed. Public warning can be given by these means:

- Structure-mounted sirens, where the public is trained to turn on the radio upon hearing the proper warning signal
- Radio and TV announcements
- Phone, such as reverse 911 or Code Red
- Word of mouth

- Mobile loudspeakers from vehicles traveling up and down blocks
- Door to door

Instructions to evacuees should be put on tape, if possible. If read by police over loudspeakers, each officer should be given the same printed copy. For the area surrounding a fixed facility such as a nuclear power plant, printed sheets can be, and in some areas have been, distributed to the public, even in advance of any incident. Some thought should be given to the instructions, which should include information regarding routes, means of transportation, articles to bring, destination, and shutting off utilities. Hazardous materials accidents may preclude anything except an urgent directive to "get out and head north as fast as possible."

Language barriers are a possibility as a greater population within the United States is non-English speaking. Although many states have adopted English as the official language, planners and response executives must prepare for a language barrier. Failure to plan for this could result in at least 15% to 20% of the population not receiving the message. Having the ability to translate messages at the EOC or the command post is essential. Many language services offer translation services via phone. Planners should arrange for this service before it is needed.

Especially in rural and suburban areas, almost everyone can travel in his own car or with a friend or relative. At Three Mile Island, 94% of households had a car, and 84% of families left in their own cars. Many leave before an order is given, and many outside the order boundaries also leave.

By using public buses, school buses, or whatever resources are available in the community, transportation must be provided for those who lack it. Pickup points can be identified in the evacuation instructions. The bedridden create an extra concern, and provisions must be made for them. If at all possible, a door-to-door check should be made, and some standard mark or sign placed on the door of evacuated homes.

Large hazmat and natural disaster incidents rarely occur in inner cities, but when they do, they present different problems. In tenement areas, the majority of inhabitants may not own cars, door-to-door notification would most likely not be effective, securing the areas would be very difficult, and getting everyone out could become impossible. In Mississauga, the evacuated population was upper-middle class; relatively well off and in private homes with small families. That is a very different profile from that of the inner city. In the evacuation that occurred during Hurricane Katrina, citizens meeting the low-income, disabled criteria were overlooked or inadequately planned for. Thus, many were forced to remain in the city of New Orleans at the Superdome, in hopes that the structure would survive. Figure 12–6 shows the line of people that were at the Superdome seeking shelter.

Institutions can be more of a problem than the general public.

- **Schools.** Should the students be allowed to return home, letting them leave with their families, or should the whole school population be transferred to another school outside the danger area? There are problems either way.
- **Hospitals.** Each should have its own plan for the transfer of patients to other hospitals. A hospital mutual aid plan with ambulances is a priority item.
- **Nursing homes.** Transportation for almost all will be required.
- **Jails.** Difficulties will occur in absorbing large numbers into other jails. Special transportation will be required, so advanced planning is a must. In one instance, state penitentiary inmates were transported to a football field.

Fig. 12–6. Refugees wait to enter the Superdome in preparation for Hurricane Katrina.

FEMA News Photo.

- **Animal shelters.** Transfer of animals to a safer location may be feasible.

After evacuation, the area must be secured against reentry.

Studies have not been done on topics such as how many police it takes to evacuate a given number of blocks, or how long it takes at different times of day or night. Such studies would probably only be valid if they were site specific, but they would contain the kinds of information a planner needs in order to make reasonable estimates. Recent hurricane evacuations offer a starting point for study, but research is still needed at the neighborhood level of evacuation.

How long?

If evacuation time is the total time from the onset of the incident until everyone has cleared the area, then it is the sum of $Td + Tn + Tm + Tt$,

where

Td is the time delay after occurrence until the evacuation order,

Tn is time required to notify the population,

Tm is the time required for people to mobilize and get underway, and

Tt is the travel time required to leave the affected area.[25]

The following can be noted about the evacuation time components:

- *Td* can be reduced by effective planning, since it includes the notification of officials, interpretation of data, decision making, etc.
- *Tn* seems to be inversely proportional to the population density.[26] The closer people are, the quicker they can be notified to evacuate. It is estimated under the best conditions of local planning to be from 15–60 minutes.
- *Tm* depends upon such things as families getting together, securing homes and farms, type of population, and institutional preparation.
- *Tt* is a consequence of the number of people, highway capacity, distance to travel, and highway congestion.

Figure 12–7 shows the mathematical equation used for total evacuation time.

Since almost all evacuees travel by private car, highway capacity is most important. Surveys during test evacuations have shown 1,100 to 4,080 cars per lane per hour. Four persons per car at 2,600 cars per lane would move 10,000 per lane per hour.[27] This figure can be expounded upon and is a result of numerous factors, such as population density and roadway capacity scores as outlined in an American Highway Users Alliance study titled *Emergency Evacuation Report Card 2006*[28] Test evacuations under ideal conditions do not necessarily mirror real life. This was the case in 2005, as Houston, Texas was evacuated. According to a Fox News article on Hurricane Rita, during the Rita evacuation, these preparations and their execution were overwhelmed by the enormous and unprecedented numbers of people fleeing from the Houston area prior to the local residents. By the time Jefferson County began its mandatory evacuation, local roads were already full of Houstonians. Designated evacuation routes slowed to a pace far worse than with any previous hurricane.

By late Thursday morning, the contraflow lanes had been ordered after it was determined that the state's highway system had become gridlocked. The Texas Department of Transportation was unprepared to execute such a large-scale evacuation. Coordination and implementation of the contraflow plan took 10 to 12 hours, as inbound traffic was forced to exit and police were stationed to assist with traffic flow. Evacuees fought traffic Wednesday afternoon through midday Friday, moving only a fraction of the normal distance expected. Many motorists ran out of gas or experienced breakdowns in the record-breaking temperatures that neared 100°F (38°C). Traffic volumes did not ease for almost 48 hours, as nearly 3 million residents evacuated the area in advance of the storm. This was the largest evacuation in Texas history.[29]

Converging roads, traffic lights, barriers, and disabled vehicles will all cause long evacuation times, and all such factors must be overcome.

It is important to remember that gas station owners jump in their cars and evacuate too, so provisions must be made for fuel. Certain stations in selected locations can be kept open, not only for the public but also for emergency vehicles.

Reentry

The longer the evacuation lasts, the more tension builds among the evacuees. Keeping the area sealed off becomes a growing police problem, especially if no danger is perceived. People have all kinds of urgent reasons for reentry, from getting their medications to feeding their pets that were left behind.

Evacuation Time Calculation

$$T_d + T_n + T_m + T_t = \text{Total Evacuation Time}$$

T_d = Time delay after occurrence until evacuation order

T_n = Time required to notify occupants

T_m = Time required for people to mobilize and start movement

T_t = Travel time required to leave affected area

Fig. 12–7. Total evacuation time equation. Failure to add all of the component times will result in an evacuation that is extended beyond the time planned for and could result in the evacuees being caught in the event.

Where to?

In past events, people have generally shifted for themselves. At Mississauga, less than 1 in 20 stayed in mass care centers. Of the others, 84% went with friends or relatives, and 12% arranged their own accommodations.[30]

Of 223,000 evacuees, between 30,000 and 40,000 registered at reception centers, but there were never more than 3,000 in all, and no more than 1,500 ever stayed overnight.

This was the exact opposite in the Hurricane Katrina evacuation. During this evacuation, most of those who fled the area, at the insistence of the government, stayed in shelters, such as the Houston Astrodome. The associate author stayed at the Astrodome in a staging position for deployment to the aftermath of Hurricane Rita. The Astrodome was nearly destroyed and the conditions were deplorable from the evacuees who stayed in the arena, just days before the arrival of FEMA teams. It was evident that those staying in the center did not bring appropriate care items during their evacuation, and that numerous evacuees were in the arena, rather than making their own arrangements.

How are shelters selected? They should be far enough away to eliminate any chance of secondary evacuation. In hazmat incidents, wind direction is almost always a factor. With the active hurricane season of 2005, this lesson was reiterated when evacuees from Hurricane Katrina needed to be moved in order to complete evacuations for Hurricane Rita. Although no one could predict hurricanes and their landfall locations, shelters should not be in the path of another potential disaster.

Schools are best for shelters, especially high schools. They have adequate bathrooms for a large population, a cafeteria, a water supply, separate classrooms for more privacy, a large hall, and an equipped nurse's room. In recent years, vacant warehouses and sports arenas have been utilized as shelters. With minimal construction, showers and restroom facilities can be erected inside both of these facilities. The Army Corps of Engineers can help in this task.

The Red Cross and Salvation Army have experience running shelters. People are registered, even if they do not remain. This helps to account for people's whereabouts and will aid in planning for showers, food, and sanitary facilities.

Pets can become a problem at a shelter and at motels or hotels. Most people seem to take their dogs along. Instructions regarding pets should be given at time of evacuation. Humane societies should assist, since in large-scale disasters, the number of pets will be in the thousands. Disaster Veterinary Assistance Teams are available for deployment by the Department of Health and Human Services. Figures 12–8 and 12–9 show the arenas that have been used as shelters. Note the large number of people in one area. Security will be needed to tend to the differences in evacuees that will become apparent as they interact in such tight quarters.

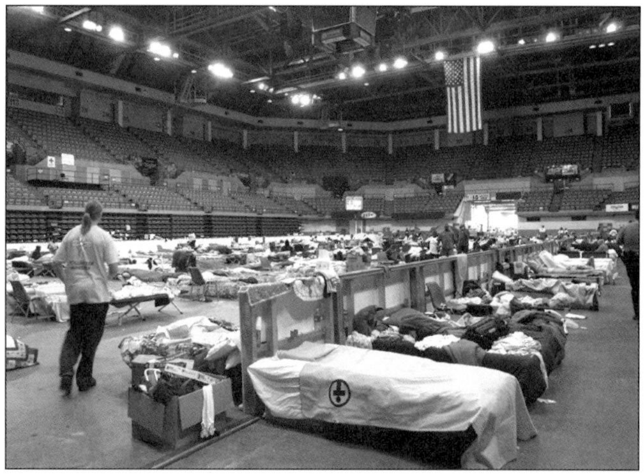

Figs. 12–8 and 12–9. Evacuation centers. Arenas are utilized for evacuees due to the large numbers they are capable of holding.

How much?

There is very little solid data on the detailed costs of past evacuations. Total approximations for an incident are sometimes stated, but breakdowns into component costs such as evacuation are not. Even if all the data could be collected on a specific past event, its applicability to a future event may be marginal.

Cost of evacuation is not considered where danger to life is obvious. Where the effects of a hazmat accident on the populace may be minute, the cost of evacuation as opposed to seeking shelter may act as a constraint on ordering the evacuation.

Costs will involve the evacuees, financial loss in farm areas, and financial loss in urban and industrial areas. The cost components are listed here:

- **Evacuee costs.** Food and housing, commercial or mass care; transportation—private, commercial, or government provided; loss of income; loss to unattended property.
- **Evacuator costs.** Food, housing, transportation; salary; extra personnel or overtime; extra or special equipment.
- **Farm loss.** Loss of finished product; harm to unattended animals; possible loss of crops.
- **Urban and industrial loss.** Value of product; damage to unattended plant; closing down and reopening costs; loss to suppliers or purchasers.

Using cost factors applicable to the area, a planner can make estimates of approximate monetary costs for evacuation of specific sites.

In summary, the cost of evacuation must be balanced against the potential cost of not evacuating.

CONCLUSION

When dealing with disasters, responders invariably deal with large numbers of persons. This has the capability of overwhelming the emergency services if planning and preparations are not completed in advance. These preparations must consider how to handle both the large number of casualties (dead and injured), known as *mass casualties*, and in handling the large number of persons who must be removed from a danger area, known as *mass evacuees*. The information and guidelines presented in this chapter could well mean the difference between order and chaos when a community is suddenly stuck with a calamity affecting large numbers of its populace.

REFERENCES

1. Collier, W. Steve. 1988. Mass casualties. Unit 12 in *Disaster and Fire Defense Planning*, Open Learning Course. Emittsburg, MD: National Fire Academy.
2. Ibid. p. 12-2.
3. *Washington Times.* 2008. Ferry travel prone to accidents. July 3. Retrieved July 17, 2008 from: http://www.washingtontimes.com/news/2008/jul/03/ferry-travel-prone-to-accidents/
4. Klaene, B., D. Kuhn, and W. Brewster. *Operations Manual.* Section 204.02. Cincinnati Fire Division. p. 7.
5. Ibid. p. 8.
6. Ibid. p. 3.
7. Ibid. p. 10.
8. Mahoney, E. L. 1986. *Disaster Medical Assistance Team Organization Guide.* Report NMDS-86. Manpower Task Force, National Disaster Medical System, U.S. Department of Health and Human Services, Public Health Service, Health

Resources and Services Administration, Bureau of Resources Development. July 25.

9. Klaene, B., D. Kuhn, and W. Brewster. Section 204.04. pp. 1–5.

10. Burkell, C. J., and H. Wood. 1999. Make the right call. *Fire Chief.* March 1. Retrieved April 15, 2007 from: http://firechief.com/mag/firefighting_right_call/index.html.

11. U.S. Department of Health and Human Services. 2007. National Disaster Medical System (NDMS). Retrieved April 15, 2007 from: http://www.hhs.fema.gov/index.html.

12. U.S. Department of Health and Human Services. Disaster Medical Assistance Teams (DMAT). Retrieved September 22, 2008 from: http://www.hhs.gov/aspr/opeo/ndms/teams/dmat.html.

13. U.S. Department of Health and Human Services. Disaster Mortuary Operational Response Teams (DMORTs). Retrieved September 22, 2008 from: http://www.hhs.gov/aspr/opeo/ndms/teams/dmort.html.

14. *CNN.com.* 2007. Three die in hell of Houston office building fire. March 28. Retrieved April 15, 2007 from: http://www.cnn.com/2007/US/03/28/fatal.fire/index.html.

15. CNN (September 18, 2008). Texans evacuate coast ahead of Ike. Retrieved November 9, 2008 from: http://www.cnn.com/2008/US/weather/09/10/hurricane.ike/index.html?iref=newssearch.

16. Federal Emergency Management Agency. *Evacuate!* HV 554.E92. Washington, DC: Federal Emergency Management Agency. p. I.

17. Weber, J. 1989. Mass evacuation. In *Disaster and Fire Defense Planning.* Emmitsburg, MD: National Fire Academy. pp. 13-1–13-18.

18. National Response Team. 2001. *Hazardous Materials Emergency Planning Guide.* NRT-1. July. Washington, DC: National Response Team. Retrieved September 23, 2008 from: http://www.epa.gov/oem/docs/chem/cleanNRT10_12_distiller_complete.pdf.

19. Ibid. pp. 58–59.

20. Perry, R. W. 1981. *Citizen Evacuation in Response to Nuclear and Nonnuclear Threats.* Washington, DC: Federal Emergency Management Agency. p. 77.

21. *Firehouse.* 1990. Fire wire. 15 (3). March. p. 67.

22. Perry, R. W. 1981. p. 52.

23. *CNN.com.* 2005. Rita scares us to death. September 21. Retrieved April 16, 2007 from http://www.cnn.com/2005/US/09/21/katrina.recovery/index.html.

24. Perry, R. W. 1981. p. 64.

25. U.S. Environmental Protection Agency. 1980. *Manual of Protective Guides and Protective Actions for Nuclear Incidents.* Washington, DC: Environmental Protection Agency. p. 1, 31.

26. Hans, J. M., Jr., and T. C. Sell. 1974. *Evacuation Risks—An Evaluation.* Las Vegas, NV: Environmental Protection Agency. p. 41.

27. Ibid. p. 42.

28. American Highway Users Alliance (2006). Emergency evacuation report card 2006. Retrieved November 9, 2008 from: http://www.highways.org/pdfs/evacuation_report_card2006.pdf.

29. Fox News. 2005. Gridlock in Houston as Texans flee. September 22. Retrieved July 17, 2008 from http://www.foxnews.com/story/0,2933,170145,00.html.

30. Sorenson, J. H., B. M. Vogt, and D. S. Mileti. 1987. *Evacuation: An Assessment of Planning and Research.* ORNL-6376. Washington, DC: Environmental Protection Agency. pp. 210–215.

5 Natural Disasters and Recovery

WEATHER-RELATED NATURAL DISASTERS 13

INTRODUCTION

Not every earthquake, tornado, hurricane, typhoon, cyclone, or flood is necessarily a natural disaster. The disaster aspects of these acts of nature occur only when they result in loss of life, injury, or property damage, or they impact the social structure.

This observation sets the tone for considering the natural disasters that are presented in this chapter. Even though all these spectacular displays of nature will not be catastrophic, the fire service, to fulfill its role of responsibility, must plan and be prepared for response to deadly natural disasters.

This chapter deals with the description and impact of tornadoes, hurricanes, typhoons, cyclones, floods, and other weather-related natural disasters, as well as the role of the emergency services in response to problems that may result. Earthquakes, volcanoes, landslides, and other "terrain-based" natural disasters are considered in chapter 14.

"Severe lightning, heavy rains, destructive winds, and heavy hail are often 'advance men' for one of the most violent storms on earth––the tornado."[1] This discussion begins with lightning and thunderstorms, proceeds to tornadoes, hurricanes, and winter storms, and concludes with floods.

LIGHTNING AND THUNDERSTORMS

Severe local storms often begin as thunderstorms. The sound of thunder is caused by the violent expansion of the air after it has been heated by lightning. By counting the time between the flash of lightning and the sound of the thunder, it is possible to tell how far away the storm is. Since sound travels about 1,100 feet per second, if it is five seconds between the flash and the sound of thunder, the storm is about a mile away.

Lightning is an electrical discharge in the sky of tremendous current—2,000 times greater than ordinary house current, which is generally less than 100 amperes. A potential difference of 100 million volts may exist between a cloud and the ground, and under certain conditions, may cause a bolt of lightning. According to the

National Weather Service (NWS), every year about 22 million strikes to ground objects occur in the United States. However, the chances of personally being struck by lightning are still 1 in 600,000.[2] Most of these will at least injure a person.

A lightning bolt is classified as *hot* or *cold* according to whether its flow of current lasts long enough to start a fire in combustible materials. About one-third of lightning strikes are hot. Because of these hot lightning bolts, the NWS reports that it is estimated that 307,000 insurance claims are filed each year due to lightning, causing $330 million worth of damage annually.

A 50-mile-wide swath of land that cuts across Florida from the Gulf Coast near Tampa to Daytona Beach on the Atlantic Ocean is called "Lightning Alley." From May through September, thunderstorms make Florida the worst state in the nation for lightning casualties. In one three-year period, more than 112 people were killed or injured in this area.[3,4]

Thunderstorms include heat energy converted into wind, electrical discharge, and violent upward motion of the air. Severe local storms may develop from heating of the ground or colder air above warm, moist, less dense air.

Storms are generally most severe when they occur along squall lines generated when cold, dense air and warm, moist air clash along borders called *fronts*. The greater the contrast in temperature and moisture across the front, the greater the energy content, and the more violent the thunderstorm. The severity of the storm can be judged by the amount of lightning, thunder, hail, and heavy rains. The more lightning, the more intense the storm; and the closer it is, the louder the thunder will be. Hail is precipitation in the form of lumps of ice, called *hailstones*. The larger they are, the more intense the thunderstorm is likely to be. Heavy rains accompany thunderstorms at their maturity.

Effects of thunderstorms

Lightning in many of its forms—forked, chain, or zigzag—is capable of great destruction. Not only are forest fires kindled, power generating stations knocked out, and buildings destroyed, but in the United States alone, lightning kills one person and injures four others every day. The NWS statistics report that one death occurs for every 345,000 flashes and one injury occurs for every 114,000 flashes.[5] Sheet lightning is light from a flash of chain lightning that takes place beyond the horizon. Heat lightning is the same, but fainter. Ball lightning has been witnessed by many observers, but does not appear to be dangerous.[6]

Hail can range from the size of peas to baseballs; the largest hailstone on record in the United States as of 2008 was nearly 19 inches in circumference.[7] Hailstorms occur most frequently near Cheyenne, Wyoming and Denver, Colorado, generally in the spring, but they are widespread in occurrence and have caused many injuries to people and animals. Severe hailstorms damage crops and greenhouses and have also produced extensive damage to vehicles and aircraft.[8]

Some transportation disasters have been attributed to severe storms. Lightning has struck aircraft not equipped with fuel tank automatic explosion prevention systems, with disastrous results. A Boeing 727 jet with 123 persons on board crashed during a thunderstorm on June 24, 1975, into sandy marshland, just short of the runway at the John F. Kennedy International Airport in New York. At least 109 persons were killed when the aircraft hit a tower and then burst into flames while descending for a runway approach. It was one of the worst domestic airplane disasters in U.S. aviation history.[9]

As reported by the Associated Press on May 20, 2007, thunderstorms that continue on the same path for numerous days can accumulate to a disaster status. During the previous week, the state of Texas saw an unprecedented number of thunderstorms, leaving five people dead.[10]

Lightning protection

Lightning protection systems can be installed, not to prevent lightning from striking, but to provide a means whereby its discharge may enter or leave the ground without passing through a nonconducting part of a building or structure.[11] Planners and municipal executives should consider the installation of lightning protection on certain key municipal infrastructures, such as communications and public safety buildings. *NFPA 1221: Standard for the Installation, Maintenance, and Use of Emergency Communications Systems* recommends their installation on all communications centers. Although not specifically recommended by code, other buildings such as firehouses and police stations should be equipped with lightning protection.

To safeguard aircraft against lightning and to help solve the problem of thunderstorm interference with aircraft communications, protection systems have been developed for antenna systems and radomes, as well as fuel systems and structural elements. Special high-energy artificial lightning discharge facilities have been set up to reproduce typical recorded flight damage and to test various methods of damage control. Little can be done to prevent lightning from striking an aircraft, but a graded resistance lightning diverter rod can be used for controlling somewhat the specific localized point at which lightning might strike.[12]

Lightning protection has become affordable for private residences and has become a popular option in the construction of newer homes, especially those located on hilltops. There are also associated nonprofit organizations, such as the Lightning Protection Institute, that offer helpful information. The Lightning Protection Institute, found on the Internet at www.lightning.org, is "dedicated to ensuring that today's lightning protection systems are the best possible quality—in design, materials, and installation."[13] This institute also has information on locating a reputable installer, discounts on homeowner's insurance, and research on lightning and protection against its destruction. The associate author's fire district saw a rash of lightning-related fires in the late 1990s. This has been reduced in recent years due to the common installation of lightning protection systems on a good portion of the homes in the area.

Lightning—how to avoid getting killed

If it is necessary to go outside in a thunderstorm, it is important to keep several precautions in mind. First, the inside of an automobile or fire truck with a metal top is a fairly safe location from lightning. Since the *NFPA 1500: Standard on Fire Department Occupational Safety and Health Program* requires all fire trucks to be equipped with enclosed crew cabs, firefighters aboard such apparatus will be relatively safe. However, many emergency service calls originate during large thunderstorms, making working within the lightning a near necessity. A risk-benefit analysis must occur on behalf of the incident commander to determine if the weather conditions permit scene operations.

The following guidelines should be considered when lightning occurs:

- If there is a choice of shelter, choose in the following order: large metal or metal-framed buildings; buildings that are protected against lightning; large unprotected buildings; small unprotected buildings.
- If outdoors, avoid the following: small sheds and shelters in an exposed location; isolated trees; wire fences; hilltops and wide open spaces.
- If on a lake in a small boat, get to shore as soon as possible; if on an ocean or lake in a ship, get inside and amidships as much as possible.
- If indoors, avoid doorways and open windows. Do not touch plumbing fixtures, electric wiring, telephones, radios (except when battery operated), and television sets. If lightning strikes the power lines leading to these appliances, they could become highly energized.

Emergency services personnel may encounter someone who has been struck by lightning, who requires medical care. If through past training, a responder has not received instructions on the care of an electrocuted victim, there are some simple steps to administering care.

If someone in the responder's presence is injured and made unconscious by lightning, it is important first to make certain that the victim is not contacting any electrical wiring or fixtures that may be retaining a residual charge. If such contact is being made, the responder should await the local power company's arrival if there is not a means of ensuring the power is not deenergized. If the objects are deenergized, a dry pole or other nonconductive object can be used to free the victim from the potentially energized object or material. It is important to be careful and quick, as electrical power could return at any time.

As soon as this has been accomplished, the responder should examine the victim to see if respiration is evident; if not, cardiopulmonary resuscitation (CPR) should be applied. The victim's clothing should be loosened to examine the skin for indication of burns. The responder should try to get medical assistance immediately, especially for those vital first five minutes when the victim's breathing may be stopped.

WIND STORMS

"Winds are by nature capricious. For example, even those of low velocity, the simple squalls and gales, have been responsible for eighty percent of all wind damage incidents and thirty-eight percent of the dollar loss reported at properties insured by Factory Mutual. But destruction by the more forceful, less frequent hurricanes and tornadoes has averaged about sixty-two percent of the total dollar loss."[14]

Buildings are destroyed and seriously damaged by high winds because it is not economically feasible to design average structures to withstand all possible wind pressures to which they may be subjected. Winds striking buildings are deflected upward and across the roof. The uplift forces are greatest at the eaves on the windward side. If several windows are broken on the windward side, or if a large door is open, additional uplift force is added inside the building to that already being exerted above the roof and alongside the building.

Wind damage prevention

Much wind damage, even near the path of tornadoes, can be prevented by good design, preventative maintenance, and emergency planning. Main roof structures should be designed to withstand wind uplift forces, and new roof coverings should be securely anchored to roof decks. Cranes and moving bridges should have brakes, clamps, and special devices to resist wind forces. Preventative maintenance should include the repair of loose window panes and frames, repair of broken or loose flashing, shingles, and roof covering, and tightening of slack guy-wires on stacks, antennas, and ventilators. It also includes strengthening of supports for exterior wiring and piping.

Due to the frequency of wind-related damage, the Federal Emergency Management Agency has developed a checklist for homeowners to prevent wind damage to their homes. The list is replicated in figure 13–1. The list includes the following measures and questions that homeowners should investigate:

- Is the roof sheathing properly installed?
- Are end gables securely attached to the rest of the roof?
- Are double entry doors secured at the top and the bottom?
- Has the garage door been properly secured?
- Are windows protected by storm shutters?
- Is the roof fastened to the walls with hurricane straps?[15]

Home & Business Owner High-Wind Preparation Checklist

☐ **Maintain Exterior Insulation Finishing System (EIFS) Walls**
- Inspect for weakened areas around doors, windows, and where they intersect with the roof
- Ensure fastening systems are rated for high winds

☐ **Protect Windows and Doors**
- Install rolldown shutters
- Replace existing glass with impact-resistant glass

☐ **Install Shutters or Plywood Covers**
- Precut the plywood covers prior to storm preparations to speed the work
- Letter/Number the cut plywood to correspond with windows/doors.
- Keep the screws with the plywood

☐ **Reinforce Double Entry Doors**
- Install slide bolt on the top and bottom of the inactive (base) door
- Install a deadbolt in the middle of the active and inactive door to secure together

☐ **Reinforce or Replace Garage Doors**
- Strengthen the glider wheel tracks
- Install horizontal girts on the backside (inside) of the garage door

☐ **Remove Trees and Potential Windborne Missiles**
- Clear trees a distance around the structure equal to the tree's height
- Anchor outbuildings, lawn furniture, trashcans, and other objects that can strike the house

☐ **Secure Metal Siding and Metal Roofs**
- Ensure sections of siding are properly fastened to the frame of the building
- Cover corners with molding to prevent wind from lifting corners/edges of siding/roof

☐ **Secure Built-Up and Single-Ply Roofs**
- Ensure flashings are reinforced with screws, concrete spikes, or a continuous cleat
- Install a continuous bar over single membrane roofs to strengthen the attachment
- Check with local building official to ensure roof is strengthened according to applicable codes

☐ **Secure Composition Shingle Roofs**
- When replacing roofs, ensure that repairs meet the following specifications
  - Each shingle is held with at least 6 nails or staples,
  - A waterproof underlayment should be installed beneath the shingles
  - The roof sheeting should be at least 15/32 thick

☐ **Brace Gable End Roof Framing**
- Consult a local building official to determine if, and how bracing should be added

For further information on natural hazard preparation for your home or business, please visit the Federal Emergency Management Agency's website at www.fema.gov/plan/prevent/howto/index.shtm

Fig. 13–1. FEMA's high-wind preparation checklist

Emergency planners should ensure distribution and access to this document prior to seasons in which wind damage is most prevalent. The placement of the document on the fire department or emergency service's Web site will allow homeowners access to the document from the convenience of their homes. In addition, any public relations events that occur during the season(s) before the windy season should focus on getting the word out to the residents. Progressive emergency services departments realize the benefit to the community that these preventative measures provide.

Emergency planning should cover the actions to take when windstorm warnings are received, procedures to follow during the danger period, and measures to clean up and restore any damaged equipment or property after the storm passes.

SANDSTORMS

High winds sometimes blow sand that can wreak havoc on structures and vehicles and can bury highways, making them impassable.

A sandstorm may result from heat of the sun on deserts or broad plains, causing heated air to rise and become quickly replaced by cooler air. The resulting winds may be strong and constant or intermittent and gusty. In the southwestern United States, sandstorms may result from the strong winds and turbulence that sometimes accompany a passing cold front. According the *Federal Meteorological Handbook No. 1*, produced by the National Weather Service, sandstorms are "particles of sand carried aloft by a strong wind. The sand particles are mostly confined to the lowest ten feet, and rarely rise more than fifty feet above the ground."[16]

Although sandstorms are common in the western portion of the United States, they are worse in China and Africa. In 2004,

> China has experienced 17 sand drifting or flowing weather systems including one strong sandstorm, five average ones and 11 drifting or flowing sand phenomena, said Zhang.
>
> A sandstorm between March 26–30, the worst in this spring, delayed more than 1,200 flights including 130 flights in Beijing Capital International Airport with 20 others forced to land at airports in Tianjin, Zhengzhou, Taiyuan and Dalian.
>
> Another dust storm between March 9 to 11, the largest this spring, swept more than 19 provinces, autonomous regions and major cities in North China and areas along the middle and lower reaches of the Yangtze River.
>
> Reviewing the frequency, feature, intensity and impact of sandstorms in China experienced this year, Zhang said, more such disasters occurred than that in the same period of last year.[17]

On the highways, the sand reduced the visibility to zero at times, and some disoriented motorists wound up driving off the road. Wind-warning signs had been posted along the highways, following advisories issued by the National Weather Service, but they were largely disregarded by vacationers with campers, motor homes, and cars pulling boats and travel trailers. The travelers apparently thought that the storm would soon blow over.

CYCLONES/TYPHOONS

Closely akin to a tornado is a cyclone, which is a storm or series of winds that rotate about a center of low atmospheric pressure, clockwise in the Southern Hemisphere and counterclockwise in the Northern Hemisphere. It advances at a speed of about 20 to 30 miles per hour and often brings abundant rain. When a cyclone occurs in the region of the Philippines or the China Sea, it is called a *typhoon*.

The term *typhoon* refers more to the origin of a storm than to its composition. Essentially it refers to storms that originate at sea. It has been used to refer to hurricanes in some parts of the world, such as in Bangladesh and India, and to tornadoes that originate at sea. It is used almost universally in this latter context. Suffice it to say that all of the destructive forces of a typhoon will be covered in the discussion of hurricanes, tornadoes, and floods.

Cyclones

A cyclone is a low-pressure area in the atmosphere in which the winds spiral inward. Cyclones may be just a few hundred miles across or they may cover areas larger than the entire United States. In the Northern Hemisphere, cyclonic winds blow counterclockwise, and in the Southern Hemisphere, they blow clockwise. Depending upon atmospheric pressure, small cyclones may move rapidly or slowly, but large ones usually move slowly, perhaps a few hundred miles per day.

A tropical cyclone occurs over the warm ocean waters in the tropics. If they form in the West Indies, they are called *hurricanes*; if they form in the Pacific Ocean, they are called *typhoons*. These storms, which may bring winds from 150 to 180 miles per hour, are called *baguios* in the Philippine Islands area.

These storms are said to be formed by the rotational effect of the Earth and the differential heating of the Sun on large masses of air.[18] Information concerning cyclones and hurricanes can be found via the Internet at the National Weather Service's National Hurricane Center located at http://www.nhc.noaa.gov/index.shtml.

TORNADOES

On April 26, 1991, at least 48 tornadoes swept Kansas and Oklahoma, killing 23 people in all. Characteristically, mobile home parks were hit hardest and suffered the most casualties. About 1,500 Andover, Kansas residents were left homeless by one tornado, the first to strike the town since 1958. This is a story that has been repeated, one place or another, every spring in the United States, and will likely occur somewhere every spring in the future. The places and death tolls may change, but tornadoes are a somber reality.[19] In 2007, Kansas was again hit with what the National Weather Service referred to as a *wedge tornado*, which covered a ¾-mile-wide path that destroyed 90% to 95% of Greensburg, Kansas. Estimates of deaths are only nine at this point, but are expected to rise as the search and rescue efforts continue. The town had a population of 1,400.[20]

No region in the world suffers so many tornadoes as the continental United States, which averages 800 per year. The areas of highest vulnerability are in those states that lie in the eastern plains and the Mississippi and Ohio river valleys. Even though some areas of the country may be relatively free from the tornado threat, tornadoes do occur in all 50 states.

Tornado season begins in February and continues into summer. About one-half occur during the months of April, May, and June. Most often, they happen in the afternoon, and the direction of travel is from southwest to the northeast. In 2005, 818 severe thunderstorm and tornado warnings were issued by the NWS.[21]

Based on this information, those communities in the areas listed would be wise to include tornadoes in their vulnerability analysis described in chapter 1 and to include the event in the exercises listed in the training chapter.

A tornado is often called the most violent of all windstorms. It is a rotating column of air pendant from a thunderstorm cloud and touching the ground. If it occurs over an ocean or lake, it is called a *waterspout*. A waterspout is not a column of water rising out of the body of water, but a whirling column of air and watery mist. Some sea or lake water may be drawn up at the base of the column, but even in ocean waterspouts, most of the water is fresh water formed by condensation.[22]

A tornado, sometimes called a *twister*, nearly always starts as a funnel-shaped attachment from the cumulonimbus clouds overhead and then builds downward to the ground. When it picks up earth and debris, it gets dark in color. Sometimes two or more funnels may form from the same thunderstorm clouds at the same time. As these clouds move along, tornadoes may form, travel along in touch with the ground for a few miles, dissipate or lift, and then touch down again––several times.

Nature of tornadoes

Tornadoes almost always form in conjunction with a thunderstorm. Heavy hail may also accompany these storms. A single thunderstorm may generate numerous tornadoes. On April 3 and April 4, 1974, 148 tornadoes developed in 11 midwestern states and in Canada. This was known as the "Super Outbreak." This outbreak was reflected in the severe rise in the death toll by year, as reported by the National Weather Service.

Author William Kramer was a firefighter on Cincinnati's Engine Co. 3 the day of the "Super Outbreak." With strong wind in his face, he remembers an eerie ride in the aisle of an open-bed pumper 20 miles to the western edge of Cincinnati. A five-alarm response was en route to the suburb of Saylor Park, where a tornado had just touched down, causing massive destruction. He and 70 other Cincinnati firefighters worked through the night helping victims and securing damaged homes.

Tornadoes are categorized from Class 1 to Class 5, with Class 5 being the most severe. A Class 1 tornado has wind speeds less than 100 miles per hour, a path width of less than 100 yards, and may stay on the ground for only 100 yards at a time. According to the NOAA, 69% of tornadoes are categorized as weak tornadoes, which have winds less than 110 miles per hour and account for about 3% of the deaths.[23]

A Class 5 tornado may have a single large vortex or multiple vortices rotating about one another. Wind speeds may reach 300 miles per hour, cutting a path up to a mile wide. These systems may stay on the ground up to an hour. According to the NOAA, "Strong tornadoes account for 29 percent of all tornadoes and approximately 27 percent of all tornado deaths. Wind speeds range from 110 to 205 mph, and these tornadoes may last 20 minutes or longer."[24] Figure 13–2 shows the destruction left behind from an F5 tornado, such as the one in Greensburg, Kansas.

Fig. 13–2. Damage from Greensburg, KS tornado. Note the large area affected.

FEMA News Photo.

According to the NOAA, "Violent tornadoes account for only 2 percent of all tornadoes but result in 70 percent of all tornado deaths. Wind speeds exceed 205 mph, and lifetimes can exceed 1 hour."[25]

The width of the touchdown path averages about 300 to 400 yards, and the length about 4 miles, though it may go as far as 300 miles. In 1917, a tornado lasted more than seven hours and traveled across Illinois and Indiana for a distance of 293 miles. The average speed that a tornado travels is from 25 to 40 miles per hour, and its sound has been described as a roaring, rushing noise, like a jet plane in flight or a train speeding through a tunnel. The horizontal wind speed in the center of a tornado may be more than 300 miles per hour, and its destructive power is due to the combined action of its strong rotary winds and the partial vacuum in the center of its vortex. As the "eye" of the tornado passes over a building, pressure builds up inside, the walls topple outward, windows explode, and debris of this destruction is driven through the air in a dangerous barrage. Heavy objects like railroad cars and machinery have been lifted and carried for considerable distances. Winds of destructive force extend for several hundred feet outward on both sides of the vortex.

What are the odds of getting hit by a tornado?

The fact that a certain town has never had a tornado does not mean that the possibility can be disregarded. Every year, places that have never been stuck before join the list of tornado disaster locations. The disaster potential is greatest where the most tornadoes hit densely populated areas. The greatest tornado threat is in Massachusetts, though the greatest number of tornadoes per 10,000 square miles occurs in Oklahoma. Texas has had more tornado-caused deaths than any other state.

The mathematical chance that a specific location will be struck by a tornado in any one year is quite small, but they have occurred with exceptional frequency at particular locations. For example, Oklahoma City has been struck 26 times since 1892. Baldwyn, Mississippi was struck twice in 25 minutes on March 16, 1942. Some cities, such as Irving, Kansas and Austin, Texas, have been struck twice in rapid succession. Codell, Kansas was struck on May 20, 1916, again on May 20, 1917, and a third time on May 20, 1918! Certain towns, such as Xenia, Ohio have experienced two tornadoes that wiped out opposite sides of the town. The first occurred on April 3, 1974. The second occurred on September 20, 2000. This will be covered in depth in the case study section of the chapter.

Examples of tornado disasters

As this chapter was being prepared, on April 26, 2007, storms with high winds rushed across three states, spawning tornadoes in Tennessee and Illinois and tossing a patrol car over a three-foot-high fence in Indiana. At least 7 people were injured. Just two days earlier, 10 people were killed when a tornado struck the U.S./Mexico border. The storm took out numerous homes and two schools.[26]

This was a far better result than the tornado that struck Enterprise, Alabama in March 2007. In the Enterprise, Alabama tornado outbreak, the Enterprise high school took a direct hit during the school day. A search and rescue effort was put into place to search for more than 1,300 schoolchildren. This search and rescue effort was conducted by local and state resources, as the lead time to receive federal search and rescue forces would have been too long to wait. The town of 23,000 was destroyed by the tornado. This outbreak of severe weather came just a month after tornadoes struck a 30-mile area in Florida, killing 21 people.[27]

In Huntsville, Alabama, a school was also struck by a tornado in November 1989. Although it was leveled, the timing was fortunate, as it hit after the school day was over, around 4:30 p.m. The twister roared through the town, killing 18 persons who ranged in age from 2 to 67 and demolishing 119 houses.[28]

These examples demonstrate two points. The first is that the dates of the disasters correspond to the National Weather Service's prediction of the times throughout the year that deadly tornadoes strike certain areas of the country. This should allow planners and response organizations to develop media and public relation opportunities in the community just prior to these periods. They can use these opportunities to educate and remind the citizens of the proper preparation, sheltering, and postdisaster instructions in order to lessen the possible death and injury rate in the community. Obviously, eliminating the tornado is not a feasible option.

The second point concerns school emergency preparedness plans. Response organizations such as the fire department must work collaboratively to ensure that these plans are in place, practiced, and understood by the response executives and incident commanders. Knowing the locations of shelters within the school buildings will save valuable time during initial search and rescue operations, thus yielding greater success in locating and rescuing victims.

There are other notable historic tornado disasters. An outbreak of 37 tornadoes on Palm Sunday, April 11, 1965, killed 257 people and injured more than 5,000 persons in the Midwest. In March 1832, a series of tornadoes in Alabama killed 268 persons and injured 1,874, with property damage of about $5 million.

The series of tornadoes that caused the greatest recorded loss of life occurred in portions of Missouri, Indiana, Kentucky, and Tennessee on the afternoon of March 18, 1925. Eight separate tornadoes were observed, one of which killed 689 persons, injured 1,890, and caused more than $16 million in property damage. An additional 41 lives were lost in the seven other tornadoes, along with added property damage.

The Xenia tornadoes—a case study of two tornadoes

Occasionally an entire city is devastated by a tornado, taxing the resources of the emergency services to the fullest and causing it to be integrated into its community to an extent never before realized. This happened in Xenia, Ohio at 4:41 p.m. on April 3, 1974, and again on September 20, 2000. This will serve as a case study showing how emergency services operations are conducted during the time of an unforeseen tornado of disastrous proportions.

The tornadoes that struck Xenia illustrate well the problems that can confront the emergency services. Xenia is a city of 27,000 people covering seven square miles. Its 36-man paid fire department operates out of two stations with an average of 10 persons on duty.

1974 tornado. For the 1974 tornado, the southwest area of Ohio had been under a tornado watch since about 1 p.m., and a tornado warning was put into effect around 3:15 p.m. At 4:40 p.m., a patrolling police officer reported that a tornado was approaching the city from the southwest, and it hit a minute later. It cut a path of total destruction about six miles long and three-fourths of a mile wide, leaving 33 persons dead and about 1,000 injured. In the 10 minutes that it took to sweep through the city, it hit the city hall, took the roof covering off one of the fire stations, seriously damaged the second, destroyed a high school, and wiped out (or damaged beyond repair) 1,347 buildings.[29,30]

Within 30 minutes, all off-duty personnel were back on duty, including five who had lost their homes; but the work of the firefighters was not without handicaps. Streets were filled with rubble, and a train that had been derailed by the tornado blocked all the major east-west streets. It was impossible to travel very far by car, so the off-duty firefighters started working in areas nearest their homes, joining any emergency unit operating in the area.

Telephone service had been knocked out, and the fire department's main and auxiliary antennas were destroyed, seriously cutting the range for use of its radio. Fire apparatus could not respond from the second fire station because cars, trees, pieces of buildings, and other debris blocked the ramp.

Fire apparatus and ambulances responded from nearby communities, without being called—since without telephone service or a radio link, there was no way to request help. Without compatible radio frequencies, there was no way to communicate with them after they arrived, making search efforts difficult to organize. According to Xenia's fire chief at the time, John Troeger, his biggest problem early in the disaster was getting incoming units to report to the disaster control center. In retrospect, he thought it might have been well to set up a second command post at the western end of the city during the early search and rescue operations, since that would have overcome the problem of the physical separation of the city caused by the derailed train.

Other problems included the fact that all power to the city was disrupted, and water pressure was low or absent due to 10 missing fire hydrants and broken pipes in about 800 buildings, including those supplying sprinkler systems. During the 32 hours it took to get the water system restored, neighboring communities supplied water tankers to help fill the need. To control natural gas leaking from the many broken pipes, it was necessary to close valves, which cut off gas to the whole community.

Ironically, the only deaths from fire during these disaster control operations were suffered by two National Guard soldiers who were trapped in their improvised basement shelter—a three-story furniture store that had become fully involved before they could escape.

The lessons learned from this tornado disaster led to some improvements. To improve the dependability of the communications system, an additional auxiliary radio antenna was added. In addition, three vehicles were equipped with 100-watt radios that could be used for a mobile base station. All fire department vehicles were equipped with four-channel radios to permit communications with all fire departments in the county and some in neighboring counties. A good disaster plan was adopted that incorporated flexibility and simplicity, and each operating division of the city developed its own operational details subject to the approval of the city manager. These plans were coordinated at countywide meetings of the emergency services chiefs, so that large-scale disaster drills could be conducted.

2000 tornado. Many of these improvements were needed and proved valuable during the 2000 tornado. Although this tornado was only an F3 and F4, it did kill one person and injured 100 others. After the 1974 tornado, town leaders decided to prepare for the next one. When this one hit, designated officials gathered in the basement of city hall, where portable generators and all the necessary supplies were available to set up a command center.

Fire Chief Price went first to a collapsed grocery store and began to sort things out. Wires were down and chaos reigned, he said. Within 10 minutes, he had a medic unit and engine on the scene. The 12 persons on duty at the two stations were out, and within 45 minutes, he had 40 of his department's 42 staff out on a general call back. The fire stations had been spared.

In the darkness, Price headed for the command center. There, the dispatch center for most of the fire departments in the county was putting calls out to 13 departments that started sending personnel and equipment. Nevertheless, they felt they were in need of more help, and issued a call to the Ohio Emergency Management Agency.

According to *Firehouse.com*,

> The Ohio Urban Search and Rescue (US&R) Task Force One had one of its first deployments that Wednesday night when it was called out to Xenia. Certified in April as one of 28 in the U.S., the unit is a highly trained and technically prepared unit of about 100 specialists from across the state, utilized in search and rescue situations.
>
> Dick Kimmins, spokesperson for the Ohio Emergency Management Agency, said the unit is self-contained with food and housing. Specialists include physicians, communication experts, heavy equipment operators and dog handlers used for victim searches.

The unit was formed by FEMA as part of the National Urban Search and Rescue (US&R) Response System.[31]

Many other jurisdictions have learned valuable lessons from the Xenia tornado. A situation where a good mutual aid plan and a countywide radio network paid off was described by the Berkshire County fire coordinator following a tornado strike in West Stockbridge, Massachusetts. The county's radio base station for the fire network had been knocked out by a thunderstorm a few minutes before the tornado hit, but portable radios were adequate to direct incoming mutual aid companies in search and rescue operations. Leaking propane gas from damaged cylinders infiltrated the piled-up cars, trucks, and structural debris. Also a gusher of fuel oil issued from a ruptured pipe from an underground tank. A "no smoking" warning had to be issued repeatedly to people in the area. Search parties had to be on the lookout for live electrical wires that might arc and ignite the leaking gas or give someone a deadly jolt.[32]

In the wake of this outbreak of tornadoes, including the Xenia tornado, municipalities and jurisdictions all across the United States installed or improved early warning systems, such as building-mounted sirens, and began testing them regularly. Senator Dennis McKinney, on the day after the Greenburg, Kansas tornado, stated that the deaths were substantially lower due to storm spotter training and advanced warning systems being in place at the time of the tornado.[33]

Tornado predictability

It is possible to predict the most probable times and places for tornadoes. The areas where tornadoes are likely to occur most often are well-known (Massachusetts, Kansas, Iowa, Texas, Oklahoma, Arkansas, Missouri, Mississippi, Ohio, and Alabama). The months of the year when at least one-half of them occur is a matter of record (April, May, and June). Also predictable is the time of day (afternoon) and the usual direction of movement (southwest to northeast). Even the type of thunderstorm clouds (cumulus and cumulonimbus), the weather conditions (heavy rain and hail), and the ominous sound (roar of planes) are long established.

The National Weather Service has provided its severe local storm forecast and warning service since 1953. Since then, the average number of deaths and injuries from tornadoes has decreased considerably. This is due in part to the help of the emergency services and individuals who watch for all the indications and make quick reports to the National Weather Service, plus the cooperation of the news media that make the information available to the public. A link to the slide presentation used in a storm spotter class is available on the Internet at http://www.wrh.noaa.gov/mso/train/STrain.html. The National Weather Service also produces a schedule of classes that are taught in areas nearby, typically in the spring, which can be accessed from http://www.weather.gov/.

Tornado preparedness planning

A tornado preparedness plan can be developed to fit any size of community or industrial plant, and such planning should not be limited to an area that has been struck before. Because tornadoes strike so suddenly and move on, it is important to have cooperative arrangements for detecting them as soon as possible and for warning those in the path. Only in this manner will people have time to get to shelter quickly. As soon as the tornado hits, planned rescue and damage control operations should be ready to start. The National Weather Service has regional offices that can be contacted, and the representatives will be glad to assist the local authorities in developing their community's tornado preparedness plan. A hazard-specific appendix sample for tornado response can be found on the FEMA Web site at http://www.fema.gov/pdf/plan/6-ch-h.pdf. This sample plan is a good starting point for any community that does not currently have a hazard appendix for a tornado.

The American Red Cross has a wealth of information available for the general public to develop plans for what to do when a tornado strikes their area. This information is found at http://www.redcross.org/services/disaster/0,1082,0_591_,00.html. Its table of contents is listed here:

- Prepare a Home Tornado Plan
- Assemble a Disaster Supplies Kit
- Stay Tuned for Storm Warnings
- What To Do When a Storm Watch Is Issued
- What To Do When a Storm Warning Is Issued
- After a Tornado Passes[34]

A major factor in preparedness is public education for the citizens of a community. They must be made aware of how advisory information is transmitted over radio and television. Many communities also utilize a special signal on horns, sirens, or whistles to alert the public. Construction of a safe room is an option that can be explored in high-risk areas.

Damage caused by tornadoes as they cut a path through a community can vary dramatically. An F1 tornado may overturn mobile homes and cause erratic and freakish damage to sign boards, store fronts, and power lines. These storms often form before warnings can be issued, but the resulting property damage and injury rate is generally low.

An F4 or F5 tornado, on the other end of the spectrum, can cause near total destruction to everything in its path, with the exception of only the sturdiest structures. Major damage to aboveground power lines and antenna towers typically occurs. Severe injuries and loss of life can be anticipated, even with an early warning. (See the NWS Web site for an in-depth explanation of the Enhanced Fujita Scale, which it adopted on February 1, 2007.)

Roadways and access routes within the touchdown area will be impassable to vehicular traffic due to fallen trees, power poles, and structural debris. Power and telephone communication will be disrupted. Water pressure may be extremely low or absent because of broken pipes throughout the area. Hazardous chemicals or flammable liquids may be released, posing special problems for both the public and emergency services.

The fire potential may be extreme following the devastating effects of a tornado. Possible fire causes include inoperative built-in protective systems (sprinklers and alarms), the drop in water pressure from broken pipes and mains, the possible delay in reporting, the delayed response, and lack of access. They all add fuel to the potential for multiple large-scale fires.

An in-depth hazard identification of a community should determine its potential for a tornado emergency and identify problems to consider.

The capability assessment should determine what resources are available and what additional resources are needed. A multiyear development plan should then be implemented to overcome capability shortfalls.

A fire department's tornado emergency planning should consider those special operational concepts (over and above standard operating procedures) that will provide effective command and control during a large-scale tornado emergency.

A tornado can be a devastating event, obliterating almost everything in its path. Loss of life, injuries, and property loss can be beyond comprehension within a community. When the fire department responds appropriately to the myriad of problems, the planning, the training, and the preparation will, to a large extent, influence the performance. At times like this, the real worth of an emergency operations plan is measured.

Emergency action—tornadoes

When a tornado approaches, a person's immediate action may mean the difference between life or death. Firefighters and civilians should be listening to radio and television stations for the latest advisory information. A *tornado watch* means that tornadoes are expected to develop, while a *tornado warning* means that a tornado has actually been sighted.

Emergency service personnel will be familiar with the advice radio and television networks broadcast to the public when a tornado watch has been issued. Whether they are at home or at the government facilities in the community, they will know what to do. The following is typical of the kind of suggestions that might be given the public:

- If you see a funnel cloud heading your way, or hear something that sounds like a freight train or a tank coming toward you, seek shelter, preferably in a tornado cellar, underground excavation, or a steel-framed or reinforced concrete building of substantial construction. Stay away from windows!
- If inside an office building, stand in an interior hallway on a lower floor, preferably in the basement. If in an industrial plant, join other workers in moving quickly to the designated location that offers the greatest protection. Stay there until the all-clear bulletin is received.
- In your home, the basement usually offers the greatest safety. Get under a heavy table or sturdy workbench, if possible. Where there is no basement, take cover under heavy furniture in the center part of the house. Keep some windows open, but stay away from them. Mobile homes are particularly vulnerable to overturning during strong winds; keep the tie-down cables' turnbuckles tightened. Get to the community shelter if one has been provided.
- Students should go to an interior hallway on the lowest floor of their school. They should get out of auditoriums and gymnasiums or other structures with wide, free-span roofs.
- In open country, move away from the tornado's path at a right angle. If there is not time to escape, lie flat in the nearest depression, such as a ditch or ravine.

This advice, though broadcast before a tornado may be formed, is intended to be acted upon after a tornado is formed and its progress is being tracked. If the situation is such that only a tornado watch has been set for the area, there may be time to activate the emergency operations center and make preparations in accordance with the tornado annex in the disaster plan.

Persons who live in an area that is susceptible to high winds could construct a safe room. FEMA has guidance on safe rooms and their construction at http://www.fema.gov/hazard/tornado/to_saferoom.shtm. The purpose of a safe room or a wind shelter is to provide a space in the home where people can seek refuge that provides a high level of protection. The safe room can be built in one of several places in the home. To protect its occupants, a safe room must be built to withstand high winds and flying debris, even if the rest of the residence is severely damaged or destroyed. According to FEMA,

Consider the following when building a safe room:

- The safe room must be adequately anchored to resist overturning and uplift.
- The walls, ceiling, and door of the shelter must withstand wind pressure and resist penetration by windborne objects and falling debris.
- The connections between all parts of the safe room must be strong enough to resist the wind.
- Sections of either interior or exterior residence walls that are used as walls of the safe room, must be separated from the structure of the residence so that damage to the residence will not cause damage to the safe room.[35]

A complete guide on the construction of the room is available to download at http://www.fema.gov/library/viewRecord.do?id=1536, or a hard copy can be requested from FEMA. Figure 13–3 shows a display at the National Emergency Management Institute that explains the construction of a safe room.

Fig. 13–3. A display explaining the benefits and construction of a safe room. In high-risk areas, these can become a good investment for a family in order to survive a high wind event if evacuation is not an option.

Photo courtesy of associate author Randall W. Hanifen.

Impact on the community

Response executives will want to make sure that their condition of readiness is such as to be able to operate in the absence of public electrical power, telephones, or normal water pressures. Their government facilities, especially fire and police stations, will be alerted to make everything secure against the force of strong winds. Off-duty personnel will know what the plan calls for in the matter of reporting to the nearest fire station or emergency services location. They could be needed in assisting with search and rescue operations, shutting off broken gas and water mains, reporting fallen power lines, and carrying out fire control and suppression activities.

Fire departments within the mutual aid areas, if not already alerted, will be notified of the possible need for assistance if a destructive tornado touches down in their community.

If a tornado hits, the officer in charge of each fire station and department facility should provide information to the emergency operations center by whatever means are available, stating damage sustained and any impairment of response capability. If station doors are obstructed by heavy debris, assistance should be obtained quickly to clear the ramp with chain saws, bulldozers, cranes, or other debris-removal equipment. If streets are obstructed by fallen walls, heaps of autos, or derailed trains, an aerial reconnaissance, if feasible, can help determine a bypass route for emergency equipment pending clearing of normal routes of travel.

Fires may break out, or there may be leaks and spills of dangerous chemicals, flammable or toxic liquids, and gases where access by fire apparatus is not readily possible. In these cases, it may be necessary to evacuate occupants of buildings in the zone of danger. If winds die down, it may be possible to fly in crews and lay lines with helicopters to handle emergencies in inaccessible locations. Measures suggested for chemical emergencies discussed in chapter 9 may be useful following tornado damage to heavy industrial occupancies, refineries, and chemical processing plants.

As soon as practical at the conclusion of emergency operations, a postdisaster inventory should be made of fire equipment lost or damaged, and steps taken to restore response readiness to a condition of normalcy. If a state of disaster was declared at either the state or federal level, this information will be needed to request reimbursement.

HURRICANES AND STORM SURGES

Hurricanes occur fairly frequently on average and have become a fearsome fact of life for coastal dwellers. A hurricane is a tropical cyclone originating in the West Indies. The name is from the Spanish *huracan*, probably derived from the Mayan storm god, Hunraken, the Quiche god of thunder and lightning, Hurkan, and numerous other Caribbean Indian terms for "evil spirit," "big wind," and the like. According to an early 1970s NOAA publication, hurricanes "are tropical children, the offspring of ocean and atmosphere, powered by heat from the sea, driven by the easterly trades and temperate westerlies, the high planetary winds and their own fierce energy. In their cloudy arms and around their tranquil core, winds blow with lethal velocity, the ocean develops an inundating surge, and as they move toward land, tornadoes now and then flutter down from the advancing wall of thunderclouds."[36]

The outer extremities of a hurricane can vary from 60 to 1,000 miles in diameter, and its center may move as much as 40 miles in a day, or it may remain fixed in one place for hours. On the average, the great spiral of a hurricane covers an area some 100 miles in diameter, with winds greater than 74 miles per hour. It spreads gale-force winds (winds over 40 miles per hour) over a 400-mile-diameter area. Their broad spiral base may dominate weather over an area of thousands of square miles. While the winds on the outer band of the storm may only be 30 miles per hour, within 30 miles of the center, winds may have velocities exceeding 150 miles per hour. Near the very center, where ascending air releases heat to drive the storm, winds may reach 200 miles per hour. Figure 13–4 shows the hurricane activity on a weather map during a recent hurricane season.

Hurricanes begin as tropical depressions. When the large counterclockwise circulating wind pattern increases in intensity to 74 miles per hour, a hurricane has developed. Whereas tornadoes have a normal duration of minutes or hours, the

Fig. 13–4. A weather map during a recent hurricane season

FEMA News Photo.

devastating forces of a hurricane may persist for days or weeks. According to the NOAA, "No other atmospheric disturbance combines duration, size and violence more destructively."[37]

The powerful winds of a hurricane may cause extensive collapse of structures and downing of trees and power poles. However, most damage from a hurricane results from flooding and the storm surge. From 6 to 12 inches of rain is typical, causing flooding in the low-lying coastal areas as well as inland. The storm surge, however, is the most lethal aspect, as the sea water level may rise 15 feet or more, along with crushing storm waves reaching heights of 50 feet or more.

In a hurricane, the winds flow toward the low pressure in the warm, comparatively calm core. Convection currents swirl the air upward around the eye of the storm, which on the average is around 14 miles in diameter, though 25 miles is not unusual. Winds in the eye may be as low as 15 miles per hour, but as the storm moves over a given location, the winds increase again at the opposite wall. However, they will be moving in the opposite direction because of the cyclonic circulation of the storm. Categories of hurricanes and their destructive effects are explained in table 13–1.

Table 13–1. Hurricane scale with wind speeds and the associated damage based on the intensity

Saffir-Simpson Hurricane Scale

Category	Wind Speed	Storm Surge	Expected Damage
1	74.95 mph (119·153 km/hr)	4·5 ft above normal	No real damage to buildings. Damage to unanchored mobile homes. Some damage to poorly constructed signs. Also some coastal flooding and minor pier damage.
2	96-110 mph (154-177 km/hr)	6-8 ft above normal	Some damage to building roofs, doors, and windows. Considerable damage to mobile homes. Flooding damages piers and small craft in unprotected moorings may break their moorings. Some trees blown down.
3	111-130 mph (178-209 km/hr)	9-12 ft above normal	Some structural damage to small residences and utility buildings. Large trees blown down. Mobile homes and poorly built signs destroyed. Flooding near coast destroys smaller structures with larger structures damaged by floating debris.
4	131-155 mph (210-249 km/hr)	13-18 ft above normal	More extensive curtain wall failures with some complete structural failure on small residences. Major erosion of beach areas. Terrain may be flooded well inland.
5	> 155 mph (>249 km/hr)	> 18 ft above normal	Complete roof failure on many residences and Industrial buildings, Some complete building failures with small utilities blown over or away. Flooding causes major damage to lower floors of all structures near the shoreline, Massive evacuation of residential areas may be required.

Effects of hurricanes and storm surges

In an average year, there will be fewer than 10 tropical cyclones, of which about 6 will develop into hurricanes. These will usually kill about 50 to 100 persons between Texas and Maine and cause property damage of more than $100 million. If it is a worse-than-average year, there will be several hundred deaths, and the property damage will run into billions of dollars. For example, 2005 was one of these worse-than-average years, with notable hurricanes Katrina and Rita, along with others, for a total of 31 named hurricanes or tropical depressions. In addition, this was a record-breaking year for tornadoes, due in part to those tornadoes spawned by hurricanes.[38]

The hurricane brings devastation by the impact of high wind pressure causing collapse of structures, breaking of windows, and knocking down trees and poles. It creates flood-producing rain, and most lethal of all, the storm surge. In a storm surge, the increase in water level, along with abnormal waves and currents, produces extensive "wave-wash" damage to coastal structures.

Floods produced by hurricane rainfall are more destructive than the winds. In 1969, Hurricane Camille produced winds of 190 miles per hour, 10 inches of precipitation on the Mississippi Gulf Coast, and waves up to 25 feet above sea level. Then sweeping inland and curving to the east, it also caused heavy flooding in Virginia.[39] Flooding is the reason that Hurricane Katrina will be remembered as one of America's worst natural disasters. The hurricane actually made landfall as a category 3 hurricane, down from the category 5; it sat off the coast of Mississippi and Louisiana. However, the tertiary flooding that accompanied the storm after the levees broke caused the most damage in terms of needed rescue assistance.

The hurricane's most dangerous aspect is due to the ocean waves. Ocean waves generated by hurricane-force winds may reach heights of 50 feet or more. The water surged in a few seconds as a result of the Galveston hurricane in 1900 and wiped out 6,000 persons—almost one-half of all the lives lost in the United States from hurricanes since that time. Because much of the United States' densely populated coastline along the Atlantic and Gulf coasts is less than 10 feet above mean sea level, the danger from storm surge is great.

Another effect of hurricanes is sudden low atmospheric pressure that produces an internal pressure inside tightly closed buildings, causing their roofs to come off or causing structural collapse. This happened to the navy hangers at Richmond, Florida, in 1945. Other effects include flying missiles from broken glass and dislodged roofs, as well as fires started by overturned stoves and electrical shorts.

Impact on the community

In addition to the direct destruction already discussed, secondary problems mount quickly. Roadways and access routes are blocked with debris. Power and telephone lines are down, causing a failure of telephone communications. All but battery-powered radios and television sets are out. There may be a tremendous impact on the emergency service communications systems with no incoming calls to report emergencies and no capability of the dispatch office to transmit radio messages. The threat from fire is intensified as the result of the hurricane's destructive forces. Many potential sources of fire are generated, reporting is delayed, access is restricted or impossible, and response is delayed. All of these add to the possibility of large-scale fires.

Low-lying population centers can be inundated by waves, causing near-total destruction in their path. (Floods are discussed later in this chapter.) The greatest loss of life in hurricanes, by a ratio of 9 to 1, is from drowning. Figure 13–5 illustrates the extreme flooding that occurred after the landfall of Hurricane Katrina. Electrocution caused by fallen power lines is another extreme danger.[40] This danger is compounded days into the response and recovery as homeowners use generators to feed their homes, and possibly back feed the electrical system. Evacuation and sheltering of thousands of people affect not only the involved

Fig. 13–5. Extensive flooding after the landfall of Hurricane Katrina and the levee failures in New Orleans

FEMA News Photo.

community but also the inland communities that act as hosts. Evacuations generally involve some holdouts who stay on and eventually present additional concerns to the emergency services.

Hurricane planning and preparation

Planning procedures for hurricanes are similar to those for tornadoes. An in-depth hazard identification of a community should determine its potential for a hurricane emergency and identify problems to consider. A capability assessment should determine what resources are available and what additional resources are needed. A multiyear development plan should be implemented to overcome capability shortfalls.

A fire department's hurricane emergency planning should consider those special operation concepts (over and above standard operating procedures) that will provide command and control during a large-scale hurricane emergency.

Hurricanes—where and when

The areas of highest vulnerability to hurricanes in the United States are those states that border the Gulf of Mexico and the entire eastern seaboard. Areas subject to flooding from the excessive rainfall include the eastern, midwestern, and New England states. Even Southern California is susceptible to severe squalls resulting from hurricanes off the Pacific Coast of Mexico.

Hurricane season for the Caribbean and Gulf of Mexico is early summer through fall. The hurricanes that form southwest of Mexico occur in the spring and summer.

Predictability of hurricanes

The National Hurricane Center (NHC) at Miami, Florida has overall responsibility for operations and meteorological analyses related to the position, intensity, and movement of hurricanes in the Atlantic, the Caribbean, and the Gulf of Mexico. The National Hurricane Center coordinates the hurricane warning offices (HWOs), which are responsible for maintaining a close watch on potential hurricane development in their respective areas and for issuing bulletins and advisories to the general public and to disaster and rescue agencies. Based upon information from the other government agencies, supplemented by weather data from ships, aircraft, satellites, and special reconnaissance flights, a detailed hurricane forecast is prepared by the NHC or by an HWO after consultation with the NHC.

Once a hurricane nears landfall, a coastal radar network extending from Maine through Texas keeps constant surveillance on the hurricane's position and movement. Forecasts and warnings are given immediate widespread distribution by e-mail, text messages, radio, television, and newspapers.

Hurricane prediction/warning

Four kinds of notices are issued through the NHC and the HWOs:

- **Advisories.** Give information on location, intensity, and movement of the storm, along with precautions to take.
- **Bulletins.** Similar to advisories, but more general or newsworthy.
- **Watches.** Announce specific areas threatened by the hurricane.
- **Warnings.** Announce that one or more of the following dangers are expected in a specified area:
  - Winds in excess of 74 miles per hour
  - High water
  - Exceptionally high waves in a specified coastal area within 24 hours or less.

The primary methods of warning the threatened areas are through radio and television. "Last appeal" types of notifications are frequently made by loud speakers from helicopters or ground patrol units.

Two major gaps have existed in prediction: the distance to be covered by the hurricane in 24 hours and the landfall location. This is further complicated by the difficulty of pinpointing the exact center of the hurricane. With respect to distance traveled, the average error is about 100 nautical miles. The same is true for the average landfall error, which could easily result from a 10° error in direction of motion (e.g., 310° instead of 320°).

Present-day hurricane warnings are issued to cover about 200 miles of coastline, while the segment identified as subject to storm surge is usually about 100 miles long. A warning is issued whenever possible to allow at least 12 hours of daylight for emergency protection of property and for population evacuation before the hurricane strikes. Officials of the NHC have reiterated their concern for safety of the increasing populations along the Atlantic and Gulf coasts, where many areas are less than 10 feet above sea level. The sledgehammer effect of pounding, surging waves over 23 feet in height, whipped by winds up to 230 miles per hour near the hurricane's eye, as in Hurricane Rita, can destroy any structure not specifically designed to resist such forces. Officials have predicted that in a hurricane emergency 10 years from now, unless proper planning is done to prevent population concentration and to provide for evacuation, the limited capacity of escape routes could cost 20,000 to 50,000 lives.

Historic hurricanes

Certainly one of the most memorable hurricanes in recent history was Hurricane Katrina, which devastated the Gulf Coast in 2005. Hurricane Katrina was the costliest and one of the deadliest hurricanes in the history of the United States. It was an extremely strong Atlantic hurricane, even though it lost its strength as it made landfall in the United States. Katrina formed on August 23 during the 2005 Atlantic hurricane season and caused devastation along much of the north-central Gulf Coast of the United States. Most notable in media coverage was the effect of Hurricane Katrina on New Orleans due to failure of the flood protection system, and also the damage in coastal Mississippi.

Due to its sheer size, Katrina devastated the Gulf Coast as far as 100 miles (160 km) from the storm's center. The storm surge caused severe damage along the Gulf Coast, devastating the Mississippi cities of Bay St. Louis, Waveland, Biloxi, Gulfport, Ocean Springs, and Pascagoula. In Louisiana, the flood protection system in New Orleans failed in 53 different places. Nearly every levee in metro New Orleans was breached as Hurricane Katrina passed east of the city, subsequently flooding 80% of the city and many areas of neighboring parishes for weeks. At least 1,836 people lost their lives in Hurricane Katrina and in the subsequent floods, making it the deadliest U.S. hurricane since the 1928 Okeechobee Hurricane. The storm is estimated to have been responsible for \$81.2 billion (2005 U.S. dollars) in damage, making it the costliest natural disaster in U.S. history.[41] Figure 13–6 shows some of the damage created by Hurricane Katrina.

Fig. 13–6. Hurricane Katrina damage. Damage was extensive throughout much of the Gulf Coast as a result of the hurricane.

FEMA News Photo.

In addition to the devastation previously described in this chapter, the hurricane was noteworthy in that it was a disaster that spread over numerous states and was followed by a large flood of an urban center. A search and rescue effort such as this had never previously occurred. The associate author witnessed a federal system that was taxed beyond its imagination. The 28 FEMA US&R teams were rostering secondary teams to be sent to either relieve or bolster the full deployment of the system. In addition, the Coast Guard rescued numerous victims via helicopter. This disaster proved that if the federal system was to be successful, numerous items had to occur and be planned for:

1. Assets within the federal system must be flexible but work within their discipline.
2. Assets must be on the ground early. This can only occur if assets are prepositioned.
3. Forethought must be used in supporting teams and assets sent to the area based on a long-term deployment.
4. Assets in the federal system must have depth in that they can roster multiple teams.
5. Coordination, as well as command and control, must be present across the system in order to deliver a full-spectrum response.

Figure 13–7 shows associate author Randall W. Hanifen shaking hands with President Bush during the federal response to Hurricane Rita.

Fig. 13–7. Associate author Randall W. Hanifen shaking hands with President George W. Bush during the response to Hurricane Rita

Photo courtesy of associate author Randall W. Hanifen.

Four decades previously, in 1969, Robert H. Simpson, then chief of the National Hurricane Center, said of Hurricane Camille, “By any yardstick, Camille was the greatest storm of any kind that has ever affected this nation.”[42] Just 40 years later, Camille would be trumped twice in one hurricane season, with the arrival of Katrina and Rita. During Camille, structures that withstood the winds risked being burned down as fires raged out of control because firefighting equipment could not reach them. Highways were impassable as the 200,000 persons who had fled inland from the storm returned to see what was left of their homes. The 600-square-mile area was deprived of means of communication and power, and its inhabitants were confronted with water and food shortages, along with the threat of a typhoid epidemic.[43]

In 1972, by the time Agnes disintegrated over western Pennsylvania, she had dumped an estimated 28.1 trillion gallons of water over the eastern seaboard, the equivalent of 25.5 cubic miles of water. Most of it was poured over the Middle Atlantic states. Streams and rivers were flooded—13,500 miles of them—and more than 500,000 people suffered losses. About 116,000 homes, 2,400 farm buildings, and 5,800 businesses were damaged or destroyed. At least $2.5 billion was lost in private and public property. Damage to highways, bridges, and public buildings amounted to more than $700 million. Considering the violence of the winds and the extent of the floods, it is amazing that only 118 lives were lost. The U.S. Army Corps of Engineers credits this to the fact that most flood control measures were adequate. Where they were insufficient, the results were devastating.[44]

Hurricane Fifi is an example of a hurricane that took a large toll in lives. On September 18 and 19, 1974, Hurricane Fifi killed between 7,500 and 8,000 persons and left more than 300,000 Hondurans homeless.[45]

In 1935, the barometer fell to 26.35 inches during a hurricane that stuck Tampa, Florida. This was the lowest reading ever recorded in the Western Hemisphere. More than 400 persons were killed in the storm, and winds reached 150 to 200 miles per hour in the Florida Keys.

In 1938 a hurricane killed about 600 persons in southern New England and on Long Island, New York. New England was badly hit again in 1955 when Hurricane Diane caused billions of dollars of damage along the Atlantic Coast, commencing with North Carolina and spreading northward. Only a year later, Hurricane Audrey struck from Texas to Alabama and killed 390 persons.

Hurricane Donna, in 1960, became the first recorded storm to strike with hurricane force winds in Florida, the Middle Atlantic states, and the New England states at the same time.

In 1963, Hurricane Flora killed about 5,000 persons in Haiti and more than 1,000 in Cuba, and left about 750,000 persons homeless.

Hurricane control

For the permanent inhabitants of the Gulf and East coasts of the United States, there is so far no practical way to avoid a hurricane. Public education and research seem to offer the most useful ways to counter this threat.

Starting in 1962, experimental work under Project STORMFURY was carried out annually until 1983 by the National Oceanic and Atmospheric Administration. Seven flights into three Atlantic hurricanes were made and confirmed that giant storms had developing cloud towers in the right places to make cloud seeding feasible. Storm seeding is based upon the premise that supercooled water (water cooled below the freezing point but still in liquid form) in a cloud can be induced to freeze by seeding it with silver iodide crystals. As the water freezes, it releases energy, causing the cloud to grow. If seeding clouds away from the storm's center would cause them to grow at the expense of the clouds around the eye, then an intense storm would tend to convert into a moderate one.[46] However, after years of experimenting, no solution came to fruition. Thus planners must ensure that codes address hurricane-resistive construction, evacuation plans are in place and function properly, and government departments are prepared to respond to and mitigate the aftermath of a hurricane. This text now examines some of these efforts.

Hazard reduction—hurricanes

The Office of Emergency Preparedness, in its report to Congress, pointed out that certain safety rules can be followed to reduce peril to human life. Anyone living in areas frequently affected by hurricanes should stock up with emergency supplies at the beginning of each hurricane season. If a hurricane is expected, windows should be boarded or protected with storm shutters, outdoor paraphernalia should be secured or placed inside, and automobiles should be fueled in readiness for swift evacuation. In addition, the progress and position of the storm should be monitored. People should stay inside until notified that the hurricane has passed and it is safe to go outdoors. They should not rush outside when the wind stops, for it may only be the eye of the storm passing, and they might get caught out in the open when the other side of the storm strikes.[47]

Each year, the National Weather Service hosts Hurricane Preparedness Week. The 2007 message for the week's campaign was "Preventing the loss of life and minimizing the damage to property from hurricanes are responsibilities that are shared by all."[48] Additional information on securing homes, steps to take when ordered to evacuate, and having a family disaster plan can be found at the National Hurricane Center's Web site at http://www.nhc.noaa.gov/.

Fire department preparation

When the warning comes, responders should review the control measures that have been adopted in the hurricane appendix of their disaster plan. The general precautions that fire departments should take do not differ materially from those outlined under the subjects of other natural disasters (floods, windstorms, and tornadoes).

Emergency action—hurricanes

The handicaps under which fire departments will have to operate following a hurricane are about the same as those encountered in a flood, windstorm, and tornado—all at once! The added danger of a storm surge in coastal communities could make shore highways impassable even before they become littered with fallen trees, demolished structures, and other debris. But on the whole, the search and rescue activities, fire and dangerous chemicals control procedures, and other emergency actions will not differ substantially from those discussed for other storms.

WINTER STORMS

Winter storms include blizzards, ice storms, and snow storms. These vary in intensity and may affect only part of one state or many states. These storms are generated from disturbances along the boundary between cold polar and warm tropical air masses. Fronts are formed where air masses of different temperatures and densities conflict. The disturbances may become intense low pressure systems, churning over tens of thousands of square miles in a great counterclockwise sweep.

Effects of winter storms

Winter storms can disrupt water, power, communications, transportation of people, and the necessities of life, in addition to causing the deaths of many persons and livestock. Buildings can collapse under heavy snow loads, and entire cities can become paralyzed under blizzard conditions, causing the cessation of all commerce. Everything may reach a standstill, including the emergency services called upon to suppress fires, provide ambulance service, apprehend criminals, release victims from stalled elevators, and deliver babies. The emergency services face difficulties not only handling all the routine emergencies, but also the new ones created by the storm.

Examples of winter storm disasters

Blizzards are possibly the most dramatic and perilous of all winter storms, characterized by low temperatures and by strong winds bearing large amounts of snow. Most of the snow in a blizzard is so fine and powdery, and in such great quantities, that it can greatly restrict visibility as it is whipped around by the wind.

The February 2007 North America winter storm (otherwise referred to as the Valentine's Day Blizzard or Valentine's Day Storm) was a massive winter storm that affected most of the eastern half of North America. It started on February 12, 2007 and peaked on Valentine's Day, February 14, 2007. The storm produced heavy snowfalls across the midwestern United States from Nebraska to Ohio and produced similar conditions across parts of the northeastern United States and into Canada in Ontario, Quebec, and New Brunswick. Significant sleet and freezing rain fell across the southern Ohio Valley and affected portions of the East Coast of the United States, including the cities of Boston, Baltimore, Washington, DC, New York City, and Philadelphia. In total, this storm system was responsible for numerous deaths across 13 U.S. states and the Canadian provinces of New Brunswick, Ontario, and Quebec. Areas of the Appalachian Mountains in northern New England

and Quebec received accumulations of more than 16 inches (40 cm). Local amounts exceeded 36 inches (90 cm) in parts of Vermont and 42 inches (105 cm) in the Adirondacks in the state of New York. In total, this storm system was responsible for 37 deaths across 13 U.S. states and three Canadian provinces.[49]

The Storm of the Century, also known as the 1993 Superstorm, the No-Name Hurricane, or the (Great) Blizzard of 1993, was a large cyclonic storm that occurred on March 12 to March 15, 1993, on the East Coast of North America. It was unique because of its intensity as well as its massive size and wide-reaching effect. At its height, the storm stretched from Canada to Central America, but its main impact was on the eastern United States and Cuba. Areas as far south as central Alabama and Georgia received 4 to 6 inches of snow. Areas as far as Birmingham, Alabama received up to 12 inches, with isolated reports of 16 inches. Snow amounts even up to 2 inches were reported on the Florida Panhandle, accompanied by hurricane-force wind gusts and record lows.

This storm complex was massive, affecting at least 26 U.S. states and much of eastern Canada. Bringing cold air along with heavy precipitation and hurricane force winds, it caused a blizzard over much of the area it affected. In addition to the snow in northern Florida, there was thundersnow from Texas to Pennsylvania and whiteout conditions. Some affected areas saw more than 3.5 feet (1.0 m) of snow, and snowdrifts were as high as 35 feet (10.0 m).[50] As previously noted, the northern states could easily deal with this snowfall, but southern states have little resources to deal with snowfall.

In January 1981, a series of devastating winter storms dipped south into the heart of California and also paralyzed New England, affecting states in between from coast to coast. When the citrus crop in California was destroyed, the ensuing high prices for produce affected all Americans, and demonstrated the far-reaching secondary, or indirect, costs of a weather-related disaster.[51] In 1977, blizzards crippled Buffalo, New York and other northeastern cities, and they brought record snowfall and record low temperatures to most of the northeast and Midwest.[52]

When a blizzard raked the West and Midwest, commencing on Good Friday, 1975, Chicago's O'Hare International Airport was closed for 24 hours and all of Chicago's schools were closed. People ate and slept in parked planes, restaurants, banquet halls, and department stores. In the Loop, hundreds of persons stayed overnight in their offices, and many stranded travelers sought refuge in churches. On a 13-mile stretch of the Kennedy Expressway, there were 350 abandoned cars and trucks either stuck in drifts, out of gasoline, or otherwise disabled. Most of the 31 storm fatalities were heart attack victims stricken while trudging through deep snow, trying to free stuck vehicles, or shoveling walks. As the storm moved eastward, it snapped power lines and tree limbs in parts of western Pennsylvania and spread freezing rain and snow across a wide area of New York. At the same time, heavy snow and high winds caused blizzard conditions in much of northern New England.[53]

Winter storms—predictability

Weather forecasters can usually tell when a blizzard is coming and warn people what to expect. But occasionally, blizzards may shift direction unexpectedly and strike a city that is unprepared. *Blizzard warnings* are issued when winds with speeds of at least 35 miles per hour are accompanied by considerable falling or blowing snow, and temperatures of 20°F or lower are expected to prevail for an extended period of time. *Severe blizzard warnings* are issued when blizzards of extreme proportions are expected and indicate wind with speeds of at least 45 miles per hour, plus a great density of falling or blowing snow and temperatures of 10°F or lower.

Other warnings include a *cold wave warning* (for a rapid fall in temperature within 24 hours) and *hazardous driving warning* (also called a *travel advisory*, indicating falling, blowing, or drifting snow, freezing rain or drizzle, sleet or strong winds). In addition, there are *stockmen's warnings* to alert ranchers to protect livestock from snow, ice, wind, etc. The National Weather Service also issues *frost warnings* to farmers in areas where grapes, citrus fruits, and other crops might be damaged by an unexpected cold wave. In certain areas, other private meteorological firms offer services to meet the needs of agricultural interests to cope with the annual threat of frosts and freezes.

Emergency preparation—winter storms

A heavy snowstorm or blizzard, with the accompanying cold temperatures made worse by the wind chill factor, can give emergency responders problems, causing delays in response and adversely affecting operations.

The amount of preparation that may be advisable depends upon the location and the kind of foul weather conditions that can be anticipated. When the fire department of London, Ontario, Canada, was confronted with a blizzard on a January morning, and visibility was reduced to 50 feet, the fire chief asked his communications division to advise all fire stations to accommodate any people who might be stranded. That night it had to fight a $100,000 fire under these severe storm conditions, and the next day the visibility was zero. These conditions continued for 36 hours. The snowfall was 20 to 25 inches, with heavy winds piling drifts up to 5 feet. As the blizzard got worse, the wind chill factor dropped to –30°F, and off-duty firefighters reported for duty in four-wheel-drive vehicles, some with plows, and others in snowmobiles and sleighs.

The storm did not abate, and it was necessary to solicit assistance over three radio stations to obtain volunteers with snowmobiles and four-wheel-drive vehicles to man each fire station. The response was excellent. The next morning the mayor declared a limited state of emergency, which brought in military personnel with tracked personnel carrier vehicles that could be equipped to fight fires and traverse difficult areas at speeds up to 40 miles per hour. The police department obtained radio-equipped four-wheel-drive jeeps from the Canadian Forces, to be manned with police and Canadian Forces personnel, and put one at each fire station. After the third day, the storm seemed to let up, but it returned again in a few hours, and the police returned with their jeeps to the fire stations. Four days after the storm started, it finally left for good, but during that interval, response had to be made to 77 alarms, and the emergencies were successfully handled.[54]

As found in many of the storms of the last two decades, building collapse is becoming prevalent. In years past, buildings were constructed of more hardy construction materials, thus a liveload factor was much higher. With today's construction, lightweight materials are used by builders to reduce construction costs. Buildings are designed to have two types of loads—live loads and dead loads. *Dead loads* are the actual structure itself, as well as items attached to the structure. *Live loads* are those loads added to the buildings, such as snow and ice. Figure 13–8 shows the heavy ice from a winter storm. Notice the weight on the trees.

Fig. 13–8. Ice accumulation as a result of a winter storm. Snow and ice weight causes power lines to snap, as well as increases the possibility of building collapse.

FEMA News Photo.

According to the Ontario Ministry of Labour, "A cubic foot of snow can weigh from seven pounds for snow that is new and dry to 30 pounds for old, compacted snow. Rain falling on accumulated snow will add even more weight. Drifting snow may put excessive loads on the areas where it piles up."[55] Structures, on the other hand, are generally only rated to hold 30 to 40 lbs per square foot of live load. As seen by these two figures, more than one foot of snow could easily cause structural collapse. In tilt-up construction, it could be even less, as witnessed by the associate author on more than one occasion.

Use of weather advisories

It might be useful to make special arrangements between the fire department and the local weather bureau office to determine as quickly and accurately as possible in advance what the storm might bring. It will be helpful to know approximately how much snow is expected to fall and how low a temperature may be expected, the duration of the storm, and whether it will be accompanied by sleet or ice. An ice storm can put heavy loads on telephone and signal wires, coat antennas and roads, and cause tree limbs to break off and block the streets. Numerous ice storms in the Ohio Valley, as well as in other northern areas of the country, have created large power outages. This puts a strain on the elderly population, who may not be able to create a heat source in their home without electrical power. Consideration should be given to opening shelters for citizens affected by the cold.

Activate the emergency operations center

Where the forecast is for heavy snow, blizzard, ice storm, or such severe cold weather that the declaration of a state of emergency is warranted, the community disaster control center will be activated. The local community will want to activate its emergency operations center and put its winter storm appendix of its disaster plan into effect. Actions might include the following:

- Hold over or recall firefighters to assist in carrying out precautionary measures, such as attaching stakes to fire hydrants that have not already been designated by markers high on poles or trees, to be able to find them when covered by snow.
- Test hydrant-thawing devices and place them on fire apparatus, ready for use.
- Mount snow tires or chains on emergency vehicles.
- Patrol schools, public buildings, and business and residential districts for possible outbreaks of fire, or roofs dangerously overloaded with snow.
- Uncover hydrants deeply buried by snow plows.
- Study the emergency snow routes established under the city's disaster plan and augment response assignments where necessary in anticipation of blocked streets and other unusual conditions that may cause delay.
- Have firefighters, paid or volunteer, who are likely to have difficulty reporting to duty in the storm, come in early and stand by until the emergency is over. Extra cots and sleeping bags can be kept for extra personnel in fire stations. Extra manpower can augment strength of in-service companies and man reserve apparatus.
- Get an extra supply of provisions for feeding members during the emergency.
- Check fire apparatus for peak conditions of readiness; batteries lose substantial power at low temperatures, so keep them fully charged. Condensation can affect air brake systems, so have condensation drained at the start of cold weather and check the air brake systems periodically. Tire pressures decrease in cold temperatures, so check tire pressures in extreme cold weather. Add bags of salt in case the apparatus becomes stuck in the snow.

- Obtain extra four-wheel-drive vehicles and snowmobiles to supplement regular fire apparatus and to carry fire equipment to emergencies not accessible by standard apparatus.
- Ask the news media—television, radio, and the press—to request the cooperation of the public. This could include instructions not to obstruct fire lanes on the emergency snow routes, not to abandon cars in main thoroughfares if they can be removed to other locations, and to remove snowdrifts that block emergency exits from places of public assembly. It could also include uncovering fire hydrants in front of their places of business to speed an attack on any fire that may break out in their buildings, along with instructions to be especially careful of fire in their homes, offices, and places of business.[56]
- Consider establishing a unified command for the local community resources that will enable city snowplows to operate in conjunction with fire and police units. This will clear the path to emergencies.
- Consider equipping fire and police support vehicles with snow-plowing equipment to aid responses when city snowplows are unavailable.

FLOODS

Dating to the earliest settlers of the North American continent, people in the United States and Canada have relied on natural water supplied by rivers. Logically, populations grew in close proximity to rivers. These rivers, which are the lifeblood of developed population centers, can, however, turn deadly in time of flood.

Although some localized flooding occurs every year, and nearly every month in some location within the United States, large-scale flooding happens every few years. This was the case in October 2006. A trio of heavy rainfall events, October 14–17, October 18–19, and again October 26–27 set the stage for moderate to significant flooding across portions of eastern Texas and Louisiana, as well as portions of Arkansas and Mississippi. During this event, up to 28 inches of rain fell in a 13-day period. Not only was this a great deal of rain, but the flat and sandy soil present in these areas compounded the problem due to the low absorption rate of these ground surfaces.

In early 1991, floodwaters threatened to spill over an area stretching from northeast Texas into the Great Lakes regions of southern Michigan and western New York. In the center of this were Indiana, Ohio, and Kentucky, which bore the brunt of floodwaters, reaching the highest levels in more than a decade. In Indiana alone, more than 3,000 residents were forced from their homes. In Cincinnati, the north-south runway at Lunken Airport was closed because of high water, forcing air traffic controllers to work with only two runways. The highest water in more than a decade left a devastating mark on the Ohio River Valley that would long be remembered.[57]

However, this flood was trumped in 1997, when most of the lower lying areas of the Cincinnati region adjacent to the Ohio River flooded again. The town of New Richmond was the hardest hit, with resulting destruction to the majority of the town of 2,200 people. The associate author spent three days traversing the area in a boat rescuing victims and gaining reconnaissance information.

Floods may originate with hurricanes, dam breaks, lengthy rainfall, or other persistent precipitation causing overflow of rivers, lakes, and seas onto the land, but this is not the case with flash floods. To have a flash flood, all that is required is a sudden heavy rain and a gully to channel it somewhere. If that gully leads to a parking lot, as it did in Las Vegas on July 3, 1975, the result can be destruction of 300 cars, dozens of homes inundated, and two people killed in less than an hour.

For action in flood waters, fire department personnel must be prepared for a variety of tasks. One of the first acts of the fire officer should be to verify which members of the fire company can swim and which ones can row or operate a power boat. In addition, the fire officer must determine what equipment will be needed to carry out rescues, fire control, and other operations while flood waters move dangerously and inexorably through the area. In these situations, hand-portable two-way radios are essential; so are rope, protective clothing, ground ladders, pike poles, and lights.

The second action should be the summoning of specialized boat rescue teams that may be available in the area. Proactive planners will have these teams listed and know an estimated time of arrival from the time of the request. Floods, especially flash floods, can be a dangerous, if not deadly, environment for even the highest trained experts in swift water rescue.

In some areas of the country, such as California, the option to reach victims in the floodwaters is to utilize helicopters. The Coast Guard utilized this tactic with great success in the aftermath of Hurricane Katrina. This is a highly specialized rescue operation, but in large-scale floods, this operation can quickly speed rescue efforts. Planners should investigate whether this is an option for their locale.

If properly trained swift water rescue teams are not available, fire officers must learn quickly where the flood currents are most dangerous and how to prepare for work in those areas. Rescue personnel must be ready to work as teams in case people are swept into the flood waters and are in danger of drowning. It will be very important for the fire department to work efficiently with all other agencies assigned to the flood area.

Finally, in the aftermath, the fire department probably will be using portable pumps to move water from submerged homes. They may also use portable generators to provide standby electrical power to important buildings like hospitals and fire stations or halls where flood victims may be housed temporarily.

Floods—where and when

Virtually every state in the country has areas that are subject to disastrous flooding conditions. Flooding occurs when the soil and vegetation are unable to absorb the falling rain or melting snow, or when the runoff exceeds the capacity of the normally adequate control channels.

Communities that are particularly vulnerable to flooding emergencies are those located in the following places:

- Low areas in proximity to rivers or flood control channels
- Watersheds of barren mountains that lack vegetation due to environment or fire
- In the path of water flow from a dam or reservoir failure
- Low coastal areas subject to storm surges

In the past several years, flooding has occurred around Lake Michigan and at the Great Salt Lake, primarily because of years of uncontrollable rise in the water level of the lakes. Figure 13–9 shows the extensive flooding around the levees in New Orleans following Hurricane Katrina.

Fig. 13–9. Major flooding present around the levees in New Orleans following Hurricane Katrina

The season for flood threat varies according to locale throughout the country. The peak time for flood threat on the West Coast and in the Pacific

Northwest occurs during winter and early spring. The Rocky Mountain area peaks in late spring. The Mississippi and Ohio River valleys and the southern states peak in the winter and early spring. The Great Lakes region and New England states peak in early springtime, while Florida is subjected to flooding in the fall.

Earthquake-related flood threats are not associated with weather or season. The potential collapse of dams or reservoirs may pose a severe flash flood threat that needs to be addressed. Earthquakes on the ocean floor may generate gigantic ocean waves called *tsunamis*. The major threat from tsunamis in the United States exists for the Hawaiian Islands and on the West Coast, including Alaska.

Nature of floods

There are generally two types of flooding: the slow-rise flood and the flash flood. The onset of a slow-rise flood, as the name implies, is a relatively slow process with several days of buildup. Long periods of heavy rainfall or melting snow combined with rain are normally the cause for rivers to overflow their banks. Flood warnings in such cases can be issued many hours or even days in advance of the flood crest.

Flash floods normally occur after torrential rains of short duration, causing violent flooding of a river, stream bed, or even a normally dry wash or gully. Heavy rainfall running off a mountain range can be the perpetrator, as in the Las Vegas situation previously mentioned. The breakup of an ice jam also presents the threat of flash flooding.

Causes of floods

Heavy rainfall causes most floods, but sometimes other factors are involved. Rain that normally soaks into a mountainside can cause a flood if timber and grass have been destroyed by fire. Rain can loosen rocks to cause a landslide that obstructs a river and causes a flood. A cloudburst can cause flash floods and fill a dam so fast that it breaks and floods the countryside. Tidal waves and high winds can cause coastal floods. Even an earthquake can cause a flood. An earthquake caused the Mississippi River to flow upstream in 1811 and 1812 and created Reelfoot Lake in Tennessee.

As an example of a landslide that obstructed a stream, on July 22, 1970, falling rocks diverted the course of the Alaknanda River in India, causing a sudden flood 200 miles northeast of New Delhi. The rushing waters carried away buses and other vehicles taking pilgrims home from Badrinath, a holy city, killing about 600 persons.

As an example of a flood caused by too much rain, about two months after the Alaknanda River flood, another 600 persons died and 20,000 were made homeless in India when heavy rains caused flooding from Bombay to Calcutta.[58]

A dam can break to produce a flood even if there has been no rainfall for months. In the Baldwin Hills area of Los Angeles on the afternoon of December 14, 1963, a small crack in the dam developed into a large break. This allowed 300 million gallons of water to gush forth and wipe out a large residential area.

A dam need not be of the earth-filled type nor as large as the one in Baldwin Hills to cause havoc upon release of its water. The coal-slag dam that gave way in February 1972 held only about 130 million gallons of water, but by the time the Buffalo Creek flood subsided, 125 West Virginians were dead and 4,000 others were homeless.[59]

In the most devastating disaster during the early 1990s, 200,000 people died in Bangladesh in early May 1991 as a result of a typhoon. This typhoon carried 20-foot waves and 145 mile-per-hour winds ashore, leaving behind acres of flooded countryside. This was just one of a series of disasters that have continued to plague the 100 million people in this densely populated country.[60]

Another cause of floods is the denuding of a watershed by fire. The western states were especially affected in 1991 after the extensive series of wildland fires previously reported left large tracks of land without forestation. A series of floods and mud slides killed more than 100 persons and made about 10,000 homeless.

Causes of seacoast floods

Floods from the sea usually are caused by high winds that pile water up against the harbor and drive the waves onto the land. In some countries, such as the Netherlands, dikes used to wall off the sea generally have been successful in preventing this type of flood. However, these are not always effective. After the great flood in Galveston, Texas, the city built a seawall, but the storms of 1913 went over it.[61] This was the case again when a one-quarter-mile breach in the levee that protected New Orleans from Lake Pontachartrain resulted after Hurricane Katrina's landfall. This breech provided the tertiary flooding in estimated 66% to 75% of the city.[62]

Tidal wave floods

A tidal wave is an ocean wave or series of waves of great height. They have no connection with true tides, but are responsible for seacoast floods. These gigantic waves are referred to as *tidal waves* when they are caused by submarine earthquakes (seaquakes), submarine volcanic action, an underwater atomic bomb burst, or by a hurricane.

The scientific name for such waves caused by submarine earthquakes is *tsunamis*, the Japanese word for storm waves. When at sea, the storm waves are about 100 to 400 miles from crest to crest, and they do not pile up until they hit shallow water. Tsunamis occur mainly in the Pacific Ocean along the west coast of Central America, northwest coast of South America, the Hawaiian Islands, and the east coast of China and Japan. However, one of the most recent and deadliest tsunamis occurred in the Indian Ocean.

On December 26, 2004, an earthquake of 9.1 on the Richter scale occurred on the western coast of Sumatra, Indonesia. This ocean floor movement created a tidal wave that measured as much as 80 feet in height. According to updated information from the U.S. Geological Survey, in total, 227,898 people were killed or were missing and presumed dead. About 1.7 million people were displaced by the earthquake and subsequent tsunami in 14 countries in South Asia and East Africa. (In January 2005, the death toll was 286,000. In April 2005, Indonesia reduced its estimate for the number missing by over 50,000.) This high toll is due to the number of coastal areas surrounding the epicenter. Countries affected by the tsunami included Indonesia, Sri Lanka, India, Thailand, Somalia, Myanmar, Maldives, Malaysia, Tanzania, Seychelles, Bangladesh, South Africa, Yemen, Kenya, and Madagascar.[63]

As a result of the Good Friday, 1964 earthquake in Alaska, with its epicenter in Prince William Sound, seismic ocean waves were created. In Valdez (50 miles distant), the harbor was emptied of water three times, with ensuing waves up to 50 feet high refilling it. Similar activity occurred in Anchorage, Seward, and Kodiak, with 40-foot waves. The impact of the wave action caused damage as far away as Crescent City, California and the Hawaiian Islands.

Tidal waves are somewhat predictable. In the areas where the contour of the ocean floor is known to be favorable, and within the rather narrow belts of the ocean where they have previously occurred, a large undersea earthquake, recorded on a seismograph ashore, may give rise to the expectancy of a tidal wave in 5 to 30 minutes. This is often the case if the tremor is followed by an abnormally low tide. However, such waves are not distinguishable from others when viewed from afar, and since they move at speeds between 300 to 500 miles per hour, they do not give much warning.

An example of a tidal wave flood caused by a volcano occurred following the eruption of Krakatoa, an Indonesian volcano that "blew its stack" in August 1883, causing one of the world's worst disasters. Much of the island was blown to bits, and the resulting tidal wave that washed over the nearby islands killed about 36,000 persons. One ship, carried 1½ miles inland, was left 39 feet above the sea.

A tidal wave caused by a hurricane that battered the island of Galveston, Texas for 18 hours in September 1900 produced the worst natural disaster ever recorded in the North American continent. Over 6,000 of the city's 37,000 inhabitants were killed, and nearly one-half of the residences—3,000 homes—were demolished, leaving the site looking like a lumberyard strewn with furniture.[64]

Tidal waves caused by submarine earthquakes create great destruction. One occurred in the Sanriku area of Japan in 1896; 27,122 people lost their lives in about 30 seconds when a wall of water, 75 feet to 100 feet high, rolled over their festival celebration.

Flash floods

A sudden, violent flooding of a river, stream bed, gully, or wash is called a *flash flood.* It is touched off by torrential rains filling natural and man-made drainage systems to overflowing with raging waters, uprooted trees, smashed structures, boulders, mud, and debris.

Flash flooding can occur many miles from the location the rain falls. This was the case on August 12, 1997, when flash floods killed 11 tourists in the Lower Antelope Canyon in Page, Arizona. The flash flooding caught the tourists by surprise, as little to no rain fell in this area throughout the day.[65]

In the summer of 1990, the small town of Shady Side, Ohio, witnessed utter devastation when the small, normally peaceful creek that flows alongside this city suddenly became a raging river, sweeping away most of the homes along its river bank. Dozens were drowned or missing, and the small Shady Side volunteer fire department could do little to help, even though it was well trained in handling water emergencies.[66]

On flash floods, warnings are transmitted via radio and television. The National Weather Service issues hydrologic forecasts and warnings using the following terminology:

- **Flash flood.** The occurrence of a dangerous rise in the water level of a stream or over a land area in a few hours or less caused by heavy rain, ice jam breakup, earthquake, or dam failure.
- **Flash flood watch.** Heavy rains occurring or expected to occur may soon cause flash flooding in certain areas. Citizens should be alert to the possibility of a flood emergency that will require immediate action.
- **Flash flood warning.** Flash flooding is occurring or is imminent on certain streams or in designated areas, and immediate precautions should be taken by those threatened.

Tsunami warnings can be issued after an earthquake has occurred if sufficient lead time exists. The speed of the surge can be calculated as well as its direction of travel, thus giving preparation time to those coastal areas that may be threatened. After the destructive tsunami of 2004, many countries, including the United States, began to examine and upgrade their tsunami warning systems.

Effects of floods on the community

Annual flooding of marshes and tidelands are sometimes beneficial to farmers and wildlife, but flash flooding situations can create utter devastation. A wall of water 10 feet to 20 feet high carrying massive boulders, trees, and large structural debris is virtually unstoppable in its destructive attack on a community. In addition to destroying houses and other structures, the raging torrents usually damage or demolish roadways, bridges, and any utilities within the path. The possibility of loss of life from drowning is always present with the swift currents.

Damage resulting from slow-rise flooding is generally caused not only by the inundation of water, but also by silt and debris passing through the affected community. In the United States, in an average year, flooding will force 75,000 people from their homes, about 80 people will be killed, and property loss will be about $1 billion.

Unexpected floods can cause drownings, render people homeless, wash away buildings, erode land from one place and deposit it elsewhere, pollute drinking waters, and spread disease. Floods can shift underground flammable liquid storage tanks, causing them to spill their contents and float away. They can also overturn aboveground flammable liquid and flammable gas storage tanks, causing fires. Flood waters may inundate storehouses containing water-reactive chemicals, causing violent reactions, explosions, and release of toxic gases.

"Flooding creates an atmosphere favorable to fire, either by causing it or by raising problems in efforts to fight it. The flood itself damages goods, equipment, buildings, fire protection equipment and machinery either through wetting or by foreign material plugging and damaging moving parts and by floating debris slamming into objects. Damage brought about by water, mud, silt, and debris is serious enough, but devastation that could be brought about by fire, when the property is unapproachable or normal protection has otherwise been interfered with, can be far worse."[67]

Industrial fires were caused by flash floods resulting from Hurricane Diane when it moved northward into the New England states more than five decades ago.[68] One fire broke out in a flooded magnesium powder plant in Putnam, Connecticut, and burning drums floated from the building exploded, setting fires in buildings downstream.

Another serious effect of floods is the restriction of activities of firefighters through inaccessibility. This was demonstrated in a warehouse fire in Dubuque, Iowa, where the fire department had to move ladders, hose, and personnel by Coast Guard boats to the fire scene. They then had to hook up to submerged hydrants, effect rescues, and after removing sand bags and dikes to get inside, extinguished the fire.[69]

The effects upon fire department response were also demonstrated when a tropical storm put much of Boundbrook, New Jersey, under flood waters. When volunteer firefighters were summoned to help fellow residents, they found that they were among the victims. Many were trapped in their homes. Two of their four pumpers were under water, and one stalled in water while leaving quarters. Fire apparatus could get only within one block of a residential fire because of the flooded streets. Flood waters reached almost six feet in depth on one street before they started to recede. Police used power boats to patrol the town to prevent looting. When the water receded, a large military wrecker was used to retrieve the submerged pumpers. Mutual aid companies assisted in pumping out hundreds of cellars, unclogging storm drains, washing down streets, and warning the public of the contaminated water supply.[70] Even if the fire apparatus is able to proceed to the area of the fire, the incident commander must plan for what the water level will do during the time it takes to extinguish the fire.

In addition to the loss of heat and power during the flood, the ripping away of bridges to which telephone cables are attached can also disrupt communications. This happened at Corning, New York during the 1972 flood. Dikes that were 25 feet above the normal water level of the Chemung River were not high enough to prevent water from flooding the computer room of the Corning Glass Works. Within an hour, the water was only 12 inches from the ceiling. Tapes, disc packs, and other items were water soaked and covered with a slick coat of mud and silt residue. Water and silt infiltrated safes in another building where backup tapes were stored for remote security. The communications center where all computer input was received was inundated, and all building services, including power, were out. The entire Corning area was without power, heat, water, and telephone and was under National Guard security, with surface travel limited to priority needs.

Flood prevention and precautions

A capability assessment should determine what resources are available to deal with the flooding problems and what additional resources are needed. A multiyear development plan should be implemented to overcome capability shortfalls.

A community's flood emergency planning should consider those special operational concepts (over and above standard operating procedures) that will provide effective command and control during a large-scale flood emergency.[71]

Measures that help prevent floods include dams, levees, dikes, storm drains, soil erosion programs, and forest fire prevention. The U.S. Department of Agriculture found that barren land lost 30% of its rainfall and 65 tons of soil per acre each year. But nearby forest land lost only about 1% of its rainfall and four pounds of soil per acre each year. Tree planting is important in modern flood control. The Soil Conservation Service warns that the size of floods is affected by the way in which farmers plow their land, the kinds of crops they plant, and how they control the drainage of their land.[72]

Other precautions include firmly securing or anchoring buried tanks, aboveground tanks, and other objects that might float away and placing valuable property on upper floors of buildings. It is also important to make sure that building foundations and retaining walls will withstand flood conditions. Having preincident plans that identify tanks and their contents in industrial facilities can aid in mitigation of hazardous materials leaks if the flood causes the tanks to become dislodged and begin to float in the water.

A flood control and evacuation plan should be prepared, selecting routes of response to areas in the community by roads that will avoid low spots and paths of torrents. The plan could provide that when off-duty members have made their own homes secure, they could report for duty to stand by while the on-duty members make their homes secure.

Flood predictability and warnings

Except for flash floods, the onset of most floods is a relatively slow process, often with adequate warning. Buildup usually takes several days, and progressive reports are available from NOAA (National Oceanic and Atmospheric Administration) through its National Weather Service river forecast centers and river district offices. Warnings are given by NOAA when rainfall is enough to cause rivers to overflow banks and when melting snow may combine with rainfall to produce similar effects.

Flash flood warnings are the most urgent type. They are transmitted to the public via radio and television, and by other means, such as sirens, horns, whistles, telephone alerts, or by means of emergency services personnel using loudspeakers on their vehicles or helicopters.

Forecasts of impending floods reach people in the form of warning messages about the severity of flooding (minor, moderate, or major), the affected river, and when and where flooding is expected to begin. Area radio and television stations usually broadcast the latest flood information and warnings. It is advisable that the local government give more specific advice and instructions over local stations, in accordance with the local disaster plan.

If a community has a history of recurring floods, continuing communication should be established with the National Weather Service. This allows local forecasts and warnings to be transmitted to local emergency services headquarters at stipulated periods. "During the flood-control planning phase, communities should coordinate closely with the nearest office of the Army Corps of Engineers, in accordance with established state procedures. Also, the Corps of Engineers can provide preventative assistance for flood control as authorized under Public Law."[73]

A flood that occurs because of a dam break is not as predictable as one that results from overflowing rivers. In the absence of heavy rains that might throw an added burden on a dam, or an earthquake that might shake its foundation, it is sometimes possible to observe trouble signs. Such signs could include leaks around an earth-filled dam on volcanic terrain riddled with caverns and crevices. These signs might cause safety officials to anticipate possible weakening and breaking. This was believed to be the situation in the Teton

Dam collapse in southern Idaho on June 5, 1976. As a result of the dam collapse, 11 people were killed, 4,000 homes were wrecked or damaged, and 100,000 acres of rich farmland were stripped of topsoil by the torrent of unchecked water that swept the dam away.[74]

Emergency action—floods

When a flood threatens, government departments or agencies should be prepared to assume their designated role under a community's flood annex to its disaster plan. When the National Weather Service has issued a flood watch or warning, and the situation calls for a 24-hour flood watch, the community should activate its emergency operations center. They should carry out planned measures for maintaining operational readiness under flood conditions. It is important to check emergency power supplies and supplemental communications systems, including the use of ham radio operators. In addition, emergency fuel and food supplies must be obtained that are sufficient to cover the need for the probable duration of the inaccessibility of roads serving the community. Fire stations and other government facilities that are apt to be flooded to a depth that would make use impractical should be rendered as impervious to damage from water as possible and then evacuated. Routes of response should be planned from temporary field headquarters established on high ground.

The depth of water in which fire trucks can safely operate should be determined in advance. Tractors and bulldozers capable of operating in deep water should be stationed where they can be readily available to tow disabled apparatus. Winches, cables, and towing eyes on fire trucks can drag stalled light vehicles out of flooded thoroughfares. However, most fire trucks are not geared sufficiently low to provide towing or pushing service to other heavy vehicles without mechanical damage to themselves.

Before water rises in the business district, colored flags on poles can be attached to fire hydrants to mark their locations, as was done in Dubuque, Iowa, and in Cincinnati, Ohio.[75] All boats, portable generators, flood lights, and emergency food, fuel, and water supplies for drinking that may be needed during search and rescue operations should be brought to the command post or field headquarters on high ground.

Emergency communications procedures should be established to ensure that National Weather Service reports are obtained by the dispatcher and operations officer. Dike patrols and helicopter reconnaissance are useful for observing conditions.

Following a flash flood, the use of handheld and airborne infrared devices to detect body heat of victims buried under silt and rubble have proven useful. In the Big Thompson River flash flood, described earlier, 12 handheld and two airborne infrared devices were used to quicken the tediously slow process of recovering bodies buried under tons of debris scattered down the 35-mile length of the canyon.

Portable generators can supply power to gasoline pumps to get fuel out of underground tanks when the power supply has been cut off. They also can be used to operate freezer equipment in food establishments, meat markets, and warehouses to prevent spoilage pending restoration of the normal power.

Using specially equipped vehicles, boats, and helicopters, firefighters can relay evacuation messages and rescue persons stranded on upper floors and roofs. They also can notify homeless persons where refuge centers are located and do other tasks related to saving lives and property.

The fire department can assist in placing salvage covers and vinyl sheeting over earth-filled dikes to prevent erosion and aid in stabilizing spongy dikes. Fire stations are often used as a supply depot for sandbags. In areas likely to suffer flooding and mud slides following a large brush fire, the public works department can dump loads of sand in the street near the threatened homes for use in filling sandbags for home protection.

Following recession of the flood waters, firefighters can restore services and salvage damaged property. The use of siphon ejectors to drain flooded basements is preferable to using pumps on first-line fire apparatus, for mud and debris will not have to go through the pump itself, thereby reducing the chance of damaging it. With chain saws, power winches, and other tools, firefighters can assist public works and street departments in clearing the streets of debris, fallen trees, and other obstructions.

CONCLUSION

This chapter has covered the major disasters caused by weather conditions—storms, tornadoes, hurricanes, and floods. While it is weather that provides fresh breezes, adequate drinking water, and other necessities of life, it is also weather unrestrained that causes some of the most serious disasters. In this chapter, the devastating effects of the weather-related types of disasters have been noted, along with the almost-impossible tasks facing firefighters who are the first line of defense against these disasters. This discussion will continue in the following chapter, where the focus will turn from the weather-related types of natural disasters to those that are terrain-based.

REFERENCES

1. U.S. Department of Commerce, National Oceanic and Atmospheric Administration. 1937. *Spotter's Guide for Identifying and Reporting Severe Local Storms.* Washington, DC: U.S. Government Printing Office. p. 5.

2. National Weather Service. 2007. Lightning facts page of the Colorado lightning resource page. Retrieved May 3, 2007 from http://www.crh.noaa.gov/pub/ltg/crh_ltg_facts.php.

3. Panati, C., and W. Schmidt. 1975. Lightning alley. *Newsweek*. October 27. p. 65.

4. National Weather Service. 2007.

5. Ibid.

6. *World Book Encyclopedia*. 1969. Vol. 12. Chicago, IL: Field Enterprises Educational Corporation. pp. 263–265.

7. NOAA. http://www.noaanews.noaa.gov/stories/s2008.htm.

8. *World Book Encyclopedia*. 1969. Vol. 9. Chicago, IL: Field Enterprises Educational Corporation. p. 8.

9. Goldman, J., and R. Dallas. 1975. Jetliner crashes in New York storm; 109 die. *Los Angeles Times*. June 25. Part I, p. 1.

10. NASA (May 207). Earth observatory: May 2007 headlines. Retrieved November 9, 2008 from: http://earthobservatory.nasa.gov/Newsroom/archive.php?m=05&y=2007&cat_id=18.

11. National Fire Protection Association. 2007. *Lightning Protection Code*. NFPA No. 780.

12. Newman, M., J. Robb, and E. Yonkers. 1962. Aircraft protection from thunderstorm electromagnetic effects. Conference Paper No. Cp 62-1133, American Institute of Electrical Engineers. p. 4.

13. Lightning Protection Institute. 2007. Welcome. Retrieved May 3, 2007 from http://www.lightning.org/?page=home.

14. *Factory Mutual Property Conservation Engineering Management Record*. 1975. Wind and buildings. 52 (6). November/December. p. 14.

15. U.S. Department of Homeland Security, Federal Emergency Management Agency. 2007. Avoiding wind damage: a checklist for homeowners. pp. 1–2. Retrieved May 4, 2007 from: http://www.ohsep.louisiana.gov/factsheets/avoidingwind-damage.pdf.

16. National Weather Service. 1995. Present weather. Chapter 8 in *Federal Meteorological Handbook No. 1.* December. Retrieved May 4, 2007 from http://www.nws.noaa.gov/oso/oso1/oso12/fmh1/fmh1ch8.htm.

17. Chao, Lang. 2004. Worst is over for sandstorms. *China Daily*. May 20. Retrieved May 4, 2007 from http://www.chinadaily.com.cn/english/doc/2004-05/20/content_332375.htm.

18. *World Book Encyclopedia*. 1969. Vol. 4. Chicago, IL: Field Enterprises Educational Corporation. p. 960.

19. Associated Press. 1991. *Cincinnati Enquirer.* April 29. p. A-3.

20. Fox News. 2007. Live coverage of tornado damage from Fox News Network, viewed by associate author, Randall W. Hanifen. May 5.

21. National Weather Service. 2006. *Shareholders' Report 2005*. 5 (1). February. Retrieved May 6, 2007 from http://www.srh.noaa.gov/ffc/html/anreport05.pdf.

22. *World Book Encyclopedia*. 1969. Vol. 20. Chicago, IL: Field Enterprises Educational Corporation. p. 116.

23. National Weather Service. Hodges, TX. *Tornadoes: Nature's Most Violent Storms.* Retrieved May 4, 2007 from http://www.crh.noaa.gov/lmk/preparedness/tornado_large/fslide16.php.

24. National Weather Service. Illinois. *Tornadoes: Nature's Most Violent Storms.* Retrieved May 4, 2007 from http://www.crh.noaa.gov/lmk/preparedness/tornado_large/fslide17.php.

25. National Weather Service. McConnell AFB. *Tornadoes: Nature's Most Violent Storms.* Retrieved May 4, 2007 from: http://www.crh.noaa.gov/lmk/preparedness/tornado_large/fslide18.php.

26. *CNN.com*. 2007. Tornado kills 10 on U.S.-Mexico border. April 25. Retrieved May 4, 2007 from: http://www.cnn.com/2007/WEATHER/04/25/storm.deaths.ap/index.html.

27. *CNN.com*. 2007. Tornado survivor: school was a complete chaos. March 2. Retrieved March 18, 2007 from: http://www.cnn.com/2007/WEATHER/03/02/enterprise.storm.ap/index.html.

28. *Time*. 1989. November 27.

29. Troeger, J. 1975. Anatomy of a major disaster. *Fire Command!* May. pp. 18–20.

30. *Fire Engineering*. 1974. Tornado hits Xenia, Ohio. Leaves 33 dead, 1,000 hurt. November. pp. 30–32.

31. Slepicka, L. 2000. Xenia tornado tests readiness. September 28. *Firehouse.com*. Retrieved May 5, 2007 from: http://www.firehouse.com/news/2000/9/28_tornado.html.

32. Villanova, R. 1974. Radio-controlled mutual aid speeds tornado rescue work. *Fire Engineering.* November. pp. 26–27.

33. Fox News. 2007.

34. American Red Cross. 2008. Tornado. Retrieved September 24, 2008 from: http://www.redcross.org/services/disaster/0,1082,0_591_,00.html.

35. Federal Emergency Management Agency. 2006. Preparing a safe room. April 6. Retrieved May 6, 2007 from: http://www.fema.gov/hazard/tornado/to_saferoom.shtm.

36. U.S. Department of Commerce, National Oceanic and Atmospheric Administration. 1971. *Hurricane, The Greatest Storm on Earth.* Washington, DC: U.S. Government Printing Office. p. 3.

37. Ibid.

38. National Weather Service. 2006.

39. Factory Insurance Association. 1971. The hurricane—its effects. *The Sentinel*. July-August. p. 7.

40. Office of Emergency Preparedness. 1972. *Disaster Preparedness*. Report to Congress. January. Vol. I. Washington, DC: U.S. Government Printing Office. p. 47.

41. U.S. Congress Bipartisan Committee. 2006. A failure of initiative: bipartisan select committee to investigate preparations for and response to Hurricane Katrina. February 15. Retrieved July 17, 2008 from: www.gpoaccess.gov/congress/index.

42. Kramer and Bahme. 1992. *Fire Officer's Guide to Disaster Control*. PennWell, Tulsa.

43. *World Book Encyclopedia*. 1990. Vol. 12. Chicago, IL: Field Enterprises Educational Corporation. p. 390.

44. This booklet, *Agnes Was No Lady*, which gives no publisher's name or date, was obtained by author Chuck Bahme from a rack outside the office of the U.S. Army Corps of Engineers in Chicago's Federal Building, where copies were available to the public without charge.

45. *World Book Encyclopedia*. 1975. Vol. 12. Chicago, IL: Field Enterprises Educational Corporation. p. 300.

46. *Los Angeles Times*. 1976. Pacific may get storm-taming project. July 6. Part IA. p. 7.

47. Bahme, C. W. 1976. *Fire Officer's Guide to Emergency Action*. Boston, MA: National Fire Protection Association.

48. U.S. Department of Commerce, National Oceanic and Atmospheric Administration, National Hurricane Center. Be prepared. Retrieved May 7, 2007 from: http://www.nhc.noaa.gov/HAW2/english/disaster_prevention.shtml.

49. U.S. Department of Commerce, National Oceanic and Atmospheric Administration. Diagnosis of precipitation efficiency from GOES, WSR-88D, and RUC analyses Northeast Storm 14 February 2007. Retrieved May 7, 2007 from: http://www.nhc.noaa.gov/HAW2/english/disaster_prevention.shtml.

50. U.S. Department of Commerce, National Oceanic and Atmospheric Administration, National Weather Service Forecast Office. Remembering the blizzard of 1993. Retrieved July 17, 2008 from: http://www.srh.noaa.gov/bmx/significant_events/1993/Blizzard_of_1993/index.php.

51. U.S. Department of Commerce, National Oceanic and Atmospheric Administration. 1990. *Winter Storms*. NOHA/PL 70018.

52. Ibid.

53. *Los Angeles Times*. 1975. 31 die as storm rakes Chicago area. April 4. Part I. p. 17.

54. Wills, O. Snow emergency in Canada. *Fire Engineering*. p. 33.

55. Ontario Ministry of Labour. 1996. Snow loading and roof failures. February. Retrieved May 7, 2007 from http://www.labour.gov.on.ca/english/hs/alerts/a17.html.

56. Bahme, C. W. 1976. *Fire Officer's Guide to Disaster Control, 1st Ed.* PennWell, Tulsa.

57. *Cincinnati Post*. 1991. January 2. p. 1.

58. *The World Book Year Book*. 1971. Chicago, IL: Field Enterprises Educational Corporation. p. 296.

59. Keerdoja, E., and E. Shammon. 1976. After the flood. *Newsweek*. July 12. p. 7.

60. *Washington Post*. 1991. May 4.

61. *World Book Encyclopedia*. 1969. Vol. 7. Chicago, IL: Field Enterprises Educational Corporation. p. 210.

62. U.S. Congress Bipartisan Committee. 2006.

63. U.S. Geological Survey. Most destructive known earthquakes on record in the world. *Earthquake Hazards Program*. Retrieved July 17, 2008 from: http://earthquake.usgs.gov/regional/world/most_destructive.php.

64. Rhodes, L. M. 1968. The great Galveston hurricane of 1900. *American Legion Magazine*. September. pp. 12–16, pp. 46–49.

65. *Luminous Landscape*. Lower Antelope Canyon. Retrieved July 17, 2008 from: http://www.

luminous-landscape.com/1photo-pages/l/lower_antelope.shtml.

66. *American Heat*. 1990. Shadyside floods. Video journal. September. St. Louis, MO.

67. *AIA Special Interest Bulletin*. 1970. Flood damage prevention. No. 323. October.

68. Babcock, C. I. 1955. Flood and fire. *NFPA Quarterly*. October. pp. 153–157.

69. Kemps, J. 1965. Flood and fire. *Firemen*. August. p. 6.

70. Newhall, R. 1972. N.J. volunteers win fight against flood waters. *Fire Engineering*. July. p. 37.

71. National Fire Academy. 1988. *Disaster and Fire Defense Planning*. Open learning course. Emmitsburg, MD: National Fire Academy.

72. *World Book Encyclopedia*. 1969. Vol. 7. Chicago, IL: Field Enterprises Educational Corporation. p. 22.

73. Defense Civil Preparedness Agency. 1974. *Disaster Planning Guide for Business and Industry*. CPG 2-5. May. Washington, DC: U.S. Government Printing Office. p. 2; pp. 525–539.

74. *Los Angeles Times*. 1976. A bureau covers its eyes. Editorial. July 20. Part II. p. 2. Based upon an analysis by *Times* staff writers Gaylord Shaw and John Kendall.

75. Kuhn, D., W. Brewster, and B. Klaene. 1992. Water emergencies. *Procedures Manual.* Section 203.17. Cincinnati Fire Division.

TERRAIN-BASED NATURAL DISASTERS 14

INTRODUCTION

Natural disasters take many forms. While weather-based natural disasters were examined in the previous chapter, the discussion will now turn to natural disasters that occur in or on land directly—*terrain-based natural disasters.* Specifically, disaster planning factors will be identified concerning earthquakes, volcanoes, and landslides.

EARTHQUAKES

Earthquakes are seen on a less frequent basis than some of the other disaster events that have been discussed, but they do occur on a frequent basis. Large, disastrous earthquakes only occur every few years, and some of these occurrences are located in countries abroad. This was the case on October 8, 2005, when the Kashmir earthquake, also known as the South Asian earthquake, struck the countries of Pakistan, India, and Afghanistan. This earthquake measured 7.6 on the Richter scale, placing it among the strongest of earthquakes, such as the 1906 earthquake in San Francisco. The South Asian earthquake killed more than 76,000 people and injured more than 106,000. The aftershocks, as many as 147 the day after the quake, continued to multiply injuries and resulting events that occurred after the initial shock.

Although this earthquake occurred on the other side of the planet, there would be many similarities if it occurred in Chicago, New York, or Miami. Therefore, it is important that disaster planners and response executives apply current business trends, which involve globalization. Studying events abroad, such as those discussed in the terrorism chapter, can show what local communities could expect in the near future.

The prevalence of earthquakes was illustrated in the United States on January 3, 1991, when at least two dozen earthquakes shook Carson City, the state capital of Nevada. These earthquakes pointed out how earthquakes can be of long duration and yet cause minimal damage. The series of earthquakes began just after 4 a.m. and continued for hours, but the largest earthquake registered only 4.5 on the Richter scale. Most are far more deadly. On June 28, 1991, 2 people died and more than 100 were injured in an earthquake that registered 6.0 on the Richter scale in Pasadena, California. Property losses were estimated at $18 million. In a geological rarity, no aftershocks were reported following the quake. More than 300 buildings were damaged in Pasadena, Monrovia, Sierra Madre, Arcadia, and other communities.[1,2]

In April 1990, aftershocks were still occurring from the earthquake that shook the San Francisco Bay area in October 1989, killing 60 people.

An earthquake is the trembling or shaking movement of the Earth's surface. Of all natural disasters, earthquakes can inflict the greatest loss of life and property, a fact that is borne out historically. As long ago as 1556, Shensi, China suffered the loss of 830,000 lives. As recently as 1990, the loss of life in an earthquake in Iran was estimated at 25,000. In 1737 in Calcutta, 300,000 were killed. In 1920, in Kansu, China, 180,000 were killed, and in 1933, 143,000 were killed in Tokyo and Yokohama.[3] If the seers and scientists are correct in their forecasts, more devastating quakes are likely to occur in the future.

Causes of earthquakes

Most earthquakes are caused by the accumulation of strain in the Earth's interior, resulting from convective movement of massive tectonic plates covering its surface. Rocks that make up the outer layers of the Earth are continually being compressed and stretched by forces within the Earth. These forces are believed by some geologists to be the result of enormous heat in the Earth's core. One theory is that the radioactivity of the Earth over a period of millions of years has produced heat inside that cannot readily escape through the crust, especially through the continental crust, so it breaks through near the sea. This could account for the fact that earthquake belts and volcano belts coincide somewhat around the perimeter of the Pacific Ocean.

When the strain becomes too great, a massive cracking of interior rocks takes place, and the Earth shakes. These fractures in the rocks are called *faults*. When slippage subsequently occurs along these faults, additional earthquakes are produced, with the intensity depending upon the length of the fault and amount of the slip. Most earthquake faults lie beneath the surface, but some, such as the San Andreas Fault in California, can be seen from the air for a distance of 600 miles.

According to some geophysicists, the combined gravitational pull of the Sun and the Moon may be strong enough to trigger an earthquake. The Tehachapi earthquake of July 21, 1952, may have been caused this way. The Sun and the Moon were almost in a perfect line when this 7.7 magnitude shock occurred, killing 14 people and causing about $60 million damage. About 20 aftershocks were of magnitude 5.0 or greater on the Richter scale.

Shrinkage of the Earth's crust, the collapse of cave roofs, volcanic action, the pumping of fluids in the crevices of subterranean rocks, and the detonation of nuclear devices beneath the Earth's surface are also said to cause earthquakes. The subterranean point of origin is called the *focus* of the earthquake, and the point on the surface directly above the focus is called the *epicenter*. The release of energy at the time of the quake produces waves in the Earth that travel along its surface; other waves pass through the Earth at a depth of about 1,800 miles.

Power of earthquakes

On December 26, 2003, just one year prior to the devastating tsunami mentioned previously, at least 15,000 deaths and an additional 10,000–50,000 injures occurred in Bam, Iran when an earthquake destroyed 70% of the town This type of destruction, which measured a relatively moderate 6.7 on the Richter scale, is representative of the tremendous power that earthquakes can unleash.[4]

The power of earthquakes is estimated in various ways. The Richter scale is used to measure energy released (magnitude) as recorded on seismographs, and the strength of an earthquake

Table 14–1. Magnitude/intensity comparison. This chart derived from the U.S. Geological Survey illustrates how the Richter scale, the Mercalli Intensity scale, and the associated damage are related. (A complete description of the systems can be found with the chart at http://earthquake.usgs.gov/learning/topics/mag_vs_int.php.)

Earthquake Magnitude vs. Intensity

Richter (magnitude)	Mercalli (intensity)
1.0-3.0	I
3.0-3.9	II-III
4.0-4.9	IV-V
5.0-5.9	VI-VII
6.0-6.9	VII-IX
7.0 and higher	X or Higher

Effects

I. Not felt except by a very few under especially favorable conditions.

II. Felt by only a few persons at rest, especially on upper floors of buildings.

III. Felt quite noticeably by persons indoors, especially on upper floors of buildings. Many people do not recognize it as an earthquake. Standing motor cars may rock slightly. Vibrations similar to the passing of a truck. Duration estimated.

IV. Felt indoors by many, outdoors by few in the day. At night some awakened. Dishes, windows, doors disturbed; walls make cracking sound. Sensation like heavy truck striking the building. Standing motor cars rocked noticeably.

V. Felt by nearly everyone; many awakened. Some dishes, windows broken. Unstable objects overturned. Pendulum clocks may stop.

VI. Felt by all, many frightened. Some heavy furniture moved; a few instances of fallen plaster. Damage slight.

VII. Damage negligible in buildings of good design and construction; slight to moderate in well-built ordinary structures; considerable damage in poorly built or designed structures; some chimneys broken.

VIII. Damage slight in specially designed buildings, considerable damage, partial collapse in ordinary buildings. Damage great in poorly built structures. Fall of chimneys, columns, walls, and monuments.

IX. Damage considerable in specially designed structures; well designed frame structures thrown out of plumb, Damage great in substantial buildings, with partial collapse. Buildings shifted off foundations.

X. Some well-built wooden structures destroyed, most masonry and frame structures destroyed with foundations. Rails bent.

XI. Few (Masonry Structures standing). Bridges destroyed. Rails bent greatly.

XII. Damage total. Objects thrown into air.

is rated in intensity scales such as the Mercalli scale. The Mercalli scale assigns a qualitative measure of damage to the terrain and structures. This could range from "not felt" to "damage nearly total."[5] No shock has been assigned a magnitude over 8.6. The intensity increases in a logarithmic fashion, and thus an increase of 1 in intensity number means the energy released is 10 times greater. Thus, an earthquake having a magnitude of 8.5 is 10 times greater than one of 7.5, and some experts say damage would be 30 times greater. A chart depicting the Richter scale, Mercalli Intensity scale, and the effects of the intensities is found in table 14–1.

In populated areas, an earthquake of 3.5 on the Richter scale can cause slight damage in the local area; 4 can cause moderate damage; 5 can cause considerable damage; and 6 can be severe. A reading of 7 is a major earthquake, capable of widespread, heavy damage. A reading of 8 is a "great" earthquake, such as the one in San Francisco in 1906, which was 8.3 on the Richter scale, and the one in Alaska in 1964, which registered between 8.3 and 8.7. The earthquake in Alaska released energy equivalent to 100 underground 100-megaton nuclear explosions on one line.[6] A shock of magnitude 9.5 would release about 40 times the energy of the largest recorded shock, and a shock of 10 would be felt over the whole Earth.[7]

The Modified Mercalli Intensity scale has 12 divisions. Moderate damage is produced at VII, and people will run out of buildings. An earthquake with an intensity of VIII will cause major damage.

The Ross-Forel scale has 10 divisions, and people will leave their dwellings on a class six earthquake.

Location of earthquakes

Violent earthquakes occur roughly in two zones of the Earth's surface. One follows a line drawn through the Mediterranean, Asia Minor, the Himalayas, and the East Indies; the other follows the western, northern, and eastern shores of the Pacific Ocean. In the United States, 80% to 90% of all earthquakes occur in the California/Arizona/Nevada region, where hundreds of them, usually of minor intensity, occur daily.

Frequency of earthquakes

In the past two decades, the United States has experienced several severe earthquakes (Alaska, San Fernando Valley, and the San Francisco Bay area in California), though thousands of small earthquakes occur in the nation every year. Seismic risk maps of the United States depict the relative damage expectancy from earthquakes and show that several population centers are located in high-risk areas.

Earthquake predictability

Anyone can predict that an earthquake will occur within an hour, someplace, and he or she would be right, because about 1 million occur every year. Of these, about 800 are in the range of 5 to 5.9 on the Richter scale; about 6,200 are in the range of 4 to 4.9; and about 49,000 are in the 3 to 3.9 range. Roughly 100,000 have a magnitude of 2.5 to 2.9. About 150,000 perceptible shocks are felt each year.[8] However, the ability to predict an earthquake of a specific magnitude at a specific location is open to question. The areas where earthquakes are likely to occur are known, but this does not preclude earthquakes from occurring elsewhere. As for the specific time when a strong earthquake will occur, there are better answers available today than the advice given to former author Charles Bahme in his college geology class, i.e., "The longer it has been since the last, the closer you are to the next."

There are scientists who have achieved moderate success in predicting earthquakes. "In localized areas small and moderate quakes are being predicted accurately today by the United States and Soviet scientists. In quake-prone mainland China, a major earthquake was predicted, and measures taken, averting a catastrophe even worse than that in Guatemala."[9] The earthquake with a 7.3 magnitude occurred on February 4, 1975, in the town of Haich'eng, and nearly every house was flattened. Since people had been warned at 6:00 p.m. to leave their homes with their animals, only those who ignored the warning were killed when the quake struck 1½ hours later. Two or three days prior to the shock, the warning signs began to multiply. Water rose in wells suddenly, farm animals refused to enter their folds, and schoolchildren skating on frozen lakes noticed that frogs were jumping through the holes in the ice.

In 1987, Tom Tugend reported on research being conducted by UCLA on the West Coast. For example, one of the country's leading geophysicists, Leon Knopoff, discovered that earthquakes do not happen at random. Over certain periods of time and geographical space, they actually occur in clusters, and by analyzing these patterns of clustering, earthquakes can be predicted.[10]

The Chinese have accumulated knowledge over the past 3,000 years about events that usually precede earthquakes. These include minute earth movements across fault lines that are sensed by lasers and slight changes recorded in tiltmeters. Changes in the electrical resistance in the Earth's crust recorded on gages (signaling changes in its density and the amount of water in the rock) have also been noted. In addition, they have noted decreases in the amount of the radioactive gas radon in well water and also variations in the speed of seismic waves.

Their system is not always successful, for there apparently was a failure to predict a strong quake in Yunan Province in May 1974, and the two quakes that struck a three-city area with a population of 15 million in July 1976. More than 700,000 were estimated to have been killed in these quakes, the first of which, measuring 8.2 in magnitude, was the strongest felt anywhere in the world since the 8.5 quake devastated Anchorage, Alaska in 1964.[11]

The first valid U.S. earthquake prediction was in 1973 by Dr. Yash P. Aggarwal, who discovered changes in speed of seismic waves from seismometers placed around Blue Mountain Lake, a region of frequent but small tremors in the Adirondack Mountains. His calculation that a 2.5 to 3 magnitude tremor would occur in two to four days in that area proved correct when one of 2.6 hit within that time interval.[12]

Other predictions, specifying place, time, and magnitude, have been made in the United States. In 1974, Dr. James H. Whitcomb, a California Institute of Technology geophysicist, came close to meeting the criteria—he predicted the time and place, but missed the magnitude. In addition, in northern California, Menlo Park scientists monitoring the network of instruments south of Hollister informally predicted an earthquake with a magnitude of 5.2 that occurred on schedule but slightly off in location. Dr. Whitcomb then made another prediction. He stated in April 1976 that there was evidence of stress building up in subsurface rock layers to the point where a 5.5 to 6.5 magnitude earthquake could be expected to occur within a roughly circular area in Southern California. This area was about 80 miles in diameter, centered in the same region as the site of the 1971 Sylmar tremor. There are at least 20 different active faults in that zone, and the temblor could occur in any one of them, he explained, but the evidence, in the form of variations in velocities of sound waves traveling those stressed layers, did not permit a more precise location. His prediction may have been realized in the modest 1986 quake in East Los Angeles, although it is a quake-prone region, regardless.

Various prediction systems are in use, some based on earthquake surveillance that is provided by seismometers and strong-motion seismographs. Comparison of surface motion on various soils and rocks in the vicinity of strong-motion stations during an earthquake can be made with seismoscopes. Several hundred seismograph stations are operating continuously throughout the world, under control of government and private agencies.

According to the United States Geological Service, the National Strong-Motion Network (NSMN) currently operates over 1,000 strong-motion devices at approximately 700 permanent stations located in 33 states and the Caribbean. Additionally, recorders are currently installed at approximately 210 structures, which include 133 buildings, 17 bridges and overcrossings and 60 dams. Over 40 instrumented buildings have 6 or more seismic sensors and several have 36 or more sensors. The NSMN is primarily the result of cooperative efforts with many the USGS and many other federal, state, and local agencies, private companies, and academic institutions[13]

When doomsayer Iben Browning predicted that a major earthquake would occur along the New Madrid Fault on December 2, 1990, city officials, school officials, and residents from north to south took his prediction seriously. There was a run on home insurance, a concerted effort to conduct school earthquake drills, and even plans to call off school in some southern counties. People generally prepared for a genuine earthquake threat. The New Madrid Fault runs about 120 miles from Marked Tree, Arkansas, to Cairo, Illinois, and includes Tennessee, Kentucky, and Missouri. One of the reasons that people took this prediction seriously is that one of the three worst earthquakes in the nation's recorded history happened in New Madrid, Missouri—named after the fault. In western Kentucky, school officials in eight districts cancelled classes December 2 and 3. Residents went on buying sprees and stocked their basements with water, food supplies, and survival gear.

Although the earthquake failed to materialize, the prediction certainly had value in raising public awareness concerning the real dangers that earthquakes present and the preparations that can and should be made in anticipation of them.[14] This fault line still remains the biggest fear for many planners and responders due to the lack of seismic-protected buildings in this region of the country. Many emergency planners and predictors have recorded documentary television shows describing what destruction would take place based on the magnitude of the earthquake. If a large-scale earthquake, such as a 7.0, were to occur, the prediction is near-total destruction for many cities in the Midwest. Predictors proclaim this would be a disaster of proportions not previously seen on this planet.

Earthquake surveillance equipment

The three types of earthquake monitoring generally employed are worldwide seismography, local seismic recording, and strong-motion seismography. In addition to a worldwide network of standard seismographs, there are microearthquake seismic networks, one of which consists of about 50 stations in the area from San Francisco to Hollister, California. This network locates about 2,000 earthquakes per year. Silent earthquakes that occur when the land along a fault slips so slowly and smoothly that it cannot be felt may be precursors to strongly felt tremors. The recording of data on this kind of "creep" may enable scientists to calculate the repeat time of large quakes. The *repeat time* is the approximate date of reoccurrence of large earthquakes, which evidence suggests is about every 100 years off Japan. Although the technique might not be applicable to Southern California because of its particular geology, it is also the time interval thought by some scientists to be applicable to large tremors along the San Andreas Fault.[15]

Dilatancy is a phenomenon believed to take place in deep underground rock layers subjected to great strain. It has been proposed as a very promising way of predicting the time, place, and magnitude of an earthquake. Dilatancy occurs when rock under extreme pressure suddenly opens with tiny cracks and expands. This process is believed to occur when two large land masses try to move past each other. Initially the total volume of these cracks is greater than the volume of underground water available to fill them, but when the cracks fill up, the water lubricates the interface between the land masses, and an earthquake occurs. Dilatancy explains other precursors—variations in electrical sensitivity of the rock, increase of radon in the well water, and uplifting of land under strain. In the Palmdale area northeast of Los Angeles, the Seismic Safety Commission has been asked to consider the possibility that a large swelling that extends for about 100 miles along the San Andreas Fault, varying from 2 to 10 inches in height over the 4,500 square miles involved, is a precursor of a coming major earthquake. Although this uplifting of the land has taken place since the early 1960s, its significance is difficult to judge, since there have been instances in which a sudden uplift of land preceded a large and destructive earthquake, and other cases where no tremors followed.[16,17]

Soviet and American scientists continue to work to achieve a reliable earthquake-forecasting system. Dr. Robert Hamilton, director of the Office of Earthquake Studies for the U.S. Geological Survey, was reported as saying that lack of funds had retarded research aimed at developing techniques that would give 75% accuracy in quake predictions.

Legislation was adopted unanimously by the U.S. Senate in the late 1970s authorizing $150 million over a three-year period to improve earthquake prediction techniques. The 1976 devastating earthquakes in Guatemala, Italy, the Soviet Union, and China, along with the discovery of the Palmdale bulge, apparently spurred the action. In addition to furthering prediction research, the bill aimed to reduce loss of life and damage caused by earthquakes. This was to be accomplished by identifying communities near fault lines or those likely to suffer from landslides or tsunamis. In addition, it called for development of building codes and construction methods to prevent future occurrences of events such as the collapse of the Sylmar Veterans Hospital in the 1971 southern California quake or the collapse of the double-decker freeway in Oakland in 1989. Both of these structures had been designed to existing earthquake-resistant standards. Finally, the bill was also aimed at development of warning capabilities and emergency services.

Currently, the National Earthquake Information Center (NEIC), a division of the United States Geological Survey, is the leader in many areas of earthquake information.

The NEIC has three main missions. First, the NEIC determines, as rapidly and as accurately as possible, the location and size of all significant earthquakes that occur worldwide. The NEIC disseminates this information immediately to concerned national and international agencies, scientists, critical facilities, and the general public. Second, the NEIC collects and provides to scientists and to the public an extensive seismic database that serves as a solid foundation for scientific research, principally through the operation of modern digital national and global seismograph networks and through cooperative international agreements. The NEIC is ***the*** national data center and archive for earthquake information. Third, the NEIC pursues an active research program to improve its ability to locate earthquakes and to understand the earthquake mechanism. These efforts are all aimed at mitigating the risks of earthquakes to mankind; and they are made possible by the fine international cooperation that has long characterized the science of seismology.[18]

An article by R. Paddock states, "Under the worst combination of an inaccurate prediction and ill-conceived public response, the prediction and quake together might be even more costly than an unpredicted quake would have been."[19,20]

It also asserts that while scientists should develop, assess, and issue predictions of earthquakes, warnings should come from elected officials. If mass evacuation is deemed necessary, it would be the decision of governmental authorities to issue such orders to residents of vulnerable areas. It was believed that all steps in the development of a prediction should be made public to prevent the possibility of inside information being used by some people to the disadvantage of the general public.

How will people react to earthquake forecasts? There has been speculation that, following prediction of a major earthquake and issuance of official warnings, the panic that would follow could be more disastrous than the physical damage. "The body of scholars headed by UCLA's Dr. [Ralph H.] Turner does not agree with that alarming idea, but their report does see some very serious problems involved in earthquake forecasting, especially if government doesn't begin to formulate strategies for action. And thus far, there's not much evidence of public policy planning for the inevitable eventuality which lies down the road."[21] The report pointed out that fewer than 10% of Americans can safely ignore the earthquake hazard, but most use or occupy buildings that are substandard for seismic safety.[22]

Often earthquakes strike in somewhat rapid succession and in close places. On January and February 13, 2001, El Salvador experienced two different earthquakes. On January 13, the 7.6 earthquake struck with the epicenter at 60 km southwest of San Miguel, El Salvador. At least 844 people were killed, 4,723 injured, 108,226 houses destroyed, and more than 150,000 buildings were damaged in El Salvador. On February 13, a 5.7 earthquake struck with the epicenter 80 km east of San Salvador, El Salvador. At least 827 people were killed, many more were injured, and there was extensive damage. The most severe damage occurred in the San Juan Tepezontes/San Vicente/Cojutepeque area.[23]

Effects of earthquakes

One important result of great earthquakes is the subsequent production of additional great earthquakes, known as *aftershocks*, sometimes in the same region, and sometimes thousands of miles away and occurring months later. It is believed now that there may be a link between the huge earthquakes that periodically erupt all around the rim of the Pacific Ocean basin. Scientists now are beginning to doubt the "local" nature of large quakes—those with a magnitude of 5.0 or more. "A huge tremor that shakes Japan, the scientists suspect, may well be the trigger that sets off another earthquake months later in Peru, Mexico, California, or some place on the edge of the Pacific basin."[24]

Scientists theorize that material from the upper mantle wells up along the mid-Atlantic Ridge, causing the movement of large areas of the Earth's surface called *plates*. These plates are thought to interact in one of three ways: spreading where new crust is formed, subduction where one plate plunges under another, or other fault action where two plates rub together.

Whether the earthquakes cause the movement of these massive tectonic plates, or vice versa, a large earthquake is one of nature's most devastating phenomena. "The energy released by a magnitude of 8.5 earthquake on the Richter scale is equivalent to 12,000 times the energy released by the Hiroshima nuclear bomb. While these cataclysms have their foci well beneath the earth's surface, cities have been destroyed and thousands of lives lost in a few seconds as the result of great earthquakes in the past."[25]

Among the primary effects of an earthquake's surface waves is the violent shaking of the ground. Shaking causes the greatest amount of damage from earthquakes in highly populated areas. Most structural failures in earthquakes likely occur in one of several ways. They may result from shaking that directly damages the structure, or that causes soil failure beneath its foundation (as by slope failure due to liquefaction of water-saturated soils), or that causes the soil beneath the foundation to densify and settle, allowing the structure to fall. Sometimes shaking will open up fissures and cause great displacement. For example, displacement in the 1906 earthquake in San Francisco was 21 feet horizontally, while the one in Yakutat Bay in 1899 was 47 feet vertically.

Shaking can cause vibrations to occur that are severe enough to cause collapse or shearing of dams, bridges, tunnels, or buildings. This shearing and collapse can be seen in figure 14–1. Chimneys, high-rise buildings, water tanks, and bridges are especially vulnerable to resonant vibration motion. Walls of high-rise buildings without adequate lateral bracing can fall outward, allowing floors to cascade one on top of the other, crushing occupants. In countries where construction is largely of unreinforced mud and adobe brick, collapse of buildings is apt to be total. This was the case in Bam, Iran. Earthquake damage from Bam, Iran can be seen in figure 14–2. The construction of buildings has a large effect on their demise when an earthquake occurs.

Fig. 14–1. Shear wall collapse from an earthquake. Note that facades built onto buildings are likely to fall away from the building during the earthquake. This is often the case in new buildings, as many of them have facades.

Fig. 14–2. Earthquake damage from Iran. This photo shows American US&R responders present at the Iran earthquake. Note the near-total destruction of the town.

FEMA News Photos.

The Earth can "break" by four different processes. These processes are: faulting (displacement of two Earth masses along the fault plane), lurching (surface cracking of soils at places other than directly along faults), liquefaction (causing soils to lose their strength), and slope failure (causing landslides).

The movement during an earthquake, lateral and vertical, severs pipes as well as underground power and communications cables, releasing flammable and toxic gases and liquids and knocking out fuel, power, sewer, and water systems. Elevators can be stalled or jammed, or their suspension system may fail.

Water damage can be caused by an oscillation known as *seiching* that occurs in lakes and bays, causing water to surge from one end to another, reaching great heights and overflowing the banks. As a result of the 1964 earthquake in Alaska, water rose six feet in Memphis, Tennessee, 5,000 miles from the epicenter, due to this type of action. Collapsed water tanks and broken dams can produce floods, the effects of which were discussed in the previous chapter.

Another secondary effect is a tsunami, discussed in the previous chapter. A tsunami is the seismic sea wave generated rather infrequently by submarine earthquakes or disturbances. Since 1900, more than 200 have been recorded in the Pacific area. It is estimated that about 600,000 people would be endangered by a tsunami in Pacific Coast states if the wave were to be 50 feet high; if the wave were 100 feet high, about 1,141,000 would endangered.[26]

Tsunamis, major and local, took most of the 122 lives in the Alaskan earthquake of 1964, including those lost in Hawaii and on the West Coast. In 1755, a seismic wave 50 feet high wrecked Lisbon, Portugal.

Local waves in confined areas generated by earthquake action can be very destructive. When a landslide caused by an earthquake fell into Lituya Bay, Alaska, a wave 1,700 feet high was created. Even without creating a water wave, landslides caused by earthquakes can result in a great loss of life. At least 70,000 persons were killed—most of them buried under a fast-moving wall of water, rock, and mud—when an earthquake struck western Peru in 1970. About 800,000 were left homeless.[27]

Earthquakes can also cause subsidence. About 30,000 square miles of Missouri sank 5 to 15 feet in 1812 during seismic activity that centered near New Madrid, and the Mississippi River flowed upstream to form Reelfoot Lake in Tennessee. Some areas in that region were raised at the same time. Part of Japan's Sagami Bay dropped 689 feet, and part raised 850 feet, creating a seismic wave 35 feet high in the 1923 earthquake that completely destroyed the city of Yokohama. It also destroyed 71% of Tokyo and 51% of Yokosuka.

Subsidence was a major cause of loss in the 1964 Alaska earthquake, when almost the entire residential subdivision of Anchorage's Turnagain Heights slid from its original site toward Turnagain Arm. Another example in the same quake was the slide on 4th Avenue in Anchorage.

Fire was the secondary effect that caused 80% of destruction in the Great San Francisco Earthquake of 1906. Of 143,000 who died in the 1923 earthquake in Kwanto, Japan, in which Tokyo lost all its records and art collections, 40,000 were residents of Tokyo who took refuge in an open space across which flames were driven, and none escaped.[28] The monetary loss from the earthquake and fire was $2.5 billion. With inflation, this would estimate to be $272 billion today.

Fires caused by overturned stoves, broken electric wiring, and spilled molten materials are fed by escaping gas and flammable liquids from ruptured fuel tanks and pipe lines, as well as debris from damaged buildings. Control of fire becomes especially difficult when water mains are broken, streets are blocked by fallen buildings, fire stations and fire apparatus are damaged, and firefighters are killed, injured, or otherwise unable to answer emergency calls. Where highways and rail systems are put out of service, not much help can be rendered by mutual aid companies responding from areas outside the zone of damage.

Property damage from earthquakes has been considerable in the United States—billions of dollars since 1865. Earthquakes have been responsible for the loss of hundreds of thousands of lives.

Earthquakes affect the economy of the nation, destroying means of production, distribution, and transportation of commercial products and disrupting communications and public utility services. Many businesses destroyed never resume, while most suffer delay until factories can be rebuilt and disruptions in the labor force and public utilities have ended.

No one knows how much damage is done to the environment by violent earthquakes, such as in Peru (1970), Alaska (1964), Montana (1959), and San Francisco (1989). However, it is believed that transient and permanent inundation and large landslides strongly affect flora and fauna. Large forests are destroyed by violent shaking, uprooting, or inundation with seawater, with adverse effects on wildlife. Crustal changes of the ocean shore can exterminate marine organisms.

Examples of major earthquakes

One of the most deadly earthquakes of the current century struck on October 8, 2005. The quake, which killed more than 74,000 and injured hundreds of thousands, registered 7.6 on the Richter scale, shaking the densely populated Balakot.

The history of three decades of other great earthquakes will now be discussed, noting that the 1970s, 1980s, and 1990s were deadly periods in earthquake history.

Managua, Nicaragua, 1972. On December 23, 1972, a series of earthquakes hit Managua, Nicaragua, killing at least 10,000 persons. Two light shocks were felt on December 22, causing the fire department to move apparatus out of fire stations. Following the second shock, all rigs were put back inside. Then the big one hit 30 minutes after midnight on December 23, collapsing the main fire station that had been built only eight years before (using earthquake-resistive standards). Four of the latest fire trucks and four rescue vehicles were destroyed with the collapse of the fire station's second floor onto the first. Two firefighters were killed at the same time. The second fire station withstood the shock, leaving about 35% of the fire equipment in the department in operable condition.

All on-duty fire department members remained on duty (unlike many government employees who left immediately to attend to their families). Within 30 minutes, 70 of the department's volunteer firefighters reported for duty, and 230 of them were on duty by 7:00 a.m., constituting the total forces for this capital city of 1.5 million people. In the absence of communications, other than vehicle radios, the fire chief had to drive about 18 miles to request help from the nearest neighboring community. As soon as word got out about their situation, additional assistance was forthcoming without further requests.

To handle fires that developed into a mass fire in an area of 9 blocks by 18 blocks in the center of the city, water had to be relayed from 3,000 feet away. Only 1½-inch lines were used to provide perimeter control, since there was insufficient water for use of master streams. With the help of the U.S. Army Corps of Engineers, heavy equipment and explosives were used to clear a firebreak around the involved area. The 13 fire trucks with 135 personnel from other towns in Nicaragua, plus 9 fire trucks from six neighboring countries, were deployed around the major fire area to prevent its spread.

The final result of the quake showed that 295 city blocks were severely damaged, with the mass fire confined to 162 blocks. Of the city's 70,000 homes, 53,000 were destroyed. Of light shops and commercial buildings, 95% were destroyed, 250,000 people were displaced, and the damage caused by the quake was over $1 billion.[29]

Soviet Asian Republic of Uzbek, 1976. The Soviet Asian Republic of Uzbek was hit by powerful earthquakes in April and May 1976. While there was said to be property damage in this area, which contained the Soviet Union's largest gas fields, no report of casualties was given. A Moscow seismologist reported by television that the Soviet earthquakes measured 7.3 and 7.0 on the Richter scale. The Moscow report said the reason that the Soviet's largest of their two quakes caused no injuries was because seismologists gave early warning of the tremor and the inhabitants were evacuated.[30]

Italy, 1976. Both of the earthquakes in Uzbek were stronger by several times than the one that devastated northeast Italy in early May 1976, killing more than 900 people. The Italian earthquake measured 6.5 to 6.9, causing death and destruction in the towns near the border of Yugoslavia—Buia, Maiano, Forgaria, and Gemona. A few chimneys were knocked down in Venice, but in western Italy, the famous Leaning Tower of Pisa, though shaken, had suffered no damage, and its tilt of 17 feet off center remained unchanged.[31]

Asia, 1976. On August 16, 1976, two major quakes struck in Asia in a two-hour period—one in central China and another in the Philippines. Seismic reporting stations said that the Chinese quake was centered near the Kansu-Szechwan province boundary in central China, about 650 miles west-southwest of Peking. This was far from the northeast China area devastated by a quake in the preceding month, which killed an estimated 750,000 people in and around Tangshan. The August Chinese earthquake was measured at 7.3 on the Richter scale; the one in July was 8.2.

The National Earthquake Information Center in Golden, Colorado reported a tremor measuring 8.0 on the Richter scale in the Moro Gulf of the southern Philippines, near Mindanao, on August 16, 1976. The quake struck more than a dozen cities, from Jolo, 550 miles south of Manila, to Legaspi City, 300 miles to the north. The Philippines National Geophysical Observatory reported the quake as 6 on the Rossi-Forel scale. More than 4,000 were killed and 90,000 made homeless in this worst quake ever to hit the Philippines Islands.[32, 33]

As has been described, the numerous earthquakes that struck during the 1970s made it a rather lethal decade.

1980s earthquakes. Moving into the 1980s, history repeated itself, with only the names of the places changing.[34] In 1980, 4,500 people died in Nigeria, and 4,800 in southern Italy. In 1981, 8,000 people died in the Iranian earthquake earlier described. In 1985, 7,000 people died in central and southwestern Mexico, followed in 1986 by 1,500 deaths in El Salvador. Then came 1988 and the devastating earthquake in Soviet Armenia that claimed 25,000 lives and taxed the relief services of the world. Both the Mexico City and Armenia incidents highlighted enormous search and rescue (SAR) problems following a major earthquake and the need for well-equipped and trained personnel for locating victims and extricating them from heavy debris.[35] This would prove to be the basis for the existence of the FEMA US&R teams in the early and mid-1990s.

The decade closed with the 1989 San Francisco Bay area earthquake, not as deadly in terms of lost lives, but certainly the most devastating to the American landscape, and taking the greatest toll in property damage. There were 60 deaths attributed to this earthquake, but property damage ran into the billions of dollars.[36]

1990s earthquakes. During 1991, earthquakes killed 115 in central China, 200 in Peru, 40,000 in Iran, and 1,650 in the Philippines.[37]

The 1990s saw a slowdown, but many earthquakes continued to occur throughout the world. On January 17, 1994, the city of Los Angeles experienced an earthquake measuring 6.7 on the Richter scale, known as the Northridge earthquake. Although the magnitude was 6.7, this became the most expensive earthquake in terms of damage due to the ground acceleration. It was the highest ever recorded in an urban area. Only 57 people were killed and 9,000 injured. The damage costs were amassed through the collapse of three different interstate and state roads. These bridge collapses resulted from a lack of structural design to withstand an earthquake. The ground acceleration of 1 g caused many well-built structures to succumb to the violent shaking.[38]

A strong earthquake with a remotely located epicenter occurred on October 16, 1999. The Hector Mine earthquake registered 7.1 on the Richter scale, but due to the epicenter being located in the middle of the Mojave Desert, with the nearest civilization 14 miles away, almost no injuries were reported. Had this earthquake's epicenter been around an urban area, the results would have differed greatly.

Can the damaging effects of earthquakes be avoided?

There appears to be no likelihood that methods will be devised to prevent earthquakes. Research was conducted on the "lubrication" of rock interfaces by pumping fluid down a borehole under pressure, but the possibility of generating sufficient microtremors to avoid a macrotremor was purely speculative at the time. Nevertheless, the hazards of earthquakes can be greatly reduced and their disastrous effects minimized by application of seismology principles in building design and construction.

In regions known to be earthquake-prone, places of public assembly should be of earthquake-resistant construction and should not be placed over faults or in areas of unstable slopes.

Building codes should be revised to require that existing structures be strengthened to resist earthquake damage. The need for such legislation to avoid the inevitable result of natural catastrophe in densely populated areas is illustrated in the estimate by the Office of Emergency Preparedness. It calculated that 500 million people are in imminent danger of death, serious injury, or substantial destruction of property due to earthquake.

In 1988, a handbook called *Rapid Visual Screening of Buildings for Potential Seismic Hazards* was developed by FEMA, dividing the country into three areas of earthquake risk: low, moderate, and high. These figures are based on geological and seismological investigations into the 1990s as well as the locations of the historical earthquakes that have been discussed in this text.[39]

In Los Angeles, engineers and scientists estimated that 200,000 people could die in the metropolitan area in the event of a major earthquake. A major earthquake could collapse about 28,000 potential "death traps"—pre-1933 unreinforced masonry structures.[40]

Since the 1933 earthquake in Long Beach, in which many schools collapsed—fortunately about 6 p.m. when they were unoccupied—the state of

California established new building standards. These required that new construction conform to minimum earthquake-resistant standards, and existing schools either conformed or by law were abandoned by 1975.

The Building and Safety Committee of the Los Angeles City Council estimated that 14,000 commercial and residential structures did not conform to seismic safety standards in the city's building code. Of these, about 300 were churches, theaters, libraries, nightclubs, restaurants, meeting halls, and assembly buildings. The Council Committee proposed that, in lieu of ordering these 14,000 buildings to be brought up to code or closed down, a requirement be adopted for posting conspicuous signs on them to warn the occupants that they were potential death traps in a severe quake. The building owners would have to notify occupants of the risk. In order to have the signs removed, they had to make improvements to bring the buildings up to code, the cost of which averaged about 80% of the cost of new building.[41] As expected, most building owners had to live with the signs, and most of these buildings survived into the 21st century.

Hazard reduction

Although building collapse is not the only problem encountered by planners and response personnel, their collapse tends to cause a majority of the injuries and deaths associated with a major earthquake. Therefore, a large push, as indicated in the Los Angeles example in the preceding paragraph, involves seismic retrofitting. However, planners and response executives must understand what options are available to retrofit a building. According to a 2004 MIT report, the following are some options available:

1. **Hybrid solution.** According to *Wikipedia*, "PRESS—Precast Seismic Structural Systems is a U.S./Japan joint program for the development of structural methods suitable for ductile reinforcement and joining of rigid concrete plate and beam structures."[42]
2. **Isolation.** This involves excavating around the building and separating the building from its foundation by placing rubber and metal isolating pads between the two structures. This allows the ground to shake independently of the building.
3. **Dampers.** Used to "absorb the energy of motion and convert it to heat, thus 'damping' resonant effects in structures that are rigidly attached to the ground."[43]
4. **Shock absorbers.** These function in the same fashion as the shock absorbers on a vehicle and are used to connect portions of the building, allowing movement during ground shaking and reducing the chance of different parts of the building colliding while shaking.
5. **Reinforcement.** This is the most common method used in low-rise buildings. It involves providing additional strength to existing structures to help them safely survive the seismic activity.[44]

Hazard reduction measures that would increase fire department usefulness in a major earthquake include the following:

1. **Increasing the ability of water systems to withstand severe earth vibrations.** In the Sylmar earthquake of February 9, 1971, the Los Angeles water system lost the use of about 2,000 fire hydrants and 25,000 individual services, and it had about 1,500 distribution main breaks. During the 14 days required to restore the system, emergency water was supplied from 53 tanks from various agencies, and by 17 fire department pumpers pumping water into the uphill systems as breaks were repaired. (This method reduces the possibility of pollution from sewer breaks near the water system breaks.)

   Speedy restoration of the water system is important. It was found that purchase of additional skid-mounted diesel pumps (capable of 350 head) that could be moved to various locations as needed, plus one-cubic-foot booster pumps for local areas, would remedy deficiencies. Increasing the amount of repair materials on hand, such as clamps for repairing small mains, is a good precaution.

Following the original San Francisco earthquake fire, approximately 150 underground tanks or cisterns were constructed at key intersections. Holding approximately 75,000 gallons or about an hour's supply for a pumper, these are maintenance free and have been so successful that in 1991, a decision was made to add 94 cisterns to additional portions of San Francisco.[45]

San Francisco has also pioneered the development of a portable water supply system (PWSS) incorporating 16,000 feet of five-inch-diameter hose carried on special hose wagons, with portable hydrants and reducing valves. This system is capable of complex gridding and can create a water curtain several city blocks in length. Cities in earthquake-prone areas might consider the development of a PWSS.[46] Many fire departments in urban and suburban areas do not concern themselves with carrying adequate amounts of hard suction hose on their pumpers to draft water from a static source, citing a hydrant program that places a fire hydrant every 300 to 500 feet. This everyday operation could be quickly tested when a small earthquake causes a catastrophic failure of the water system. Redundancy is vital in disaster planning.

2. **Increasing the earthquake resistance of public utility systems to lessen the chance of power outages from violent earth movement.** This is another measure that would help the fire department. Broken power lines are a life and fire hazard, and the absence of regular power supplies makes rescue work, especially at nighttime, more difficult. In such conditions, portable generators are necessary to operate some of the rescue tools and to provide needed light.

    The experience gained in the Sylmar quake, which had a magnitude of 6.6 and caused an area of approximately 200 square miles in the San Gabriel Mountains to move southward and rise permanently several feet, is illustrative of the time required to restore public utilities after a moderate quake. (It released only 1/100th as much energy as the 1964 Alaska earthquake.) Power was restored in 2 days to all areas with the exception of the evacuated area below the San Fernando Dam, which regained power in 3 days. Gas service was restored to most customers in 5 days, and to the remainder in 11 days. The telephone system was back in service for 8,000 customers in 6 days, with complete restoration of service in 21 days. Repair of the sewer system took longer because of the difficulty in locating the many breaks, but emergency repairs were virtually completed within two weeks.[47]

3. **Increasing earthquake resistance of streets, tunnels, bridges, and freeway overpasses.** This effort will improve the fire department's response capability. Even if highways are not damaged, the fire service will have difficulty trying to respond on streets jammed with vehicles carrying people evacuating the area after a major earthquake. With only three or four ways for residents of Los Angeles County to escape from the metropolitan basin, the highways following a magnitude 8 earthquake would become a seven-million-car parking lot. In the Northridge earthquake, four bridges collapsed; streets were also ruptured by gas line breaks and washed out by major water main ruptures. With urban areas often having double-decker bridges, cars can be easily crushed, creating extensive rescue operations as well as delaying emergency vehicles.

4. **Increasing fire protection measures for hazardous occupancies.** This will lighten the work of the fire department in case of a major tremor.

5. **Increasing earthquake resistance of structures whose use is critically needed after a disaster.** This includes fire and police stations, hospitals, disaster control, and emergency communications centers. "Communication centers are often surprisingly vulnerable, due to non-secured computers, inadequately raised floors, unsecured battery racks, etc."[48]

6. **Increasing the earthquake resistance of schools, churches, theaters, large hotels, and other high-rise buildings housing large numbers of people.** This will reduce the amount and difficulty of rescuing the occupants of large assemblage buildings. In addition, it will be help to provide safe places of refuge for displaced persons following a disaster.

Emergency action in earthquakes

An earthquake is like a terrorist attack in many respects—since it is not known when it will occur, there is no warning. Planners and responders do not know where the main attack will hit, or with how much force, so they cannot tell how much damage has been caused until it is over and they have had a chance to look around. They may be only in the periphery of the heaviest area of damage. They do not know if or when they will experience another attack, perhaps greater than the first.

If a plan of action has not already been created and widely disseminated and understood by the community's emergency forces, the problems will be multiplied. Any plan adopted after the earthquake—perhaps sitting on the curb in front of a collapsed headquarters building—will be difficult to promulgate if the personal computers, telephones, and radio base station are knocked out. If the fire stations were not earthquake resistive before the strike, it will be too late to do anything about that afterward. Those that have not collapsed may do so in the aftershocks that follow.

Assuming that the quake has not deprived a community of its fire department and all of their response capability, and there are still some uninjured firefighters and apparatus to cope with problems, what should be done first?

There will be people trapped under fallen buildings, fires at broken gas mains, overturned molten materials, and fires started by building owners who have no earthquake insurance, but do have fire insurance. There will also be fires set by looters to cover up their thefts. It is also likely that there will be one or two large fires that could develop into a conflagration.

A dam could break, or a dam could be weakened enough to require removal of everyone, including fire department personnel, from the zone of danger, before a second major shock completes the destruction. If the community is a coastal city, there could be a tsunami on the way capable of wiping out the response forces and the people in the harbor area and along the coast.

The fire department is not likely to be flooded with calls for help, for its communications will probably consist of whatever portable and vehicle radios they still have available. Unless they have a common radio frequency with other agencies and other communities, and enough signal strength to reach them, the only way they may be able to request help is to send someone on an all-terrain vehicle or in a helicopter. The expressways will be blocked by fallen overpasses, bridges will be down, and roads will be blocked with debris from collapsed structures. For several hours, local responders are apt to be on their own. It will take some time for the emergency management organization to get its disaster control center in operation. In any case, the mayor or whoever is the head of it probably is going to ask the fire department executive what steps should be taken, in what order, to save lives and property.

To make a decision, the fire department executive needs facts. The earthquake disaster annex should include provision for activating and staffing the emergency operations center. There should be no need to telephone off-duty members to notify them to report, since the violent shaking will serve as better notification than a dead telephone system.

Facts will be needed concerning response capability and the extent of damage to the community. After learning what personnel and equipment can respond and assessing the total situation, the fire department executive can evaluate the need for deploying forces to the most critical locations. In the Sylmar earthquake, three Los Angeles County fire stations were destroyed, and in the station at Olive View Hospital, it took 15 minutes to break down the station's door to free the apparatus.[49]

The fire department executive must use his or her best judgment in establishing priorities for emergency services. Some structures may be so badly damaged or so well involved in fire that they will have to be written off. Efforts must be directed where there is a chance of stopping fire spread. Triage will take a different form. The disaster events occurring will have to be triaged, and it will be necessary to choose to mitigate the events that will provide the best outcome for the overall area affected. Utilization of an area command will help ensure that adequate resources are applied to those incidents identified through the triage process as requiring immediate attention.

If there is a helicopter available, the damage should be surveyed to plan for deployment of forces. Nevertheless, whether the survey is by air or vehicle, or only based on reports received from field supervisors, it will still be necessary to plan for concerted action. If the executive fire officer leaves discretion to his or her subordinate supervisors, they may commit units to a local rescue or fire suppression action. However, the chief fire officer may realize that such action will be futile if a sweeping fire headed their way is not stopped.

The biggest problem may involve rescue of victims trapped inside buildings that have partially or completely collapsed, though some fire chiefs feel that the primary goal of the fire department is to extinguish fires and prevent development of a conflagration.[50] The OEP report to Congress, *Disaster Preparedness*, seemed to concur with this view when it disclosed its findings as follows, "The capability to mount effective search and rescue (SAR) operations in an earthquake disaster is marginal."[51] As evident from the San Fernando case in 1971 and the San Francisco case in 1989, there is a need for readily available special equipment and special procedures to locate and extricate buried persons. This report helped form the development of the FEMA US&R Teams. These teams, 28 throughout the country, have the specialized equipment and training to shore and provide SAR activities in the urban environment. Although they have become a WMD/terrorist attack resource, their primary mission is to rescue people trapped in collapsed structures.

As with most of the major obligations facing the emergency services, preparation for an earthquake begins with properly sizing up the potential for earthquake in a given jurisdiction. Cities with buildings constructed before 1970 could easily face simultaneous building collapses, major fires, and compounding emergency events, all without warning. Size-up starts with estimating the impact on the community's public safety resources. What will happen to the fire stations, apparatus, and personnel? It should be noted that fire stations are often vulnerable due to large spans and the openness of apparatus floors. When San Francisco conducted structural surveys of all of its fire stations in 1991, more than 40% were judged "high collapse hazards."[52] Another study pointed out that fire departments need to pay more attention to SAR capabilities involving heavy debris.[53] All firefighters should be given minimal training, and in high-risk areas, special teams are advised. These reports, as well as subsequent earthquakes in the 1990s, have driven FEMA and many state EMAs to develop resources of equipment and personnel that can rapidly respond to large-scale building collapse incidents and initiate SAR operations. Many states have become self-sufficient in terms of highly trained rescuers. This was the case in Hurricane Ophelia, which struck North Carolina. The state assets for SAR were so advanced that the federal-level teams were not utilized.

The advice in the OEP report was referring to the 1971 Sylmar earthquake rescue operations at the collapsed Veterans Administration Hospital, where 49 persons lost their lives. Personnel of 18 engine companies, using three bulldozers, one motor grader, a demolition truck, ladder trucks, wreckers, and four helicopters accomplished the tasks. The fire departments also used power saws, acetylene torches, hydraulic jacks, spreaders, "jaws," and similar tools during the four days it took to recover the last body in the wreckage. In addition, the U.S. Army Corps of Engineers supplied dump trucks, loaders, backhoes, cranes, jackhammers, and concrete-cutting equipment. A field kitchen and a morgue were set up, and medical supplies were brought in by four teams of firefighter paramedics who took EKGs of patients removed from demolished buildings. Two patients in cardiac arrest were defibrillated, and their lives were saved. Doctors and nurses were flown in helicopters to and from Harbor Hospital 40 miles away, and helicopters also transported casualties.

The small town fire chief who has been training volunteers for years will be on familiar ground if it becomes necessary to put citizen volunteers to work on a moment's notice. Even firefighters who are unable to report to their regular places of duty may be called upon to form volunteer companies to assist in search and rescue work and to extinguish fires.[54]

Three years prior to the Bay Area earthquake, San Francisco Fire Chief Emmett Condon stressed the need for community self-reliance. "When it occurs, the damage, injuries, and death will overwhelm our capability to deal with them. That's why we teach self-help. People who can take care of themselves and families can do more to save lives than all the emergency equipment in the state."[55]

The Bay Area, due to its extensive population and the relatively recent earthquake with aftershocks felt into 1990, has served as a focal point for study of earthquakes. This will continue into the future, hopefully evaluating the seismic retrofitting of many buildings and the increased training and cache available to the emergency services to respond to and mitigate earthquake damage.

VOLCANOES

In 2002 and 2003, Mount Etna had the biggest series of eruptions for many years, which threw up a huge column of ash that could easily be seen from space and fell as far away as Libya, on the far side of the Mediterranean Sea. Seismic activity in this eruption caused the eastern flanks of the volcano to slip by up to two meters, and many houses on the flanks of the volcano experienced structural damage. The eruption also completely destroyed the Rifugio Sapienza on the southern flank of the volcano.[56]

After more than 600 years of dormancy, the Mt. Pinatubo volcano in the Philippines began erupting on June 9, 1991. After a series of massive explosions, tremendous ash storms, and ongoing tremors, 310 people had died, and the nearby Clark Air Base and Subic Bay Naval Base used by the U.S. military were both heavily damaged.

Approximately one week after the eruptions started, they grew to such cataclysmic proportions that this was considered among the major volcanic events of recorded history. According to scientists at NOAA, the volcano pumped enough ash and gas into the air to color sunsets and lower temperatures slightly throughout the Northern Hemisphere, perhaps for years.[57]

In May 1990, a lava flow 2,000 feet wide crept toward the heart of the village of Kalapana, Hawaii, having destroyed 120 homes since 1983, including 44 after a new lava flow began in April 1990. A small historical church was moved by the townspeople before their village was overrun with lava. This is an example of the massive destruction and disastrous consequences that can result from the powerful release of molten energy from volcanoes.[58]

What is a volcano?

A volcano is an opening or number of openings in the surface of the Earth at the top of, and sometimes on the sides of, a cone-shaped hill or mountain connected to the interior of the Earth by a funnel or crater. Through the funnel or crater, steam, gases, ashes, rocks, and sometimes molten materials are expelled in periods of activity.

The word *volcano* comes from Volcanus, the name of the Roman god of fire. The name "Volcano" was first used to identify one of the Lipari Islands in the Mediterranean Sea where the god was supposed to have lived.

History of volcanoes

From ancient times, volcanoes have terrified people with the violence and roar of their eruptions, which have been responsible for many disasters. The close correlation of locations of volcanic activity with the epicenters of earthquakes makes their location and activity significant. According to Dr. Charles Richter, it is possible that massive subsidence and volcanic activity of the sort experienced off Iceland is a manifestation of tectonic forces that may sooner or later generate earthquakes. It may be significant, in view of Guatemala's disastrous earthquake, that more than one-half of the 80 volcanoes in Central America are located in Guatemala.[59]

The OEP report states that debris, pyroclastic flows, mudflows, floods, and lava flows could create danger for people and property in the immediate vicinity of volcanoes. Fallout of airborne ash and flood-transported debris could affect people within a radius of 50 to 75 miles. Flying ash could cause people serious respiratory problems or even suffocate people. Ash and debris ruin crops and reduce the productivity of the land for many years. If fallout is sufficiently great, buildings collapse, and people and animals may be trapped. "In short, all life and property is endangered by volcanic eruptions, and the extent of the disaster is chiefly a function of the magnitude of the particular event."[60]

Some relatively recent adverse effects from volcanoes include the pelting of an airliner with fist-size rocks, cracking cockpit windows and spraying ash into engine intakes. This caused an Al Nippon Lockheed Tri-Star Airliner carrying 317 passengers over Japan to make a forced landing at Sapporo. Since volcanoes are often located close to the oceans, when their lava flow reaches the sea, it creates an acid-laden mist known as *laze* (short for "lava haze").[61]

Cause of volcanoes

The cause of volcanoes is not clearly identified, but in the last 100 years, scientists have gained a better understanding of their origin. Since they are generally located near the sea, it was once thought that the entrance of seawater into the supposedly molten interior of the Earth was responsible; this is not the accepted view today. A more recent theory asserts that weakness in the Earth's crust combined with pressure exerted in the formation of folds and faults is responsible for starting volcanoes. It is believed that chambers of extremely hot magma (molten material) exist from tens to hundreds of miles below the volcano, and a passage called the *conduit* carries lava gases and rock to the Earth's surface.

The extreme heat and magma contribute to the process of mountain building (*orogeny*). Shifting rocks are forced to great depth, where increased temperature and pressure cause them to be melted and converted to magma (possibly by *anatexis*, which is partial melting or melting of certain constituents in the rock). Once formed, magma exerts pressure on the surrounding solid material. Where a fissure in the solid material extends to the surface, the molten material surges upward, and a volcano is born.

Explosive eruptions occur due to gas pressures that build up when the conduit becomes plugged and is finally blown out. As the lava (molten rock) flows out of the Earth, it cools and forms a cone, which can eventually become a high mountain. Cotopaxi in Ecuador rises to a majestic 19,347 feet, and Argentina's Aconcagua reaches 22,831 feet.

Occasionally when a volcano erupts, its majestic height can be lessened considerably. For example, when Mount St. Helens erupted in the state of Washington during May 1980, 1,200 feet—or 1/7th of its height—was blown away.[62]

Volcanic emissions

The gaseous materials ejected from a volcano include steam, carbon dioxide, hydrogen chloride, hydrogen fluoride, hydrogen sulfide, sulfur dioxide, and hydrogen. In the presence of water vapor, most of these gases form toxic, irritating, and in some cases, foul-smelling acid vapors. Other emissions include pyroclastic ash, pyroclastic flows, and lava flows. *Pyroclastic flows* consist of rocks thrown into the air that are too heavy to remain aloft. Thus, they flow downhill like a hot liquid at speeds up to 60 mph, like an avalanche, setting everything on fire in their path. *Lava flows* are streams of molten rock that rise from the vent, explosively or quiescently, and flow down the side of the mountain at speeds up to 30 mph.

Solfataras and *fumaroles* are volcanic vents that expel only gases and magmatic vapors. A volcano may have a number of such openings in addition to the main conduit.

Kinds of volcanoes

Volcanoes have been classified into four groups: Hawaiian, Strombolian, Vulcanian, and Peléan. Hawaiian volcanoes have a fluid magma that usually erupts without violence. The Strombolian volcanoes have thicker lavas from which gases are released with explosions. The Vulcanian volcanoes have magmas almost as thick as tar; they also have violent explosions that produce heavy clouds of volcanic dust. The Peléan volcanoes are the most violent of all, and the clouds above their eruptions glow as though on fire.[63]

Some volcanoes are constantly active, such as Stromboli, and erupt so regularly they are like lighthouses on their little islands. Mt. Etna is an intermittent volcano that erupts at fairly regular intervals. This type of volcano erupts in cycles, with the duration depending on the length of time it takes to build up enough heat and pressure to reactivate.

Dormant volcanoes are those not active at present, but no one knows when they will erupt again. They have not been quiet long enough to be designated "extinct," as have those that have not erupted since the beginning of recorded history.

Another grouping of volcano types offers three classifications:

1. Shield volcanoes
2. Cinder-cone volcanoes
3. Composite volcanoes

The first of these is a *shield volcano*, formed by free-flowing lava spilling from a vent and spreading widely. This produces a low, broad, dome-shaped mountain. The second type, a *cinder-cone volcano*, builds up when tephra (cinders) erupts from a vent and returns to the ground near the event area. The third type, a *composite volcano*, is formed by a combination of lava and tephra from a single vent. The materials accumulate in alternate layers around the vent and form a towering, cone-shaped mountain.[64]

Location of volcanoes

Most volcanoes are distributed in belts of weakness in the Earth's crust near the great mountain chains. About 79% of known volcanoes are contained in the Circum-Pacific System, though their distribution is irregular, with 45% in the western Pacific Islands and only 17% in the Pacific coastal region of North and South America. The United States has 27 active volcanoes in the Aleutian chain of Alaska and 2 in Hawaii. Along the Pacific Coast of the United Stares is a chain of 12 volcanoes, located in Washington, Oregon, and California. They are all potentially dangerous to the urban and agricultural centers of the Pacific Northwest.

Effects of volcanoes

A violent eruption can produce a lava fountain one-half mile high. Airborne ash, containing cinders and pyroclastic rocks, is blown into the air and remains suspended for long periods before eventually falling to Earth. About 1 million cubic feet of these materials were produced every second in the first one-half hour of the 1947 eruption of Hekla in Iceland. Before it subsided, it ejected 225 trillion cubic feet of material, enough to cover a 100,000 square-mile-area with a one-inch deposit. In addition, the eruption of Mt. Vesuvius in 1779 caused a river of lava 65 feet wide that flowed at a rate of 25 to 30 feet per second.

When solid particles are suspended in a gas, a pyroclastic flow occurs that acts like a liquid, capable of moving with great speed down the mountain. In the Katmai, Alaska eruption of 1912, 300 million cubic feet of *ignimbrite*, or rock layers formed from the deposit of these materials, were produced in 60 hours. These ignimbrite sheets may be as deep as 1,600 feet.

The choking fine ash thrown into the air from the Crater Lake volcano about 6,600 years ago covered central Oregon with a layer more than six inches deep. "If similar eruptions were to take place today and the wind happened to be blowing in an adverse direction, Mt. Ranier or Mt. Hood would shower a layer of debris more than six inches deep over downtown Seattle or Portland. In such an eventuality, the loss of life and property would, of course, be enormous."[65]

Most of the largest eruptions have occurred in uninhabited areas, such as the Valley of Ten Thousand Smokes (1912) and Bezymianny, Kamchatka (1956). However, as the OEP report points out, such an eruption in a densely populated area would produce a "catastrophe of unheard-of proportions."[66] Catastrophic enough was the Mount St. Helens eruption on May 18, 1980.

When the Mount St. Helens volcano erupted, a stratospheric cloud of grit, 10 miles up, fanned out across the country and dumped ash as far east as Montana. In all, 57 people died in the Mount St. Helens eruption. Shock waves leveled 44,000 acres of fir trees, and boiling mud traveling 30 miles per hour overflowed local riverbanks, smashed bridges, swallowed homes, and killed millions of salmon and trout. Some rivers got so hot that fish actually jumped out. So much mud poured into one stretch of the Columbia River, normally 40 feet deep and 600 feet wide, that the passage shrank to 14 feet deep and 200 feet wide—trapping 31 ocean-going ships in Portland, Oregon and Vancouver, Washington.[67]

Examples of volcanic activity

The most active volcanoes in the United States that are located in populated areas are Haleakala on Maui and the Hualalai, Mauna Loa, and Kilauea volcanoes on the island of Hawaii. Since 1750, Kilauea has been constantly active, with 32 eruptions on the record, but Mauna Loa has erupted only about once every 3.6 years on the average since 1832. Haleakala last erupted in 1750.

Eruptions of many volcanoes have been sufficiently close to inhabited areas to cause disasters. Mount Katmai in Alaska has been made a national monument since its last eruption in 1912. In 1963, Mt. Agung on the Indonesian island of Bali erupted and killed about 1,500 persons. In the same year, a lava flow started building the new island of Surtsey about 20 miles south of Iceland. By the time the lava stopped flowing in 1967, it had risen to a height of 567 feet and covered more than one square mile.

After 25 years of inactivity, Mauna Loa in Hawaii started spouting molten rock in 1975, creating a fountain over 40 feet high along the northeast flank of this 13,680-foot mountain. Two major lava flows issued from the northeast rift area: one moving north, and the other in the direction of Hilo, the island's principal city of 35,000 persons, a distance of 35 miles from Mauna Loa's summit. In a 1942 eruption, lava flows stopped 12 miles from Hilo. Prior to the 1975 eruption, its previous eruption was in 1950, when it put on a 23-day display of fountaining, sending 600 million cubic yards of lava into the sea on the southwest side of the island.[68]

Predictability of volcanic eruptions

Despite regular monitoring of volcanoes in the Cascade Range, geologists admit that they cannot predict when one might erupt. Volcanic eruptions usually are preceded by earthquake activity. One U.S. Geological Survey report indicated that increased volcanic activity was detected on Mount Baker in Washington. The report predicts that if this mountain erupts, it could send millions of cubic yards of rock, mud, and ice hurtling down the mountainside. "The mudflow could move at speeds of more than thirty miles per hour, engulfing hundreds of acres of surrounding forest. If it reached Baker Lake, the mudflow would create a twenty-foot-high wave that might completely wash out the surrounding beaches, campgrounds, and resorts."[69]

In regions where volcanic activity poses a threat to safety of residents, the U.S. Geological Survey is carrying out programs to warn of imminent danger. In Hawaii, the prediction and warning of volcanic hazards is performed by the Hawaiian Volcano Observatory (HVO) on the rim of the Kilauea Crater. It measures the earthquake activity, long-term tilt, and bulging (deformation). The HVO also takes regular temperature measurements, conducts chemical analysis of the gases and lava of the volcanoes, and assesses the activity to establish normal patterns. The detection of any unusual activity can then be the basis for preparing predictions and warnings.

According to the OEP report, the study of volcanoes in the Cascade Range of the Pacific Northwest is largely concentrated on Mt. Ranier and Mount St. Helens in Washington, and Lassen Peak in California. It has consisted mostly of mapping and evaluating existing hazards, with brief periods of microearthquake monitoring of these three volcanoes.[70]

No one can be sure a volcano is extinct. Some thought to be extinct have become active again. In 1973, a volcano that had been dormant for more than 5,000 years erupted on the island of Heimaey near Iceland. It poured so much ash over the island's only town of Vestmannaeyjar that all residents were forced to escape.[71]

In 1990, the *Los Angeles Times* reported that most persons are willing to take a fatalistic view of dormant volcanoes, assuming they will not erupt. This is the same attitude that creates the urban/wildland interface problem. Most people assume that disasters will not happen to them and want to have a property location of their liking rather than concern themselves with the disaster possibilities.

Damage control—volcanoes

Since the eruption of volcanoes cannot be prevented, volcanoes will always be a hazard to persons who live in the vicinity. Two types of damage limitation action can be taken. The areas that are likely to be in the path of molten lava flows can be plotted, and zoning restrictions could then be instituted to control the use of land in such areas.

Emergency action

Emergency services organizations in areas where volcanic eruptions pose a threat must maintain communications with agencies that provide warning of impending danger and activate the disaster plan that deals with the volcano danger. Such a plan should provide for rescue of persons in structures in the path of the lava flows and containment of fires started by molten rock and hot cinders. Breathing apparatus should be used in areas where the air is heavily contaminated with sulfur dioxide, hydrogen sulfide, and acid vapors while conducting rescue and fire suppression operations.

If a volcano "blows its top" with no warning and showers the city with burning cinders and suffocating ash, the fire department will encounter many simultaneous fires. If the eruption is accompanied by strong earthquakes, which is likely, the problems will be compounded.

An aerial survey might be hazardous due to flying rocks and cinders, and the sky would be so filled with ash that visibility would be greatly restricted. Debris fallout can make surface travel hazardous and eventually might bury the community completely. If this prospect appears likely, the fire department has no alternative but to assist with and participate in the evacuation of the populace.

LANDSLIDES

One secondary effect of an earthquake or volcanic eruption is a landslide. Independent landslides, large and small, also occur. Fire departments sometimes respond to emergencies involving the burial of people, autos, and even buildings under debris that slides off hills and mountains. Sometimes structures slide and collapse, or a mass of soil or rock may break from a mountain and carry structures along. Sometimes land falls away from under a building, leaving it teetering on the edge of a cliff.

As states become more populated, there is a tendency to subdivide hillsides and build terraced residences without thorough investigation of the geological conditions of the region. To add to the problem, natural vegetation is often cleared away to such an extent that soil erosion occurs, and the burden on the land is increased by many tons in the form of swimming pools or other structures built near the edge of the slope or cliff. A ground cover may be planted to replace brush and trees that have been removed (sometimes to reduce the fire hazard) and then watered so frequently as to saturate the soil, with the result that the risk of landslide is increased.

It has been estimated that the cost of landslides includes damage to highways, railroads, utility company structures, public works, industrial property, residences, and personal property.[72]

Causes of landslides

Landslides occur when a mass of soil or rocks slips down the side of a slope. They may be falls, slides, slumps, or flows. Water seepage, cracks in rocks, earthquakes, and volcanic action are common causes of landslides. Heavy rains, snow accumulation, and erosion are additional causes. The action of waves on a coastline also may start a landslide.

Falls of soil or rock occur from natural cliffs and overly steepened river banks, or from cuts or excavations made for highways, railroad roadbeds, and construction sites.

Earth blocks that slide on fracture surfaces may be a few hundred to several thousand feet across, and the fracture surfaces may be a considerable distance below the surface of the Earth. Such slides usually involve a slow downward movement of an entire slope due to a subsurface shearing. When it moves as a unit, it is called *slumping*. If it breaks into small units, it is called *avalanching*. These landslides usually move a few inches to feet per day, but some move considerably faster.

Flows may be dry or wet and involve greater masses of material that continue downhill far beyond the base of the slope from which they originate. Mudflows usually begin on steep slopes when shallow soil layers are transformed into a liquid state, usually by rainfall, though leaking water systems might be the water source. When this happens, liquefied soil flows like streams of soft plastic down canyons, arroyos, and valleys, spreading on the plain below. Large flows may move as fast as 20 mph, carrying along houses, barns, fences, trees—everything on top of the ground and in the ground, including sewer pipes, gas lines, and other utilities. Figure 14–3 shows how mudslides affect the urban interface as people continue to seek homes with a view by placing them on hillsides.

Following a brush fire that denuded California hillsides, and a subsequent long period of steady rain, huge amounts of jelly-like mud flowed through windows and doors of beautiful homes at the foot of a hill. This river of mud continued into the street, leaving a deposit several feet deep that later congealed into a hard mass of clay, encasing the furniture like cement. Houses were pushed from foundations, causing them to topple, collapse, break, or become buried completely.

Slumps occur when the strength of the ground is exceeded by forces that tend to move the ground downhill. "These forces arise from the weight of earth materials and from ground water seepage. Landslides are triggered by both natural and man-made processes that reduce the strength

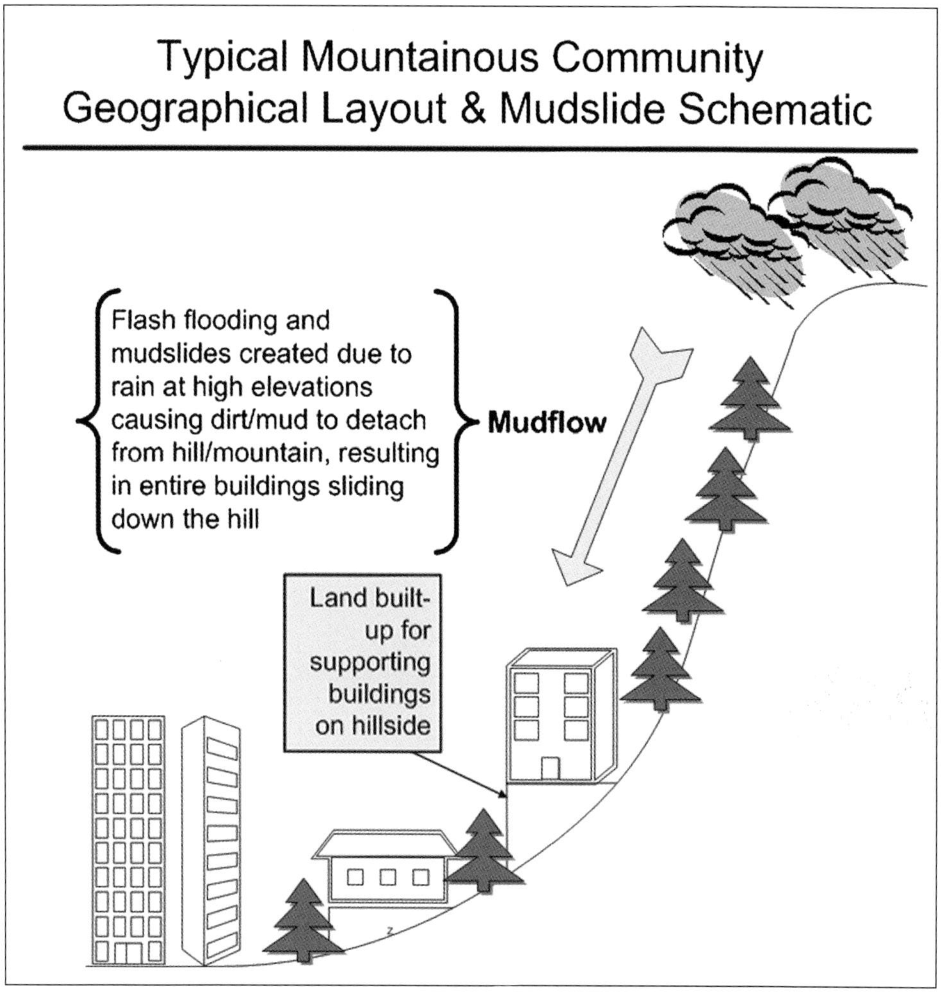

Fig. 14–3. Mudslides that involve structural collapse will increase as Americans continue to build structures on hillsides. This is a two-fold problem. The first problem is the reduced area to absorb the rain. The second is the inability to maintain structural stability of buildings when such large flows of mud occur.

of ground or increase the forces associated with weight and seepage. Such processes include rainfall, erosion, earthquakes, and various modifications such as cutting and filling that are made to ground by man."[73]

An OEP report lists five causes for landslides:

1. Slow erosion, which results in rock falls due to gradual wearing away from supporting materials and avalanches that sweep down a mountainside, gathering momentum and masses of loose debris.
2. Water, causing marginally stable slopes to give way in avalanches, slumps, or mudflows.
3. Glacial action, melting and removing the support that glacial ice gives to canyon walls, resulting in landslides and mudflows.
4. Earthquakes and volcanic action, shaking material loose and forming an avalanche, such as the one that formed Earthquake Lake in 1959 at Hegben, Montana. Earthquakes create vibrations that cause soil to act like liquid, with resulting ground failure and slumping over wide areas. This type of action took place in the Van Norman Dam during the 1971 earthquake in Sylmar, California, necessitating evacuation of 80,000 residents for several days while the reservoir was being drained.

5. Engineering defects, such as miscalculating slope stability when excavating cuts through a slide area for dams, bridge abutments, and buildings.[74]

Effects of landslides

"Landslides occur in every state and U.S. territory. The Appalachian Mountains, the Rocky Mountains and the Pacific Coastal Ranges and some parts of Alaska and Hawaii have severe landslide problems".[75] "In the United States, it is estimated that the total dollar losses from landslides is between one and two billion dollars (1.6 and 3.2 billion, year 2000 dollars). This figure is a conservative estimate, as there is a uniform method or overall agency that keeps track of or reports landslide losses.[76] Many foreign countries experience a higher frequency of landslides, partially due to lesser engineering requirements in these countries.

Examples of landslide disasters

January 2001 proved to be a disastrous month for El Salvador. On January 13, at least 844 people were killed, 4,723 injured, 108,226 houses destroyed, and more than 150,000 buildings damaged in El Salvador due to a 7.6 earthquake. About 585 of the deaths were caused by large landslides in Nueva San Salvador and Comasagua. Utilities and roads were damaged by more than 16,000 landslides.[77]

On May 1, 1991, a Soviet town vanished and 81 people died when the village of Khakhieti was obliterated by a mountain slide. The village was buried so completely that not one of its two-story houses was left visible. This slide, triggered by an earthquake, left no ruins, no bodies, and no signs of the farming village that had once been there. It was described as "just dirt, rocks, and fallen oak trees."[78]

When Peru experienced its 7.75 magnitude earthquake in 1970, most of the 70,000 persons who lost their lives died as a result of landslides. Yungay, a town of 25,000 people, was completely destroyed when the earthquake loosened the northern face of Mt. Huascaran and an avalanche of water, rocks, and mud roared down upon the town. "A few people reached high ground in time. Rescuers found them huddled in a cemetery. Everyone and everything else had been obliterated."[79]

In the middle of the night on August 17, 1959, an eight-state area felt the first heavy jolt of the Hebgen Lake earthquake, and a huge landslide consisting of about 38 million cubic yards of rock slid into the Madison River Canyon at a speed of about 100 mph. People, cars, and trailers were crushed and buried. Some of the survivors were engulfed by a huge wave of water from the displaced Madison River. The slide blocked the canyon so that hundreds of vacationers asleep in their tents, campers, trailers, and lodges near Hebgen Lake and along the river had their escape routes cut off.

By dawn, rescue units were on the way from many agencies in Idaho and Montana. Aerial reconnaissance was made and the Forest Service parachuted smokejumpers into the canyon to give first aid and to set up communications. Others came in on horseback and by helicopter to assist the survivors. By evening, bulldozer roads had been built around highway slipouts, and all who wished to leave were able to get out of the canyon. However, 28 men, women, and children lost their lives in the disaster.

The slide was about a mile long and three-quarters of a mile wide, and it formed a dam about 225 feet thick. After the valley was blocked by the mass of displaced soil and rock, the waters that poured over the dam as a result of the seiche, plus the waters normally released through the spillway at the dam, began to fill the canyon and form "Earthquake Lake."

Landslide prevention

Little can be done to prevent earthquakes or heavy rains, which are the cause of many landslides. However, the danger of landslides can be greatly lessened in hillside developments if construction practices are regulated by grading ordinances and building codes and if geologic and engineering advice is followed. Also, the laying of rights-of-way for dams, bridge abutments, and other structures should be done with special consideration for the landslide characteristics of the area.

There are several steps that can be taken to reduce the landslide danger. Where the slope cannot be altered, it is important to maintain good surface and subsurface drainage. Other suggestions include the use of retaining walls and planting of certain types of vegetation on loose slopes. FEMA also suggests that homeowners "minimize home hazards by having flexible pipe fittings [professionally] installed to avoid gas or water leaks, as flexible fittings are more resistant to breakage."[80] There are private consulting companies that specialize in addressing landslide potential and offering more technical solutions to the hazard.

Once a landslide begins, little can halt its progress. In the Palo Verdes movement near San Pedro, California, residents of homes below the main scarp attempted to continue living in the houses by connecting flexible pipe and hoses to utility lines before they became severed by the movement. Tunnels were dug back into the hillside to drain ground water, and gas heaters were installed in them to dry out the soil. Steel rods and retaining walls proved useless, as did the other measures tried, and eventually the homes had to be moved, disassembled, or abandoned, as the subdivision gradually slid down to the ocean below.

Although it might be very costly, highways, railways, or aqueducts that must traverse unstable areas where control of slumps is not possible can sometimes be bridged over the area, allowing movement to continue without disturbing the construction project.

Emergency action—landslides

From the emergency services standpoint, landslides that involve the eventual destruction of buildings through the slow process of slump are not problems calling for emergency response. This is true unless a gas line is sheared off during the movement, fallen power lines start a brush fire, or some similar emergency dictates an immediate need.

When part of an embankment fell away from beneath a residential area in the Pacific Palisades a number of years ago, Los Angeles firefighters assisted persons to escape from homes left precariously perched on the edge of the bluff. In some cases it was possible to bridge a gap with ladders and shore up the overhanging building sufficiently to permit the occupants to remove some of their personal property before the house fell or slid any further. These response activities are noted in figure 14–4.

Fig. 14–4. Responders examine structures after a landslide. Entire structures can be buried during these events. Utilizing zoning records to ensure that all structures are accounted for can prevent missed rescues.

FEMA News Photo.

In another instance in which a series of homes situated on the top of a bluff in the Santa Monica Mountains started to slump, many of the homes had swimming pools. Fire department pumpers were called, and using submersible pumps and siphon ejectors, the pools were emptied quickly to lighten the burden on the affected terrain. In a few weeks, the homes were beyond saving, but the occupants had ample time to remove possessions and find other quarters.

In mud flows that struck many homes following brush fires in California and other western states in the early 1990s, fire departments dispatched companies to assist in filling sand bags to use in diversion of flowing mud and water. An attempt was made to keep homes from being inundated with the putty-like effluent that was oozing out of the gullies and arroyos as a result of the heavy rains. The efforts of many firefighters were concentrated on the digging of channels to provide a bypass for homes being crushed, pushed, and infiltrated by thick mudflows slowly moving down the slopes, but it was often a losing battle. Bulldozers and skip loaders were put to work in an attempt to divert the mud streams to the streets, where similar equipment would keep it moving. Between rains, hastily bulldozed bulwarks were constructed in gullies that posed the greatest threat from mudflows. However, these were soon filled after the next rain commenced and provided only temporary relief.

Search and rescue efforts can take days or weeks due to the time-consuming task of digging for survivors or the deceased. Use of plywood will prevent rescuers from becoming entrapped in the mud themselves during rescue operations. The utilization of military 6X4s will aid in traversing the mud in order to deliver personnel and equipment to the site of the rescue. Safety of response personnel during a landslide response should focus on accountability and the use of shoring to prevent subsequent slides that could bury the rescuers.

CONCLUSION

This chapter has covered the terrain-based types of natural disasters that are awesome and devastating in their effects. Earthquakes that strike without warning have the power to level even the most earthquake-resistant structures and wreak havoc on an unsuspecting community. Volcanoes on the West Coast and in Hawaii and Alaska have potential atmospheric repercussions that could stretch across the United States and Canada. They also have economic repercussions that could touch every part of the North American continent.

Finally, because they occur independently, landslides are found everywhere. A brief study of these three major types of disasters would easily lead one to believe that the term *terra firma* is a conflict in terminology. In the next chapter, the discussion turns from natural disasters themselves to focus on the aftermath and recovery from disasters.

REFERENCES

1. Associated Press. 1991. *Cincinnati Enquirer.* January 3. p. A-2.
2. Associated Press. 1991. *Cincinnati Enquirer.* June 30. p. A-8.
3. Webre, A. L., and P. H. Liss. 1974. *The Age of Cataclysm*. New York: Berkeley Publishing Corporation, distributed by G. P. Putnam's Sons. p. 11.
4. BBC News. 2003. Iran earthquake kills thousands. December 26. Retrieved July 18, 2008 from: http://news.bbc.co.uk/1/hi/world/middle_east/3348613.stm.
5. *Encyclopedia Britannica*. 2007. Earthquake. Retrieved May 11, 2007 from: http://www.britanica.com/eb/article-9363313.
6. Office of Emergency Preparedness, Executive Office of the President. 1972. *Disaster Preparedness*. Report to Congress. January. Washington, DC: U.S. Government Printing Office. Vol. 1. p. 73.
7. Gutenberg, B., and C. F. Richter. 1949. *Seismicity of the Earth*. Princeton, NJ: Princeton University Press.
8. Ibid.

9. Canby, T. 1976. Can we predict quakes? *National Geographic*. 149 (6). June. p. 830.

10. *UCLA Monthly*. 1987. 17 (3). January/February. pp. 10–11.

11. Carrol, R., et al. "China's Killer Quake," Newsweek, August 9, 1976, p. 30.

12. Canby, T. 1976. p. 831.

13. United States Geological Service (2008). NSMP stations. Retrieved November 10, 2008 from: http://nsmp.wr.usgs.gov/stations.html.

14. *Eastsider*. 1991. The big seismic scare. Winter. Cincinnati.

15. Alexander, G. 1976. Forecast role seen for 'silent quakes.' *Los Angeles Times*. March 31. Part II. p. 2.

16. Alexander, G. 1975. Concern voiced over bulge in land along San Andreas Fault. *Los Angeles Times*. March 12. Part II. p. 1.

17. Alexander, G. 1975. Cracks in quake forecast seen. *Los Angeles Times*. December 2. Part II. p. 1.

18. U.S. Geological Survey. 2007. National Earthquake Information Center. January 10. Retrieved May 12, 2007 from: http://earthquake.usgs.gov/regional/neic/who_we_are.php.

19. Paddock, R. 1976. Senate OK's bill to perfect U.S. quake warnings. *Los Angeles Times*. May 25. Part I. p. 1 and p. 3.

20. Cranston, A. (Senator). 1976. Report to Californians. No. 34. July. p. 3.

21. *UCLA Monthly*. 1975. Quake forecasts: how will we react? 6 (2). November/December.

22. Ibid.

23. *CNN.com*. 2001. Quake aftershock frightens Salvadorians. February 13. Retrieved July 18, 2008 from: http://archives.cnn.com/2001/WORLD/americas/02/13/salvador.quake/index.html.

24. Alexander, G. 1975. Possible link between great quakes studied. *Los Angeles Times*. April 7. Part II. p. 1.

25. Office of Emergency Preparedness, Executive Office of the President. 1972. *Disaster Preparedness*. Report to Congress. January. Vol. 3. Washington, DC: U.S. Government Printing Office. p. 72.

26. Kramer, W. M., and C. W. Bahme. 1992. *Fire Officer's Guide to Disaster Control*. 2nd ed. Saddle Brook, NJ: Fire Engineering Books and Videos. p. 307; citing: G. Alexander. 1975. Possible link between great quakes studied. *Los Angeles Times*. April 7. Part II. p. 1.

27. *The World Book Year Book*. 1971. Chicago, IL: Field Enterprises Educational Corporation. p. 294.

28. Heck, N. 1936. *Earthquakes*. Princeton, NJ: Princeton University Press.

29. Office of Emergency Services, Fire and Rescue Division, State of California. 1973. *Report of the Fire and Rescue Problem—Earthquake Disaster, Managua, Nicaragua, December 23, 1972*. p. 12.

30. *Los Angeles Times*. 1976. Large region of Soviet Asia hit by strong quake. May 18. Part I. p. 1.

31. *Los Angeles Times*. 1976. Death toll climbs in Italian quake. May 8. Part I. p. 1.

32. *Los Angeles Times*. 1976. 2 major earthquakes strike Asia during 2-hour period. August 17. Part I. p. 1.

33. *Los Angeles Times*. 1976. Philippines quake toll tops 4,000. August 19. Part I. p. 6.

34. Earthquakes of the 1980s taken from the respective yearbooks of the *World Book Encyclopedia*.

35. Blackburn, F. T., C. Scathorn, and J. Neil. 1989. Earthquake preparedness for fire departments. *Firehouse.* 14 (12). December. p. 37.

36. Scathorn, C. 1989. First report on the quake. *Firehouse.* 14 (12). December. p. 31.

37. *World Book Year Book.* 1991. Chicago, IL: Field Enterprises Educational Corporation. p. 274.

38. U.S. Geological Survey. 2005. *USGS response to an urban earthquake—Northridge 1994.* November 16. Retrieved July 18, 2008 from: http://pubs.usgs.gov/of/1996/ofr-96-0263/introduc.htm#impacts.

39. Federal Emergency Management Agency. 1988. *Rapid Visual Screening of Buildings for Potential Seismic Hazards: A Handbook.* FEMA-154 Earthquake Hazards Reduction Series 41. July.

40. Hume, E. 1975. Unsafe buildings endanger 200,000. *Los Angeles Times.* April 7. Part II. p. 1.

41. Baker, E. 1976. Panel OK's shift in quake safety policy. *Los Angeles Times.* August 28.

42. *Wikipedia.* 2008. Seismic retrofit. Retrieved October 6, 2008 from: http://en.wikipedia.org/wiki/Seismic_retrofit#Types_of_modifications.

43. Ibid.

44. IST Group. 2004. Methods for seismic retrofitting of structures. Retrieved July 18, 2008 from: http://web.mit.edu/istgroup/ist/documents/earthquake/Part5.pdf

45. Blackburn, F. T., et al. 1989. supra. p. 48.

46. Ibid. p. 52.

47. Los Angeles County Board of Supervisors. 1971. *Report of the Los Angeles County Earthquake Commission—San Fernando Earthquake, February 9, 1971.* November. p. 20.

48. Blackburn, F. T., et al. 1989. supra. p. 48.

49. Dektar, C. 1971. Extensive rescue operations follow Los Angeles quake. *Fire Engineering.* June. pp. 38–40.

50. Long, K. The fire department's fire problem (resulting from the unexpected structural damage). Earthquake Fire Seminar. p. 9.

51. Office of Emergency Preparedness, Executive Office of the President. 1972.

52. Blackburn, Frank T., et al. 1989. supra. p. 37.

53. Ibid. p. 49.

54. Connick, W. B. 1975. An isolated fire department's approach to earthquake caused problems. Earthquake Fire Seminar. June 5.

55. *Press Democrat.* 1986. Santa Rosa, CA. April 19.

56. *Italy's Volcanoes.* 2003. The 2002–2003 NE and S flank eruption. November 7. Retrieved July 18, 2008 from http://boris.vulcanoetna.it/ETNA_2002.html.

57. Enquirer News Service. 1991. *Cincinnati Enquirer.* June 23. p. A-6.

58. *USA Today.* 1990. May 1.

59. Webre, A., and P. Liss. 1974. *Age of Cataclysm.* Berkeley Publishing Company, distributed by G. P. Putnam's Sons, NY, citing Richter, C. F. 1958. *Elementary Seismology.* San Francisco: W. H. Freeman and Co. p. 407.

60. Office of Emergency Preparedness, Executive Office of the President. 1972.

61. *USA Today.* 1990.

62. *Newsweek.* 1980. June 2. p. 25.

63. *World Book Encyclopedia.* 1991.

64. Ibid.

65. Kramer, W. M., and C. W. Bahme. 1992.

66. The OEP report (Office of Emergency Preparedness, Executive Office of the President.

1972. *Disaster Preparedness*. Report to Congress. January. Washington, DC: U.S. Government Printing Office) cites the following: Cliff Ollier. 1969. *Volcanoes*. Cambridge, MA: MIT Press. p. 154.

67. *Newsweek*. 1980. June 2. p. 25.

68. *Los Angeles Times*. 1975. Volcano erupts on Hawaii Island. July 7. Part I. p. 8.

69. Panati, C., with M. Kellogg. 1975. Thunder mountain. *Newsweek*. August 18. p. 74.

70. Office of Emergency Preparedness, Executive Office of the President. 1972. Vol. 1. p. 103.

71. *World Book Encyclopedia*. 1976. Vol. 10. p. 18.

72. Office of Emergency Preparedness, Executive Office of the President. 1972. Vol. I. p. 87.

73. U.S. Department of the Interior, U.S. Geological Survey. *Landslides*. 1974. Washington, DC: U.S. Government Printing Office. p. 1.

74. *World Book Encyclopedia*. 1976. Vol. 3. p. 89.

75. United States Geological Society (2008). Landslide hazard program. Retrieved November 14, 2008 from: http://landslides.usgs.gov/learning/ls101.php?phpsessid=6a7hlv6611csk5t5be486qjp5.

76. United States Geological Society (2008). Landslides 101. Retrieved November 14, 2008 from: http://landslides.usgs.gov/learning/ls101.php?phpsessid=6a7hlv6611csk5t5be486qjp5.

77. *CNN.com*. 2001. Quake aftershock frightens Salvadorians. February 13. Retrieved July 18, 2008 from: http://archives.cnn.com/2001/WORLD/americas/02/13/salvador.quake/index.html.

78. Associated Press. 1991. Earthquake buries Soviet village, killing all. *New York Times*. May 2. Retrieved October 6, 2008 from: http://query.nytimes.com/gst/fullpage.html?res=9D0CE2DF163AF931A35756C0A967958260 .

79. *World Book Year Book*. 1971. p. 294.

80. U.S. Department of Homeland Security, Federal Emergency Management Agency. 2008. Landslides and debris flow (mudslides). In *Are You Ready?* May 30. Retrieved October 10, 2008 from: http://www.fema.gov/areyouready/landslide.shtm.

AFTERMATH AND RECOVERY 15

INTRODUCTION

As conditions are brought under control in a major emergency or disaster, the emergency services and community enter the recovery phase. This is a period in which families, neighborhoods, and government agencies attempt to return to normal. The recovery period can be emotionally intense as well as crucial to the future economic well-being of the community. Although community recovery efforts are not a primary responsibility of the fire service, fire departments are involved.

This chapter will examine the key features of the recovery phase:

- Presidential declaration of disaster
- Overview of the federal disaster relief process
- Psychological and sociological effects in victims and responders
- Community recovery process
- Evaluation of emergency response operations

The chapter focuses first on the correct process to request federal-level assistance and the various levels of government assistance. Next, the authors examine how individuals are affected by disasters.

PRESIDENTIAL DECLARATION OF DISASTER AREAS

The governmental structure of the United States focuses on allowing local communities to care for their citizens. Thus, the federal government will not come in and take over a disaster area. Because of this freedom allowed to the states and the local communities, a request for assistance by the federal government must occur in order to activate federal resources.

The FEMA document *A Guide to the Disaster Declaration Process and Federal Assistance* explains that according to the Stafford Act,

> "All requests for a declaration by the President that a major disaster exists shall be made by the governor of the affected State." A State also includes the District of Columbia, Puerto Rico, the Virgin Islands, Guam, American Samoa, and the Commonwealth of the Northern Mariana Islands...
>
> The Governor's request is made through the regional FEMA/EPR office. State and Federal officials conduct a preliminary damage assessment (PDA) to estimate the extent of the disaster and its impact on individuals and public facilities. This information is included in the Governor's request to show that the disaster is of such severity and magnitude that effective response is beyond the capabilities of the State and the local governments and that Federal assistance is necessary. Normally, the PDA is completed prior to the submission of the Governor's request. However, when an obviously severe or catastrophic event occurs, the Governor's request may be submitted prior to the PDA. Nonetheless, the Governor must still make the request.
>
> As part of the request, the governor must take appropriate action under State law and direct execution of the State's emergency plan. The Governor shall furnish information on the nature and the amount...and severity of damage and the impact on the private and public sector, and provide an estimate of the type and amount of assistance needed under the Stafford Act. In addition, the Governor will need to certify that, for the current disaster, State and local government obligations and expenditures...will comply with all applicable cost-sharing requirements.
>
> Based on the Governor's request, the President may declare that a major disaster or emergency exists, thus activating an array of Federal programs to assist the response and recovery effort.[1]

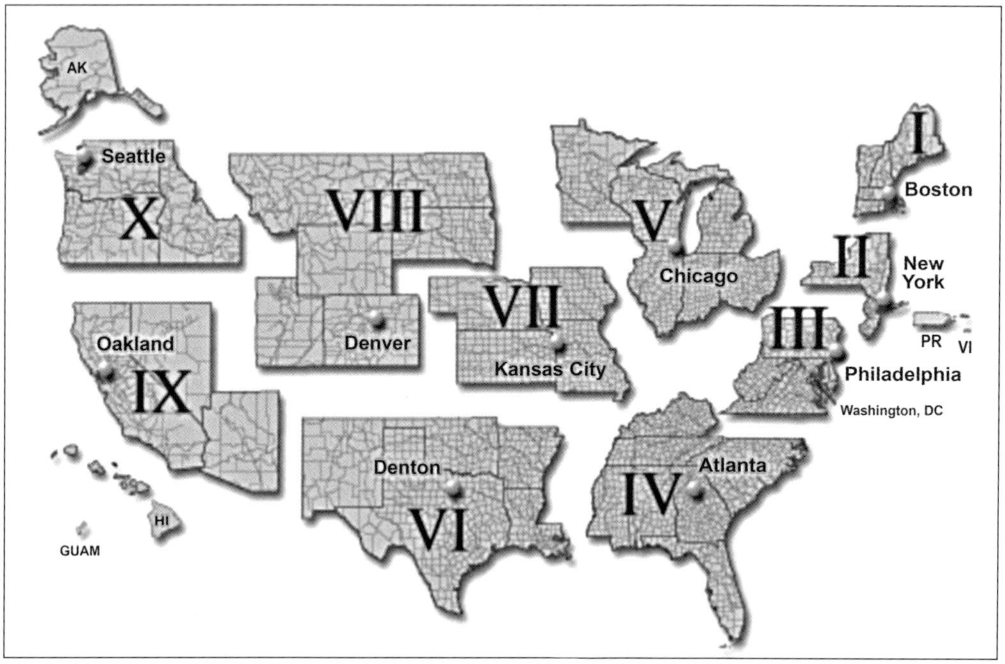

Fig. 15–1. The 10 FEMA regions of the U.S.

Figure 15–1 shows the states that are covered by the 10 FEMA regions. This regional approach allows FEMA to accomplish operational and administrative tasks more efficiently. Knowledge of the regional boundaries can aid in planning and official requests.

FEDERAL RESPONSE, FEDERAL/STATE INTERFACE

In Executive Order 12148, the president delegated the primary responsibility for administering the Disaster Relief Act (Public Law 93-288) to the Federal Emergency Management Agency (FEMA). Since 1979, this agency has served as the central point of contact within the federal government for a wide range of emergency response activities.

As described above, a governor's request for federal aid, addressed to the president, is forwarded to the appropriate FEMA regional director, who evaluates the on-scene damage assessments. The regional director's analysis and recommendations are forwarded to FEMA headquarters for review. The FEMA director's recommendation is then forwarded to the president.

It should be understood that the analysis and review process could take a few hours or a few days. At all levels of FEMA, responsible officials charged with accountability for federal disaster funds—taxpayers' dollars—must be certain that there truly is a need for supplemental federal aid.

For this reason, it should not be assumed that action on a request for federal aid would be either quick or affirmative. In fact, history shows that more requests for assistance are turned down by FEMA than are approved. A key question must be answered in the affirmative: Is the magnitude of the crisis beyond state and local capability?

Following 9/11 and recent major hurricanes, this process has become streamlined, and federal assets have been mobilized prior to the request from a governor, based on expected damage. However, as discovered in Hurricane Katrina, the process must still take place in order to fully activate the federal response.

As soon as possible following a presidential declaration, FEMA designates, on the basis of damage assessments, the counties or political subdivisions that are eligible for federal aid. The president's declaration also names a federal coordinating officer (FCO) who, as the president's representative, will direct federal recovery and coordination activities. FEMA then signs a disaster assistance agreement with the governor of the affected state outlining what the state's role will be and its commitment to the recovery operation.

FEMA's responsibility is to supplement disaster assistance available through state and local governments, not to supersede it. FEMA is also responsible for coordinating disaster assistance provided by all other federal agencies, the military, state and local governments, and, with their consent, those of private relief agencies as well.

Basic disaster assistance from the federal government falls into three possible categories:

- **Individual assistance.** This can provide help to individuals and families, and coordination of aid to businesses, including farmers.
- **Public assistance.** This provides funding assistance and technical expertise to aid local and state governments and certain private, nonprofit organizations to repair essential facilities. (Reporters often call this assistance *civil* or *public works*.)
- **Hazard mitigation assistance.** This provides funding for measures designed to reduce future losses to public and private property.

Whether these categories of assistance are made available depends on the needs identified during damage assessments by teams usually made up of federal, state, and local officials.

FEMA's first concern is for families and individuals adversely affected by the disaster. The degree of suffering and deprivation plays a major role in determining the need for federal assistance.

There are no hard-and-fast rules within FEMA requiring that a specified number of people must be affected before a request can be considered. FEMA has recommended approval of federal assistance in situations involving only a few dozen disaster victims. The critical consideration is whether those impacted by the disaster have needs that state, local, and/or volunteer recovery efforts cannot meet for one reason or another.

While most major disaster declarations are aimed at providing assistance to individuals and families, there are occasions when the only needs determined to be beyond state and local capabilities are those addressed in the public assistance categories.

Public assistance is designed to provide federal aid to repair damaged or destroyed facilities essential to the overall integrity of the affected political jurisdiction and to protect public health, safety, and welfare. This can include repairs to roads, bridges, parks, city halls, public libraries, publicly held hospitals and utilities, and other facilities of this type. It can also be used to clear debris.

A governor's request may seek individual assistance, public assistance, or both. FEMA may recommend that the president authorize either or both, depending on the needs found to be beyond the state and local recovery efforts.

APPLYING FOR FEDERAL FUNDS

Application for federal disaster assistance involves specific damage assessment and application procedures. The official damage assessment is conducted by a team consisting of officials from the regional office of FEMA, the state emergency management agency, and representatives from local government. Assessments are detailed, including specific numbers of damaged dwellings, business and public structures, assessed valuations, and percent of insured damages. Funding applications normally must be submitted within 30 days of the presidential declaration. According to the FEMA Web site, the following are deadlines associated with the filing of claims:

- An applicant must submit a Request for Public Assistance within 30 days of the date that the area was designated a disaster area.
- Project information must be submitted to FEMA within 60 days of the Kickoff Meeting.
- FEMA will take action on granting funds for a specific project within 45 days of receiving all of the information for that project.
- A subgrantee may appeal FEMA's decision to the grantee within 60 days of being notified of that decision...

All time frames are set by regulation; however, if extenuating circumstances or unusual project conditions exist, a time extension may be requested through the grantee [state]. The grantee has the authority to extend the time frames for completion of debris removal and emergency work by 6 months and permanent work by 30 months. For all other extensions, the grantee must request the extension from FEMA.[2]

Figure 15–2 lists the available programs for disaster assistance from the federal government.

Checklist of Potential Federal Funds for Disaster Recovery

Individual Assistance

- ☐ Temporary Housing
- ☐ Disaster Loans for the following purposes (maximum loan payouts)
  - -Homes ($200,000.00)
  - -Personal Property ($40,000.00)
  - -Business ($1,500,000.00)

    Note: all loan maximums increased 20% for mitigation activities
- ☐ Individual and Family Grants
- ☐ Agricultural Assistance (80% of production up to $500,000.00)
- ☐ Disaster Unemployment Assistance
- ☐ Food Stamp/commodity assistance
- ☐ Legal Services
- ☐ Crisis Counseling
- ☐ Federal Tax Assistance
- ☐ Elderly Aid via the Social Security Administration
- ☐ Volunteer Assistance via American Red Cross

Public Assistance

- ☐ Federal Aid for Highways
- ☐ Restoration of Public Schools
- ☐ Fire Suppression Assistance

Hazard Mitigation Assistance

- ☐ Property Acquisitions
- ☐ Pre-disaster Mitigation Program
- ☐ Flood Mitigation Assistance

Fig. 15–2. Federal disaster assistance options. Response executives are encouraged to work closely with local, state, and federal officials in order to accurately inform victims of the assistance available. Often the local fire, police, and administrative officials will have to field these questions.

Eligible applicants

According to FEMA, there are four types of entities eligible to apply for public assistance. If an entity meets the requirements of one of the types, the applicant may be eligible to receive federal disaster assistance.

1. **State government agencies**
2. **Local governments and special districts** Any county, city, village, town, district, or other political subdivision of any state, including any rural community, unincorporated town or village, or other public entity for which an application for assistance is

made by a state or political subdivision thereof. Other state and local political subdivisions may be eligible if they are formed in accordance with state law as a separate entity and having taxing authority. These include, but are not limited to, school districts, irrigation districts, fire districts, and utility districts.

3. **Private nonprofit organizations** These organizations are private nonprofit organizations or institutions that own or operate facilities that are open to the general public and that provide certain services otherwise performed by a government agency. These services include:

- Education
  - Colleges and universities.
  - Parochial and other private schools.
- Utility
  - Systems of energy, communication, water supply, sewage collection and treatment, or other similar public service facilities.
- Emergency
  - Fire protection, ambulance, rescue, and similar emergency services.
- Medical
  - Hospital, outpatient facility, rehabilitation facility, or facility for long-term care for mental or physical injury or disease.
- Custodial care
  - Homes for the elderly and similar facilities that provide institutional care for persons who require close supervision but do not require day-to-day medical care.
- Other essential governmental services
  - Museums, zoos, community centers, libraries, homeless shelters, senior citizen centers, rehabilitation facilities, shelter workshops, and facilities that provide health and safety services of a governmental nature. Health and safety services are essential services that are commonly provided by all local governments and directly affect the health and safety of individuals. Low-income housing, alcohol and drug rehabilitation, programs for battered spouses, transportation to medical facilities, and food programs are examples of health services.

4. **Federally recognized Indian tribes, Alaskan Native tribal governments, Alaskan Native village organizations or authorized tribal organizations.** This does not include Alaska Native corporations, which are owned by private individuals.

The application is submitted through the state office of emergency services to the regional office of FEMA and then to the director of FEMA. The FEMA director will make a determination on whether the application meets federal guidelines. If the decision is affirmative, a recommendation will be made to the president to make the appropriate declaration. If the application is rejected, the director will notify the state.

DISASTER APPLICATION CENTERS

To make it easier for disaster victims to obtain individual assistance, the FCO establishes one or more disaster application centers (DACs). The DACs allow representatives of federal agencies, local and state governments, and voluntary relief agencies to counsel and accept applications for aid from individuals, families, and business firms, all in a centralized location.

These federal-state DACs are different in nature and purpose from assistance centers operated by the American Red Cross and other volunteer groups. While both usually are referred to as *centers*, the volunteer center is established to dispense emergency aid—food, clothing, medical supplies, and shelter. A DAC, on the other hand, is a centralized location where the application process for long-term recovery aid—grants, loans, and so forth—is begun.

Disaster application centers are established close to the stricken area in large facilities, often in schools or armories, to accommodate many workers and the applicants they will serve. Following an initial interview, each applicant is directed to the desks of one or more of the government or private assistance programs most likely to meet the needs of that applicant. Since this is the beginning of the application process, no decision is made in the DAC whether the applicant is eligible for one type of aid or another. Such decisions are made by the various agencies on the basis of information provided by the applicant on forms obtained during this interview. Officials will attempt to properly mark the DAC with a large sign and bold lettering to ensure that the sign is visible and distinctive.

DACs remain open as long as the FCO and his or her counterpart, the state coordinating officer (SCO), deem necessary. New DACs may open in the disaster area, and mobile DACs may be sent to remote areas if needed.

When DACs do close, the FCO makes available the forwarding addresses and phone numbers of the participating agencies where individuals may make applications or obtain needed assistance or information.

GETTING OUT THE WORD

Prior to the opening of the DAC, the FCO's staff conducts a brief but intense public information campaign. This campaign is designed to advise disaster victims about the types of federal aid that will be available, eligibility requirements, and how to apply. The objective is to prepare people for the disaster relief application process, to keep expectations at realistic levels, and to encourage only the most seriously affected individuals to visit DACs on opening day. This information effort, with appropriate media support, can save individuals much time, trouble, and disappointment at DACs, and may save them an unnecessary trip altogether.

The fire department can play a pivotal role in public information. A *Newsweek* article pointed out that the profession Americans trust most is that of a firefighter.[3] They should be willing to cooperate with FEMA in providing public information.

The FEMA public information campaign makes these important points:

- A visit to a DAC is the first of several steps in the recovery process. Applicants will receive information and applications at these centers, not goods or cash, although recent disasters proved that individuals could receive a $2,000 check card from FEMA.
- Those not severely affected by the disaster can avoid long lines on opening day by waiting a day or two before visiting a DAC.
- Basic information will be required from individuals at the DAC, such as the applicant's name, current and predisaster address (if appropriate), proof of residence, current phone number, insurance coverage, policy number(s), and agent's name. Applications will be taken without this information, but the process will move more quickly if this information is readily available.
- Applicants have a responsibility to contact their insurance company or companies and file necessary claims. Government assistance does not cover damages or losses already covered by private insurance.

- Federal disaster aid does not guarantee total recovery from all disaster losses. Assistance is limited to that necessary to return damaged or lost property to tenable condition.
- Federal and state assistance is long-range in nature. Therefore, it should be understood that recovery will not occur overnight.

DISASTER HOTLINES

Another effective tool the FCO may use to communicate with disaster victims is the disaster hotline, a telephone question-and-answer service that helps both disaster victims and the federal response officials. In all disaster operations where communications links can be established, FEMA sets up a special telephone number—usually toll-free 1-800 lines—that victims may use to obtain basic information. Through this service, victims can obtain such information as where the DACs are, operating hours, what documents (if any) will be needed at the DAC, and so forth. Hotline operators, trained by FEMA, maintain a daily log of the types of questions asked. An unusually large number of calls relating to one program or another suggests possible confusion about the program and may indicate a need for greater emphasis in this area by public information officers and the media.

As with almost every other aspect of life in the 21st century, disaster assistance has moved to an online environment. Through the www.fema.gov Web site, applicants can investigate their eligibility for the particular assistance programs, file the needed paperwork, and check the status of their application. Although this is a great addition to the application process, disaster victims are not likely to have Internet access in the middle of the disaster. However, this process can be completed by victims who have been evacuated to nondisaster areas.

HOW MUCH WILL IT COST?

Monetary disaster losses must eventually be calculated, but this process is not easy. It is virtually impossible to assign a dollar value to the amount of property lost or to the amount that will be needed to restore those losses. Even if it could be done, it is unlikely that the two figures—loss and cost—would balance neatly.

A governor's request for federal assistance almost always includes a dollar estimate of the damages. However, as the record shows, these estimates can be far too high or too low. Such estimates are developed quickly, in most cases, and their only role in the decision process is to determine whether the recovery effort is beyond state and local response capability.

To illustrate the problem with assigning dollar values on loss and cost, consider an example of a disaster that involves a city hall. Anybody with eyes can see that the city's $1 million city hall was destroyed. The mayor cites the $1 million loss, and he is correct. But then it is determined that the city government insured the structure for $800,000.

Thus, the city lost its $1 million city hall, but insurance will reimburse the local government $800,000. The net loss to the city is $200,000.

But if there is a presidential declaration, the city would be eligible to seek federal reimbursement from FEMA to rebuild the structure. FEMA can reimburse 75% of the uninsured loss, or $150,000. The state and/or local government pays the remaining $50,000.

So what is the actual loss? Actually, it varies with each viewpoint:

Mayor: $1 million

Insurance company: $800,000

FEMA: $150,000

State/local government: $50,000

Factors such as these can make damage estimates and repair costs confusing and seemingly contradictory.

In most cases, the actual dollar loss will not be known until the last damaged bridge is rebuilt, the final roofing shingle nailed in place, the final recovery loan disbursed, and the last insurance claim paid. Until then, there can be only estimates.

In addition, it is extremely unlikely that the FCO or the disaster public information officer on the scene will make estimates. They certainly would not do so unless there is solid information that such estimates would be accurate.

There will be some figures available, however, that may be useful. For instance, there will be daily reports on the number of disaster victims who have visited a disaster assistance center and what kinds of assistance they sought. The American Red Cross and other volunteer organizations will be able to say how many victims they have helped and the kind of help provided. Local governments or the U.S. Army Corps of Engineers may let repair contracts, and these details will be available.

Once the DACs have been operating for a few days, federal officials may be able to make projections. These can be developed on the basis of the number of aid applicants visiting a center, the needs they have, and the average cost of providing these needs. However, once again, these will be only estimates. For short durations, the mobile disaster response center pictured in figure 15–3 allows the DAC to become operational in a limited time frame.

Fig. 15–3. A mobile disaster recovery center operated by FEMA

FEMA News Photo.

One of the major programs usually involved in disaster recovery is the disaster loan program. Figures generated by applications to repair or replace homes, personal property, and businesses can make headlines, and they do represent losses. The record may show, for instance, that 500 homeowners and businesses applied for $60 million in disaster loans. This clearly represents at least part of the disaster loss, but is it a cost? Remember that these are loans, virtually all of which will be repaid to the federal government with interest.

Another important point to be remembered is that the federal funds needed for response and recovery in a presidentially declared major disaster are available no matter how much is required. This is because the President's Disaster Relief Fund, a line item in the federal budget, is replenished as needed.

None of this is intended to suggest that reporters will never get those all-important dollar figures that are so much a part of telling the story of the disaster. It is intended only to illustrate the problem facing the disaster information officer who tries to respond to the question, "How much will it cost?"

DISASTER RECOVERY PROGRAMS

Presidential approval of a governor's request for federal supplemental assistance activates a wide range of disaster recovery programs from a variety of federal agencies. Representatives of these agencies will move into the disaster area to provide the long-term recovery assistance not available through emergency response groups such as the American Red Cross, the Salvation Army, and others.

While each federal agency is responsible for administering its programs, management and coordination of the overall recovery operation are the responsibilities of FEMA through the FCO, which is dictated by the National Response Framework. In addition, FEMA, under the Department of Homeland Security, and in cooperation with state officials, arranges for establishment of disaster application centers. FEMA also sets up appropriate communications and provides personnel to ensure that the needs of the disaster victims are met in a timely and efficient manner.

While there is a wide range of programs that can aid disaster victims, it should not be assumed that all of them will be activated in a given situation. The determination of which programs will be used is made by the actual need found during the damage assessment and any subsequent information that may be developed.

The following is a brief explanation of the major provisions of the various disaster recovery programs. There is also an outline of the capabilities of the various volunteer agency programs. These explanations are general because of space limitations. However, they are suitable for publication or broadcast should an area be declared by the president to be a major disaster area.

INDIVIDUAL AND BUSINESS ASSISTANCE

Temporary housing

FEMA, or an agency designated by the state, can provide housing assistance to those whose homes are damaged, destroyed, or otherwise rendered uninhabitable by the disaster. The assistance may be provided for up to one year, but the applicant's need is reviewed at least quarterly.

Temporary accommodations may be uninhabited government-owned or subsidized housing units, commercial rental property, or as a last resort, mobile homes or trailers. Mobile homes are seen as a last resort because of the time required to move them to the area of need, development of suitable sites, and construction of necessary utilities. In recent years, FEMA has made a large investment in mobile homes and trailers. These trailers are deployed to disaster areas to provide temporary housing to the victims.

Funds may also be provided for use in making habitable those homes that sustained only minor damage.

Disaster loans

The Small Business Administration (SBA) can make federally subsidized loans to repair or replace homes, personal property, or businesses that sustained damages not covered by insurance. Maximum loans are $200,000 for homes, $40,000 for personal property, and $1,500,000 on businesses. These maximums will be increased by 20% if the improvements involve preventing damage from future disasters. Over the years, the interest rates on disaster loans have varied in relation to the cost to the federal government to borrow money. Since interest rates can vary from time to time, it is recommended that the media confer with the local SBA office or the FEMA regional public affairs officer to determine the applicable rates at the time of the declaration. Repayment of loans may go to 30 years, but actual terms are based upon the applicant's ability to repay. The application deadline is 60 days from the date of the president's declaration. (*Note:* The administrator of SBA is empowered to make the disaster loan program available in a designated area without presidential declaration of a major disaster.)

Individual and family grants

Upon request of the governor, the individual and family grant program may be made available. This program is administered by the state. However, it is underwritten with 75% FEMA funding and can provide grants up to $5,000 through an agency designated by the state to meet disaster-related serious needs and necessary expenses not provided before by other programs, insurance, or other means. One test for grant eligibility requires, among other things, a decision by the SBA that the applicant is ineligible for a disaster loan. Among needs that can be met with grant funds are housing repairs, home cleaning and sanitation, necessary household items, replacement of tools of the applicant's trade, clothing, transportation, and medical and dental expenses, among others. The application deadline is 60 days from the declaration date.

Agricultural assistance

The Farmers Home Administration (FmHA) may make emergency loans to farmers and ranchers—owners or tenants—who were operating and managing a farm or ranch at the time of the disaster. These loans are limited to the amount necessary to compensate for actual losses to essential property and/or production capacity. Loans are limited to 80% of the actual production loss of up to $500,000. As with SBA, FmHA interest rates have varied through the years. It is recommended that the media contact FmHA or the FEMA regional public affairs officer to determine the applicable rates at the time of the declaration. Disaster loans from FmHA also are available to agriculturists in counties that are contiguous to those counties designated as a result of a presidential declaration. This contiguous rule applies across state lines as well.

The U.S. Department of Agriculture has other programs that may come into play in varying degrees during recovery operations. Among these programs are those that include assistance in removing debris from agriculture lands, fence repair, crop insurance, and other. (*Note*: The Secretary of Agriculture is empowered to make disaster loans available without presidential declaration of a major disaster.)

Disaster unemployment assistance

This program is administered by the state's employment commission in cooperation with the U.S. Department of Labor. It may provide weekly benefit payments to those out of work because of a major disaster, and who have no entitlement to regular state unemployment insurance programs. Examples include self-employed persons or farm and ranch workers, among others. The president must declare a major disaster for this program to be activated.

Food stamps/commodity assistance

Food stamps and commodities are made available by the Food and Nutrition Service of the U.S. Department of Agriculture (USDA) after designation by the Secretary of Agriculture following a presidential declaration. Assistance includes:

- A 30-day allotment of food stamps to low-income households in areas affected by the disaster.
- Food commodities to recognized disaster organizations for mass feeding.
- Food commodities for household distribution in areas served by the Food Stamp Program whenever commercial food sources remain disrupted as a result of the disaster.
- In order to determine eligibility for food stamp assistance, victims need to call 1-888-312-4567. This is in addition to the immediate food and water supplies, similar to the ones in figure 15–4, that are handed out by FEMA.

Fig. 15–4. FEMA responds with water and food to aid disaster victims. In recent years, FEMA has increased this portion of its response to disasters.

FEMA News Photo.

Legal services

This program, usually supported by the Young Lawyers Division of the American Bar Association, can be activated by the appropriate FEMA regional director to provide free representation to low-income disaster victims requiring such assistance. Fee-generating cases—cases that a local attorney likely would accept—are not accepted by these disaster counselors.

Crisis counseling

This program, funded by FEMA through the National Institute of Mental Health, may be implemented if requested by the governor. Its purpose is to relieve those possible mental health problems caused or aggravated by the disaster or its aftermath. The assistance is provided by state and local health agencies. According to FEMA,

> There are two separate portions of the CCP that can be funded: immediate services and regular services. A State may request either or both types of funding.
>
> The **immediate services** program is intended to enable the State or local agency to respond to the immediate mental health needs with screening, diagnostic, and counseling techniques, as well as outreach services such as public information and community networking.
>
> The **regular services** program is designed to provide up to nine months of crisis counseling, community outreach, and consultation and education services to people affected by a Presidentially declared disaster. Funding for this program is separate from the immediate services grant.
>
> To be eligible for crisis counseling services funded by this program, the person must be a resident of the designated area or must have been located in the area at the time the disaster occurred. The person must also have a mental health problem which was caused by or aggravated by the disaster or its aftermath, or he or she must benefit from services provided by the program.[4]

Federal tax assistance

The Internal Revenue Service makes available advisors to assist disaster victims who wish to deduct uninsured casualty losses from the current tax year's return or the prior year's return. This action can result in an expedited tax refund useful to a disaster victim's recovery effort. The Bureau of Alcohol, Tobacco and Firearms helps businesses recover excise taxes and customs duties paid on distilled spirits, wines, beer, tobacco, etc.

Aid to elderly

The Social Security Administration provides assistance to annuitants who require address changes and can expedite benefit check delivery. It also can assist disaster victims in applying for disability, death, and survivor benefits.

Assistance to veterans

The Veterans Administration provides assistance in applying for VA death benefits, pensions, insurance settlements, and adjustments to VA-guaranteed home mortgages made appropriate because of financial burdens caused by disaster-related losses or damages.

Volunteer assistance

The American Red Cross provides individual and mass shelters, feeding, vouchers, clothing or funds for clothing, cleaning supplies, supplemental medical care, and blood and blood products. It also can provide funds for certain essential home furnishings, minor repairs to homes, rent, equipment, eyeglasses, prescription medicines, lost dentures, and prosthetic devices.

The Salvation Army provides emergency food, clothing, furniture, and financial assistance for immediate needs.

Other emergency aid

Emergency assistance programs also may be available through private relief organizations such as Church World Services, Seventh Day Adventists, Southern Baptist Disaster Services, local churches, private organizations, and others.

Mennonite Disaster Services may provide skilled and unskilled labor for debris removal, home repair and reconstruction, and home cleaning. Many disaster relief funds were created as a result of 9/11 and the destructive hurricanes of 2005.

PUBLIC ASSISTANCE

At the direction of the president, FEMA can provide funding and technical assistance to the state and local government and certain private, nonprofit institutions for the repair or replacement of public properties damaged or destroyed by the disaster. Among these can be restoration of public-owned facilities (roads, bridges, buildings, parks, etc.) and other facilities considered necessary to reestablish the integrity of the disaster area's infrastructure.

FEMA also can provide assistance for emergency protective measures, debris removal, emergency communications, and emergency public transportation.

The agency may also provide assistance to the state government for fire suppression for any fire that threatens such destruction as would constitute a major disaster. This can be accomplished by FEMA without a presidential declaration.

In addition, the agency may make limited loans to local governments that have suffered a substantial loss from their tax rolls and other revenues as a result of the disaster.

Federal aid for highways

The U.S. Department of Transportation's Federal Highway Administration can provide financial and technical assistance for repair or reconstruction of highways on the federal aid system, forest highways, park roads and trails, Indian reservation roads, and other similar projects damaged by the disaster.

Restoration of public schools

The Office of Elementary and Secondary Education, U.S. Department of Education, can provide grants to restore public schools damaged or destroyed by disasters that are declared by the president. The agency also may provide grants to construct minimum public school facilities to take the place of a destroyed nonpublic school that will not be replaced.

Fire suppression assistance

Chapter 7 discussed the ever-present threat of wildland/urban interface fires. These occur often in many parts of the country, particularly during extended periods of drought. Careless campers, lightning, and arson are the most common causes of such fires.

In most cases, state and local resources—often augmented by trained firefighters brought in by the state and local volunteers—are adequate to contain, control, and ultimately extinguish these fires. However, there are times when a governor foresees a need for federal assistance from FEMA.

The Disaster Relief Act in effect since 1974 contains a provision that authorizes the agency to provide grants, equipment, supplies, and personnel to a state government, but not specifically to local governments. Though the agency can provide any or all of these forms of assistance to a state, the most common is financial reimbursement for a portion of its actual cost to battle the fire.

Financial reimbursement is determined by a formula based on the state's average annual expenditure to fight these types of fires. A *floor cost*, based on a five-year average of state expenditures, is used to determine the amount of extraordinary costs incurred by the state to fight a specific fire.

If the request for federal aid is approved and there are found to be extraordinary costs beyond the state's floor cost for the year, FEMA can provide reimbursement of 75%. It should be noted that fires on federal national forests and Bureau of Land Management lands are the responsibility of the federal agency concerned.

Whether or not assistance can be provided by FEMA depends upon a key phrase in the enabling legislation. According to the law, the specific fire—on nonfederal public or private lands—must "threaten destruction as would constitute a major disaster." In other words, the fire must pose a present or continuing threat to public safety or to property necessary to the public welfare.

The authority to grant fire suppression assistance has been delegated to FEMA. That means no specific action by the president is required.

PHYSIOLOGICAL AND PSYCHOLOGICAL DIMENSIONS OF DISASTERS

Major emergencies are stressful experiences, not only for victims, but also for responders and many others in the immediate community. Many persons associated with such episodes will exhibit or experience some degree of psychological trauma.

Even today, many of the responders and victims of 9/11 and Hurricane Katrina are still suffering from depression and stress-related illnesses. As the science of psychology develops, many recent reports show a link between illness and stress. Providing mental health treatment to responders and victims is an important aspect of returning the community to a predisaster state.

PSYCHOLOGICAL EFFECTS OF THE COMMUNITY: MYTHS AND REALITY

Victims of disasters tend to go through a series of phases in reacting to the incident. If there has been warning, preincident activities focus on attempts to validate the severity of the threat. This threat validation process can have many factors, such as the effect of previous disaster predictions that were not valid or were hyped. On the other hand, people can learn from not reacting properly to a disaster, which occurred post-Katrina in 2005.

As emergency conditions strike, immediate reactions turn to protecting self and family. An immediate inventory phase then follows in which people anxiously attempt to account for their homes and loved ones. Highly altruistic behavior is often seen as victims and unaffected volunteers work to assist those in need.

Once the emergency needs have been brought under control, those who are unhurt begin an energetic process of securing possessions, if their homes have been damaged or destroyed, and cleaning up the mess. Psychological reactions in the early recovery phase focus very strongly on returning to some semblance of a normal routine.

In South Carolina, in the aftermath of Hurricane Hugo, hundreds of families went without heat or electricity for days, and yet they improvised in varying ways and were able to carry on a semblance of reasonable, normal living. This, by the way, prompted Congressman Curt Weldon to proclaim "Can you imagine? A nation that can send 400,000 troops halfway around the world in four months cannot get generators to families that are without heat and electricity."[5] This type of response was again the case in 2005 with Hurricane Katrina, which drew sharp criticism by the public. With the great level of reporting available today via media outlets, many families not affected by the disaster can see and experience the devastation via television and the Internet. Thus, even citizens not affected by the disaster have feelings of betrayal when the government does not care for citizens in a timely manner.

DISASTER SYNDROME

Victims and survivors of the immediate impact area frequently exhibit a typical behavior pattern called *disaster syndrome* that psychologists have characterized as a defensive reaction. The behavior appears in several phases, as described by Anthony Wallace, an early researcher in disaster behavior:

> During the first stage, the person displaying it appears to the observer to be "dazed," "stunned," "apathetic," "passive," "immobile," or "aimlessly puttering around." This stage presumably varies in duration from person to person, depending on circumstances and individual character, from a few minutes to hours; apparently severely injured people remain "dazed" longer than the uninjured although this emotionally dazed condition is no doubt often overlaid by wound shock. The second stage is one of extreme suggestibility, altruism, gratitude for help, and anxiousness to perceive that known persons and places have been preserved; personal loss is minimized, concern is for the welfare of family and community. This stage may last for days. In the third stage, there is a mildly euphoric identification with the damaged community, and enthusiastic participation in repair and rehabilitation enterprises; it sometimes appears to observers as if a revival of neighborhood spirit has occurred. In the final stage, the euphoria wears off, and "normally" ambivalent attitudes return, with the expression of criticism and complaints, and awareness of the annoyance of the long-term effects of the disaster. The full course of the syndrome may take several weeks to run.[6]

Research that is more recent suggests that the disaster syndrome may not be a universal occurrence. Furthermore, actual disaster behavior may vary considerably depending on the type of disaster and the social and cultural characteristics of the community and particular people.

In an article published in the *American Family Physician*, Dr. John R. Freedy and Dr. William M. Simpson address the effects of disaster-related illnesses, which are likely to present to the local family physician. According to the article,

> The use of primary health care services typically increases for 12 or more months following major disasters. A conceptual framework for assisting disaster victims involves understanding the individual and environmental risk factors that influence post-disaster physical and mental health. Victims of disaster will typically present to family physicians with acute physical health problems such as gastroenteritis or viral syndromes. Chronic problems often require medications and ongoing primary care. Some victims may be at risk of acute or chronic mental health problems such as post-traumatic stress disorder, depression, or alcohol abuse. Risk factors for post-disaster mental health problems include previous mental health problems and high levels of

exposure to disaster-related stresses (e.g., fear of death or serious injury, exposure to serious injury or death, separation from family, prolonged displacement).[7]

"COUNTER-DISASTER" SYNDROME, PANIC, AND CONVERGENCE

Several years ago, Steve Collier (previously introduced as coauthor of the National Fire Academy's course *Disaster and Fire Defense Planning*) was reviewing the recruit training curriculum at a major law enforcement academy. In doing so, he came across a section on disaster response. He noted that the section consisted of only a few paragraphs that described the principal law enforcement role at such incidents as the control of looting and public panic. Fortunately a substantial body of research indicates that actual occurrences of either are relatively rare and take place only under specific conditions. *Panic*, characterized as totally irrational fear, generally occurs only when individuals perceive themselves to be trapped or threatened with no option for escape, such as being trapped in a burning building. However, panic is clearly not a characteristic response to stress or loss of possessions.[8]

Convergence

Convergence refers to the tremendous influx of people and equipment that typically occurs moments after a major incident, and it may continue for a period of hours or days. Convergence behavior can be divided into three sources: the general public, emergency response agencies, and the news media. This behavior is so prevalent that many recent emergency operations plans have added sections to specifically deal with this issue based on recent disasters such as 9/11 and Hurricane Katrina.

Convergence by the general public

Disasters that are particularly violent or spectacular tend to draw tremendous numbers of people even in the first few moments. These people may be motivated by one or more of the following factors:

- Curiosity about the nature of the incident
- Desire to be part of or involved in the incident; "to have been there"
- Intent to volunteer assistance, particularly if they have medical or other emergency skills
- Desire to learn the status of friends or relatives who may be involved

The convergence of looters, as accepted notions hold, had not been extensively documented in the immediate aftermath period until recently in New Orleans. This looting by the victims and others has shocked many response planners and executives and required some new thinking about response to national disasters. In post-Katrina disasters, responders felt the need to arm themselves, which proves a quandary when trying to balance a look of helping with the safety of the responders.

Looting, of course, is a major factor in civil disturbances or social unrest, since it is a tactic in itself.

Convergence by emergency responders

Convergence is also a phenomenon among emergency responders. Virtually every major incident attracts law enforcement officers, firefighters, and emergency medical personnel from miles away who can usually gain access to the site because of their vehicles and uniforms. In some cases this convergence is in response to general calls for assistance, but in many cases, it is simply a desire to be part of the action. The convergence often creates serious problems for jurisdictional agencies, since many of the outside responders are not familiar with local plans and procedures.

Convergence by responders has created such problems in recent disasters that the Department of Homeland Security, through the U.S. Fire Administration, has put out mass e-mails to emergency responders urging responders not to self-dispatch. The associate author participated in the monthly *Fire Engineering* Roundtable discussion on self-deployment, in which he wrote a response to the question, "What should be the consequences of self-deployment to major disasters?" His response stated,

> Ultimately, the consequence should be the return of the firefighters or EMTs to their home jurisdiction without using their services. This may be difficult given the severe need for help directly following a major event. I would propose that the fire service develop a firefighter version (FERT) of the Community Emergency Response Teams (CERT) model. The fire service has provided personnel to US&R and DMAT, but much of the work needed directly after a disaster involves logistical support and humanitarian aid. Firefighters and EMTs are excellent for both positions, but they only add to the problem when they show up unequipped to care for themselves and without a dedicated assignment. Through a FERT program, firefighters could be assembled at a predetermined location. They could be deployed with the proper self-care items (clothes, PPE, and so on) and transported to designated locations with designated jobs. An online class could cover the basics to obtain the certification for the team. This program would allow the fire service to show up in force but in a professional and coordinated effort.[9]

Convergence by the news media

Any significant incident, particularly a spectacular one, will attract local news media from up to several hundred miles away, as well as network broadcast and cable news crews. This type of convergence intensifies the feeling that the community is in a fishbowl being watched by millions. The frantic competition of outside, unfamiliar reporters and the influx of high-tech broadcast equipment can be particularly unnerving to incident commanders and response agencies.

Techniques for dealing with convergence

Fire, law enforcement, and other emergency response agencies can minimize the effects of convergence by doing the following:

- Anticipate convergence problems at major incidents and address them in plans so that field personnel will be prepared.
- Work together to develop major incident plans that prevent congestion, thus preventing the need to clear it out after it occurs.
- Provide a responsive media briefing center so that the news media can get valid information. Release technical and factual information only through designated public information officers. News media representatives should not be permitted to randomly interview agency officials.
- Try to release information as quickly as possible so that those converging on the scene simply to find out what happened will have valid information from an informed

source. This was one of the most successful operations of Mayor Rudolph Giuliani in the days after 9/11. He provided timely assertive information to the public via the news media. This instilled confidence in the public during a time of panic from a terrorist attack.

CRISIS COUNSELING FOR VICTIMS

Various phases of the disaster syndrome may last for several days or weeks depending on the type of incident. In the airplane crash in Indianapolis, when the jet fighter struck the hotel, damage was so extensive that the hotel closed for months. Most of the 10 people killed were employees of the hotel, as were many of the affected survivors. These survivors experienced severe difficulties because they had lost fellow employees and friends. They had lost their jobs because the hotel was closed, and they had narrowly missed being killed themselves. To some extent, they had experienced guilt feelings because they survived and their friends did not. Some of these employees experienced reactions so severe that they could not function well enough to take advantage of available help or to seek employment.

Many of these victims eventually received counseling assistance from local mental health counselors and from an outside counseling team sent in by the National Organization for Victim Assistance. This team works with local mental health practitioners to counsel victims and to set up longer-term counseling programs.[10]

Most adverse psychological reactions can be dealt with effectively by early intervention that focuses on sharing feelings and receiving reassurance that these feelings and emotions are completely normal. Local mental health associations are routinely involved with crisis counseling, but the stigma of these agencies as dealing primarily with the mentally ill prevents many victims from seeking their assistance. Agency complaints and local or outside victim assistance groups can act as intermediaries in getting those who need help into the system.

CRITICAL INCIDENT STRESS IN RESPONDERS

Psychological reactions are also seen frequently in responders involved in major incidents. Certain types of incidents commonly result in problems. Such incidents include those causing significant loss of life, particularly in the following situations:

- When it is violent (such as an airplane crash).
- When it involves children (San Diego McDonald's sniping incident).
- When protracted rescue efforts are involved (Kansas City Hyatt Regency collapse).
- When rescuers are exposed to sustained personal danger (e.g., pinned down by sniper fire).

These so-called critical incidents can lead to *critical incident stress*. Figure 15–5 contains a critical incident stress checklist.

Critical Incident Stress Checklist

The following is a sample list of possible signs and symptoms from Critical Incident Stress
Contact a CISM resource or medical attention if these symptoms are present

Incident Characteristics

- ☐ Large-Scale Event
- ☐ Death(s) as a result of the event
- ☐ Characteristics of event closely correspond to caregivers current life situations such as
  - -Child same age
  - -Same age as caregiver
  - -Within same social setting, i.e., hobbies, sports
- ☐ Failed rescue event
- ☐ First Event of type, i.e., first death, first rescue, first extrication, first dismemberment

Physical Symptoms

- ☐ Shock Symptoms
- ☐ Dizziness
- ☐ Chest Pain
- ☐ Profuse Sweating
- ☐ Weakness

Mental Symptoms

- ☐ Nightmares
- ☐ Blaming Others
- ☐ Heightened or Lowered Alertness
- ☐ Poor Concentration
- ☐ Confusion
- ☐ Uncertainty

Emotional Symptoms

- ☐ Anxiety
- ☐ Agitation
- ☐ Depression
- ☐ Emotional Outbursts

Behavioral Symptoms

- ☐ Loss or Increase in Appetite
- ☐ Change in Social Activity
- ☐ Increased Alcohol Consumption

Note: Not an all inclusive list.

Fig. 15–5. Critical incident stress checklist. Although not an all-inclusive list, supervisors should refer to the list during and after the incident to ensure rescue workers receive proper psychological care to remain healthy and productive.

Even when there is only one victim, the same psychological reactions are manifested in the emergency responders, and for that matter, in the public at large. As an example, recall how the nation watched and waited for the rescue of little Jessica McClure. "The eyes of the nation were glued to the television as rescuers tunneled frantically for more than two days to save the life of a little girl trapped in an abandoned well."[11] Consider, then, how the emotional involvement of emergency responders can multiply with multiple similar victims.

The symptoms of critical incident stress include one or more of the following:

- Compulsion to recount details of the incident for days, weeks, or months
- Difficulty in concentrating
- Difficulty in making decisions
- Disciplinary problems at work
- Sleep problems or recurring nightmares

- Dependence on alcohol or drugs
- Marital or sexual difficulties
- Feelings of guilt ("Isn't there something more I could have done?")
- Suicide (This is rare but well documented in the 9/11 responders.)

Critical incident stress (CIS) has been around for a long time. However, the fire service until the past two decades failed to recognize it or failed to provide a program to deal with it, primarily because there was little direct research or literature regarding the problem. As the 1990s began, research was conducted regarding the effects of CIS on emergency service providers. Literature also began to be published on the subject. One of the leading CIS authorities is Dr. Jeffrey T. Mitchell, an assistant professor with the Emergency Health Sciences Department at the University of Maryland, Baltimore County. Dr. Mitchell and other experts who have studied the effects of CIS on emergency service providers agree that when a person has been stressed beyond his or her stress threshold, normal stress coping mechanisms no longer are effective in reducing stress levels. At that point, other ways of dealing with this level of stress are needed.[12]

Most experts agree that an effective program for dealing with critical stress must have several component phases. There must be an education component that allows all persons in an emergency response unit to learn what critical stress is and how it could affect them. This is such an important component that the FEMA response teams now require a critical incident stress management (CISM) orientation in order to become deployable on the teams. Fire chiefs and officers need to learn how to recognize the signs of stress in fire department personnel. There needs to be a peer support unit of individuals (coworkers) who have been professionally trained. These peers need to be available at all times and must be well versed in the importance of confidentiality. Another part of an effective program is the availability of a CISM team, including professionals from outside the fire department. The CISM team performs two functions. First, it defuses individuals or small groups of individuals immediately after an incident. Next, it performs a full-scale debriefing about 48 hours after an incident. This is discussed in more detail in the section on debriefing.

Educational programs

Information on CIS is increasingly available for fire departments. Educational programs and literature are available to educate firefighters. Group lectures presented by mental health professionals, employee assistance counselors, and chaplains can be effective. Many articles on the subject have been published in the *Journal of Emergency Medical Services (JEMS)*, and books on the subject are available. Existing CISD teams, eager to reach out to new fire agencies, will usually be glad to make a presentation on the subject. The educational portion of the program is so important that it should be conducted by qualified professionals. CISM teams will even customize a presentation on their services for a particular organization, as well as visit the site to deliver the presentation. The Cincinnati chapter is now spearheading a program to deal with alcoholism in responders, which studies have shown is a coping activity related to CIS.

Peer support

The next consideration is peer support. The people selected for the program should be volunteers who are carefully screened to ensure that they support the CIS concept, are compassionate, can be available, and most importantly, will keep information they receive confidential. The confidentiality of the program must be emphasized to all members so that in time of need, individuals will feel free to share their feelings. Peer supporters should be trained in active listening skills as well as what to say and what not to say to victims of critical stress. They should be taught to read nonverbal signs and body language. Training is ongoing, so peer supports should meet on a regular basis to share information and techniques.

They should develop knowledge of the experiences of one another because victims may wish to talk to someone who has been through a similar traumatic experience.[13]

Debriefing

A necessary component of the CIS program is the critical incident stress management (CISM) team (formerly, CISD teams). "CISM teams are made up of dedicated and trained mental health professionals, who combine their expert knowledge and talents with specially trained peer support personnel drawn from the emergency service's ranks. The CISM team is essentially a partnership between the two groups with a common goal—the reduction of critical incident stress in emergency personnel."[14] There are three general areas in which critical incident stress debriefing (CISD) teams function: preincident, incident, and postincident.

CISM or CISD functions

Preincident functions would include education of personnel, command staff, and the families of emergency personnel. They would also include development of stress management protocols to be used in the field. Functions of CISM teams during an incident would include the following:

- On-the-scene support services (to provide assistance to obviously distressed personnel)
- Defusing (a short, usually unstructured debriefing that would encourage discussions of the events and reduce acute stress)
- Demobilizations (used on large-scale incidents in place of defusing)

Demobilizations take place away from the scene and are held in a large meeting facility where mental health professionals can address large numbers of emergency services providers about signs and symptoms of critical stress. All personnel being demobilized are given the opportunity to have something to eat and drink. They are also provided with the opportunity to talk to mental health professionals or chaplains before they return to duty or go home.

After demobilization or defusing, emergency personnel usually enter a phase lasting about 24 hours where they try to sort out events and compare their actions with accepted procedures and protocols. During this period, a full-scale debriefing would probably not be very effective, as fire service personnel are usually not ready to deal with whatever feelings have been generated during the incident. After about 48 hours, the postincident functions, the third phase, takes place.

Postincident functions

The postincident functions include the formal critical incident stress debriefing (CISD). Formal CISD is a carefully designed process that progresses through seven different phases and provides important stress reduction information. According to Dr. J. T. Mitchell,

> Events that require a CISD include the following: Any event that has significant emotional power to overwhelm usual coping mechanisms. Line-of-duty deaths, serious line-of-duty injuries, emergency worker suicide, disasters, unusually tragic deaths to children, significant events where the victims are relatives or friends of emergency personnel, events that attract excessive media attention, and events that seriously threaten the lives of the responders. Because overuse of CISDs dilutes their effectiveness, they are reserved for only those events that overwhelm the usual coping methods of emergency personnel.[15]

SEVEN BASIC STEPS IN DEBRIEFING

Seven basic steps are involved in a debriefing period. These are summarized in the following:

1. Selection of a place, notification of participants, review of videos and reports, and occasionally a site visit.
2. Collection of facts from personnel who are asked to describe what happened at the scene.
3. Thought phase in which people are asked their thoughts regarding the incident.
4. Reaction phase in which personnel are asked to describe the most mentally traumatic parts of the events, including the venting of thoughts and feelings.
   - The symptom phase in which the group is asked to describe stress symptoms felt at three times:
     - During the incident
     - Three to five days after the incident
     - Those stress symptoms still remaining
5. Teaching phase in which CIS team members furnish stress reduction information to the group.
6. Reentry phase where the group is encouraged to ask questions and resolve doubts.
7. A summary and end to the critical incident stress debriefing.

As a first step, the CISM team selects a place, notifies the personnel who are going to participate, and reviews the incident by viewing video tapes and reading reports of the incident. Many teams also visit the scene of the incident. Only the personnel involved in the incident should attend the debriefing, and they should not be required to respond to calls during the debriefing. At this point, the peer support people from the emergency response agency involved in the incident become important. Personnel will feel more comfortable with and are more likely to trust fellow workers. It should be made plain from the beginning that everything discussed is strictly confidential, and it should be stressed that debriefing does not involve a critique of operations at the incident. At the outset, the attendees need to be told what to expect in the debriefing, what the rules are, and that "the main objectives of CISM are to mitigate the impact of a critical incident and accelerate the return of personnel to routine functions after the incident."[16]

The second phase of the CISM is the fact phase, in which personnel are asked to describe what happened at the scene.

In the third phase, the team leader will then lead the discussion into the thought phase, in which people are asked what their thoughts were. In the fourth phase, the reaction phase, personnel are asked to describe the worst part of the events for them. This phase allows for venting of thoughts and feelings associated with the incident, and it prepares the participants for useful stress reduction information.

According to Dr. J. T. Mitchell, "The fifth phase of the CISM process is the symptom phase. At this point the group is asked to describe stress symptoms felt at three different times: first, those symptoms experienced during the incident; second, those that appeared three to five days after the incident; and third, those symptoms that might still remain at the time of debriefing."[17]

The sixth phase is the teaching phase, wherein team members furnish stress-reduction information to the group. The last phase is the reentry phase, in which the group is encouraged to ask questions and resolve any doubts. There is a summary of the debriefing, and the CISD is ended. After the debriefing, team members make themselves available to individuals who may require more assistance. If necessary, referrals are made for counseling. After all the emergency personnel have left the debriefing site, the CISD team has a postdebriefing meeting to review the process and to discuss ways to improve future debriefings.

Follow-up services

There should be follow-up services after all defusing, demobilizations, and debriefings. Some of the ways that follow-up can be accomplished include telephone calls to the individuals, discussions with fire company officers or commanders, and visits to stations. Other follow-up can be done through peer supporters, continuing education programs, support services for spouses, and counseling sessions with individuals. Follow-up services usually begin about 24 hours after the debriefing and are very important to the success of the program.

CISM training

An important consideration for a CISM team is the amount of training required for its members. There should be a minimum of two days of training, in conjunction with continuing education on a regular basis. CISM teams should have good peer support personnel and mental health professionals in order to be effective. Without mental health professionals, CISM teams will not only be ineffective but also dangerous. Mental health professionals are needed for their skills in supervision and leadership, and for their diagnostic skills. Undetected symptoms of critical stress could contribute to the suicide of an emergency worker.

It may take six months to a year to organize a CISM team, and it is very important that all members of the team receive the same training. Operating procedures should be established in writing before the team starts to do formal debriefings. Competency should be the most important consideration when selecting team members.

Dr. J. T. Mitchell notes that "'critical incident stress debriefing teams have experienced phenomenal growth in recent years, overcoming many problems and achieving many successes. They assume an important place within emergency services organizations and are likely to continue their support services in the future."[18]

OVERALL FIRE DEPARTMENT RECOVERY

In disastrous situations, the fire department may be severely affected, at times almost catastrophically. Damaged apparatus and equipment may need to be replaced without funds available. Fire stations, maintenance facilities, and communications centers may require repair or reconstruction; and most seriously of all, firefighters may have been killed or injured. Imagine the recovery process that was needed after 9/11 for the FDNY. Not only did they need to replace a large amount of their fleet, but also 343 of the world's finest firefighters were now no longer present in the department. This disaster crippled the FDNY, but a good plan to put the department back to its previous state needed to be implemented. The fire department may well have difficulty solving its own problems because other agencies in the community may be in the same or worse predicaments, and adequate government funds may be unavailable. In fact, the local tax base may be severely affected by decreased revenues due to business closings, failures, or relocations. This was the case in post-9/11 New York City. Articles in the fire service journals and Internet-based journals relayed information on the FDNY and their closing of firehouses due to financial constraints.

The fire department may also be affected by damage to lifelines or the infrastructure it depends on, such as roadways, water supply systems, utilities, and telephone communications services. If state or federal disaster funds do not become available, complete fire department recovery may be a long process during which operation at previous levels may not be possible.

RESPONSE DOCUMENTATION

One very important activity during both the response and recovery phases of a disaster is that of response documentation. In applying for federal disaster funds, discussed later in this chapter, it is critical that all costs associated with fire department operations be documented. These include the following:

- Manpower overtime
- Apparatus damages
- Apparatus usage above normal
- Fuel
- Equipment and supplies used or damaged
- Administrative costs associated with response, such as those involved with purchasing, vendor payment, legal advice, etc.

Many fire departments are becoming quite skilled at accounting for their costs—even at relatively small-scale events, such as hazardous materials spills—and billing the responsible parties. Equipment, manpower, and materials expenses can be recovered. Volunteer fire departments can also bill for services, and according to a well-researched article by Michael Ward, can charge the same rates for labor as comparable career departments.[19]

A rate chart depicting the amount of cost per hour for response and recovery equipment can be found via the FEMA Web site at http://www.fema.gov/government/grant/pa/eqrates.shtm. An example is a 1,500 gpm fire truck, which is billed at a rate of $95.00 per hour. This does not include the personnel costs.

COMMUNITY RECOVERY

Community recovery operations generally divide into short-term and long-term activities that take place more or less concurrently. Short-term activities are those focusing on the return to normal life and may last from several days to several weeks. Examples of short-term and long-term recovery activities are listed in the following discussion.

Short-term recovery activities include the following:

- Development of a short-term action plan
- Preliminary damage assessment
- Debris clearance
- Temporary housing
- Immediate financial assistance to individuals
- Expedient repair of lifelines and utilities
- Crisis counseling

Long-term recovery activities include the following:

- Development of a long-term action plan
- Acquisition of financial aid
- Reconstruction of lifelines and utilities
- Reconstruction of community business section
- Preventive legislation, zoning, building codes
- Economic development programs
- Improvement of emergency response capabilities

Combined short-term and long-term activities to accomplish include the following:

- Review recovery elements of emergency operations plans
- Determine if an emergency has been declared
- Review preliminary damage assessments and prepare comprehensive assessments
- Identify and prioritize recovery needs
- Identify sources of assistance
- Identify specific funding sources
- Develop and implement a detailed recovery action plan
- Submit documentation/application for state and federal funding
- Document recovery effort
- Monitor entire emergency responses and recovery operation

The Web site www.ready.gov is a joint venture between the Department of Homeland Security and the Citizen Corps, which helps individuals, businesses, and kids prepare for and deal with a disaster. Some of the publications available from the Web site include the *Are You Ready?* manual, which is a 204-page in-depth guide to citizen preparedness. Another useful publication is a brochure explaining what supplies citizens should have on hand. This is in the publication *Emergency Supply List*. This one-stop Web site provides the average citizen the needed information to prepare for and deal with a disaster. Planners should be familiar with these documents so that they have an understanding of the recommended supplies and useful actions on the part of citizens. This can then be factored into emergency response and recovery plans.

THE RESPONSE PROCESS

People expect their government to protect them, assist them, and inform them when a disaster threatens or occurs.

They have these expectations for good reasons. Thousand of hours and many dollars—local, state and federal—have gone into the development of comprehensive plans of action to be followed before, during, and after disaster strikes. Fire service personnel and police at the local level, and emergency management personnel at all levels of government, have undergone extensive training. Communications equipment and procedures are in place, and duties and responsibilities are assigned. In late 1989, when Hurricane Hugo hit and the earthquake occurred in San Francisco in rapid succession, many of the nation's lawmakers questioned the speed and thoroughness of the FEMA response, no doubt feeling pressure from their constituents. The agency was, however, extremely strapped as it was spread across two major disasters.[20]

Should the local government need outside assistance, the governor has the authority to activate the state's emergency plan. The governor may find it appropriate to declare a state of emergency in the affected area, thus freeing state resources to provide the appropriate response and recovery activities.

Should combined local and state resources be inadequate, the governor may turn to the federal government with a request that the president declare a major disaster or an emergency.

Thus, there is a clear-cut chain of events leading from the local level, through the state, and, if necessary, to the federal level. Each step to higher authority is predicated upon the inability of the preceding authority to adequately meet emergency needs.

Local government

A local government's disaster authority and responsibility is centered in the mayor, city manager, or county executive's office. Most cities and populous counties direct disaster relief activities from emergency operating centers (EOCs), discussed in chapter 4. Fire department and local government EOC priorities in the early phases of a disaster are to warn and evacuate citizens and protect life and property.

Immediately following the emergency, local government must ensure public order and security. Vital services such as water, power, communications, transportation, shelter, and medical care must be continued uninterrupted if possible or be restored quickly if disrupted. Debris removal also must begin, as noted in chapter 2.

Local officials rely on help from the private sector and individual volunteers and charitable organizations to cope with immediate postdisaster tasks. Disaster plans and annexes should include liaison with such organizations as medical care facilities, gas and electric companies, telephone companies, transportation companies, and chemical companies. They also must work with firms with heavy equipment such as cranes and bulldozers that can supplement city equipment.

Kansas City, Missouri responded in the wake of several disasters, including the Kansas City Country Club Plaza flood of 1977, which killed 18 people, and the Hyatt Regency Hotel disaster in 1981, which killed 113 people. The city has put together an agreement with the Heavy Constructors Association of the Greater Kansas City Area (HCA), Mid-American Regional Council, and various local governmental bodies. A catalog is published annually by the "Plan Bulldozer" Committee of the HCA, which provides lists of hundreds of resources that can be used in the event of a disaster. Contractor names, equipment, and key figures are provided. This could serve as an excellent model for other communities that have yet to establish an effective link to the private sector and its numerous invaluable resources.[21] The NIMS resources typing section will only enhance these programs, as it provides a basis for what constitutes a "bulldozer," along with its size. This common language allows commanding officers to order the correct resource the first time.

Invaluable assistance is provided by charitable institutions such as the American Red Cross, the Salvation Army, church groups, and others. These private organizations remove a great burden from the local government during an emergency by being partly or totally responsible for distributing food, medicine, and supplies, and by providing emergency shelter.

State government

If local officials need state help, state governors can provide several valuable services. These services could include state police to help ensure public safety and protection and state national guard troops to provide security, transportation, medical, food, and temporary shelter services. Other services include state transportation services, including highway equipment and personnel, and state funds for disaster relief and recovery.

At any time in the disaster or emergency, local government or state officials may turn to a number of available federal agencies for assistance. Most can act quickly under their own statutory authorities without a presidential declaration. These agencies include the U.S. Coast Guard and armed forces for search and rescue operations, the U.S. Army Corps of Engineers for emergency flood protection, and the Small Business Administration for homeowner and business loans. Other agencies included are the Federal Highway Administration for road and bridge repairs and the U.S. Department of Agriculture for emergency loans or grants to farmers and ranchers. Finally, the Department of Health and Human Services can assist in public health and welfare measures.

Federal government

When a disaster situation is beyond the capabilities of local and state forces, supplemented by private and volunteer organizations and limited assistance of federal agencies on the scene, the governor—and only a governor—may request that the president declare a major disaster or an emergency.

Major disaster is defined in the Stafford Act (PL 93-288) as

> any natural catastrophe (including any hurricane, tornado, storm, high water, winddriven water, tidal wave, tsunami, earthquake, volcanic eruption, landslide, mudslide, snowstorm, or drought), or, regardless of cause, any fire, flood, or explosion, in any part of the United States, which in the determination of the President causes damage of sufficient severity and magnitude to warrant major disaster assistance under this Act to supplement the efforts and available resources of States, local governments, and disaster relief organizations in alleviating the damage, loss, hardship, or suffering caused thereby.[22]

EVALUATION OF DISASTER RESPONSE OPERATIONS: IMPROVING THE EMERGENCY

Management Program

An important step in the recovery process that normally takes place near the end of the short-term recovery phase is an evaluation of the emergency response effort. This evaluation is a systematic comparison of the outcome of emergency response efforts to the response goals, objectives, and priorities established in the emergency operations plan. It is intended to identify specific changes indicated in the preparedness program, including the hazard identification and emergency operations plans, and in the capabilities of carrying out those plans.

Evaluation methods

There are a number of methods for evaluating response plans and operations that are listed as follows:

- Review the hazard analysis. (Was this emergency anticipated?)
- Review documentation compiled during the response phase.
- Obtain outside input (local, state, federal officials, researchers).
- Evaluate public reaction or criticism.
- Hold critique sessions among emergency responders.
- Interview involved agency heads.
- Review other after action reports and see if similarities occurred during the response.

Evaluation criteria

The major areas to be evaluated should follow the major elements in the emergency operations plans. Those categories and examples of specific aspects to evaluate are as follows:

- **Direction, control, and coordination**
  Were the strategy and tactics effective?
  Was the command organization effective?
  Were control preparations effective?
  Were resources adequate?
  Were mutual aid procedures adequate?
  Was coordination among external agencies adequate?

- **Alerting and warning**
  Was warning information heeded?
  Were there preventable injuries or fatalities?

- **Communications**
  Were there major hardware failures?
  Were emergency notifications made?
  Were field tactical communications adequate?
  Were EOC/field/area command communications adequate?
  Were there problems in receiving requests for emergency services from the public?

- **Emergency Public Information**
  Were there complaints from the public?
  Were there complaints from the news media?
  Were there excessive rumors or misinformation?
  Was conflicting information distributed by government agencies?
  Were emergency information bulletins/recommendations heeded?
  Was the public response appropriate?
  Did the public receive sufficient information?

- **Assessment of hazards to responders**
  Were personnel injured or resources lost unnecessarily?

- **Size-up and situation status**
  Was available information sufficient for action planning?
  Did recon personnel deliver relevant information?

- **Evacuation and sheltering**
  Was evacuation conducted in time?
  Were transportation resources adequate?
  Were shelter facilities adequate?
  Was temporary housing (if long-term) adequate?
  Were food and support adequate?

- **Law enforcement**
  Was perimeter/traffic control adequate?
  Was control of criminal activity adequate?
  Were evacuation efforts of law enforcement agencies adequate?
  Was a coordinated investigation of the event or disaster conducted, if warranted?
  Did police personnel have the proper PPE needed?

- **Fire suppression**
  Were fires contained and controlled?
  Were there adequate firefighting sources (including water)?
  Were search and rescue efforts effective?
  Were hazardous material leaks controlled?
  Were potential WMD/CBRNE products identified and mitigated?
  Were mutual aid plans effective?

- **Medical/mass casualty**
  Was patient triage adequate?
  Was field treatment appropriate and adequate?
  Was distribution of casualties to hospitals adequate?
  Were there sufficient resources, including manpower, ambulances, equipment, and supplies?
  Was management of fatalities appropriate and adequate?
  Were casualty lists prepared and released appropriately?

Were hospitals able to handle the disaster?
Did decontamination of victims occur if needed?

- **Public works**
  Was debris cleared adequately?
  Were public facilities protected (including sewers, waste treatment, dams, flood control facilities, reservoirs, aqueducts, cisterns)?
  Was support in assessing damages adequate?
  Did public works personnel have appropriate PPE for their tasks?

- **Transportation**
  Were transportation resources adequate?
  Was maintenance/repair of transportation facilities (airports, roads, bridges, railways, mass transit systems, pipelines) adequate?

- **Human services and distance relief**
  Were the following items available from relief agencies? Food, counseling, clothing, sheltering/temporary housing, nonemergency medical needs (medicine, oxygen, appliances), money (for rent, food, etc.)

- **Utilities**
  Was service returned in a reasonable time?
  Were resources (regular, standby, emergency alternate) adequate?

- **Administration**
  Were purchasing procedures adequate?
  Was legal advice timely and adequate?
  Were critical government records preserved?
  Were overall emergency response costs adequately documented?

- **Recovery**
  Was there adequate funding available for necessary recovery activities?
  Did recovery include provisions to prevent or minimize future emergencies?
  Was there prompt response to requests for funds?

- **Hazard identification**
  Was this hazard identified as a serious threat before the emergency?
  Was the analysis of primary and secondary effects accurate?
  Was the hazard corrected to prevent recurrence, if possible?

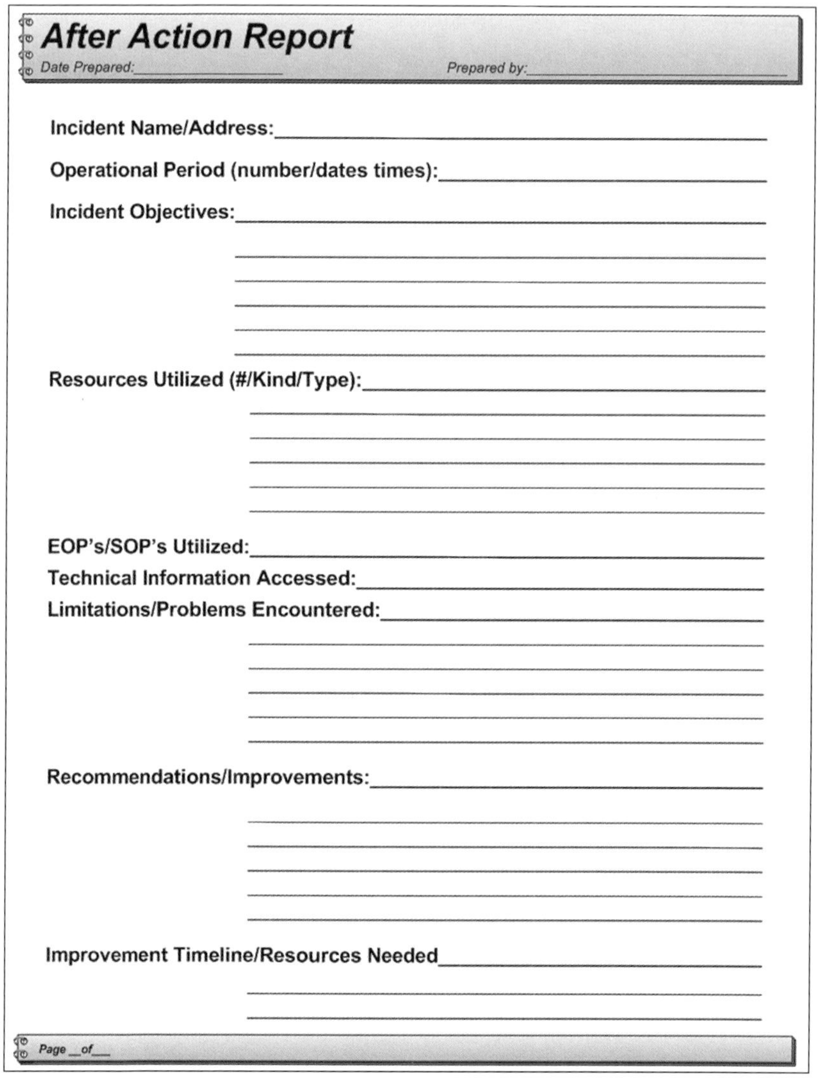

After Action Report

Date Prepared:__________ Prepared by:__________

Incident Name/Address:__________

Operational Period (number/dates times):__________

Incident Objectives:__________

Resources Utilized (#/Kind/Type):__________

EOP's/SOP's Utilized:__________

Technical Information Accessed:__________

Limitations/Problems Encountered:__________

Recommendations/Improvements:__________

Improvement Timeline/Resources Needed__________

Page __of__

Fig. 15–6. Sample after action report (AAR). It is important for responders to learn from their experiences. In order to record the lessons learned and the improvements needed, responders at all levels should participate in the creation of AARs.

Incident case study

As a final step in the evaluation process, it is often useful to summarize and document the response to the incident in a case study or after action report. A great deal will be learned about what actually happened as participants are interviewed and reports, documents, and photographs are collected. Figure 15–6 contains a sample after action report (AAR).

Such a report is also helpful in sharing experiences and lessons learned with other agencies. The Learning Resource Center at the National Emergency Training Center maintains a comprehensive library of case studies of major incidents from around the world. The Lessons Learned Information Sharing Web site found at www.llis.gov has a comprehensive list of disaster and disaster exercise after action reports that all emergency responders can view and learn from at any time.

In compiling a case study, it is important to identify problems and potential solutions. Generally it is more productive to focus on improving plans and response systems rather than to highlight poor individual decisions.

Improving the emergency management program

After the conclusion of the evaluation process, efforts turn to preventing or mitigating future emergencies and to improving response. The hazard analysis should be reviewed and updated if the incident was unexpected, or if it caused effects that were different from those expected.

Plans at the departmental and community level should be reviewed and revised as necessary. Resources that were needed but were unavailable should be identified, and arrangements should be made for acquiring them in the future. Figure 15–7 shows the never-ending cycle of disaster management. Executives and responders must participate in all of the steps present in the process in order to ensure the steps will function as planned.

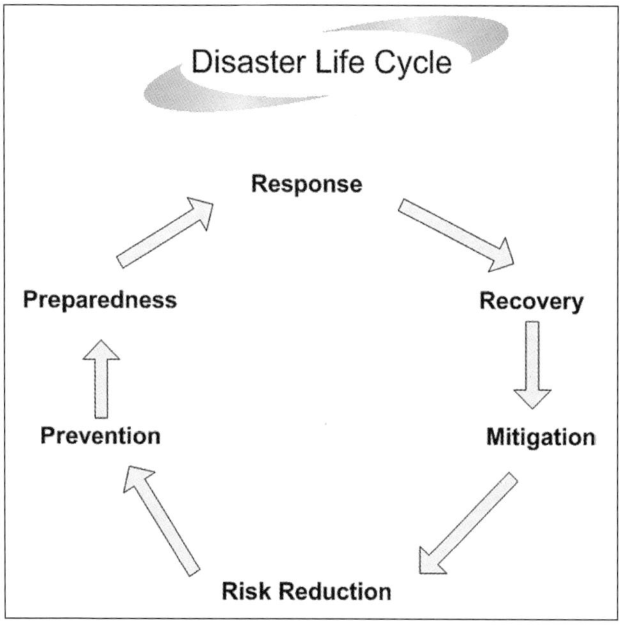

Fig. 15–7. Disaster cycle. If any portion of the cycle is overlooked, it will hamper the success of the other portions.

Finally, responder training programs and exercises should be developed to address the problems that invariably occur in large-scale incidents. The response will never be as the written plans describe, but substantial improvements in both the plans and in capabilities for carrying them out can be made.

CONCLUSION

Throughout this book, the seemingly endless complexities of disaster planning and control were examined. Regardless of where a disaster strikes, and regardless of the type of disaster, one fact remains clear: the greater the degree of advanced planning, the less serious the negative impact on the community will be. Those emergency response organizations that have adopted a vigorous forward approach to disaster planning will be those most capable of serving their communities when disaster strikes. But keep in mind that despite every effort to anticipate all possible contingencies that could arise in every conceivable form of disaster, "Murphy's Law" may affect the best-laid plans, and unforeseen problems may arise. If this happens, consolation may be found in the old adage, "This, too, shall pass!"

REFERENCES

1. U.S. Department of Homeland Security, Federal Emergency Management Agency. 2003. *A Guide to the Disaster Declaration Process and Federal Disaster Assistance*. March. Retrieved October 26, 2008 from: http://www.fema.gov/pdf/rrr/dec_proc.pdf.

2. U.S. Department of Homeland Security, Federal Emergency Management Agency. 2007. Deadlines and timelines. Retrieved April 25, 2007 from: http://www.fema.gov/government/grant/pa/re_deadlines.shtm.

3. *Newsweek*. 1991. The moral minority. May 6. p. 63.

4. U.S. Department of Homeland Security, Federal Emergency Management Agency. 2006. Additional assistance. March 21. Retrieved April 27, 2007 from: http://www.fema.gov/assistance/process/additional.shtm#0.

5. Weldon, C. 1991. Weldonís wireófinding a common ground with a new friend. May. *Firehouse*. 16 (5). p. 21.

6. Wallace, A. F. C. 1956. *Tornado in Worcester: An Exploratory Study of Individual and Community Behavior in an Extreme Situation*. Washington, DC: National Academy of Sciences-National Research Council. pp. 109–110. Retrieved October 13, 2008 from: http://openlibrary.org/details/tornadoinworcest00wallrich.

7. Freedy, J. R., and W. M. Simpson, Jr. 2007. Disaster-related physical and mental health: a role for the family physician. *American Family Physician*. 75 (6). March 15. Retrieved April 27, 2007 from: http://www.aafp.org/afp/20070315/841.html.

8. Collier, S. 1989. Disaster recovery. In *Disaster and Fire Defense Planning*. Emmitsburg, MD: National Fire Academy.

9. Hanifen, R. W. 2006. Roundtable: the consequences of self-deployment. *Fire Engineering*. 159 (3). March. p. 56. Accessed October 7, 2008 from: http://www.fireengineering-digital.com/fireengineering/200603/?pg=56.

10. Quarantelli, E. L., and B. Baisden. 1979. The recovery period in the U.S. disasters: problems and opportunities. In *Proceedings: National Public Policy Forum for Disaster Relief*. Racine, WI: National Voluntary Organizations Active in Disasters.

11. Moore, J. 1991. Self-evaluation. *Firehouse*. 16 (3). March. p. 79.

12. Mitchell, J. T. 1988. Using the REAPER model to learn to cope. *Journal of Emergency Medical Services (JEMS)*. September.

13. Mitchell, J. T. 1988. The history, status, and future of critical incident stress debriefing team. *Journal of Emergency Medical Services (JEMS)*. November.

14. Mitchell, J. T. 1988. Stress: development and functions of a critical incident stress debriefing team. *Journal of Emergency Medical Services (JEMS)*. December. pp. 43–46.

15. Ibid.

16. Ibid.

17. Ibid.

18. Ibid.

19. Ward, M. 1987. Recovering costs for better service. *Chief Fire Executive*. 2 (1). May/June. pp. 41–43.

20. Weldon, C. 1991.

21. Williams, K. 1990. Plan Bulldozer. *Firehouse*. 15 (10). October. p. 42.

22. Federal Emergency Management Agency. 2007. *Robert T. Stafford Disaster Relief and Emergency Assistance Act, Public Law 93-288, as Amended, 42 USC 5121-5207, and Related Authorities.* United States Code. Title 42. The Public Health and Welfare. Chapter 68. Disaster relief. p. 2. Retrieved October 7, 2008 from http://www.fema.gov/pdf/about/stafford_act.pdf.

GLOSSARY

ALS: Advanced Life Support

Base: That location at which the primary logistics functions are coordinated and administered. (Incident name or other designator will be added to the term "Base.") The Incident Command Post may be co-located with the base. There is only one base per incident.

BLEVE: Boiling Liquid Expanding Vapor Explosion.

BLS: Basic Life Support.

Branch: That organizational level having functional/geographic responsibility for major segments of incident operations. The branch level is organizationally between section and division group.

CAD: See Computer-Aided Dispatch.

CAMEO: See Computer-Aided Management of Emergency Operations.

Camp: A geographical site, within the general incident area, separate from the base, equipped and staffed to provide food, water, and sanitary services to incident personnel.

CDC: Civil Defense Corps.

Cellular Mobile Telephone: A recently implemented mobile telephone system which utilizes trunked radio technology to virtually eliminate waiting time by efficiently utilizing a limited number of radio frequencies.

CEPP: Chemical Emergency Preparedness Program.

CERCLA: Comprehensive Environmental Response and Liability Act.

Chemical Transportation Emergency Center: (CHEMTREC) Offers guidance of the dangers and precautions associated with the uncontrolled release of hazardous materials.

CHEMTREC: See Chemical Transportation Emergency Center.

CHRIS: Chemical Hazards Response Information System.

Civil Disturbance: A demonstration of popular unrest that may manifest itself in acts of violence and destruction against human life and property.

Clear Text: The use of plain English in radio communications transmissions. No ten codes or specific codes are used when using clear text.

Command Staff: The command staff consists of the information officer, the safety officer, and the liaison officer, and reports directly to the incident commander.

Company: Any piece of equipment having a full complement of personnel.

Computer-Aided Dispatch: (CAD) Computer systems designed to enhance the speed and accuracy of dispatching equipment to emergencies.

Computer-Aided Management of Emergency Operations: (CAMEO) A computer program, developed by the federal government, which can be used in the field to provide real-time emergency information for managing chemical incidents.

Conflagration: Large continuous fire involving buildings on more than one city block.

COPE: Communities Organized for Potential Emergencies.

Disaster Medical Assistance Team: (DMAT) An organizational unit developed under the National Disaster Medical System.

Division: That organizational level having responsibility for operations within a defined geographic area or with functional responsibility. The division level is organizationally between the strike team and the branch. See also Group.

DMAT: See Disaster Medical Assistance Team.

DOD: Department of Defense.

EEM: Earthquake Emergency Mode.

EIS: Emergency Information System.

EMA: Emergency Management Assistance.

Emergency: As it relates to the federal disaster assistance program, a declaration made by the president of the United States authorizing certain specific assistance from one or more federal agencies. See Major Disaster.

EMS: Emergency Management Assistance.

EOC: See Emergency Operations Center.

Emergency Operations Center: (EOC) A facility that serves as a multi-agency focal point for coordination activities during disasters.

Engine Company: Any ground vehicle providing specified levels of pumping, water, hose capacity, and personnel.

EOP: Emergency Operations Plan.

EPA: Environmental Protection Agency.

Facsimile: (FAX) Electronic devices which can transmit copies or facsimiles of written documents via ordinary telephone lines.

FAX: See Facsimile.

Federal Coordinating Officer: A federal officer appointed to coordinate with state agencies following the establishment of a FEMA/state agreement for disaster assistance to communities in that state.

FEMA: Federal Emergency Management Agency.

FHA: Federal Housing Administration.

FIRESCOPE: FIre REources of Southern California Organized for Potential Emergencies. (More recently—FIre RESources of California Organized for Potential Emergencies).

General Staff: The group of incident management personnel comprised of the incident commander, the operations chief, the planning chief, the logistics chief, and the finance chief.

GHQ: General Headquarters.

Group: A functional division (e.g., air support, medical, salvage, etc.), having no geographical boundaries.

Group fire: Large fires involving several structures on the same city block.

HA MANEUVER: Sudden flight of personnel (Haul ass!).

Hazardous Material: Any substance that poses an unreasonable risk to health, safety, property, and the environment.

Hazardous Materials Response Team: A group of individuals whose abilities, training, and equipment inventory represent an asset in the handling of a hazardous materials incident.

HAZMAT: Hazardous Materials.

HEAR: See Hospital Emergency Administrative Radio.

Hertz: A unit of frequency of a periodic process equal to one cycle per second.

HICA/MYDP: Hazardous Identification, Capability Assessment, Multi-Year Development Plan.

Hospital Emergency Administrative Radio: (HEAR) A radio system developed to permit coordination among hospitals during disasters.

HWOs: Hurricane Warning Offices.

IAFC: International Association of Fire Chiefs.

IC: See Incident Commander.

Icon: A picture symbol utilized by certain computer programs to simplify use of the program by reducing the number of written terms which must be remembered.

ICP: See Incident Command Post.

ICS: See Incident Command System.

IEMS: Integrated Emergency Management System.

Incident Action Plan: The basic operational plan for the incident. The plan which is initially prepared at the first planning meeting containing general control objectives reflecting the overall incident strategy and specific suppression and rescue action plans for the next operational period.

Incident Command Post: (ICP) That location at which the primary command functions are executed.

Incident Command System: (ICS) The combination of facilities, equipment, personnel, procedures, and communications operating within a common organizational structure with responsibility for the management of assigned resources to effectively accomplish stated objectives pertaining to the incident.

Incident Commander: (IC) The individual responsible for the management of all incident operations.

MACS: Multi-Agency Coordination System.

MAJOR Disaster: As it relates to the federal disaster assistance program, a declaration made by the president of the United States authorizing broad federal assistance and funds as delineated by the Federal Disaster Relief Act of 1974. See Emergency.

MHz: Megahertz: a unit of frequency equal to one million hertz.

Multi-Hazard Approach: An approach to emergency planning in which one response plan is adapted for several types of emergencies.

National Disaster Medical System: (NDMS) A system being developed by the federal government to deal with very large mass casualty disasters.

NCP: National Contingency Plan.

NDMS: See National Disaster Medical System.

NHC: National Hurricane Center.

NIMS: National Interagency Management System.

NOAA: National Oceanic and Atmospheric Administration.

NOVA: National Organization for Victims Assistance.

NRT: National Response Team.

OCC: See Operations Control Center.

OES: Office of Emergency Services.

On Scene Coordinator: Federal official pre-designated to coordinate and direct federal response to environmental disasters.

Operations Control Center: A term used by some police and fire departments to designate the alarm office or dispatch center.

RACES: See Radio Amateur Civil Emergency Service.

Radio Amateur Civil Emergency Service: (RACES) Groups of amateur (HAM) radio operators formally affiliated with local and state emergency management agencies.

Repeater: An electronic device that allows 3-way communication on 2-frequency radio channels.

RESTAT: Resource Status.

Riot: The most serious variant of crowd behavior, characterized by the threat of destruction of lives and property.

SARA: Superfund Amendment and Reauthorization Act.

SBA: Small Business Administration.

Sector: Division under Incident Command.

Sectorization: Dividing the incident scene into more manageable command areas called sectors.

Single Hazard Approach: An approach to emergency planning in which each type of emergency has its own response plan.

SITSTAT: Situation Status.

Staging Area: That location where incident personnel and equipment are assigned on a three-minute available status.

State Coordinating Officer: An officer normally appointed by the state emergency management agency to coordinate federal disaster assistance with the federal coordinating officer and local officials.

Strike Team: Specified combinations of the same kind and type of resources, with common communications and a leader.

Task Force: A group of resources with common communications and a leader, temporarily assembled for a specific mission.

Terminal Node Controller: (TNC) An electronic device which divides and encodes computer data into "packets" and transmits it via regular two-way radio. The data are received and decoded by another TNC. The technology permits error-free transmission of data with minimal interruption of voice communications.

Terrorism: The use of non-military violence by individuals and organizations for the attainment of political ends.

TMI: Three Mile Island.

TNC: See Terminal Node Controller.

Trunked Radio: An advanced technology designed to maximize the use of a limited number of frequencies by allowing them to be efficiently shared. Users of the system are linked on channels controlled by a computer rather than communication on dedicated radio frequencies.

UHF: See Ultra High Frequency.

USDA: United States Department of Agriculture.

USFS: United States Forest Service.

VA: Veterans Administration.

VHF: Very High Frequency; a segment of the radio spectrum at approximately 155 MHz.

Warning: A term developed by the National Weather Service meaning that a certain type of severe weather condition has actually been detected or spotted locally. Often follows the issuance of a severe weather "watch." Used in conjunction with specific weather types.

Watch: A term developed by the National Weather Service meaning that conditions are favorable to the formation of a severe weather condition.

Wildland Fire: Large unfriendly fires in which the natural cover of the fire area constitutes the primary fuel supply.

ACRONYMS

AAR	after action report
AFFF	aqueous film-forming foam
ALOHA	aerial location of hazardous atmospheres
ALS	advanced life support
ANFO	ammonium nitrate/fuel oil
AOC	Army Operations Center
APCO	Association of Public-Safety Communications Officials
AVLs	automatic vehicle locators
BERT	basic emergency rescue technician
BLEVE	boiling liquid expanding vapor explosion
BLS	basic life support
CAD	computer-assisted dispatch; computer-aided dispatch
CAI	Common Air Interface
CAMEO	Computer-Aided Management of Emergency Operations (database)
CBRNE	chemical, biological, radiological, nuclear, and explosive
CERCLA	Comprehensive Environmental Response, Compensation, and Liability Act
CERT	Community Emergency Response Team
CIA	Central Intelligence Agency
CIS	critical incident stress
CISD	critical incident stress debriefing
CISM	critical incident stress management
CMCHS	Civilian-Military Contingency Hospital System
CPR	cardiopulmonary resuscitation
CSA	The Covenant, the Sword, and the Arm of the Lord
CSG	Counterterrorism Security Group
CST	Civil Support Team
C-TPAT	Customs-Trade Partnership Against Terrorism
CVAT	Community Vulnerability Assessment Tool
DACs	disaster application centers
DCO	designated coordinating officer
DEST	Domestic Emergency Support Team
DHS	Department of Homeland Security
DMAT	Disaster Medical Assistance Team

DMORT	Disaster Mortuary Operational Response Team
DOT	Department of Transportation
DPMU	Disaster Portable Morgue Units (Team)
DRG	Domestic Readiness Group
EMAC	emergency mutual aid compact; Emergency Management Assistance Compact
EMI	Emergency Management Institute
EMS	emergency medical services
EMT	emergency medical technician
EOC	emergency operations center
EPR	Emergency Preparedness and Response (Directorate)
ERRI	Emergency Response & Research Institute
ESF	emergency support functions
FAA	Federal Aviation Administration
FBI	Federal Bureau of Investigation
FCC	Federal Communications Commission
FCO	federal coordinating officer
FDC	fire department connection
FDIC	Fire Department Instructors Conference
FEMA	Federal Emergency Management Agency
FmHA	Farmers Home Administration
FRC	federal resource coordinator
FTOs	foreign terrorist organizations
GIA	Armed Islamic Group
GIS	geographic information systems
GPS	global positioning systems
HEAR	Hospital Emergency Administrative Radio
HIPAA	Health Insurance Portability and Accountability Act
HLT	Hurricane Liaison Team
HSC	Homeland Security Council
HSEEP	Homeland Security Exercise and Evaluation Program
HVO	Hawaiian Volcano Observatory
HWOs	hurricane warning offices
IACET	International Association of Continuing Education and Training
IACP	International Association of Chiefs of Police
IAEA	International Atomic Energy Agency
IAFC	International Association of Fire Chiefs
IAP	incident action plan
ICP	incident command post
ICTAP	Interoperable Communications Technical Assistance Program
IEMS	Integrated Emergency Management System
IMAS	Intrastate Mutual Aid System
IMATS	Incident Management Assistance Teams
IMBE	Improved Multi-Band Excitation (vocoder)
IMP	incident medical plan
IMS	incident management system
ITP	incident traffic plan
JFO	joint field office
JIC	joint information center
JIS	Jamiyyat Ul Islam Is Saheeh
JOC	joint operations center
JTF	joint task force
JTTF	Joint Terrorism Task Force
LEPC	local emergency planning committee
LNG	liquefied natural gas
LPG	liquefied petroleum gas
MASTF	Mutual Aid System Task Force
MAT	Medical Assistance Team
MDCs	mobile data computers
MERS	Mobile Emergency Response Support
MGRS	Military Grid Reference System
MSDS	material safety data sheets
NASTD	National Association of State Telecommunications Directors
NCP	National Contingency Plan
NCS	National Communications System
NCTC	National Counterterrorism Center
NDMS	National Disaster Medical System
NEIC	National Earthquake Information Center
NFPA	National Fire Protection Association
NHC	National Hurricane Center
NIC	National Integration Center
NICC	National Infrastructure Coordinating Center
NIFC	National Interagency Fire Center
NIMS	National Incident Management System
NNRT	National Nurse Response Team
NOAA	National Oceanic and Atmospheric Administration
NOC	National Operation Center
NPRT	National Pharmacy Response Team

NRC	National Response Center
NRCC	National Response Coordination Center
NRF	National Response Framework
NRP	National Response Plan
NRT	National Response Team
NSC	National Security Council
NTSB	National Transportation Safety Board
NWS	National Weather Service
OSHA	Occupational Safety and Health Administration
PASS	personal alert safety system
PDA	preliminary damage assessment
PIO	public information officer
PPE	personal protective equipment
PWSS	portable water supply system
RDD	radiological dispersion device
RMS	records management system
RNA	rapid needs assessment
RRCC	Regional Response Coordination Center
RRT	Regional Response Team
SAR	search and rescue
SARA	Superfund Amendment and Reauthorization Act
SBA	Small Business Administration
SCBA	self-contained breathing apparatus
SCO	state coordinating officer
SFLEO	senior federal law enforcement official
SFPE	Society of Fire Protection Engineers
SIOC	Strategic Information and Operations Center
SOP	standard operating procedure
SWAT	special weapons and tactics
TEC	Technical Escort Center (U.S. Army)
TIA	Telecommunications Industry Association
TICP	Tactical Interoperable Communications Plan
UCG	Unified Coordination Group
UCS	unified command structure
US&R	Urban Search and Rescue (Team)
USNG	U.S. National Grid
UTM	Universal Transverse Mercator
VBIED	vehicle-borne improvised explosive device
VMAT	Veterinary Medical Assistance Team
WMD	weapons of mass destruction

INDEX

A

B

C

D

E

F

G

H

I

J

K

L

M

N

O

P

Q

R

S

T

U

V

W

X–Z